5G/5G-ADVANCED

The New Generation Wireless Access Technology

5G/5G-ADVANCED

The New Generation Wireless Access Technology

Third Edition

ERIK DAHLMAN

STEFAN PARKVALL

JOHAN SKÖLD

ACADEMIC PRESS

An imprint of Elsevier

ELSEVIER

Academic Press is an imprint of Elsevier
125 London Wall, London EC2Y 5AS, United Kingdom
525 B Street, Suite 1650, San Diego, CA 92101, United States
50 Hampshire Street, 5th Floor, Cambridge, MA 02139, United States
The Boulevard, Langford Lane, Kidlington, Oxford OX5 1GB, United Kingdom

ISBN 978-0-443-13173-8

For information on all Academic Press publications
visit our website at https://www.elsevier.com/books-and-journals

Publisher: Mara Conner
Acquisitions Editor: Tim Pitts
Editorial Project Manager: Teddy A Lewis
Production Project Manager: Anitha Sivaraj
Cover Designer: Greg Harris

Typeset by STRAIVE, India

Working together
to grow libraries in
developing countries

www.elsevier.com • www.bookaid.org

Contents

Preface

Deployment of 5G wireless access network based on the 3GPP NR ("New Radio") technology is now well under way. Compared to 4G networks based on the LTE technology, 5G provides substantially better mobile-broadband performance including data rates of many 10 Gbps and substantially lower latency. But 5G goes beyond this, also addressing services requiring new levels of reliability in combination with very lower latency. The evolution of 5G also includes extensive support for various verticals and new deployment scenarios.

This book describes NR as of Spring 2023. Chapter 1 provides a brief introduction, followed by a description of the standardization process and relevant organizations such as the aforementioned 3GPP and ITU in Chapter 2. The frequency bands available for mobile communication are covered in Chapter 3 together with a discussion on the process for finding new frequency bands.

An overview of LTE and its evolution is found in Chapter 4. Although the focus of the book is NR, a brief overview of LTE as a background to the coming chapters is relevant. One reason is that both LTE and NR are developed by 3GPP and hence have a common background and share several technology components. Many of the design choices in NR are also based on experience from LTE. Furthermore, LTE remains important for the overall 5G radio access, not only as important component for certain IoT applications but also in terms of operating together with 5G/NR in dual-connectivity scenarios.

Chapter 5 provides an overview of NR. It can be read on its own to get a high-level understanding of NR or as an introduction to the subsequent chapters.

Chapter 6 outlines the overall protocol structure in NR, followed by a description of the overall time/frequency structure of NR in Chapter 7.

Multiantenna processing and beamforming are integral parts of NR. The channel sounding tools to support these functions are outlined in Chapter 8, followed by the overall transport-channel processing in Chapter 9 and the associated control signaling in Chapter 10. How the functions are used to support different multi-antenna schemes and beamforming functions is the topic of Chapters 11 and 12.

Retransmission functionality and scheduling are the topics of Chapters 13 and 14, followed by power control in Chapter 15, cell search in Chapter 16, and random access in Chapter 17.

Coexistence and interworking with LTE is an essential part of NR, especially in the non-standalone version that relies on LTE for mobility and initial access, and is covered in Chapter 18.

Chapters 19–27 focus on some of the major new features brought by the evolution of NR in release 16, release 17, and release 18. Remote interference management for TDD networks is discussed in Chapter 19. Accessing unlicensed spectrum is treated in Chapter 20. Enhancements for machine-type communication are discussed in Chapters 21 and 22, with Chapter 21 focusing on ultra-reliable, low-latency communication and industrial internet-of-things and Chapter 22 treating RedCap devices and small-data transmission. Chapter 23 describes the enhancements for multicast and broadcast services. Chapter 24 describes integrated access and backhaul where NR is used not only for the access link but also for backhauling purposes. Chapter 25 covers the introduction of NR support for so-called non-terrestrial networks, in practice 5G access via satellites and high-altitude platforms. Vehicular-to-anything communication and the NR sidelink design is the scope of Chapter 26. Positioning is treated in Chapter 27.

Radio-frequency (RF) requirements, taking into account spectrum flexibility across large frequency ranges and multistandard radio equipment, are the topic of Chapter 28. Chapter 29 discusses the RF implementation aspects for higher frequency bands in the mm-wave range.

Finally, the book is concluded by Chapter 30 with an outlook to future NR releases and the upcoming 6G technologies.

Acknowledgments

We thank all our colleagues at Ericsson for assisting in this project by helping with contributions to the book, giving suggestions and comments on the contents, and taking part in the huge team effort of developing NR and the next generation of radio access for 5G.

The standardization process involves people from all parts of the world, and we acknowledge the efforts of our colleagues in the wireless industry in general and in 3GPP RAN in particular. Without their work and contributions to the standardization, this book would not have been possible.

Finally, we are immensely grateful to our families for bearing with us and supporting us during the long process of writing this book.

Abbreviations and acronyms

3GPP	third generation partnership project
5GCN	5G core network
AAS	active antenna system
ACIR	adjacent channel interference ratio
ACK	acknowledgment (in ARQ protocols)
ACLR	adjacent channel leakage ratio
ACS	adjacent channel selectivity
ADC	analog-to-digital converter
AF	application function
AGC	automatic gain control
AI	artificial intelligence
AM	acknowledged mode (RLC configuration)
AM	amplitude modulation
AMF	access and mobility management function
A-MPR	additional maximum power reduction
AMPS	advanced mobile phone system
AoA	angle of arrival
AP	application protocol
AR	augmented reality
ARI	acknowledgment resource indicator
ARIB	Association of Radio Industries and Businesses
ARQ	automatic repeat-request
AS	access stratum
ATIS	Alliance for Telecommunications Industry Solutions
AUSF	authentication server function
AWGN	additive white Gaussian noise
BAP	backhaul adaptation protocol
BC	band category
BCCH	broadcast control channel
BCH	broadcast channel
BiCMOS	bipolar complementary metal oxide semiconductor
BPSK	binary phase-shift keying
BS	base station
BW	bandwidth
BWP	bandwidth part
CA	carrier aggregation

CACLR	cumulative adjacent channel leakage ratio
CBG	Codeblock group
CBGFI	CBG flush information
CBGTI	CBG transmit indicator
CC	component carrier
CCCH	common control channel
CCE	control channel element
CCSA	China Communications Standards Association
CDM	code-division multiplexing/code-domain multiplexing
CDMA	code-division multiple access
CEPT	European Conference of Postal and Telecommunications Administration
CFR	common MBS frequency resource
CITEL	Inter-American Telecommunication Commission
CHO	conditional handover
CLI	cross-link interference
C-MTC	critical machine-type communications
CMAS	commercial mobile alert service
CMOS	complementary metal oxide semiconductor
CN	core network
CoMP	coordinated multi-point transmission/reception
CORESET	control resource set
COT	channel occupancy time
CP	cyclic prefix
CP	compression point
CQI	channel-quality indicator
CRB	common resource block
CRC	cyclic redundancy check
C-RNTI	cell radio-network temporary identifier
CS	capability set (for MSR base stations)
CSI	channel-state information
CSI-IM	CSI interference measurement
CSI-RS	CSI reference signal
CS-RNTI	configured scheduling RNTI
CSS	common search space
CU	centralized unit
CW	continuous wave
D2D	device-to-device
DAC	digital-to-analog converter
DAI	downlink assignment index
D-AMPS	digital AMPS
DAPS	dual active protocol stacks

DC	dual connectivity
DC	direct current
DCCH	dedicated control channel
DCH	dedicated channel
DCI	downlink control information
DFT	discrete Fourier transform
DECT	Digital Enhanced Cordless Telecommunications
DFTS-OFDM	DFT-spread OFDM (DFT-precoded OFDM, see also SC-FDMA)
DL	downlink
DL-SCH	downlink shared channel
DM-RS	demodulation reference signal
DR	dynamic range
DRB	data radio bearer
DRX	discontinuous reception
DTX	discontinuous transmission
DU	distributed unit
ECC	Electronic Communications Committee (of CEPT)
EDGE	enhanced data rates for GSM evolution, enhanced data rates for global evolution
EESS	earth exploration satellite systems
eIMTA	enhanced interference mitigation and traffic adaptation
EIRP	effective isotropic radiated power
EIS	equivalent isotropic sensitivity
eMBB	enhanced MBB
EMF	electromagnetic field
eMTC	enhanced machine-type communication support, see LTE-M
eNB	eNodeB
EN-DC	E-UTRA NR dual-connectivity
eNodeB	E-UTRAN NodeB
EPC	evolved packet core
ETSI	European Telecommunications Standards Institute
ETWS	earthquake and tsunami warning system
EUHT	enhanced ultra high throughput
E-UTRA	evolved UTRA
EVM	error vector magnitude
FBE	frame-based equipment
FCC	Federal Communications Commission
FD	frequency domain
FDD	frequency-division duplex
FDM	frequency-division multiplexing/frequency-domain multiplexing
FET	field-effect transistor
FDMA	frequency-division multiple access

FFT	fast Fourier transform
FoM	figure-of-merit
FPLMTS	future public land mobile telecommunications systems
FR1	frequency range 1
FR2	frequency range 2
FR2-1	frequency range 2-1
FR2-2	frequency range 2-2
GaAs	gallium arsenide
GaN	gallium nitride
GEO	geostationary equatorial orbit
GERAN	GSM/EDGE radio access network
gNB	gNodeB (in NR)
gNodeB	generalized NodeB
GNSS	global navigation satellite system
GPS	global positioning system
GSA	Global mobile Suppliers Association
GSM	Global System for Mobile Communications
GSMA	GSM Association
GSO	geostationary orbit
GTP	GPRS tunneling protocol
HAP	high-altitude platform
HAPS	high-altitude platform station/system
HIBS	high-altitude IMT base station
HARQ	hybrid ARQ
HBT	heterojunction bipolar transistor
HEMT	high electron-mobility transistor
HFN	hyperframe number
HRNN	human readable network name
HSPA	high-speed packet access
IAB	integrated access backhaul
IC	integrated circuit
ICNIRP	International Commission on Non-Ionizing Radiation
ICS	in-channel selectivity
IEEE	Institute of Electrical and Electronics Engineers
IFFT	inverse fast Fourier transform
IIoT	industrial IoT
IL	insertion loss
IMD	inter modulation distortion
IMT-2000	International Mobile Telecommunications 2000 (ITU's name for the family of 3G standards)
IMT-2020	International Mobile Telecommunications 2020 (ITU's name for the family of 5G standards)

IMT-Advanced	International Mobile Telecommunications Advanced (ITU's name for the family of 4G standards)
InGaP	indium gallium phosphide
IoT	Internet of things
IP	Internet protocol
IP3	third order Intercept point
IR	incremental redundancy
IRDS	International Roadmap for Devices and Systems
ITRS	International Telecom Roadmap for Semiconductors
ITU	International Telecommunications Union
ITU-R	International Telecommunications Union-Radiocommunication Sector
ITS	intelligent transportation systems
KPI	key performance indicator
L1-RSRP	Layer 1 reference signal received power
LC	inductor(L)-capacitor
LAA	license-assisted access
LBE	load-based equipment
LBT	listen before talk
LCID	logical channel index
LDPC	low-density parity check code
LEO	low earth orbit
LO	local oscillator
LNA	low-noise amplifier
LPWA	low power wide area
LTCC	low temperature co-fired ceramic
LTE	long-term evolution
LTE-M	see eMTC
LTM	L1/L2-triggered mobility
MAC	medium access control
MAC-CE	MAC control element
MAN	metropolitan area network
MBB	mobile broadband
MB-MSR	multi-band multi-standard radio (base station)
MBS	multicast and broadcast services
MCCH	multicast control channel
MCG	master cell group
MCS	modulation and coding scheme
MIB	master information block
ML	maximum likelihood/machine learning
MMIC	monolithic microwave integrated circuit
MIMO	multiple-input multiple-output

ML	machine learning
mMTC	massive machine-type communication
MPR	maximum power reduction
MSR	multi-standard radio
MT	mobile terminal (in IAB context)
MTC	machine-type communication
MTCH	multicast traffic channel
MU-MIMO	multi-user MIMO
NAK	negative acknowledgment (in ARQ protocols)
NB-IoT	narrow-band Internet-of-things
NCJT	non-coherent joint transmission
NCR	network controlled repeater
NCR-Fwd	forwarding unit of NCR
NCR-MT	control unit ("mobile terminal") of NCR
NDI	new-data indicator
NEF	network exposure function
NF	noise figure
NG	the interface between the gNB and the 5G CN
NG-c	the control-plane part of NG
NGMN	next generation mobile networks
NG-u	the user-plane part of NG
NMT	Nordisk MobilTelefon (Nordic Mobile Telephony)
NodeB	NodeB, a logical node handling transmission/reception in multiple cells. Commonly, but not necessarily, corresponding to a base station
NOMA	nonorthogonal multiple access
NPN	non-public networks
NR	new radio
NRF	NR repository function
NS	network signaling
NSA	non-standalone
NTN	non-terrestrial network
NZP-CSI-RS	non-zero-power CSI-RS
OAM	operation and maintenance
OBUE	operating band unwanted emissions
OCC	orthogonal cover code
OFDM	orthogonal frequency-division multiplexing
OOB	out-of-band (emissions)
OSDD	OTA sensitivity direction declarations
OTA	over-the-air
OTDOA	observed time difference of arrival
PA	power amplifier
PAE	power-added efficiency

PAPR	peak-to-average power ratio
PAR	peak-to-average ratio (same as PAPR)
PBCH	physical broadcast channel
PCB	printed circuit board
PCCH	paging control channel
PCell	primary cell
PCF	policy control function
PCG	project coordination group (in 3GPP)
PCH	paging channel
PCI	physical cell identity
PDC	personal digital cellular
PDCCH	physical downlink control channel
PDCP	packet data convergence protocol
PDSCH	physical downlink shared channel
PDU	protocol data unit
PEI	paging early indicator
PHS	personal handy-phone system
PHY	physical layer
PLL	phase-locked loop
PM	phase modulation
PMI	precoding-matrix indicator
PN	phase noise
PRACH	physical random-access channel
PRB	physical resource block
P-RNTI	paging RNTI
PRS	positioning reference signal
PSBCH	physical sidelink broadcast channel
PSCCH	physical sidelink control channel
PSD	power spectral density
PSFCH	physical channel feedback channel
PSS	primary synchronization signal
PSSCH	physical sidelink shared channel
PTM	point-to-multipoint
PTP	point-to-point
PUCCH	physical uplink control channel
PUCCH-sCell	PUCCH switching cell
PUSCH	physical uplink shared channel
QAM	quadrature amplitude modulation
QCL	quasi co-location
QoS	quality-of-service
QPSK	quadrature phase-shift keying
RACH	random-access channel

RAN	radio access network
RA-RNTI	random-access RNTI
RAT	radio access technology
RB	resource block
RE	resource element
RedCap	reduced capability
RF	radio frequency
RFIC	radio frequency integrated circuit
RI	rank indicator
RIB	radiated interface boundary
RIM	remote interference management
RIT	radio interface technology
RLC	radio link control
RLF	radio link failure
RMSI	remaining minimum system information
RNTI	radio-network temporary identifier
RoAoA	range of angle of arrival
ROHC	robust header compression
RRC	radio resource control
RRM	radio resource management
RS	reference symbol
RSPC	radio interface specifications
RSRP	reference signal received power
RTT	roundtrip time
RV	redundancy version
RX	receiver
SA	standalone
SAI	sidelink assignment index
SBCCH	sidelink broadcast control channel
SCell	secondary cell
SCG	secondary cell group
SCCH	sidelink control channel
SCI	sidelink control information
SCG	secondary cell group
SCS	sub-carrier spacing
SCTP	stream control transmission protocol
SDAP	service data application protocol
SDL	supplementary downlink
SDMA	spatial division multiple access
SDO	standards developing organization
SDT	small data transmission
SDU	service data unit

SEM	spectrum emissions mask
SFI	slot format indicator
SFI-RNTI	slot format indicator RNTI
SFN	system frame number (in 3GPP)/single-frequency network
SI	system information message
SIB	system information block
SIBn	system information block n
SiGe	silicon germanium
SINR	signal-to-interference-and-noise ratio
SIR	signal-to-interference ratio
SiP	system-in-package
SI-RNTI	system information RNTI
SL-BCH	sidelink broadcast channel
SL-SCH	sidelink Shared Channel
SL-SSB	sidelink SSB
SMF	session management function
SNDR	signal-to-noise-and-distortion ratio
SNR	signal-to-noise ratio
SoC	system-on-chip
S-PSS	sidelink primary synchronization signal
SR	scheduling request
SRB	signaling radio bearer
SRI	SRS resource indicator
SRIT	set of radio interface technologies
SRS	sounding reference signal
SS	synchronization signal
SSB	synchronization signal block
SSS	secondary synchronization signal
SSSG	search-space-set group
S-SSS	sidelink secondary synchronization signal
SMT	surface-mount assembly
STCH	sidelink traffic channel
SUL	supplementary uplink
SU-MIMO	single-user MIMO
TAB	transceiver-array boundary
TACS	total access communication system
TBoMS	transport block over multiple slots
TCI	transmission configuration indication
TCP	transmission control protocol
TC-RNTI	temporary C-RNTI
TDD	time-division duplex
TDM	time-division multiplexing/time-domain multiplexing

TDMA	time-division multiple access
TD-SCDMA	time-division-synchronous code-division multiple access
TIA	Telecommunication Industry Association
TMGI	temporary mobile group identity
TR	technical report
TRP	total radiated power/transmission and reception point
TRS	tracking reference signal
TS	technical specification
TSDSI	Telecommunications Standards Development Society, India
TSG	Technical Specification Group
TSN	time-sensitive networks
TTA	Telecommunications Technology Association
TTC	Telecommunications Technology Committee
TTI	transmission time interval
TX	transmitter
UAV	unmanned aerial vehicle
UCI	uplink control information
UDM	unified data management
UE	user equipment, the 3GPP name for the mobile terminal
UL	uplink
UMTS	Universal Mobile Telecommunications System
UPF	user plane function
URLLC	ultra-reliable low-latency communication
USS	user-specific search space
UTC	coordinated universal time
UTRA	universal terrestrial radio access
V2X	vehicular-to-anything
V2V	vehicular-to-vehicular
VCO	voltage-controlled oscillator
VR	virtual reality
WARC	World Administrative Radio Congress
WCDMA	wideband code-division multiple access
WG	working group
WiMAX	Worldwide Interoperability for Microwave Access
WHO	World Health Organization
WP5D	Working Party 5D
WRC	World Radiocommunication Conference
XR	refers to all of AR, VR, and MR
Xn	the interface between gNBs
ZC	Zadoff-Chu
ZP-CSI-RS	zero-power CSI-RS

CHAPTER 1

What is 5G?

1.1 The evolution of mobile communication – From 1G to 5G

Over the last 40+ years, the world has witnessed five generations of mobile communication, see Fig. 1.1.

The first generation of mobile communication, emerging around 1980, was based on analog transmission with the main technologies being AMPS (Advanced Mobile Phone System) developed within North America, NMT (Nordic Mobile Telephony) jointly developed by the Nordic countries, and TACS (Total Access Communication System) used in, for example, the United Kingdom. The mobile communication systems based on first-generation technology were limited to voice services and, for the first time, made mobile telephony accessible to ordinary people.

The second generation of mobile communication, emerging in the early 1990s, saw the introduction of digital transmission on the radio link. Although the target service was still voice, the use of digital transmission allowed for second generation mobile-communication systems to also provide limited data services. There were initially several different second-generation technologies, including GSM (Global System for Mobile communication) jointly developed by a large number of European countries, D-AMPS (Digital AMPS), PDC (Personal Digital Cellular) developed and solely used in Japan, and the CDMA-based IS-95 technology. As time went by, GSM spread from Europe to other parts of the world and eventually came to completely dominate among the second-generation technologies. Primarily due to the success of GSM, the second-generation systems also turned mobile telephony from something still being used by only a relatively small fraction of people to a communication tool being a necessary part of life for a large majority of the world's population.

The third generation of mobile communication, often just referred to as 3G, was introduced in the early 2000s, based on the work in the newly established Third Generation Partnership Project (3GPP). With 3G and, especially, with the 3G evolution known as HSPA (High Speed Packet Access) [19], the first true step toward high-quality mobile broadband was taken. In addition, while earlier mobile-communication technologies were only designed for operation in paired spectrum (separate spectrum for network-to-device and device-to-network links) using Frequency-Division Duplex (FDD), see Chapter 7, 3G also saw the first introduction of mobile technologies targeting unpaired spectrum using Time Division Duplex (TDD).

5G/5G-Advanced
https://doi.org/10.1016/B978-0-443-13173-8.00008-6

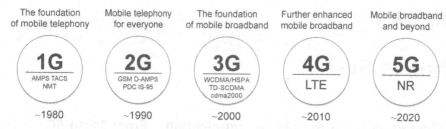

Fig. 1.1 Generations of mobile communication.

Around 2010, fourth-generation (4G) mobile communication was introduced in form of the LTE technology [26]. LTE followed in the steps of HSPA, with higher efficiency and further enhanced mobile-broadband experience by enabling higher achievable data rates and lower latency. This was provided by means of an OFDM-based transmission technology enabling wider transmission bandwidths and more advanced multi-antenna technologies. Furthermore, while 3G allowed for mobile communication in unpaired spectrum by means of a specific radio-access technology different from the one used in paired spectrum, LTE supported both FDD and TDD operation, that is, operation in both paired and unpaired spectrum, within the same radio-access technology. By means of LTE the world thus converged into a single global technology for mobile communication. As discussed in somewhat more detail in Chapter 4, the later evolution of LTE has extended mobile communication into unlicensed spectrum.

Wireless communication has now evolved well into the fifth generation (5G) era represented by the NR radio-access technology. 5G further enhances the mobile-broadband experience, beyond 4G and also expands mobile communication to address a wide range of new use cases beyond mobile broadband.

1.2 3GPP and the standardization of mobile communication

Agreeing on multi-national technology specifications and standards has been key to the success of mobile communication. This has allowed for the deployment and interoperability of devices and infrastructure of different vendors and enabled devices and subscriptions to operate on a global basis.

As already mentioned, already the first-generation NMT technology was created on a multinational basis, allowing for devices and subscription to operate over the national borders between the Nordic countries. The next step in multinational specification/standardization of mobile-communication technology took place when GSM was jointly developed between a large number of European countries within CEPT, later moved to the newly created ETSI (European Telecommunications Standards Institute). As a consequence of this, GSM devices and subscriptions were already from the beginning able to operate over a large number of countries, covering a very large number of potential users. This large common market had a profound impact on device availability, leading to an unprecedented number of different device types and substantial reduction in device cost.

However, the final step to true global standardization of mobile communication came with the specification of the 3G technologies, especially WCDMA. Work on 3G technology was initially also carried out on a regional basis, that is, separately within Europe (ETSI), North America (TIA, T1P1), Japan (ARIB), etc. However, the success of GSM had shown the importance of a large technology footprint, especially in terms of device availability and cost. It also become clear that although work was carried out separately within the different regional standard organizations, there were many similarities in the underlying technology being pursued. This was especially true for Europe and Japan which were both developing different but very similar flavors of wideband CDMA (WCDMA) technology.

As a consequence, in 1998, the different regional standardization organizations came together and jointly created the Third-Generation Partnership Project (3GPP) with the task of finalizing the development of 3G technology based on WCDMA. A parallel organization (3GPP2) was somewhat later created with the task of developing an alternative 3G technology, cdma2000, as an evolution of second-generation IS-95. For a number of years, the two organizations (3GPP and 3GPP2) with their respective 3G technologies (WCDMA and cdma2000) existed in parallel. However, over time 3GPP came to completely dominate and has, despite its name, continued into the development of 4G (LTE), and 5G (NR) technologies. Today, 3GPP is the only significant organization developing technical specifications for mobile communication.

1.3 The new generation – 5G NR

Discussions on fifth-generation (5G) mobile communication began around 2012. In many situations, the term 5G is used specifically for the new 5G radio-access technology referred to as NR or "New Radio". However, 5G is also often used in wider context, referring to the range of new use cases toward which mobile communication is evolving in the 5G era.

1.3.1 The 5G use cases

In the context of 5G, one is often talking about three distinctive classes of use cases: enhanced mobile broadband (eMBB), massive machine-type communication (mMTC), and ultra-reliable and low-latency communication (URLLC) (see also Fig. 1.2).

- eMBB corresponds to a more or less straightforward evolution of the mobile broadband services of 4G, enabling even larger data volumes and further enhanced user experience, for example, by supporting even higher end-user data rates.

eMBB
High data rates, High traffic volumes

mMTC
Massive number of devices,
Low cost, low energy consumption

URLLC
Very low latency,
Very high reliability and availability

Fig. 1.2 High-level 5G use-case classification.

- mMTC corresponds to services that are characterized by a massive number of devices, for example, remote sensors, actuators, and the monitoring of various equipment. Key requirements for such services include very low device cost and very low device energy consumption, allowing for very long device battery life of up to at least several years. Typically, each device consumes and generates only a relatively small amount of data, that is, support for high data rates and high spectral efficiency is of less importance.
- URLLC type-of-services are envisioned to require very low latency and extremely high reliability. Examples hereof are traffic safety, automatic control, and factory automation.

It is important to understand that the classification of 5G use cases into these three distinctive classes is somewhat artificial, primarily aiming to simplify the definition of requirements for the technology specification. There will be many use cases that do not fit exactly into one of these classes. Just as an example, there may be services that require very high reliability but for which the latency requirements are not that critical. Similarly, there may be use cases requiring devices of very low cost but where the possibility for very long device battery life may be less important.

1.3.2 Evolving LTE to 5G capability

The first release of the LTE technical specifications was introduced in 2009. Since then, LTE has gone through several steps of evolution providing enhanced performance and extended capabilities. This has included features for enhanced mobile broadband, including means for higher achievable end-user data rates as well as higher spectrum efficiency. However, it has also included important steps to extend the set of use cases to which LTE can be applied. Especially, there have been important steps to enable truly low-cost devices with very long battery life, in line with the characteristics of massive MTC applications. Thus, LTE-derived technologies for massive MTC will continue to play an important role also in the 5G era and can be seen as part of an overall 5G system. LTE is also an important part of the overall 5G solution when operating in so-called non-standalone mode as discussed in Chapter 6. Although not being the main aim of this book, an overview of LTE and its evolution is provided in Chapter 4.

1.3.3 NR – The new 5G radio-access technology

Despite LTE being a very capable technology, it was quite clear around 2015 that the full set of use cases envisioned for 2020 and beyond required performance and capabilities beyond what could efficiently be met by just evolving LTE. Also, the general technology advancements since the introduction of LTE paved the way for new technology solutions that were difficult or even impossible to efficiently introduce in a direct LTE evolution. To ensure that its technologies could meet the requirements of future use cases and fully exploit the potential of new technologies, 3GPP therefore initiated the development of a new 5G radio-access technology known as NR (New Radio). A workshop setting the scope for NR was held in the fall of 2015 and technical work began in the spring of 2016. The first version of the NR specifications was available by the end of 2017 to meet commercial requirements on early 5G deployments already in 2018.

NR reuses many of the structures and features of LTE. However, being a new radio-access technology means that NR, unlike the LTE evolution, is not restricted by a need to retain backward compatibility. The requirements on NR are also broader than what was the case for LTE, motivating a partly different set of technical solutions.

Chapter 2 discusses the standardization activities related to NR, followed by a spectrum overview in Chapter 3 and a brief summary of LTE and its evolution in Chapter 4. The main part of this book (Chapters 5–30) then provides an in-depth description of the current stage of the NR technical specifications, finishing with an outlook of the future NR evolution in the last chapter.

1.3.4 5GCN – The new 5G core network

In parallel to NR, that is, the new 5G radio-access technology, 3GPP also developed a new 5G core network referred to as 5GCN. The new 5G radio-access technology connects to the 5GCN. However, 5GCN is also able to provide connectivity for the evolution of LTE. At the same time, NR may connect via the legacy core network EPC when operating in so-called *non-stand-alone mode* together will LTE, as will be further discussed in Chapter 6. For a detailed description of the 5G core network, see [84].

CHAPTER 2

5G standardization

The research, development, implementation and deployment of mobile-communication systems is performed by the wireless industry in a coordinated international effort by which common industry specifications that define the complete mobile-communication system are agreed. The work depends heavily on global and regional regulation, in particular for the spectrum use that is an essential component for all radio technologies. This chapter describes the regulatory and standardization environment that has been, and continues to be, essential for defining the mobile-communication systems.

2.1 Overview of standardization and regulation

There are a number of organizations involved in creating technical specifications and standards as well as regulation in the mobile-communications area. These can loosely be divided into three groups: Standards Developing Organizations, regulatory bodies and administrations, and industry forums.

Standards Developing Organizations (SDOs) develop and agree on technical standards for mobile communications systems, in order to make it possible for the industry to produce and deploy standardized products and provide interoperability between those products. Most components of mobile-communication systems, including base stations and mobile devices, are standardized to some extent. There is also a certain degree of freedom to provide proprietary solutions in products, but the communications protocols rely on detailed standards for obvious reasons. SDOs are usually non-profit industry organizations and not government controlled. They often write standards within a certain area under mandate from governments(s) however, giving the standards a higher status.

There are nationals SDOs, but due to the global spread of communications products, most SDOs are regional and also cooperate on a global level. As an example, the technical specifications of GSM, WCDMA/HSPA, LTE and NR are all created by 3GPP (Third Generation Partnership Project) which is a global organization from seven regional and national SDOs in Europe (ETSI), Japan (ARIB and TTC), United States (ATIS), China (CCSA), Korea (TTA) and India (TSDSI). SDOs tend to have a varying degree of transparency, but 3GPP is fully transparent with all technical specifications, meeting documents, reports and e-mail reflectors publicly available without charge even for non-members.

5G/5G-Advanced
https://doi.org/10.1016/B978-0-443-13173-8.00011-6

Regulatory bodies and administrations are government-led organizations that set regulatory and legal requirements for selling, deploying and operating mobile systems and other telecommunication products. One of their most important tasks is to control spectrum use and to set licensing conditions for the mobile operators that are awarded licenses to use parts of the *Radio Frequency* (RF) spectrum for mobile operations. Another task is to regulate "placing on the market" of products through regulatory certification, by ensuring that devices, base stations and other equipment is type approved and shown to meet the relevant regulation.

Spectrum regulation is handled both on a national level by national administrations, but also through regional bodies in Europe (CEPT/ECC), the Americas (CITEL) and Asia (APT). On a global level, the spectrum regulation is handled by the *International Telecommunications Union* (ITU). The regulatory bodies regulate what services the spectrum is to be used for and in addition set more detailed requirements such as limits on unwanted emissions from transmitters. They are also indirectly involved in setting requirements on the product standards through regulation. The involvement of ITU in setting requirements on the technologies for mobile communication is explained further in Section 2.2.

Industry forums are industry led groups promoting and lobbying for specific technologies or other interests. In the mobile industry, these are often led by operators, but there are also vendors creating industry forums. An example of such a group is GSMA (GSM Association) which is promoting mobile-communication technologies based on GSM, WCDMA, LTE and NR. Other examples of industry forums are *Next Generation Mobile Networks* (NGMN) *Alliance* which is an open forum founded by leading operators defining requirements on the evolution of mobile systems, and *5G Americas* which is a regional industry forum that advocates for the advancement and transformation of 5G and beyond throughout the Americas.

Fig. 2.1 illustrates the relation between different organizations involved in setting regulatory and technical conditions for mobile systems. The figure also shows the mobile industry view, where vendors develop products, place them on the market and negotiate with operators who procure and deploy mobile systems. This process relies heavily on the technical standards published by the SDOs, while placing products on the market relies on certification of products on a regional or national level. Note that in Europe, the regional SDO (ETSI) is producing the so called *harmonised standards* used for product certification (through the "CE"-mark), based on a mandate from the regulators, in this case the European Commission. These standards are used for certification in many countries also outside of Europe. In Fig. 2.1, full arrows indicate formal documentation such as technical standards, recommendations and regulatory mandates that define the technologies and regulation. Dashed arrows show more indirect involvement through for example liaison statements and white papers.

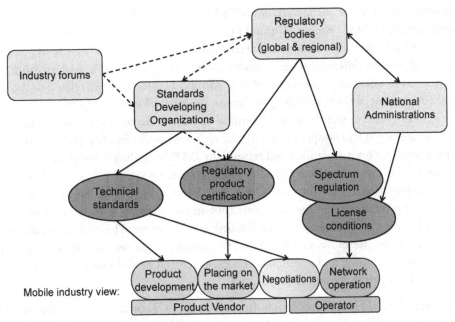

Fig. 2.1 Simplified view of relation between Regulatory bodies, standards developing organizations, industry forums and the mobile industry.

2.2 ITU-R activities from 3G to 6G

2.2.1 The role of ITU-R

ITU-R is the Radio communications sector of the International Telecommunications Union. ITU-R is responsible for ensuring efficient and economical use of the radio-frequency (RF) spectrum by all radio communication services. The different study groups and working parties produce reports and recommendations that analyze and define the conditions for using the RF spectrum. The quite ambitious goal of ITU-R is to "ensure interference-free operations of radiocommunication systems", by implementing the *Radio Regulations* and regional agreements. The Radio Regulations [46] is an international binding treaty for how RF spectrum is used. A *World Radiocommunication Conference* (WRC) is held every 3–4 years. At WRC the Radio Regulations are revised and updated, resulting in revised and updated use of the RF spectrum across the world.

While the technical specification of mobile-communication technologies, such as NR, LTE and WCDMA/HSPA is done within 3GPP, there is a responsibility for ITU-R in the process of turning the technologies into global standards, in particular for countries that are not covered by the SDOs that are partners in 3GPP. ITU-R defines the spectrum for different services in the RF spectrum, including mobile services, and some of that spectrum is particularly identified for so-called International Mobile Tele-communications (IMT) systems. Within ITU-R, it is *Working Party 5D* (WP5D) that has the responsibility for the overall radio system aspects of IMT systems, which, in practice, corresponds to the different generations of mobile communication systems from 3G onwards. WP5D has the prime responsibility within ITU-R for issues related to the ter-restrial component of IMT, including technical, operational and spectrum-related issues.

WP5D does not create the actual technical specifications for IMT, but has kept the roles of defining IMT in cooperation with the regional standardization bodies and main-taining a set of recommendations and reports for IMT, including a set of detailed spec-ifications of the radio interfaces. These recommendations contain "families" of *Radio Interface Technologies* (RITs) for each IMT generation, all included on an equal basis. For each radio interface, the specifications contain an overview of that radio interface, followed by a list of references to the detailed specifications. The actual specifications are maintained by the individual SDO and the specifications provide references to the specifications transposed and maintained by each SDO. The following detailed specifi-cations for IMT are in existence or planned:

- For IMT-2000: ITU-R Recommendation M.1457 [47] containing six different RITs including the 3G technologies such as WCDMA/HSPA.
- For IMT-Advanced: ITU-R Recommendation M.2012 [43] containing two differ-ent RITs where the most important one is 4G/LTE.
- For IMT-2020: ITU-R Recommendation M.2150 [123] containing two RITs and two SRITs, where the most important ones are the 5G NR RIT and SRIT.
- For IMT-2030: A new ITU-R Recommendation, containing the RITs for 6G tech-nologies is planned for completion in 2030.

Each set of detailed specifications are continuously updated to reflect new developments in the referenced detailed specifications, such as the 3GPP specifications for WCDMA and LTE. Input to the updates is provided by the SDOs and the Partnership Projects, nowadays primarily 3GPP.

2.2.2 IMT-2000 and IMT-Advanced

Work on what corresponds to third generation of mobile communication started in the ITU-R in the 1980s. First referred to as *Future Public Land Mobile Systems* (FPLMTS) it was later renamed IMT-2000. In the late 1990s, the work in ITU-R coincided with the work in different SDOs across the world to develop a new generation of mobile systems. Detailed specifications for IMT-2000 was first published in 2000 and included WCDMA from 3GPP as one of the RITs.

The next step for ITU-R was to initiate work on IMT-Advanced, the term used for systems that include new radio interfaces supporting new capabilities of systems beyond IMT-2000. The new capabilities were defined in a framework recommendation

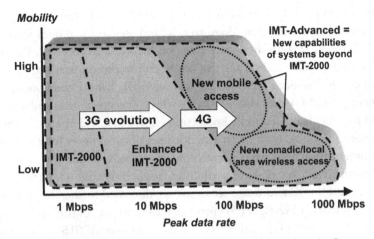

Fig. 2.2 Illustration of capabilities of IMT-2000 and IMT-Advanced, based on the framework described in ITU-R Recommendation M.1645 [39].

published by the ITU-R [39] and were demonstrated with the "van diagram" shown in Fig. 2.2. The step into IMT-Advanced capabilities by ITU-R coincided with the step into 4G, the next generation of mobile technologies after 3G.

An evolution of LTE as developed by 3GPP was submitted as one candidate technology for IMT-Advanced. While actually being a new release (Release 10) of the LTE specifications and thus an integral part of the continuous evolution of LTE, the candidate was named LTE-Advanced for the purpose of ITU-R submission and this name is also used in the LTE specifications from Release 10. In parallel with the ITU-R work, 3GPP set up its own set of technical requirements for LTE-Advanced, with the ITU-R requirements as a basis [10].

The target of the ITU-R process is always harmonization of the candidates through consensus building. ITU-R determined that two technologies would be included in the first release of IMT-Advanced, those two being LTE-Advanced and WirelessMAN-Advanced [35] based on the IEEE 802.16m specification. The two can be viewed as the "family" of IMT-Advanced technologies. Note that, among these two technologies, LTE has emerged as the dominating 4G technology by far.

2.2.3 5G and IMT-2020 in ITU-R WP5D

Starting in 2012, ITU-R WP5D set the stage for the next generation of IMT systems, named IMT-2020. As a further development of the terrestrial component of IMT beyond the year 2020, it corresponds to what is more commonly referred to as "5G", the fifth generation of mobile systems. The framework and objective for IMT-2020 is outlined in ITU-R Recommendation M.2083 [45], often referred to as the "Vision" recommendation. The recommendation provided the first step for defining the new developments of IMT, looking at the future roles of IMT and how it can serve society, looking at market, user and technology trends, and spectrum implications. The user trends for IMT together with the future role and market lead to a set of *usage scenarios* envisioned for both human-centric and machine-centric communication. The usage scenarios identified are *Enhanced Mobile Broadband* (eMBB), *Ultra-Reliable and Low Latency Communications* (URLLC) and *Massive Machine-Type Communications* (mMTC).

The need for an enhanced mobile broadband experience, together with the new and broadened usage scenarios, leads to an extended set of capabilities for IMT-2020. The Vision recommendation [45] gave a first high-level guidance for IMT-2020 requirements by introducing a set of key capabilities, with indicative target numbers. The key capabilities and the related usage scenarios are discussed further in Section 2.3.

As a parallel activity, ITU-R WP5D produced a report on "Future technology trends of terrestrial IMT systems" [41], with a focus on the time period 2015–20. It covers trends of IMT technology aspects by looking at the technical and operational characteristics of IMT systems and how they are improved with the evolution of IMT technologies. In this way, the report on technology trends relates to LTE in 3GPP Release 13 and beyond, while the Vision recommendation looked further ahead and beyond 2020. A new aspect considered for IMT-2020 is that it is capable of operating in potential new IMT bands above 6 GHz, including mm-wave bands. With this in mind, WP5D produced a separate report studying radio wave propagation, IMT characteristics, enabling technologies, and deployment in frequencies above 6 GHz [42].

At WRC-15, potential new bands for IMT were discussed and an agenda item 1.13 was set up for WRC-19, covering possible additional allocations to the mobile services and for future IMT development. These allocations were identified in a number of frequency bands in the range between 24.25 and 86 GHz. At WRC-19, several new bands identified for IMT emerged as an outcome of agenda item 1.13. The specific bands and their possible use globally are further discussed in Chapter 3.

After WRC-15, ITU-R WP5D continued the process of setting requirements and defining evaluation methodologies for IMT-2020 systems, based on the Vision recommendation [45] and the other previous study outcomes. This step of the process was completed in mid-2017, as shown in the IMT-2020 work plan in Fig. 2.3. The result was three documents published late in 2017 that further define the performance and characteristics for IMT-2020 and that is applied in the evaluation phase:

- **Technical requirements:** Report ITU-R M.2410 [49] defines 13 minimum requirements related to the technical performance of the IMT-2020 radio interface(s). The requirements are to a large extent based on the key capabilities set out in the Vision recommendation [45]. This is further described in Section 2.3.
- **Evaluation guideline:** Report ITU-R M.2412 [48] defines the detailed methodology to use for evaluating the minimum requirements, including test environments, evaluation configurations, and channel models. More details are given in Section 2.3.

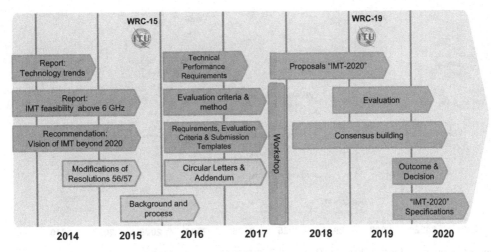

Fig. 2.3 Work plan for IMT-2020 in ITU-R WP5D [38].

- **Submission template:** Report ITU-R M.2411 [50] provides a detailed template to use for submitting a candidate technology for evaluation. It also details the evaluation criteria and requirements on service, spectrum and technical performance, based on the two previously mentioned ITU-R reports M.2410 and M.2412.

External organizations were informed of the IMT-2020 process through a circular letter. After a workshop on IMT-2020 was held in October 2017, the IMT-2020 process was open for receiving candidate proposals. A total of seven candidates were submitted from six proponents. These are presented in Section 2.3.4.

The work plan for IMT-2020 in Fig. 2.3 shows the complete timeline starting with technology trends and "Vision" in 2014, continuing with the submission and evaluation of proposals in 2018, and resulting in an outcome with the detailed specifications for IMT-2020 published late in early 2021 [123].

2.2.4 IMT-2030 and ITU-R work towards 6G

The first steps towards the next generation were taken in ITU-R WP5D in 2021 and will around the year 2030 lead to new detailed specifications for IMT-2030, more commonly called 6G. The present time plan for IMT-2030 in ITU-R is detailed in Fig. 2.4 and describes a process that has similar steps and targets as the one for IMT-2020. The main difference is that the total time span is ~3 years longer, giving more time to develop the evaluation process including requirements, and for the final evaluation and consensus building.

As a first step, a report on "Future technology trends of terrestrial IMT systems towards 2030 and beyond" was published in 2022 [122]. The report first gives an overview of emerging services and applications for IMT, and then goes into more details of

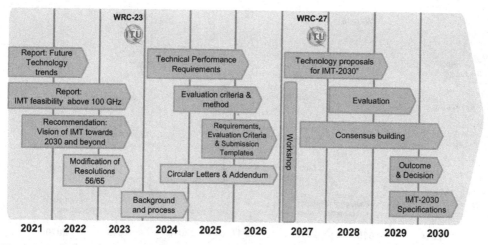

Fig. 2.4 Work plan for IMT-2030 in ITU-R WP5D [124].

emerging general technology trends and enablers, continuing with specific technologies to enhance the radio interface and the radio network.

The framework and objective for IMT-2030 will be outlined in a new ITU-R recommendation that is targeted for completion in 2023. It will cover motivation and societal considerations for IMT-2030, user and application trends, technology trends and spectrum implications. It also describes an expanded set of usage scenarios and capabilities for IMT-2030, where use cases are grouped in terms of the capabilities needed from the system.

Based on the needed capabilities, continued work on the technical performance requirements and evaluation criteria will continue after WRC-23, leading to submission and evaluation of technology proposals to be completed after WRC-27, and resulting in detailed specifications of IMT-2030 by the year 2030.

2.3 5G and IMT-2020

The detailed ITU-R process for developing IMT-2020 was presented above with the most important steps summarized in Fig. 2.3. The ITU-R activities on IMT-2020 started with development of the "vision" recommendation ITU-R M.2083 [45], outlining the expected usage scenarios and corresponding required capabilities of IMT-2020. This was followed by definition of more detailed requirements for IMT-2020, requirements that candidate technologies are then to be evaluated against, as documented in the evaluation guidelines. The requirements and evaluation guidelines were finalized mid-2017.

The candidate technologies submitted to ITU-R were evaluated both through a self-evaluation by the proponent and by independent evaluation groups, based on the IMT-2020 requirements. The technologies that fulfill the requirements are approved and published as part of the IMT-2020 specifications [123]. Further details on the ITU-R process can be found in Section 2.2.3.

2.3.1 Usage scenarios for IMT-2020

With a wide range of new use cases being one principal driver for 5G, ITU-R defined three usage scenarios that form a part of the IMT Vision recommendation [45]. Inputs from the mobile industry and different regional and operator organizations were taken into the IMT-2020 process in ITU-R WP5D, and were synthesized into the three scenarios:

- **Enhanced Mobile Broadband (eMBB):** With mobile broadband today being the main driver for use of 3G and 4G mobile systems, this scenario points at its continued role as the most important usage scenario. The demand is continuously increasing, and new application areas are emerging, setting new requirements for what ITU-R calls *Enhanced Mobile Broadband*. Because of its broad and ubiquitous use, it covers a range of use cases with different challenges, including both hotspots and wide-area coverage, with the first one enabling high data rates, high user density and a need for very high capacity, while the second one stresses mobility and a seamless user experience, with lower requirements on data rate and user density. The Enhanced Mobile Broadband scenario is in general seen as addressing human-centric communication.
- **Ultra-reliable and low latency communications (URLLC):** This scenario is intended to cover both human- and machine-centric communication, where the latter is often referred to as critical machine type communication (C-MTC). It is characterized by use cases with stringent requirements for latency, reliability, and high availability. Examples include vehicle-to-vehicle communication involving safety, wireless control of industrial equipment, remote medical surgery, and distribution automation in a smart grid. An example of a human-centric use case is 3D gaming and "tactile internet," where the low-latency requirement is also combined with very high data rates.
- **Massive machine type communications (mMTC):** This is a pure machine-centric use case, where the main characteristic is a very large number of connected devices that typically have very sparse transmissions of small data volumes that are not delay-sensitive. The large number of devices can give a very high connection density locally, but it is the total number of devices in a system that can be the real challenge and stresses the need for low cost. Due to the possibility of remote deployment of mMTC devices, they are also required to have a very long battery lifetime.

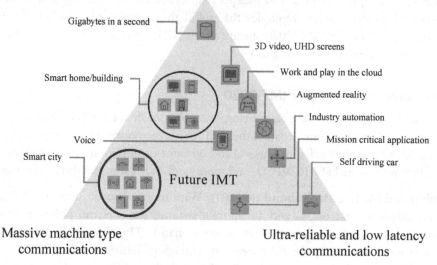

Fig. 2.5 IMT-2020 use cases and mapping to usage scenarios. *From ITU-R, IMT Vision — Framework and overall objectives of the future development of IMT for 2020 and beyond, Recommendation ITU-R M.2083, September 2015, used with permission from the ITU.*

The usage scenarios are illustrated in Fig. 2.5, together with some example use cases. The three scenarios above are not claimed to cover all possible use cases, but they provide a relevant grouping of a majority of the presently foreseen use cases and can thus be used to identify the key capabilities needed for the next generation radio interface technology for IMT-2020. There will most certainly be new use cases emerging, which we cannot foresee today or describe in any detail. This also means that the new radio interface must have a high flexibility to adapt to new use cases and the "space" spanned by the range of the key capabilities supported should support the related requirements emerging from evolving use cases.

2.3.2 Capabilities of IMT-2020

As part of developing the framework for IMT-2020 as documented in the IMT Vision recommendation [45], ITU-R defined a set of capabilities needed for an IMT-2020 technology to support the 5G use cases and usage scenarios identified through the inputs from regional bodies, research projects, operators, administrations and other organizations. There is a total of 13 capabilities defined in ITU-R [45], where eight were selected as *key capabilities*. Those eight key capabilities are illustrated through two "spider web" diagrams, see Figs. 2.6 and 2.7.

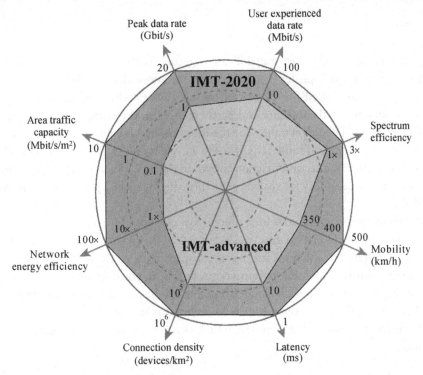

Fig. 2.6 Key capabilities of IMT-2020. *From ITU-R, IMT Vision – Framework and overall objectives of the future development of IMT for 2020 and beyond, Recommendation ITU-R M.2083, September 2015, used with permission from the ITU.*

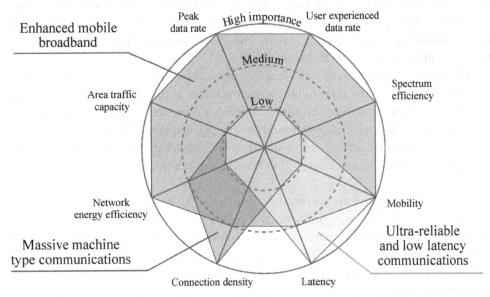

Fig. 2.7 Relation between key capabilities and the three usage scenarios of ITU-R. *From ITU-R, IMT Vision – Framework and overall objectives of the future development of IMT for 2020 and beyond, Recommendation ITU-R M.2083, September 2015, used with permission from the ITU.*

Fig. 2.6 illustrates the key capabilities together with indicative target numbers intended to give a first high-level guidance for the more detailed IMT-2020 requirements that were later used for evaluation of candidate technologies. As can be seen the target values are partly absolute and partly relative to the corresponding capabilities of IMT-Advanced. The target values for the different key capabilities do not have to be reached simultaneously and some targets are to a certain extent even mutually exclusive. For this reason, there is a second diagram shown in Fig. 2.7 which illustrates the "importance" of each key capability for realizing the three high-level usage scenarios envisioned by ITU-R.

Peak data rate is a number on which there is always a lot of focus, but it is in fact quite an academic exercise. ITU-R defines peak data rate as the maximum achievable data rate under ideal conditions, which means that the impairments in an implementation or the actual impact from a deployment in terms of propagation, etc. does not come into play. It is a dependent *key performance indicator* (KPI) in that it is heavily dependent on the amount of spectrum available for an operator deployment. Apart from that, the peak data rate depends on the peak spectral efficiency, which is the peak data rate normalized by the bandwidth:

$$\text{Peak data rate} = \text{System bandwidth} \times \text{Peak spectral efficiency}$$

Since large bandwidths are really not available in any of the existing IMT bands below 6 GHz, it is expected that really high data rates will be more easily achieved at higher frequencies. This leads to the conclusion that the highest data rates can be achieved in indoor and hotspot environments, where the less favorable propagation properties at higher frequencies are of less importance.

The *user experienced data rate* is the data rate that can be achieved over a large coverage area for a majority of the users. This can be evaluated as the 95th percentile from the distribution of data rates between users. It is also a dependent capability, not only on the available spectrum but also on how the system is deployed. While a target of 100 Mbit/s is set for wide area coverage in urban and suburban areas, it is expected that 5G systems could give 1 Gbit/s data rate ubiquitously in indoor and hotspot environments.

Spectrum efficiency gives the average data throughput per Hz of spectrum and per "cell," or rather per unit of radio equipment (also referred to as *Transmission Reception Point*, TRP). It is an essential parameter for dimensioning networks, but the levels achieved with 4G systems are already very high. The target for IMT-2020 was set to three times the spectrum efficiency target of 4G, but the achievable increase strongly depends on the deployment scenario.

Area traffic capacity is another dependent capability, which depends not only on the spectrum efficiency and the bandwidth available, but also on how dense the network is deployed:

$$\text{Area Traffic Capacity} = \text{Spectrum efficiency} \cdot \text{BW} \cdot \text{TRP density}$$

By assuming the availability of more spectrum at higher frequencies and that very dense deployments can be used, a target of a 100-fold increase over 4G was set for IMT-2020.

Network energy efficiency is, as already described, becoming an increasingly important capability. The overall target stated by ITU-R is that the energy consumption of the radio access network of IMT-2020 should not be greater than IMT networks deployed today, while still delivering the enhanced capabilities. The target means that the network energy efficiency in terms of energy consumed per bit of data therefore needs to be reduced with a factor at least as great as the envisaged traffic increase of IMT-2020 relative to IMT-Advanced.

These first five key capabilities are of highest importance for the Enhanced Mobile Broadband usage scenario, although mobility and the data rate capabilities would not have equal importance simultaneously. For example, in hotspots, a very high user-experienced and peak data rate, but a lower mobility, would be required than in wide area coverage case.

The next key capability is *latency*, which is defined as the contribution by the radio network to the time from when the source sends a packet to when the destination receives it. It will be an essential capability for the URLLC usage scenario and ITU-R envisions that a 10-fold reduction in latency from IMT-Advanced is required.

Mobility is in the context of key capabilities only defined as mobile speed and the target of 500 km/h is envisioned in particular for high-speed trains and is only a moderate increase from IMT-Advanced. As a key capability, it will, however, also be essential for the URLLC usage scenario in the case of critical vehicle communication at high speed and will then be of high importance simultaneously with low latency. Note that high mobility and high user-experienced data rates are not targeted simultaneously in the usage scenarios.

Connection density is defined as the total number of connected and/or accessible devices per unit area. The target is relevant for the mMTC usage scenario with a high density of connected devices, but an eMBB dense indoor office can also give a high connection density.

In addition to the eight capabilities given in Fig. 2.6 there are five additional capabilities defined in [45]:

- *Spectrum and bandwidth flexibility*:
  Spectrum and bandwidth flexibility refers to the flexibility of the system design to handle different scenarios, and in particular to the capability to operate at different frequency ranges, including higher frequencies and wider channel bandwidths than in previous generations.
- *Reliability*:
  Reliability relates to the capability to provide a given service with a very high level of availability.
- *Resilience*:
  Resilience is the ability of the network to continue operating correctly during and after a natural or man-made disturbance, such as the loss of mains power.
- *Security and privacy*:
  Security and privacy refer to several areas such as encryption and integrity protection of user data and signaling, as well as end user privacy preventing unauthorized user tracking, and protection of network against hacking, fraud, denial of service, man in the middle attacks, etc.
- *Operational lifetime*:
  Operational life time refers to operation time per stored energy capacity. This is particularly important for machine-type devices requiring a very long battery life (for example more than 10 years) whose regular maintenance is difficult due to physical or economic reasons.

Note that these capabilities are not necessarily less important than the capabilities of Fig. 2.6 despite the fact that the latter are referred to as "key capabilities". The main difference is that the "key capabilities" are more easily quantifiable while the remaining five capabilities are more of qualitative capabilities that cannot easily be quantified.

2.3.3 IMT-2020 performance requirements

Based on the usage scenarios and capabilities described in the Vision recommendation [45], ITU-R developed a set of minimum technical performance requirements for IMT-2020. These are documented in ITU-R report M.2410 [49] and served as the baseline for the evaluation of IMT-2020 candidate technologies (see Fig. 2.3). The report describes 14 technical parameters and the corresponding minimum requirements. These are summarized in Table 2.1.

The evaluation guideline of candidate radio interface technologies for IMT2020 is documented in ITU-R report M.2412 [48] and follows the same structure as the previous evaluation done for IMT-Advanced. It describes the evaluation methodology for the 14 minimum technical performance requirements, plus two additional requirements: support of a wide range of services and support of spectrum bands.

Table 2.1 Overview of minimum technical performance requirements for IMT-2020.

Parameter	Minimum technical performance requirement
Peak data rate	Downlink: 20 Gbit/s Uplink: 10 Gbit/s
Peak spectral efficiency	Downlink: 30 bit/s/Hz Uplink: 10 bit/s/Hz
User experienced data rate	Downlink: 100 Mbit/s Uplink: 50 Mbit/s
5th percentile user spectral efficiency	$3\times$ IMT-Advanced
Average spectral efficiency	$3\times$ IMT-Advanced
Area traffic capacity	10 Mbit/s/m^2 (Indoor hotspot for eMBB)
User plane latency	4 ms for eMBB 1 ms for URLLC
Control plane latency	20 ms
Connection density	1 000 000 devices per km^2
Energy efficiency	Related to two aspects for eMBB: **(a)** Efficient data transmission in a loaded case **(b)** Low energy consumption when there is no data The technology shall have the capability to support a high sleep ratio and long sleep duration.
Reliability	$1\text{--}10^{-5}$ success probability of transmitting a layer 2 PDU (Protocol Data Unit) of 32 bytes within 1 ms, at coverage edge in Urban Macro for URLLC
Mobility	Normalized traffic channel data rates defined for 10, 30 and 120 km/h at $\sim1.5\times$ IMT-Advanced numbers. Requirement for High speed vehicular defined for 500 km/h (compared to 350 km/h for IMT-Advanced).
Mobility interruption time	0 ms
Bandwidth	At least 100 MHz and up to 1 GHz in higher frequency bands. Scalable bandwidth shall be supported.

The evaluation is done with reference to five *test environments* that are based on the usage scenarios from the Vision recommendation [45]. Each test environment has a number of *evaluation configurations* that describe the detailed parameters that are to be used in simulations and analysis for the evaluation. The five test environments are:

- **Indoor Hotspot-eMBB:** An indoor isolated environment at offices and/or in shopping malls based on stationary and pedestrian users with very high user density.
- **Dense Urban-eMBB:** An urban environment with high user density and traffic loads focusing on pedestrian and vehicular users.
- **Rural-eMBB:** A rural environment with larger and continuous wide area coverage, supporting pedestrian, vehicular and high speed vehicular users.

- **Urban Macro–mMTC:** An urban macro environment targeting continuous coverage focusing on a high number of connected machine type devices.
- **Urban Macro–URLLC:** An urban macro environment targeting ultra-reliable and low latency communications.

There are three fundamental ways that requirements are evaluated for a candidate technology:

- **Simulation:** This is the most elaborate way to evaluate a requirement and it involves system- or link-level simulations, or both, of the radio interface technology. For system-level simulations, deployment scenarios are defined that correspond to a set of test environments, such as indoor, dense urban, etc. Requirements that are evaluated through simulation are average and fifth percentile spectrum efficiency, connection density, mobility and reliability.
- **Analysis:** Some requirements can be evaluated through a calculation based on radio interface parameters or be derived from other performance values. Requirements that are evaluated through analysis are peak spectral efficiency, peak data rate, user experienced data rate, area traffic capacity, control and user plane latency, and mobility interruption time.
- **Inspection:** Some requirements can be evaluated by reviewing and assessing the functionality of the radio interface technology. Requirements that are evaluated through inspection are bandwidth, energy efficiency, support of wide range of services and support of spectrum bands.

Once candidate technologies are submitted to ITU-R and have entered the process, the evaluation phase starts. Evaluation is done by the proponent ("self-evaluation") or by an external evaluation group, doing partial or complete evaluation of one or more candidate proposals.

2.3.4 IMT-2020 candidates and evaluation

As shown in Fig. 2.3, the IMT-2020 work plan spanned over seven years and was completed in 2020. The details for submitting candidate technologies for IMT-2020 is described in detail in the WP5D agreed process [92].

Submissions for candidates were either as an individual Radio Interface Technology (RIT) or a Set of Radio Interface Technologies (SRIT). The following were the criteria for submission, and defines what can be a RIT or SRIT in relation to the IMT-2020 minimum requirements:

- A RIT needs to fulfil the minimum requirements for at least three test environments; two test environments under eMBB and one test environment under mMTC or URLLC.
- An SRIT consists of a number of component RITs complementing each other, with each component RIT fulfilling the minimum requirements of at least two test

environments and together as an SRIT fulfilling the minimum requirements of at least four test environments comprising the three usage scenarios.

A number of IMT-2020 candidates were submitted to the ITU-R up until the formal deadline on 2 July 2019. Each submission contained characteristics template, compliance template, link-budget template and self-evaluation report. A total of seven submissions were made from six different proponents.

Based on the outcome of the evaluation and some merging of proposals, the following RIT/SRIT were included in the detailed specification ITU-R recommendation M.2150 [123], as updated in 2022:

- **3GPP 5G-SRIT:** The SRIT from 3GPP consists of NR and LTE as component RITs. Both the individual RIT components as well as the complete SRIT fulfils the evaluation criteria above. The self-evaluation is contained in 3GPP TR 37.910 [93].
- **3GPP 5G-RIT:** The RIT from 3GPP consists of NR. It fulfils all test environments for all usage scenarios. The same self-evaluation document is used [93] as for the 3GPP 5G-SRIT
- **5Gi:** The RIT from TSDSI is based on the 3GPP 5G NR technology, with a limited set of changes applied to the specifications. An independent evaluation report was submitted for the RIT.
- **DECT 5G-SRIT:** The SRIT from ETSI/DECT Forum consists of DECT-2020 as one component RIT and 3GPP 5G NR as a second component RIT. References are made to the 3GPP NR submission and 3GPP self-evaluation in [93] for aspects related to NR.

It should be noted that all of the final RIT/SRITs included are either based directly on 5G NR or have 5G NR as one component RIT.

An additional RIT was submitted consisting of the Enhanced Ultra High Throughput (EUHT) technology. After a final evaluation completed in 2022, it was concluded that the candidate technology could not be declared as a qualified Radio Interface Technology and it did therefore not go forward for inclusion in the detailed specifications.

2.4 3GPP standardization

With a framework for IMT systems set up by the ITU-R, with spectrum made available by the WRC and with an ever-increasing demand for better performance, the task of specifying the actual mobile-communication technologies falls on organizations like 3GPP. More specifically, 3GPP writes the technical specifications for 2G GSM, 3G WCDMA/HSPA, 4G LTE and 5G NR. 3GPP technologies are the most widely deployed in the world, with more than 95% of the world's 9 billion mobile subscriptions in 2022 [28]. In order to understand how 3GPP works, it is important to also understand the process of writing specifications.

2.4.1 The 3GPP process

Developing technical specifications for mobile communication is not a one-time job; it is an ongoing process. The specifications are constantly evolving, trying to meet new demands for services and features. The process is different in the different fora, but typically includes the four phases illustrated in Fig. 2.8:

- *Requirements*, where it is decided what is to be achieved by the specification.
- *Architecture*, where the main building blocks and interfaces are decided.
- *Detailed specifications*, where every interface is specified in detail.
- *Testing and verification*, where the interface specifications are proven to work with real-life equipment.

These phases are overlapping and iterative. As an example, requirements can be added, changed, or dropped during the later phases if the technical solutions call for it. Likewise, the technical solution in the detailed specifications can change due to problems found in the testing and verification phase.

The specification starts with the *requirements* phase, where it is decided what should be achieved with the specification. This phase is usually relatively short.

In the *architecture* phase, the architecture is decided – that is, the principles of how to meet the requirements. The architecture phase includes decisions about reference points and interfaces to be standardized. This phase is usually quite long and may change the requirements.

After the architecture phase, the *detailed specification* phase starts. It is in this phase that the details for each of the identified interfaces are specified. During the detailed specification of the interfaces, the standards body may find that previous decisions in the architecture or even in the requirements phases need to be revisited.

Finally, the *testing and verification* phase starts. It is usually not a part of the actual specification, but takes place in parallel through testing by vendors and interoperability testing between vendors. This phase is the final proof of the specification. During the testing and verification phase, errors in the specification may still be found and those errors may change decisions in the detailed specification. Albeit not common, changes may also need to be made to the architecture or the requirements. To verify the specification, products are needed. Hence, the implementation of the products starts after (or during) the detailed specification phase. The testing and verification phase ends when there are stable

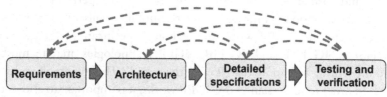

Fig. 2.8 The standardization phases and iterative process.

test specifications that can be used to verify that the equipment is fulfilling the technical specification.

Normally, it takes approximately one year from the time when the specification is completed until commercial products are out on the market.

3GPP consists of three *Technical Specifications Groups* (TSGs) – see Fig. 2.9 – where TSG RAN (*Radio Access Network*) is responsible for the definition of functions, requirements, and interfaces of the Radio Access. TSG RAN consists of six working groups (WGs):

- RAN WG1, dealing with the physical layer specifications.
- RAN WG2, dealing with the layer 2 and layer 3 radio interface specifications.
- RAN WG3, dealing with the fixed RAN interfaces – for example, interfaces between nodes in the RAN – but also the interface between the RAN and the core network.
- RAN WG4, dealing with the *radio frequency* (RF) and *radio resource management* (RRM) performance requirements.

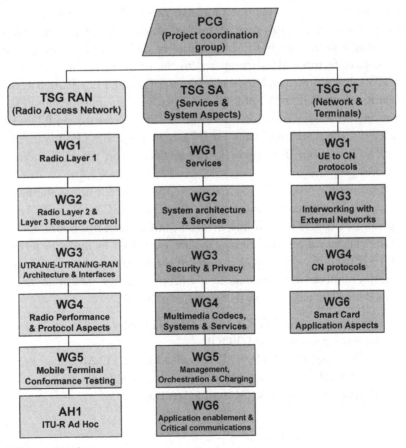

Fig. 2.9 3GPP organization.

- RAN WG 5, dealing with the device conformance testing.
- RAN AHG ITU, is a permanent ad hoc group that acts as a focal point to ensure a proper flow of information and contributions between 3GPP and ITU-R

Until 2020, there was an additional group RAN WG6, dealing with standardization of GSM/EDGE (previously in a separate TSG called GERAN) and HSPA (UTRAN). Any remaining activity in those areas has become a direct TSG RAN responsibility. 3GPP standardization activities are now focused on 4G and 5G.

The work in 3GPP is carried out with relevant ITU-R recommendations in mind and the result of the work is also submitted to ITU-R as being part of IMT-2000, IMT-Advanced, and now also part of IMT-2020. The organizational partners are obliged to identify regional requirements that may lead to options in the standard. Examples are regional frequency bands and special protection requirements local to a region. The specifications are developed with global roaming and circulation of devices in mind. This implies that many regional requirements in essence will be global requirements for all devices, since a roaming device has to meet the strictest of all regional requirements. Regional options in the specifications are thus more common for base stations than for devices.

The specifications of all releases can be updated after each set of TSG meetings, which occur four times a year. The 3GPP documents are divided into releases, where each release has a set of features added compared to the previous release. The features are defined in Work Items agreed and undertaken by the TSGs. LTE is defined from Release 8 and onwards, where Release 10 of LTE is the first version approved by ITU-R as an IMT-Advanced technology and is therefore also the first release named LTE-Advanced. From Release 13, the marketing name for LTE is changed to LTE-Advanced Pro. An overview of LTE is given in Chapter 4. Further details on the LTE radio interface can be found in [26].

The first release for NR is in 3GPP Release 15, marketed under the name "5G". From Release 18, the marketing name for NR is changed to 5G-Advanced. An overview of NR is given in Chapter 5, with further details throughout this book.

The 3GPP Technical Specifications (TS) are organized in multiple series and are numbered TS XX.YYY, where XX denotes the number of the specification series and YYY is the number of the specification within the series. The following series of specifications define the radio access technologies in 3GPP:

- 25-series: Radio aspects for UTRA (WCDMA/HSPA)
- 45-series: Radio aspects for GSM/EDGE
- 36-series: Radio aspects for LTE, LTE-Advanced and LTE-Advanced Pro
- 37-series: Aspects relating to multiple radio access technologies
- 38-series: Radio aspects for NR (5G and 5G-Advanced)

2.4.2 Specification of 5G in 3GPP as an IMT-2020 candidate

In parallel with the definition and evaluation of the next-generation access initiated in ITU-R, 3GPP started to define the next-generation 3GPP radio access. A workshop on 5G radio access was held in 2015 and a process to define the evaluation criteria for 5G was initiated with a second workshop in early 2016. The evaluation follows the same process that was used when LTE-Advanced was evaluated and submitted to ITU-R and approved as a 4G technology as part of IMT-Advanced. The evaluation and submission of NR follows the ITU-R timeline described in Section 2.2.3.

The scenarios, requirements, and evaluation criteria to use for the new 5G radio access are described in the 3GPP report TR 38.913 [94], which is in general aligned with the corresponding ITU-R reports [48,49]. As for the case of the IMT-Advanced evaluation, the corresponding 3GPP evaluation of the next-generation radio access has a larger scope and may have stricter requirements than the ITU-R evaluation of candidate IMT-2020 radio interface technologies that is defined by ITU-R WP5D.

The standardization work for NR started with a study item phase in Release 14 and continued with development of a first set of specifications through a work item in Release 15. A first set of the Release 15 NR specifications was published in December 2017 and the full specifications were available in mid-2018. With the continuing work on NR, Release 16 specifications were published starting mid-2019. Further details on the time plan and the content of the NR releases is given in Chapter 5.

3GPP made a first submission of NR as an IMT-2020 candidate to the ITU-R WP5D meeting in February 2018. NR was submitted both as a RIT by itself and as an SRIT (set of component RITs) together with LTE. The following candidates were submitted, all including NR as developed by 3GPP:

- 3GPP submitted a candidate named "5G", containing two submissions: the first submission was an SRIT containing two component RITs, these being NR and LTE. The second submission was a separate RIT being NR.
- Korea submitted NR as a RIT, with reference to 3GPP.
- China submitted NR as a RIT, with reference to 3GPP.

Further submissions were made during 2018 and 2019 of characteristics templates, compliance templates, link-budget templates and self-evaluation reports, as part of the process described in Section 2.3.4. The self-evaluation performed by 3GP PTSG RAN is documented in 3GPP TR 37.910 [93].

During 2020, further inputs were made by 3GPP to ITU-R WP5D of as input to the detailed specification for IMT-2020 being developed by the ITU-R. In the global specification for IMT-2020 published by ITU-R [123], the 3GPP submission is included as 3GPP 5G-SRIT and 3GPP 5G-RIT. See Section 2.3.4 for more details.

CHAPTER 3

Spectrum for 5G

3.1 Spectrum for mobile systems

Historically, the bands for the first and second generation of mobile services were assigned at frequencies around 800–900 MHz, but also in a few lower and higher bands. When 3G (IMT-2000) was rolled out, focus was on the 2 GHz band and with the continued expansion of IMT services with 3G and 4G, new bands were added at both lower and higher frequencies, presently spanning from 450 MHz to around 6 GHz. While new, previously unexploited, frequency bands are continuously defined for new mobile generations, the bands used for previous generations are used for the new generation as well. This was the case when 3G and 4G were introduced and it is also the case for 5G.

Bands at different frequencies have different characteristics. Due to the propagation properties, bands at lower frequencies are good for wide area coverage deployments, in urban, suburban and rural environments. Propagation properties of higher frequencies make them more difficult to use for wide-area coverage and for this reason, higher frequency bands have to a larger extent been used for boosting capacity in dense deployments.

With the introduction of 5G, the demanding eMBB usage scenario and related new services requires even higher data rates and high capacity in dense deployments. While many early 5G deployments are in bands already used for previous mobile generations, frequency bands above 24 GHz are specified as a complement to the frequency bands below 6 GHz. With the 5G requirements for extreme data rates and localized areas with very high area traffic capacity demands, deployment using even higher frequencies, even above 60 GHz, are being considered. Referring to the wavelength, these bands are often called mm-wave bands.

New bands are defined continuously by 3GPP, both for NR and LTE specifications. Many new bands are defined for NR operation only, which is always the case for mm-wave operation. Both paired bands, where separated frequency ranges are assigned for uplink and downlink, and unpaired bands with a single shared frequency range for uplink and downlink, are included in the NR specifications. Paired bands are used for Frequency Division Duplex (FDD) operation, while unpaired bands are used for Time Division Duplex (TDD) operation. The duplex modes of NR are described further in Chapter 7. Note that some unpaired bands are defined as *Supplementary Downlink* (SDL) or *Supplementary Uplink* (SUL) bands. These bands are paired with the uplink or downlink of other bands through *carrier aggregation*, as described in Section 7.6.

5G/5G-Advanced
https://doi.org/10.1016/B978-0-443-13173-8.00021-9

3.1.1 Spectrum defined for IMT systems by the ITU-R

The ITU-R identifies frequency bands to use for mobile service and specifically for IMT. Many of these were originally identified for IMT-2000 (3G) and new ones came with the introduction of IMT-Advanced (4G). The identification is however technology and generation "neutral," since the identification is for IMT in general, regardless of generation or Radio Interface Technology. The global designations of spectrum for different services and applications are done within the ITU-R and are documented in the ITU Radio Regulations [46] and the use of IMT bands globally is described in ITU-R Recommendation M.1036 [44].

The frequency listings in the ITU Radio Regulations [46] do not directly list a band for IMT, but rather allocate a band for the mobile service with a footnote stating that the band is identified for use by administrations wishing to implement IMT. The identification is mostly by region, but is in some cases also specified on a per-country level. All footnotes mention "IMT" only, so there is no specific mentioning of the different generations of IMT. Once a band is assigned, it is therefore up to the regional and local administrations to define a band for IMT use in general or for specific generations. In many cases, regional and local assignments are "technology neutral" and allow for any kind of IMT technology. This means that all existing IMT bands are potential bands for IMT-2020 (5G) deployment in the same way as they have been used for previous IMT generations.

The *World Administrative Radio Congress* WARC-92 identified the bands 1885–2025 and 2110–2200 MHz as intended for implementation of IMT-2000. Out of these 230 MHz of 3G spectrum, 2×30 MHz were intended for the satellite component of IMT-2000 and the rest for the terrestrial component. Parts of the bands were used during the 1990s for deployment of 2G cellular systems, especially in the Americas. The first deployments of 3G in 2001–2002 in Japan and Europe were done in this band allocation, and for that reason it is often referred to as the IMT-2000 "core band".

Additional spectrum for IMT-2000 was identified at the World Radio-communication Conference[a] WRC-2000, where it was considered that an additional need for 160 MHz of spectrum for IMT-2000 was forecasted by the ITU-R. The identification includes the bands used for 2G mobile systems at 806–960 and 1710–1885 MHz, and "new" 3G spectrum in the bands at 2500–2690 MHz. The identification of bands previously assigned for 2G was also a recognition of the evolution of existing 2G mobile systems into 3G. Additional spectrum was identified at WRC-07 for IMT, encompassing both IMT-2000 and IMT-Advanced. The bands added were 450–470, 698–806, 2300–2400, and 3400–3600 MHz, but the applicability of the bands varied on a regional and national basis. At WRC-12 there were no additional spectrum allocations identified for IMT, but the issue was put on the agenda for WRC-15. It was also

---

[a] The World Administrative Radio Conference (WARC) was reorganized in 1992 and became the World Radio-communication Conference (WRC).

determined to study the use of the band 694–790 MHz for mobile services in Region 1 (Europe, Middle East and Africa).

WRC-15 was an important milestone setting the stage for 5G. First a new set of bands were identified for IMT, where many were identified for IMT on a global, or close to global, basis:

- 470–694/698 MHz (600 MHz band): Identified for some countries in Americas and the Asia-Pacific. For Region 1, it was considered for a new agenda item for IMT at WRC-23.
- 694–790 MHz (700 MHz band): This band was identified fully for Region 1 and is thereby a global IMT band.
- 1427–1518 MHz (L-band): A new global band identified in all countries.
- 3300–3400 MHz: Global band identified in many countries, but not in Europe or North America
- 3400–3600 MHz (C-band): A global band identified for all countries. The band was already allocated in Europe.
- 3600–3700 MHz (C-band): Global band identified in in many countries, but not in Africa and some counties in Asia-Pacific. In Europe, the band has been available since WRC-07.
- 4800–4990 MHz: New band identified for a few countries in Asia-Pacific.

Especially the frequency range from 3300 to 4990 MHz is of interest for 5G, since it is new spectrum in higher frequency bands. This implies that it fits well with the new usage scenarios requiring high data rates and is also suitable for massive MIMO implementation, where arrays with many elements can be implemented with reasonable size. Since it is new spectrum with no widespread previous use for mobile systems, it is easier to assign this spectrum in larger spectrum blocks, thereby enabling wider RF carriers and ultimately higher end user data rates.

At WRC-15, an agenda item 1.13 was appointed for WRC-19, to identify high-frequency bands above 24 GHz for IMT. Based on the studies conducted by the ITU-R following WRC-15, a set of new bands were identified for IMT at WRC-19, targeting mainly IMT-2020 and 5G mobile services. The new bands were assigned to the mobile service on a primary basis, in most bands together with fixed and satellite services. They consist of a total of 13.5 GHz in the following band ranges, where mobile services now have a primary allocation:

- 24.25–27.5 GHz
- 37–43.5 GHz
- 45.5–47 GHz
- 47.2–48.2
- 66–71 GHz

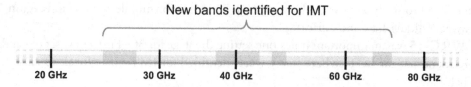

Fig. 3.1 New bands identified for IMT by WRC-19 are shown in blue (gray in print version).

Specific technical conditions were agreed at WRC-19 for some of these bands, specifically protection limits of the Earth Exploration Satellite Systems (EESS) in the frequency ranges 23.6–24.0 GHz and 36.0–37.0 GHz were defined. The new bands are illustrated in Fig. 3.1.

Agenda items were also created for the coming WRC-23 to consider further bands for IMT identification. The first two have identification for IMT in some countries and regions as mentioned above, but further consideration will now be made for those in other regions:

- 3300–3400 MHz
- 3600–3800 MHz,
- 6425–7025 MHz
- 7025–7125 MHz
- 10.0–10.5 GHz

It should be noted that there are also a large number of other frequency bands identified for *mobile services*, but not specifically for IMT. These bands are often used for IMT on a regional or national basis.

The somewhat diverging arrangement between regions of the frequency bands assigned to IMT means that there is not one single band that can be used for roaming worldwide. Large efforts have, however, been put into defining a minimum set of bands that can be used to provide truly global roaming. In this way, multiband devices can provide efficient worldwide roaming. With many of the new bands identified at WRC-15 and WRC-19 being global or close to global, global roaming is made possible for devices using fewer bands and it also facilitates economy of scale for equipment and deployment.

3.1.2 Global spectrum situation for 5G

There is a considerable interest globally to make spectrum available for 5G deployments. This is driven by operators and industry organizations such as the Global mobile Suppliers Association [33] and DIGITALEUROPE [27], but is also supported by regulatory bodies in different countries and regions. In standardization, 3GPP has focused its activities on bands where a high interest is evident (the full list of bands is in Section 3.2 below). The spectrum of interest can be divided into bands at low, medium, and high frequencies:

Low frequency bands correspond to existing LTE bands below 2 GHz, which are suitable as a coverage layer, providing wide and deep coverage, including indoor. The bands with highest interest here are the 600 and 700 MHz bands, which correspond to 3GPP NR bands n12, n13, n14, n28, n71, n83 and n105 (see Section 3.2 for further details). Since the bands are not very wide, a maximum of 20 MHz channel bandwidth is defined in the low frequency bands.

For early deployment, the 600 MHz band is considered for NR in the US, while the 700 MHz band is defined as one of the so-called pioneer bands for Europe. In addition, a number of additional LTE bands in the below 2 GHz range are identified for possible "re-farming" and have been assigned NR band numbers. Since the bands are in general already deployed with LTE, NR is expected to be deployed gradually at a later stage.

Medium frequency bands are in the range 2–6 GHz and can provide coverage, capacity, as well as high data rates through the wider channel bandwidth possible. The highest interest globally is in the range 3300–4200 MHz, where 3GPP has designated NR bands n77 and n78. Due to the wider bands, channel bandwidths up to 100 MHz are possible. Up to 200 MHz per operator may be assigned in this frequency range in the longer term, where carrier aggregation could then be used to deploy the full bandwidth.

The range 3300–4200 MHz is of global interest, with some variations seen regionally; and 3400–3800 MHz is a pioneer band in Europe, while China and India are planning for 3300–3600 MHz and in Japan 3600–4200 MHz is considered. Similar frequency ranges are considered in North America (3550–3700 MHz and initial discussions about 3700–4200 MHz), Latin America, the Middle East, Africa, India, Australia, etc. A total of 45 countries signed up to the IMT identification of the 3300–3400 MHz band at WRC-15. There is also a large amount of interest for a higher band in China (primarily 4800–5000 MHz) and Japan (4400–4900 MHz). In addition, there are a number of potential LTE refarming bands in the 2–6 GHz range that have been identified as NR bands.

In the range 5125–5925 MHz and 5925–7125 MHz, bands are defined for unlicensed operation with NR-U, see also Chapter 20.

High frequency bands are in the mm-wave range above 24 GHz. They will be best suited for hotspot coverage with locally very high capacity and can provide very high data rates. The highest interest is in the range 24.25–29.5 GHz, with 3GPP NR bands n257 and n258 assigned. Channel bandwidths up to 400 MHz are defined for these bands, with even higher bandwidths possible through carrier aggregation.

The mmWave frequency range is new for IMT deployment as discussed above. The band 27.5–28.35 GHz was identified at an early stage in the US, while 24.25–27.5 GHz, also called the "26 GHz band", is a pioneer band for Europe. Different parts of the larger range 24.25–29.5 GHz are being considered globally. The range 27.5–29.5 GHz is the first range considered for Japan and 26.5–29.5 GHz in Korea. Overall, this band can be seen as global with regional variations. The range 37–40 GHz is also defined in the US and similar ranges around 40 GHz are considered in many other regions too, including China.

There are not yet any frequency bands identified for IMT in the frequency range from 6 to 24 GHz. There is however now an agenda item for WRC-23 to consider the band 10–10.5 GHz and there are also some regional and national consideration for other bands in this range. 3GPP has made a thorough technical study in Release 16 [106] of how NR can be implemented for operation in the bands 7–24 GHz.

For the frequency range 57–71 GHz is, NR-U operation is available in unlicensed spectrum as band n263, see also Chapter 20.

3.2 Frequency bands for NR

5G NR can be deployed both in existing IMT bands used by 3G UTRA and 4G LTE, in the new bands defined for IMT at WRC-19 and in bands that may be identified at future WRC, or in regional bodies. The possibility of operating a radio-access technology in different frequency bands is a fundamental aspect of global mobile services. Most 2G, 3G, 4G and 5G devices are multiband capable, covering bands used in the different regions of the world to provide global roaming. From a radio-access functionality perspective, this has limited impact and the physical layer specifications such as those for NR do not assume any specific frequency band. Since NR however spans such a vast range of frequencies, there are certain provisions that are intended only for certain frequency ranges. This includes how the different NR numerologies can be applied (see Chapter 7).

While many RF requirements are band agnostic, some are specified with different requirements across bands. This is certainly the case for NR, but also for previous generations. Examples of RF requirements that may be band specific are the allowed maximum transmit power, requirements/limits on out-of-band (OOB) emission and receiver blocking levels. Reasons for such differences are varying external constraints, often imposed by regulatory bodies, in other cases differences in the operational environment that are considered during standardization.

The differences between bands are more pronounced for NR due to the very wide frequency range. For NR operation in the new mm-wave bands above 24 GHz, both devices and base stations will be implemented with partly novel technology and there will be a more widespread use of massive MIMO, beam forming and highly integrated advanced antenna systems. This creates differences in how RF requirements are defined, how they are measured for performance assessment and ultimately also what limits are set for the requirements. Frequency bands in 3GPP are for this reason divided into frequency ranges:

- Frequency range 1 (FR1) includes all existing bands in the range 410–7125 MHz.
- Frequency range 2 (FR2) includes bands in the range 24.25–71 GHz and is divided into two subranges, FR2-1 (24.25–52.6 GHz) and FR2-2 (52.6–71 GHz).

These frequency ranges may be extended or complemented with new ranges in future 3GPP releases. The impact of the frequency ranges on the RF requirements is further discussed in Chapter 28.

The frequency bands where NR will operate are in both paired and unpaired spectrum, requiring flexibility in the duplex arrangement. For this reason, NR supports both FDD and TDD operation. Some ranges are also defined for SDL or SUL. These features are further described in Section 7.7.

3GPP defines *operating bands*, where each operating band is a frequency range for uplink and/or downlink that is specified with a certain set of RF requirements. The operating bands are each associated with a number. When the same frequency range is defined as an operating band for different radio access technologies, the same number is used, but written in a different way. 4G LTE bands are written with Arabic numerals (1, 2, 3, etc.), while 3G UTRA bands are written with Roman numerals (I, II, II, etc.). LTE operating bands that are used with the same arrangement for NR are often referred to as "LTE re-farming bands."

Release 17 of the 3GPP specifications for NR includes 61 operating bands in frequency range 1 and seven in frequency range 2. The bands for NR are assigned numbers from n1 to n512 using the following rules:

- For NR in LTE refarming bands, the LTE band numbers are reused for NR, just adding an "n".
- New bands for NR are assigned the following numbers:
  - The range n65 to n256 is reserved for NR bands in frequency range 1 (some of these bands can be used for LTE in addition)
  - The range n257 to n512 is reserved for new NR bands in frequency range 2

The scheme "conserves" band numbers and is backward compatible with LTE (and UTRA) and does not lead to any new LTE numbers above 256, which is the present maximum possible. Any new LTE-only bands can also be assigned unused numbers below 65. In release 17, the operating bands in frequency range 1 are in the range n1 to n104 as shown in Table 3.1. The bands in frequency range 2 are in the range from n257 to n263, as shown in Table 3.2. All bands for NR are summarized in Figs. 3.2–3.5, which also show the corresponding frequency allocation defined by the ITU-R.

Some of the frequency bands are partly or fully overlapping. In most cases this is explained by regional differences in how the bands defined by the ITU-R are implemented. At the same time, a high degree of commonality between the bands is desired to enable global roaming. Originating in global, regional, and local spectrum developments, a first set of bands was specified as bands for UTRA. The complete set of UTRA bands was later transferred to the LTE specifications in 3GPP Release 8. Additional bands have been added in later releases. In release 15, many of the LTE bands were transferred to the NR specifications.

Table 3.1 Operating bands defined by 3GPP for NR in Frequency range 1 (FR1).

NR band	Uplink range (MHz)	Downlink range (MHz)	Duplex mode	Main region(s)
n1	1920–1980	2110–2170	FDD	Europe, Asia
n2	1850–1910	1930–1990	FDD	Americas (Asia)
n3	1710–1785	1805–1880	FDD	Europe, Asia (Americas)
n5	824–849	869–894	FDD	Americas, Asia
n7	2500–2570	2620–2690	FDD	Europe, Asia
n8	880–915	925–960	FDD	Europe, Asia
n12	699–716	729–746	FDD	US
n13	777–787	746–756	FDD	US
n14	788–798	758–768	FDD	US
n18	815–830	860–875	FDD	Japan
n20	832–862	791–821	FDD	Europe
n24	1626.5–1660.5	1525–1559	FDD	US
n25	1850–1915	1930–1995	FDD	Americas
n26	814–849	859–894	FDD	
n28	703–748	758–803	FDD	Asia/Pacific
n29	N/A	717–728	N/A	Americas
n30	2305–2315	2350–2360	FDD	Americas
n34	2010–2025	2010–2025	TDD	Asia
n38	2570–2620	2570–2620	TDD	Europe
n39	1880–1920	1880–1920	TDD	China
n40	2300–2400	2300–2400	TDD	Europe, Asia
n41	2496–2690	2496–2690	TDD	US, China
n46	5150–5925	5150–5925	TDD	(NR-U)
n48	3550–3700	3550–3700	TDD	US
n50	1432–1517	1432–1517	TDD	Europe
n51	1427–1432	1427–1432	TDD	Europe
n53	2483.5–2495	2483.5–2495	TDD	
n54	1670–1675	1670–1675	TDD	
n65	1920–2010	2110–2200	FDD	Europe
n66	1710–1780	2110–2200	FDD	Americas
n67	N/A	738–758	SDL	
n70	1695–1710	1995–2020	FDD	Americas
n71	663–698	617–652	FDD	Americas
n74	1427–1470	1475–1518	FDD	Japan
n75	N/A	1432–1517	SDL	Europe
n76	N/A	1427–1432	SDL	Europe
n77	3300–4200	3300–4200	TDD	Europe, Asia

Table 3.1 Operating bands defined by 3GPP for NR in Frequency range 1 (FR1)—cont'd

NR band	Uplink range (MHz)	Downlink range (MHz)	Duplex mode	Main region(s)
n78	3300–3800	3300–3800	TDD	Europe, Asia
n79	4400–5000	4400–500	TDD	Asia
n80	1710–1785	N/A	SUL	
n81	880–915	N/A	SUL	
n82	832–862	N/A	SUL	
n83	703–748	N/A	SUL	
n84	1920–1980	N/A	SUL	
n85	698–716	728–746	FDD	
n86	1710–1780	N/A	SUL	Americas
n89	824–849	N/A	SUL	
n90	2496–2690	2496–2690	TDD	US
n91	832–862	1427–1432	FDD	
n92	832–862	1432–1517	FDD	
n93	880–915	1427–1432	FDD	
n94	880–915	1432–1517	FDD	
n95	2010–2025	N/A	SUL	
n96	5925–7125	5925–7125	TDD	(NR-U)
n97	2300–2400	N/A	SUL	
n98	1880–1920	N/A	SUL	
n99	1626.5–1660.5	N/A	SUL	
n100	874.4–880	919.4–925	FDD	
n101	1900–1910	1900–1910	TDD	
n102	5925–6425	5925–6425	TDD3	(NR-U)
n104	6425–7125	6425–7125	TDD	

Table 3.2 Operating bands defined by 3GPP for NR in frequency range 2 (FR2).

NR band	Uplink and downlink range (MHz)	Duplex mode	Main region(s)
n257	26,500–29,500	TDD	Asia, Americas, (global)
n258	24,250–27,500	TDD	Europe, Asia, (global)
n259	39,500–43,500	TDD	Global
n260	37,000–40,000	TDD	Americas, (global)
n261	27,500–28,350	TDD	Americas
n262	47,200–48,200	TDD	
n263	57,000–71,000	TDD	(NR-U)

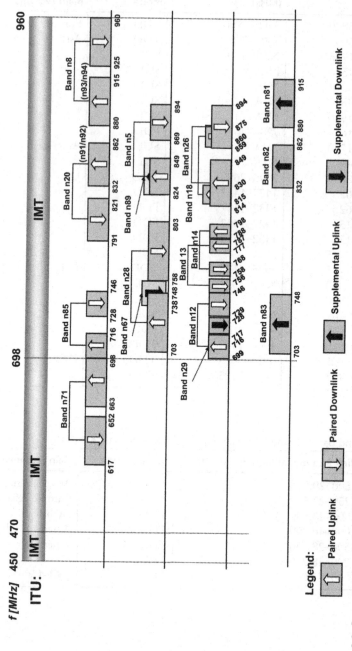

Fig. 3.2 Operating bands specified in 3GPP Release 17 for NR below 1 GHz (in FR1), shown with the corresponding ITU-R allocation. Not fully drawn to scale.

Legend:

Paired Uplink   Paired Downlink   Supplemental Uplink   Supplemental Downlink

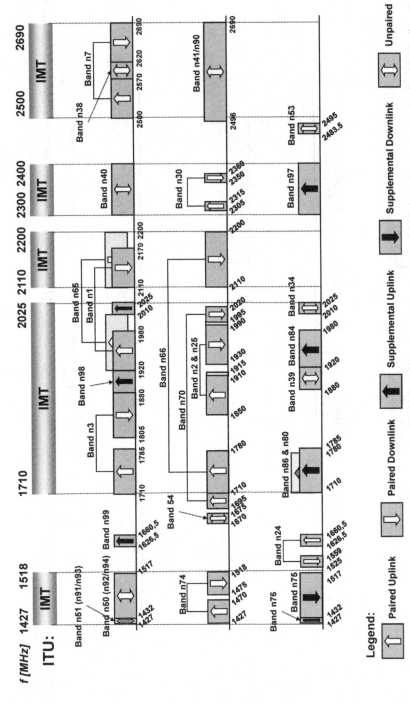

Fig. 3.3 Operating bands specified in 3GPP Release 17 for NR between 1 GHz and 3 GHz (in FR1), shown with the corresponding ITU-R allocation. Not fully drawn to scale.

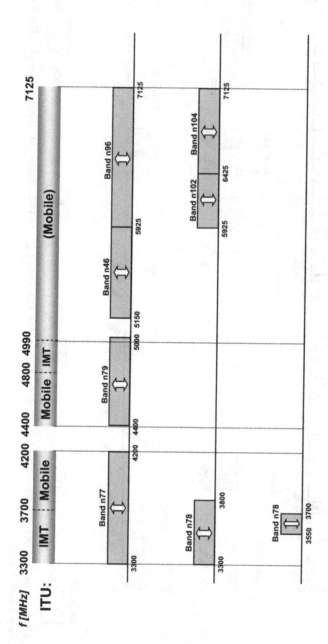

Fig. 3.4 Operating bands specified in 3GPP Release 17 for NR between 3 GHz and 7 GHz (in FR1), shown with the corresponding ITU-R allocation. Not fully drawn to scale.

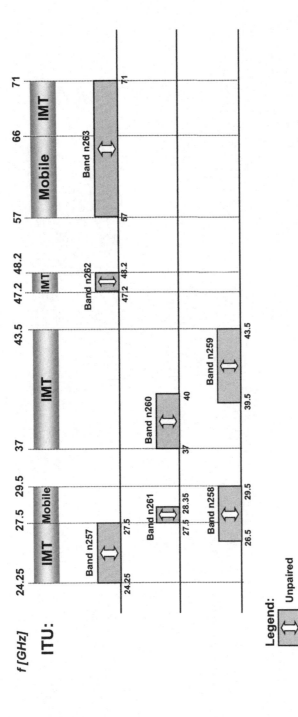

Fig. 3.5 Operating bands specified in 3GPP Release 17 for NR above 24 GHz (in FR2), shown with the corresponding ITU-R allocation. Not fully drawn to scale.

CHAPTER 4

LTE overview

The focus of this book is NR, the new 5G radio access. Nevertheless, a brief overview of LTE as background to the coming chapters is relevant. One reason is that both LTE and NR have been developed by 3GPP and hence have a common background and share several technology components. Many of the design choices in NR are also based on experience from LTE. For a detailed description of LTE see [26].

The work on LTE was initiated in late 2004 with the overall aim of providing a new radio-access technology focusing on packet-switched data only. The first release of the LTE specifications, release 8, was completed in 2009 and commercial network operation began in late 2009. Release 8 has been followed by subsequent LTE releases, introducing additional functionality and capabilities in different areas, as illustrated in Fig. 4.1. Releases 10 and 13 are particularly interesting. Release 10 is the first release of LTE Advanced, and release 13, finalized in early 2016, is the first release of LTE Advanced Pro. Note that neither of these two names imply a break of backward compatibility. Rather they represent steps in the evolution where the amount of new features was considered large enough to merit a new name. Currently, as of this writing, 3GPP has completed release 17 and is working on release 18. The main focus in these releases is on NR, but there are also some enhancements related to massive machine-type communication for LTE-derived technologies.

4.1 LTE release 8 – Basic radio access

Release 8 is the first LTE release and forms the basis for all the following LTE releases. In parallel with the LTE radio access scheme, a new core network, the Evolved Packet Core (EPC) was developed [60].

One important requirement imposed on the LTE development was spectrum flexibility. A range of carrier bandwidths up to and including 20 MHz is supported for carrier frequencies from below 1 GHz up to around 3 GHz. One aspect of spectrum flexibility is the support of both paired *and* unpaired spectrum using *Frequency-Division Duplex* (FDD) and *Time-Division Duplex* (TDD), respectively, with a common design albeit two different frame structures. The focus of the development work was primarily wide-area macro networks with above-rooftop antennas and relatively large cells. For TDD, the uplink-downlink allocation is therefore in essence static with the same uplink-downlink allocation across all cells.

5G/5G-Advanced
https://doi.org/10.1016/B978-0-443-13173-8.00007-4

43

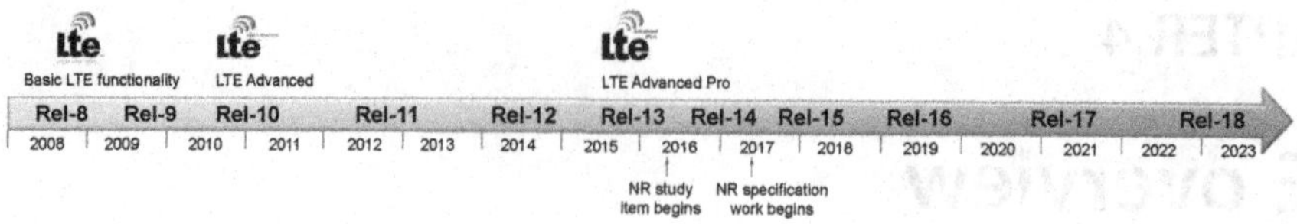

lte   lte   lte
Basic LTE functionality   LTE Advanced   LTE Advanced Pro

Rel-8	Rel-9	Rel-10	Rel-11	Rel-12	Rel-13	Rel-14	Rel-15	Rel-16	Rel-17	Rel-18
2008	2009	2010	2011	2012	2013	2014	2015	2016	2017	2018	2019	2020	2021	2022	2023

NR study   NR specification
Item begins   work begins

Fig. 4.1 LTE and its evolution.

The basic transmission scheme in LTE is *orthogonal frequency-division multiplexing* (OFDM). This is an attractive choice due to its robustness to time dispersion and ease of exploiting both the time and frequency domain. Furthermore, it also allows for reasonable receiver complexity also in combination with spatial multiplexing (MIMO) which is an inherent part of LTE. Since LTE was primarily designed with macro networks in mind with carrier frequencies up to a few GHz, a single subcarrier spacing of 15 kHz and a cyclic prefix of approximately 4.7 μs[a] was found to be a good choice. In total 1200 subcarriers are used in a 20 MHz spectrum allocation.

For the uplink, where the available transmission power is significantly lower than for the downlink, the LTE design settled for a scheme with a low peak-to-average ratio to provide a high power-amplifier efficiency. DFT-precoded OFDM, with the same numerology as in the downlink, was chosen to achieve this. A drawback with DFT-precoded OFDM is the larger complexity on the receiver side, but given that LTE release 8 does not support spatial multiplexing in the uplink this was not seen as a major problem.

In the time domain, LTE organizes transmissions into 10 ms frames, each consisting of ten 1 ms subframes. The subframe duration of 1 ms, which corresponds to 14 OFDM symbols, is the smallest schedulable unit in LTE.

Cell-specific reference signals is a cornerstone in LTE. The base station continuously transmits one or more reference signals (one per layer), regardless of whether there are downlink data to transmit or not. This is a reasonable design for the scenarios which LTE was designed for—relatively large cells with many users per cell. The cell-specific reference signals are used for many functions in LTE: downlink channel estimation for coherent demodulation, channel-state reporting for scheduling purposes, correction of device-side frequency errors, initial access, and mobility measurements to mention just a few. The reference signal density depends on the number of transmission layers set up in a cell, but for the common case of 2 × 2 MIMO, every third subcarrier in four out of 14 OFDM symbols in a subframe are used for reference signals. Thus, in the time domain there are around 200 μs between reference signal occasions, which limits the possibilities to switch off the transmitter to reduce power consumption.

Data transmission in LTE is primarily scheduled on a dynamic basis in both uplink and downlink. To exploit the typically rapidly varying radio conditions, channel-dependent

---

[a] There is also a possibility for 16.7 μs extended cyclic prefix but that option is rarely used in practice.

scheduling can be used. For each 1-ms subframe, the scheduler controls which devices are to transmit or receive and in what frequency resources. Different data rates can be selected by adjusting the code rate of the Turbo code as well as varying the modulation scheme from QPSK up to 64-QAM. To handle transmission errors, fast hybrid ARQ with soft combining is used in LTE. Upon downlink reception the device indicates the outcome of the decoding operation to the base station, which can retransmit erroneously received data blocks.

The scheduling decisions are provided to the device through the Physical Downlink Control Channel (PDCCH). If there are multiple devices scheduled in the same subframe, which is a common scenario, there are multiple PDCCHs, one per scheduled device. The first up to three OFDM symbols of the subframe are used for transmission of downlink control channels. Each control channel spans the full carrier bandwidth, thereby maximizing the frequency diversity. This also implies that all devices must support the full carrier bandwidth up to the maximum value of 20 MHz. Uplink control signaling from the devices, for example hybrid-ARQ acknowledgments and channel-state information for downlink scheduling, is carried on the Physical Uplink Control Channel (PUCCH), which has a basic duration of 1 ms.

Multi-antenna schemes, and in particular single-user MIMO, are integral parts of LTE. A number of transmission layers are mapped to up to four antennas by means of a precoder matrix of size $N_A \times N_L$, where the number of layers N_L, also known as the transmission rank, is less than or equal to the number of antennas N_A. The transmission rank, as well as the exact precoder matrix, can be selected by the network based on channel-status measurements carried out and reported by the terminal, also known as *closed-loop spatial multiplexing*. There is also a possibility to operate without closed-loop feedback for precoder selection. Up to four layers is possible in the downlink although commercial deployments often use only two layers. In the uplink only single-layer transmission is possible.

In case of spatial multiplexing, by selecting rank-1 transmission, the precoder matrix, which then becomes an $N_A \times 1$ precoder vector, performs a (single-layer) beamforming function. This type of beamforming can more specifically be referred to as codebook-based beamforming as the beamforming can only be done according to a limited set of predefined beamforming (precoder) vectors.

Using the basic features discussed above, LTE release 8 is in theory capable of providing peak data rates up to 150 Mbit/s in the downlink using two-layer transmission in 20 MHz and 75 Mbit/s in the uplink. Latency-wise LTE provides 8 ms roundtrip time in the hybrid-ARQ protocol and (theoretically) less than 5 ms one-way delay in the LTE RAN. In practical deployments, including transport and core network processing, an overall end-to-end latency of some 10 ms is not uncommon in well-deployed networks.

Release 9 added some smaller enhancements to LTE such as multicast/broadcast support, positioning, and some multi-antenna refinements.

4.2 LTE evolution

Release 8 and 9 form the foundation of LTE, providing a highly capable mobile-broadband standard. However, to meet new requirements and expectations, the releases following the basic ones provide additional enhancements and features in different areas. Fig. 4.2 illustrates some of the major areas in which LTE has evolved over the more than 10 years since its introduction with details provided in the following. Additional information about each release can be found in the release descriptions which 3GPP prepares for each new release.

Release 10 marks the start of the LTE evolution. One of the main targets of LTE release 10 was to ensure that the LTE radio-access technology would be fully compliant with the IMT-Advanced requirements, thus the name LTE Advanced is often used for LTE release 10 and later. However, in addition to the ITU requirements, 3GPP also defined its own targets and requirements for LTE Advanced [10]. These targets/requirements extended the ITU requirements both in terms of being more aggressive as well as including additional requirements. One important requirement was backward compatibility. Essentially this means that an earlier-release LTE device should be able to access a carrier supporting LTE release-10 functionality, although obviously not being able to utilize all the release-10 features of that carrier. The principle of backward compatibility is important and has been kept for all LTE releases, but also imposes some restrictions on the enhancements possible; restrictions that are not present when defining a new standard such as NR.

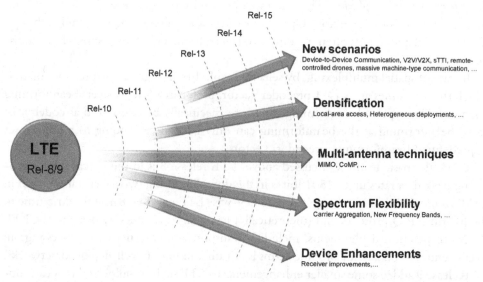

Fig. 4.2 LTE evolution.

LTE release 10 was completed in early 2011 and introduced enhanced LTE spectrum flexibility through carrier aggregation, further extended multiantenna transmission, support for relaying, and improvements around intercell interference coordination in heterogeneous network deployments.

Release 11 further extended the performance and capabilities of LTE. One of the most notable features of LTE release 11, finalized in late 2012, is radio-interface functionality for *coordinated multi-point* (CoMP) transmission and reception. Other examples of improvements in release-11 are carrier-aggregation enhancements, a new control-channel structure, and performance requirements for more advanced device receivers.

Release 12 was completed in 2014 and focused on small cells with features such as dual connectivity, small-cell on/off, and (semi-)dynamic TDD, as well as on new scenarios with introduction of direct device-to-device communication and provisioning of complexity-reduced devices targeting massive machine-type communication.

Release 13, finalized at the end of 2015, marks the start of *LTE Advanced Pro*. It is sometimes in marketing dubbed 4.5G and seen as an intermediate technology step between 4G defined by the first releases of LTE and the 5G NR air interface. License-assisted access to support unlicensed spectra as a complement to licensed spectra, improved support for machine-type communication, and various enhancements in carrier aggregation, multi-antenna transmission, and device-to-device communication are some of the highlights from release 13. Massive machine-type communication support was further enhanced and the narrow-band internet-of-things (NB-IoT) technology was introduced.

Release 14 was completed in the spring of 2017. Apart from enhancements to some of the features introduced in earlier releases, for example enhancements to operation in unlicensed spectra, it introduced support for vehicle-to-vehicle (V2V) and vehicle-to-everything (V2X) communication, as well as wide-area broadcast support with a reduced subcarrier spacing. There are also a set of mobility enhancements in release 14, in particular make-before-break handover and RACH-less handover to reduce the handover interruption time for devices with dual receiver chains.

Release 15 was completed in the middle of 2018. Significantly reduced latency through the so-called sTTI feature, as well as communication using aerials are two examples of enhancements in this release. The support for massive machine-type communication has been continuously improved over several release and release 15 also included enhancements in this area.

Release 16, completed at the end of 2019, brought enhancements in multi-antenna support with increased uplink sounding capacity, enhanced support for terrestrial broadcast services, and even further enhancements to massive machine-type communication. Improved mobility through enhanced make-before-break handover, also known as dual active protocol stack (DAPS), was introduced where the device maintains the source-cell radio link (including data flow) while establishing the target-cell radio link. Conditional handover can also be configured, where device-initiated handover to a set of preconfigured cells is triggered by rules configured by the network. In general, expanding LTE to new use cases beyond traditional mobile broadband has been in focus for the later releases and the evolution will continue also in the future. This is also an important part of 5G overall and exemplifies that LTE remains important and a vital part of the overall 5G radio access.

Release 17, completed at the end of 2021, contained a smaller amount of LTE enhancements as 3GPP primarily focused on the NR evolution. Massive machine-type communication over satellites and a set of new bandwidths for LTE-based broadcast are the main areas.

4.3 Spectrum flexibility

Already the first release of LTE provides a certain degree of spectrum flexibility in terms of multi-bandwidth support and a joint FDD/TDD design. In later releases this flexibility was considerably enhanced to support higher bandwidths and fragmented spectra using carrier aggregation and access to unlicensed spectra as a complement using license-assisted access (LAA).

4.3.1 Carrier aggregation

As mentioned earlier, the first release of LTE already provided extensive support for deployment in spectrum allocations of various characteristics, with bandwidths ranging from roughly 1 MHz up to 20 MHz in both paired and unpaired bands. With LTE release 10 the transmission bandwidth can be further extended by means of *carrier aggregation* (CA), where multiple *component carriers* are aggregated and jointly used for transmission to/from a single device. Up to five component carriers, possibly each of different bandwidth, can be aggregated in release 10, allowing for transmission bandwidths up to 100 MHz. All component carriers need to have the same duplex scheme and, in the case of TDD, the same uplink-downlink configuration. In later releases, these restrictions were relaxed. The number of component carriers possible to aggregate was increased to 32, resulting in a total bandwidth of 640 MHz. Backward compatibility was ensured as each component carrier uses the release-8 structure. Hence, to a release-8/9 device each component carrier will appear as an LTE release-8 carrier, while a carrier-aggregation-capable device can exploit the total aggregated bandwidth, enabling higher data rates. In the general case, a different number of component carriers can be aggregated for the downlink and uplink. This is an important property from a device complexity point-of-view where aggregation can be supported in the downlink where very high data rates are needed without increasing the uplink complexity.

Component carriers do not have to be contiguous in frequency, which enables exploitation of fragmented spectra; operators with a fragmented spectrum can provide

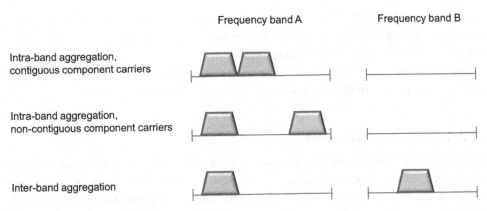

Fig. 4.3 Carrier aggregation.

high–data-rate services based on the availability of a wide overall bandwidth even though they do not possess a single wideband spectrum allocation.

From a baseband perspective, there is no difference between the cases in Fig. 4.3 and they are all supported by LTE release 10. However, the RF-implementation complexity is vastly different with the first case being the least complex. Thus, although carrier aggregation is supported by the basic specifications, not all devices will support it. Furthermore, release 10 has some restrictions on carrier aggregation in the RF specifications, compared to what has been specified for physical layer and related signaling, while in later releases there is support for carrier-aggregation within and between a much larger number of frequency bands.

Release 11 provided additional flexibility for aggregation of TDD carriers. Prior to release 11, the same downlink–uplink allocation was required for all the aggregated carriers. This can be unnecessarily restrictive in the case of aggregation of different bands as the configuration in each band may be given by coexistence with other radio access technologies in that particular band. An interesting aspect of aggregating different downlink-uplink allocations is that the device may need to receive and transmit simultaneously in order to fully utilize both carriers. Thus, unlike previous releases, a TDD-capable device may, similar to an FDD-capable device, need a duplex filter. Release 11 also saw the introduction of RF requirements for inter-band and non-contiguous intra-band aggregation, as well as support for an even larger set of inter-band aggregation scenarios.

Release 12 defined aggregations between FDD and TDD carriers, unlike earlier releases that only supported aggregation within one duplex type. FDD-TDD aggregation allows for efficient utilization of an operator's spectrum assets. It can also be used to improve the uplink coverage of TDD by relying on the possibility for continuous uplink transmission on the FDD carrier.

Release 13 increased the number of carriers possible to aggregate from 5 to 32, resulting in a maximum bandwidth of 640 MHz and a theoretical peak data rate around

25 Gbit/s in the downlink. The main motivation for increasing the number of subcarriers is to allow for very large bandwidths in unlicensed spectra as will be further discussed in conjunction with license-assisted access below.

Carrier aggregation is one of the most successful enhancements of LTE to date with new combinations of frequency band added in every release.

4.3.2 License-assisted access

Originally, LTE was designed for licensed spectra where an operator has an exclusive license for a certain frequency range. A licensed spectrum offers many benefits since the operator can plan the network and control the interference situation, but there is typically a cost associated with obtaining the spectrum license and the amount of licensed spectrum is limited. Therefore, using unlicensed spectra as a *complement* to offer higher data rates and higher capacity in local areas is of interest. One possibility is to complement the LTE network with Wi-Fi, but higher performance can be achieved with a tighter coupling between licensed and unlicensed spectra. LTE release 13 therefore introduced *license-assisted access*, where the carrier-aggregation framework is used to aggregate downlink carriers in unlicensed frequency bands, primarily in the 5 GHz range, with carriers in licensed frequency bands as illustrated in Fig. 4.4. Mobility, critical control signaling and services demanding high quality-of-service rely on carriers in the licensed spectra while (parts of) less demanding traffic can be handled by the carriers using unlicensed spectra. Operator-controlled small-cell deployments is the target. Fair sharing of the spectrum resources with other systems, in particular Wi-Fi, is an important characteristic of LAA which therefore incudes a listen-before-talk mechanism. In release 14, license-assisted access was enhanced to address also uplink transmissions and in release 15, further enhancements in the area of autonomous uplink transmissions were added. Although the LTE technology standardized in 3GPP supports license-assisted access only, where a

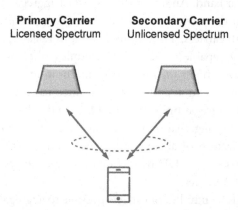

Fig. 4.4 License-assisted access.

licensed carrier is needed, there has been work outside 3GPP in the MulteFire alliance resulting in a stand-alone mode-of-operation based on the 3GPP standard.

4.4 Multi-antenna enhancements

Multi-antenna support has been enhanced over the different releases, increasing the number of transmission layers in the downlink to eight and introducing uplink spatial multiplexing of up to four layers. Full-dimension MIMO and two-dimensional beam-forming are other enhancements, as is the introduction of coordinated multipoint transmission.

4.4.1 Extended multi-antenna transmission

In release 10, downlink spatial multiplexing was expanded to support up to eight trans-mission layers. This can be seen as an extension of the release-9 dual-layer beamforming to support up to eight antenna ports and eight corresponding layers. Together with the support for carrier aggregation this enables downlink data rates up to 3 Gbit/s in 100 MHz of spectrum in release 10, increased to 25 Gbit/s in release 13 using 32 carriers, eight layers spatial multiplexing, and 256QAM.

Uplink spatial multiplexing of up to four layers was also introduced as part of LTE release 10. Together with the possibility for uplink carrier aggregations this allows for uplink data rates up to 1.5 Gbit/s in 100 MHz of spectrum. Uplink spatial multiplexing consists of a codebook-based scheme under the control of the base station, which means that the structure can also be used for uplink transmitter-side beamforming.

An important consequence of the multi-antenna extensions in LTE release 10 was the introduction of an enhanced downlink *reference-signal structure* that more extensively sep-arated the function of channel estimation and the function of acquiring channel-state information. The aim of this was to better enable novel antenna arrangements and new features such as more elaborate multi-point coordination/transmission in a flexible way.

In release-13, and continued in release 14, improved support for massive antenna arrays was introduced, primarily in terms of more extensive feedback of channel-state information. The larger degrees of freedom can be used for, for example, beamforming in both elevation and azimuth and massive multi-user MIMO where several spatially sep-arated devices are simultaneously served using the same time-frequency resource. These enhancements are sometimes termed full-dimension MIMO and form a step into massive MIMO with a very large number of steerable antenna elements. Further enhancements were added in release 16 where the capacity and coverage of the uplink sounding refer-ence signals were improved to better address massive MIMO for TDD-based LTE deployments.

4.4.2 Multi-point coordination and transmission

The first release of LTE included specific support for coordination between transmission points, referred to as *Inter-Cell Interference Coordination* (ICIC), to control the interference between cells. However, the support for such coordination was significantly expanded as part of LTE release 11 including the possibility for much more dynamic coordination between transmission points.

In contrast to release 8 ICIC, which was limited to the definition of certain messages between base stations to assist (relatively slow) scheduling coordination between cells, the release 11 activities focused on radio-interface features and device functionality to assist different coordination means, including the support for channel-state feedback for multiple transmission points. Jointly these features and functionality go under the name *Coordinated Multi Point* (CoMP) transmission/reception. Refinement to the reference-signal structure was also an important part of the CoMP support, as was the enhanced control-channel structure introduced as part of release 11, see below.

Support for CoMP includes *multi-point coordination* – that is, when transmission to a device is carried out from one specific transmission point but where scheduling and link adaptation are coordinated between the transmission points, as well as *multi-point transmission* in which case transmission to a device can be carried out from multiple transmission points either in such a way that transmission can switch dynamically between different transmission points (*Dynamic Point Selection*) or be carried out jointly from multiple transmission points (*Joint Transmission*), see Fig. 4.5.

A similar distinction can be made for uplink where one can distinguish between (uplink) multi-point coordination and multi-point *reception*. In general, uplink CoMP is mainly a network implementation issue and has very little impact on the device and very little visibility in the radio-interface specifications.

The CoMP work in release 11 assumed 'ideal' backhaul, in practice implying centralized baseband processing connected to the antenna sites using low-latency fiber connections. Extensions to relaxed backhaul scenarios with non-centralized baseband processing were introduced in release 12. These enhancements mainly consisted of defining new X2 messages between base stations for exchanging information about so-called CoMP hypotheses, essentially a potential resource allocation, and the associated gain/cost.

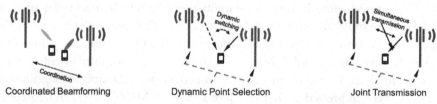

Coordinated Beamforming   Dynamic Point Selection   Joint Transmission

Fig. 4.5 Different types of CoMP.

4.4.3 Enhanced control channel structure

In release 11, a new complementary control channel structure was introduced to support inter-cell interference coordination and to exploit the additional flexibility of the new reference-signal structure not only for data transmission, which was the case in release 10, but also for control signaling. The new control-channel structure can thus be seen as a prerequisite for many CoMP schemes, although it is also beneficial for beamforming and frequency-domain interference coordination as well. It is also used to support narrow-band operation for MTC enhancements in release 12 and onward.

4.5 Densification, small cells, and heterogeneous deployments

Small cells and dense deployments have been in focus for several releases as means to provide very high capacity and data rates. Relaying, small-cell on/off, dynamic TDD, and heterogeneous deployments are some examples of enhancements over the releases. License-assisted access, discussed in Section 4.3.2, is another feature primarily targeting small cells.

4.5.1 Relaying

In the context of LTE, *relaying* implies that the device communicates with the network via a *relay node* that is *wirelessly connected* to a *donor cell* using the LTE radio-interface technology (see Fig. 4.6). From a device point of view, the relay node will appear as an ordinary cell. This has the important advantage of simplifying the device implementation and making the relay node backward compatible – that is, LTE release-8/9 devices can also access the network via the relay node. In essence, the relay is a low-power base station wirelessly connected to the remaining part of the network.

4.5.2 Heterogeneous deployments

Heterogeneous deployments refer to deployments with a mixture of network nodes with different transmit power and overlapping geographical coverage (Fig. 4.7). A typical example is a pico node placed within the coverage area of a macro cell. Although such deployments were already supported in release 8, release 10 introduced new means to

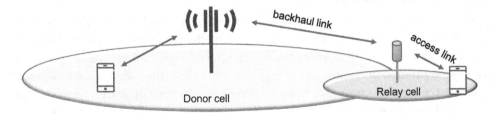

Fig. 4.6 Example of relaying.

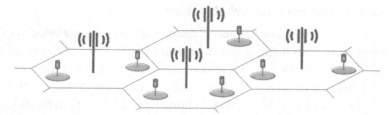

Fig. 4.7 Example of heterogeneous deployment with low-power nodes inside macro cells.

handle the inter-layer interference that may occur between, for example, a pico layer and the overlaid macro. The multi-point-coordination techniques introduced in release 11 further extend the set of tools for supporting heterogeneous deployments. Enhancements to improve mobility between the pico layer and the macro layer were introduced in release 12.

4.5.3 Small-cell on-off

In LTE, cells are continuously transmitting cell-specific reference signals and broadcasting system information, regardless of the traffic activity in the cell. One reason for this is to enable idle-mode devices to detect the presence of a cell; if there are no transmissions from a cell there is nothing for the device to measure upon and the cell would therefore not be detected. Furthermore, in a large macro-cell deployment there is a relatively high likelihood of at least one device being active in a cell motivating continuous transmission of reference signals.

However, in a dense deployment with many relatively small cells, the likelihood of not all cells having devices to serve at a certain point in time can be relatively high in some scenarios. The downlink interference scenario experienced by a device may also be more severe with devices experiencing very low signal-to-interference ratios due to interference from neighboring, potentially empty, cells, especially if there is a large amount of line-of-sight propagation. To address this, release 12 introduced mechanisms for turning on/off individual cells as a function of the traffic situation to reduce the average inter-cell interference and reduce power consumption.

4.5.4 Dual connectivity

Dual connectivity implies a device is simultaneously connected to two cells at different sites, see Fig. 4.8, as opposed to the baseline case with the device connected to a single site only. User-plane aggregation where the device is receiving data transmission from multiple sites, separation of control and user planes, and uplink-downlink separation where downlink transmissions originate from a different site than the uplink reception site are some examples of the potential benefits with dual connectivity. To some extent it can be seen as carrier aggregation extended to the case of non-ideal backhaul. The dual

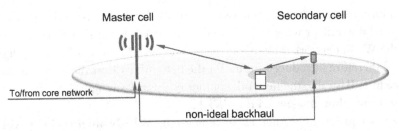

Fig. 4.8 Example of dual connectivity.

connectivity framework has also turned out to be useful for integrating other radio-access schemes such as WLAN into 3GPP networks. It is also essential for NR when operating in non-standalone mode with LTE providing mobility and initial access as will be described in the following chapters.

4.5.5 Dynamic TDD

In TDD, the same carrier frequency is shared in the time domain between uplink and downlink. The fundamental approach to this in LTE, as well as in many other TDD systems, is to statically split the resources into uplink and downlink. Having a static split is a reasonable assumption in larger macro cells as there are multiple users and the aggregated per-cell load in uplink and downlink is relatively stable. However, with an increased interest in local-area deployments, TDD is expected to become more important compared to the situation for wide-area deployments to date. One reason is unpaired spectrum allocations being more common in higher frequency bands less suitable for wide-area coverage. Another reason is that many problematic interference scenarios in wide-area TDD networks are not present with below-rooftop deployments of small nodes. An existing wide-area FDD network could be complemented by a local-area layer using TDD, typically with low output power per node, to boost capacity and data rates.

To better handle the high traffic dynamics in a local-area scenario, where the number of devices transmitting to/receiving from a local-area access node can be very small, dynamic TDD is beneficial. In dynamic TDD, the network can dynamically use resources for either uplink or downlink transmissions to match the instantaneous traffic situation, which leads to an improvement of the end-user performance compared to the conventional static split of resources between uplink and downlink. To exploit these benefits, LTE release 12 includes support for dynamic TDD, or *enhanced Interference Mitigation and Traffic Adaptation* (eIMTA) as it the official name for this feature in 3GPP.

4.5.6 WLAN interworking

The 3GPP architecture allows for integrating non-3GPP access, for example WLAN but also cdma2000 [12]. Essentially, these solutions connect the non-3GPP access to the EPC and are thus not visible in the LTE radio-access network. One drawback of this way of

WLAN interworking is the lack of network control; the device may select Wi-Fi even if staying on LTE would provide a better user experience. One example of such a situation is when the Wi-Fi network is heavily loaded while the LTE network enjoys a light load. Release 12 therefore introduced means for the network to assist the device in the selection procedure. Basically, the network configures a signal-strength threshold controlling when the device should select LTE or Wi-Fi.

Release 13 provided further enhancements in WLAN interworking with more explicit control from the LTE RAN on when a device should use Wi-Fi and when to use LTE. Furthermore, release 13 also includes LTE-WLAN aggregation where LTE and WLAN are aggregated at the PDCP level using a framework very similar to dual connectivity. Additional enhancements were introduced in release 14.

4.6 Device enhancements

Fundamentally, a device vendor is free to design the device receiver in any way as long as it supports the minimum requirements defined in the specifications. There is an incentive for the vendors to provide significantly better receivers as this could be directly translated into improved end-user data rates. However, the network may not be able to exploit such receiver improvements to their full extent as it might not know which devices have significantly better performance. Network deployments therefore need to be based on the minimum requirements. Defining performance requirements for more advanced receiver types to some extent alleviates this as the minimum performance of a device equipped with an advanced receiver is known. Both releases 11 and 12 saw a lot of focus on receiver improvements with cancellation of some overhead signals in release 11 and more generic schemes in release 12, including network-assisted interference cancellation (NAICS) where the network can provide the devices with information assisting inter-cell interference cancellation.

4.7 New scenarios

LTE was originally designed as a mobile broadband system, aiming at providing high data rates and high capacity over wide areas. The evolution of LTE has added features improving capacity and data rates, but also enhancements making LTE highly relevant also for new use cases. Massive machine-type communication, where a large number of low-cost devices, for example sensors, are connected to a cellular network is a prime example of this. Operation in areas without network coverage, for example in a disaster area, is another example, resulting in support for device-to-device commination being included in LTE. V2V/V2X and remote-controlled drones are yet other examples of new scenarios.

4.7.1 Machine-type communication

Machine-type communication (MTC) is a very wide term, basically covering all types of communication between machines. Although spanning a wide range of different applications, many of which are yet unknown, MTC applications can be divided into two main categories, massive MTC and ultra-reliable low-latency communication (URLLC).

Examples of massive MTC scenarios are different types of sensors, actuators, and similar devices. These devices typically have to be of very low cost and have very low energy consumption enabling very long battery life. At the same time, the amount of data generated by each device is normally very small and very low latency is not a critical requirement. URLLC, on the other hand, corresponds to applications such as traffic safety/control or wireless connectivity for industrial processes, and in general scenarios where very high reliability and availability is required, combined with low latency.

To better support massive MTC, the 3GPP specifications provide two parallel and complementing technologies – eMTC and NB-IoT.

Addressing the MTC area started with release 12 and the introduction of a new, low-end device category, category 0, supporting data rates up to 1 Mbit/s. A power-save mode for reduced device power consumption was also defined. These enhancements are often referred to as enhanced MTC (eMTC) or LTE-M. Release 13 further improved the MTC support by defining category-M1 with enhanced coverage and support for 1.4 MHz device bandwidth, irrespective of the system bandwidth, to further reduce device cost. From a network perspective these devices are normal LTE devices, albeit with limited capabilities, and can be freely mixed with more capable LTE devices on a carrier. The eMTC technology has been evolved further in subsequent releases to improve spectral efficiency and to reduce the amount of control signaling.

Narrow-band Internet-of-Things (NB-IoT) is a parallel track starting in release 13. It targets even lower cost and data rates than category-M1, 250 kbit/s or less, in a bandwidth of 180 kHz, and even further enhanced coverage. Thanks to the use of OFDM with 15 kHz subcarrier spacing, it can be deployed inband on top of an LTE carrier, outband in a separate spectrum allocation, or in the guard bands of LTE, providing a high degree of flexibility for an operator. In the uplink, transmission on a single tone is supported to obtain very large coverage for the lowest data rates. NB-IoT uses the same family of higher-layer protocols (MAC, RLC, and PDCP) as LTE, with extensions for faster connection setup applicable to both NB-IoT and eMTC, and can therefore easily be integrated into existing deployments.

Both eMTC and NB-IoT have been evolving over several releases and play an important role in 5G networks for massive machine-type communication. With the introduction of NR, the broadband traffic will gradually shift from LTE to NR. However, massive machine-type communication is expected to rely on eMTC and NB-IoT for many years to come. Special means for deploying NR on top of an already existing carrier used for massive machine-type communication has therefore been included (see Chapter 18). Furthermore, the focus of the LTE evolution in release 17 was primarily on massive machine-type communication, confirming the trend for the last few releases. In release 17, NTN support is introduced for eMTC and NB-IoT, that is, connecting these devices via satellites, support that is further improved in release 18.

Improved support for URLLC has been added in the later LTE releases. Examples hereof are the sTTI feature (see below) and the general work on the reliability part of URLLC in release 15.

4.7.2 Latency reduction – sTTI

In release 15, work on reducing the overall latency has been carried out, resulting in the so-called *short TTI* (sTTI) feature. The target with this feature is to provide very low latency for use cases where this is important, for example factory automation. It uses similar techniques as used in NR, such as a transmission duration of a few OFDM symbols and reduced device processing delay, but incorporated in LTE in a backward-compatible manner. This allows for low-latency services to be included in existing networks, but also implies certain limitations compared to a clean-slate design such as NR.

4.7.3 Device-to-device communication

Cellular systems, such as LTE, are designed assuming that devices connect to a base station to communicate. In most cases this is an efficient approach as the server with the content of interest is typically not in the vicinity of the device. However, if the device is interested in communicating with a neighboring device, or just detecting whether there is a neighboring device that is of interest, the network-centric communication may not be the best approach. Similarly, for public safety, such as a first responder officer searching for people in need in a disaster situation, there is typically a requirement that communication should also be possible in the absence of network coverage.

To address these situations, release 12 introduced network-assisted device-to-device communication using parts of the uplink spectrum (Fig. 4.9). Two scenarios were

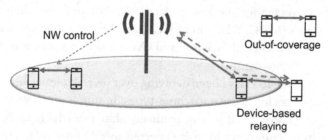

Fig. 4.9 Device-to-device communication.

considered when developing the device-to-device enhancements, in coverage as well as out-of-coverage communication for public safety, and in coverage discovery of neighboring devices for commercial use cases. In release 13, device-to-device communication was further enhanced with relaying solutions for extended coverage. The device-to-device design also served as the basis for the V2V and V2X work in release 14.

4.7.4 V2V and V2X

Intelligent transportation systems (ITSs) refer to services to improve traffic safety and increase efficiency. Examples are vehicle-to-vehicle communication for safety, for example to transmit messages to vehicles behind when the car in front breaks. Another example is platooning where several trucks drive very close to each other and follow the first truck in the platoon, thereby saving fuel and reducing CO_2 emissions. Communication between vehicles and infrastructure is also useful, for example to obtain information about the traffic situation, weather updates, and alternative routes in case of congestion (Fig. 4.10).

In release 14, 3GPP specified enhancements in this area, based on the device-to-device technologies introduced in release 12 and quality-of-service enhancements in the network. Using the same technology for communication both between vehicles and between vehicles and infrastructure is attractive, both to improve the performance but also to reduce cost.

4.7.5 Aerials

The work on aerials in release 15 covers communication via a drone acting as a relay to provide cellular coverage in an otherwise non-covered area, but also remote control of drones for various industrial and commercial applications. Since the propagation conditions between the ground and an airborne drone are different than in a terrestrial network, new channel models were developed as part of release 15. The interference

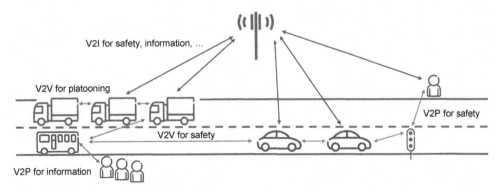

Fig. 4.10 Illustration of V2V and V2X.

situation for a drone is different than for a device on the ground due to the larger number of base stations visible to the drone, calling for interference-mitigation techniques such as beamforming as well as enhancements to the power-control mechanism.

4.7.6 Multicast/broadcast

Multimedia Broadcast Multicast Services (MBMS), where the same content can be delivered simultaneously to several devices with a single transmission, has been part of LTE since release 9. The focus of release 9 was support for single-frequency networks using the original LTE subcarrier spacing of 15 kHz where the same signal is transmitted across multiple cells in a semi-static and coordinated manner, sometimes referred to as *Multimedia Broadcast Single-Frequency Network* (MBSFN).

In release 13, an additional mode, *single-cell point-to-multipoint* (SC-PTM) was added as a complement to MBSFN for services of interest in a single cell only. All transmissions are dynamically scheduled but instead of targeting a single device, the same transmission is received by multiple devices simultaneously.

To improve the support for broadcast-only MBSFN carriers over larger areas, an additional numerology of 1.25 kHz to obtain a longer cyclic prefix was introduced in release 14. This is formally known as *enhanced MBMS* (eMBMS) but sometimes also referred to as LTE broadcast. Further enhancements, known as *LTE-based 5G terrestrial broadcast*, were added in release 16. Additional subcarrier spacing of 2.5 kHz and 0.37 kHz with a corresponding cyclic prefix of 100 μs and 400 μs were introduced, thereby supporting transmission over very wide areas in high-power/high-tower scenarios. In release 17, support for bandwidths of 6, 7, and 8 MHz were added.

CHAPTER 5

NR overview

The technical work on NR was initiated in the spring of 2016 as a study item in 3GPP release 14, based on a kick-off workshop in the fall of 2015, see Fig. 5.1. During the study item phase, different technical solutions were studied, but given the tight time schedule, some technical decisions were taken already in this phase. The work continued into a work item phase in release 15, resulting in the first version of the NR specifications available by the end of 2017, before the closure of 3GPP release 15 in mid-2018. The reason for the intermediate release of the specifications, before the end of release 15, was to meet commercial requirements on early 5G deployments in some markets.

The first specification from December 2017 is limited to non-standalone NR operation (see Chapter 6), implying that NR devices rely on LTE for initial access and mobility. The final release 15 specifications support stand-alone NR operation as well. The difference between stand-alone and non-stand-alone primarily affects higher layers and the interface to the core network; the basic radio technology is the same in both cases.

Once release 15 was completed, more features have been added to NR in subsequent releases to improve performance and to address new use cases and deployment scenarios. Of the releases following the introduction of NR, release 18 is particularly interesting as it marks the start of 5G-Advanced. This name does not imply a break in backwards compatibility – release 18 is no different of any other NR release in this respect. Rather, the amount of functionality added since release 15 was considered large enough to motivate a new name for marketing reasons.

In parallel to the work on the NR radio-access technology, a new 5G core network was developed in 3GPP, responsible for functions not related to the radio access but needed for providing a complete network. However, it is possible to connect the NR radio-access network also to the legacy LTE core network known as the *Evolved Packet Core* (EPC). In fact, this is the case when operating NR in non-stand-alone mode where LTE and EPC handle functionality like connection establishment and mobility, and NR primarily provides a data-rate and capacity booster.

The remaining part of this chapter provides an overview of NR radio access including basic design principles and the most important technology components of NR release 15, as well as the evolution of NR in release 16 and onwards. The chapter can either be read on its own to get a high-level overview of NR, or as an introduction to the subsequent Chapters 6–29 which provide a detailed description of NR.

5G/5G-Advanced
https://doi.org/10.1016/B978-0-443-13173-8.00017-7

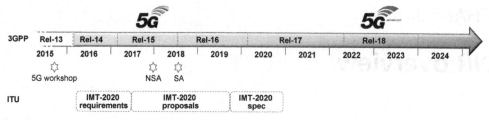

Fig. 5.1 3GPP timeline.

5.1 NR basics in release 15

NR release 15 is the first version of NR. During the development, the focus was primarily on eMBB and (to some extent) URLLC services. For massive machine-type communication (mMTC), LTE-derived technologies such as eMTC and NB-IoT [26,55] can be used with excellent results. The support for LTE-derived massive MTC on a carrier overlapping with an NR carrier has been accounted for in the design of NR (see Chapter 18) resulting in an integrated overall system capable of handling a very wide range of services.

Compared to LTE, NR provides many benefits. Some of the main ones are:

- exploitation of much higher frequency bands as a mean to obtain additional spectra to support very wide transmission bandwidths and the associated high data rates;
- ultra-lean design to enhance network energy performance and reduce interference;
- forward compatibility to prepare for future, yet unknown use cases and technologies;
- low latency to improve performance and enable new use cases; and
- a beam-centric design enabling extensive usage of beamforming and a massive number of antenna elements not only for data transmission (which to some extent is possible in LTE) but also for control-plane procedures such as initial access.

The first three can be classified as design principles (or requirements on the design) and will be discussed first, followed by a discussion of the key technology components applied to NR.

5.1.1 Higher-frequency operation and spectrum flexibility

One key feature of NR is a substantial expansion in terms of the range of spectra in which the radio-access technology can be deployed. Unlike LTE, where support for licensed spectra at 3.5 GHz and unlicensed spectra at 5 GHz were added at a relatively late stage, NR supports licensed-spectrum operation from below 1 GHz up to 52.6 GHz[a] already from its first release, with extension to unlicensed spectra in release 16 and the upper frequency limit extended to 71 GHz in release 17.

---

[a] The upper limit of 52.6 GHz is due to some very specific spectrum situations.

Operation at higher frequencies in the mm-wave band offers the possibility for large amounts of spectrum and associated very wide transmission bandwidths, thereby enabling very high traffic capacity and extreme data rates. However, higher frequencies are also associated with higher radio-channel attenuation, limiting the network coverage. Although this can partly be compensated for by means of advanced multiantenna transmission/reception, which is one of the motivating factors for the beam-centric design in NR, a substantial coverage disadvantage remains, especially in non-line-of-sight and outdoor-to-indoor propagation conditions. Thus, operation in lower-frequency bands will remain a vital component for wireless communication also in the 5G era. Especially, joint operation in lower and higher spectra, for example 2 GHz and 28 GHz, can provide substantial benefits. A higher-frequency layer, with access to a large amount of spectra, can provide service to a large fraction of the users despite the more limited coverage. This will reduce the load on the more bandwidth-constrained lower-frequency spectrum, allowing the use of this to focus on the worst-case users [62].

Another challenge with operation in higher frequency bands is the regulatory aspects. For non-technical reasons, the rules defining the allowed radiation changes around 6 GHz, from a SAR-based limitation to a more EIRP-like limitation. Depending on the device type (handheld, fixed, etc), this may result in a reduced transmission power, making the link budget more challenging than what propagation conditions alone may indicate and further stressing the benefit of combined low-frequency/high-frequency operation.

5.1.2 Ultra-lean design

An issue with current mobile-communication technologies is the amount of transmissions carried by network nodes regardless of the amount of user traffic. Such signals, sometimes referred to as "always-on" signals, include, for example, signals for base-station detection, broadcast of system information, and always-on reference signals for channel estimation. Under the typical traffic conditions for which LTE was designed, such transmissions constitute only a minor part of the overall network transmissions and thus have relatively small impact on the network performance. However, in very dense networks deployed for high peak data rates, the average traffic load per network node can be expected to be relatively low, making the always-on transmissions a more substantial part of the overall network transmissions.

The always-on transmissions have two negative impacts:

- they impose an upper limit on the achievable network energy performance; and
- they cause interference to other cells, thereby reducing the achievable data rates.

The *ultra-lean-design* principle aims at minimizing the always-on transmissions, thereby enabling higher network energy performance and higher achievable data rates.

In comparison, the LTE design is heavily based on cell-specific reference signals, signals that a device can assume are always present and use for channel estimation, tracking, mobility measurements, and so on. In NR, many of these procedures have been revisited and modified to account for the ultra-lean design principle. For example, the cell-search procedures have been redesigned in NR compared to LTE to support the ultra-lean paradigm. Another example is the demodulation reference-signal structure where NR relies heavily on reference signals being present only when data are transmitted but not otherwise.

5.1.3 Forward compatibility

An important aim in the development of the NR specification was to ensure a high degree of *forward compatibility* in the radio-interface design. In this context, forward compatibility implies a radio-interface design that allows for substantial future evolution, in terms of introducing new technology and enabling new services with yet unknown requirements and characteristics, while still supporting legacy devices on the same carrier.

Forward compatibility is inherently difficult to guarantee. However, based on experience from the evolution of previous generations, 3GPP agreed on some basic design principles related to NR forward compatibility as quoted from [3]:

• *Maximizing the amount of time and frequency resources that can be flexibly utilized or that can be left blank without causing backward compatibility issues in the future;*
• *Minimizing transmission of always-on signals;*
• *Confining signals and channels for physical layer functionalities within a configurable/allocable time/frequency resource.*

According to the third bullet one should, as much as possible, avoid having transmissions on time/frequency resources fixed by the specification. In this way one retains flexibility for the future, allowing for later introduction of new types of transmissions with limited constraints from legacy signals and channels. This differs from the approach taken in LTE where, for example, a synchronous hybrid-ARQ protocol is used, implying that a retransmission in the uplink occurs at a fixed point in time after the initial transmission. The control channels are also vastly more flexible in NR compared to LTE in order not to unnecessarily block resources.

Note that these design principles partly coincide with the aim of ultra-lean design as described here. There is also a possibility in NR to configure *reserved resources*, that is, time-frequency resources that, when configured, are not used for transmission and thus available for future radio-interface extensions. The same mechanism is also used for LTE-NR coexistence in case of overlapping LTE and NR carriers.

5.1.4 Transmission scheme, bandwidth parts, and frame structure

Similar to LTE [26], OFDM was found to be a suitable waveform for NR due to its robustness to time dispersion and ease of exploiting both the time and frequency domains when defining the structure for different channels and signals. However, unlike LTE where DFT-precoded OFDM is the sole transmission scheme in the uplink, NR uses conventional, that is, non-DFT-precoded OFDM as the baseline uplink transmission scheme due to the simpler receiver structures in combination with spatial multiplexing and an overall desire to have the same transmission scheme in both uplink and downlink. Nevertheless, DFT-precoding can be used as a complement in the uplink for similar reasons as in LTE, namely to enable high power-amplifier efficiency on the device side by reducing the *cubic metric* [57]. Cubic metric is a measure of the amount of additional power back-off needed for a certain signal waveform.

To support a wide range of deployment scenarios, from large cells with sub-1 GHz carrier frequency up to mm-wave deployments with very wide spectrum allocations, NR supports a flexible OFDM numerology with subcarrier spacings ranging from 15 kHz up to 960 kHz (in release 15 and 16, the maximum subcarrier spacing in 240 kHz) with a proportional change in cyclic prefix duration. A small subcarrier spacing has the benefit of providing a relatively long cyclic prefix in absolute time at a reasonable overhead while higher subcarrier spacings are needed to handle, for example, the increased phase noise at higher carrier frequencies and to support wide bandwidths with a reasonable number of subcarriers. Up to 3300 subcarriers are used, resulting in maximum carrier bandwidths of 50/100/200/400/1600/2000 MHz for subcarrier spacings of 15/30/60/120/480/960 kHz, respectively. Note that, for the largest subcarrier spacing of 960 kHz, not all 3300 subcarriers are used (in theory a carrier bandwidth of 3200 MHz would be possible). If larger bandwidths than what is possible to support with one carrier are needed, carrier aggregation can be used.

Although the NR physical-layer specification is band-agnostic, not all supported numerologies are relevant for all frequency bands. For each frequency band, radio requirements are therefore defined for a subset of the supported numerologies as illustrated in Fig. 5.2. The frequency range 0.45–7.125 GHz is commonly referred to as

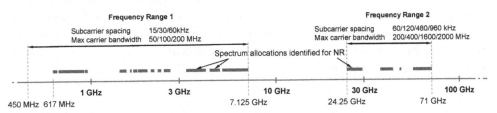

Fig. 5.2 Spectrum identified for NR and corresponding subcarrier spacings.

frequency range 1 (FR1)[b] in the specifications while the range 24.25–71 GHz is known as FR2. With the extension of FR2 up to 71 GHz in release 17, FR2 is sometimes divided into FR2-1 covering the original 24.25–52.6 GHz range and FR2-2 for the 52.6–71 GHz range. Currently, there is no NR spectrum identified between 7.125 GHz and 24.25 GHz. However, the basic NR radio-access technology is spectra agnostic and the NR specifications can easily be extended to cover additional frequencies, for example, spectra from 7.125 GHz up 24.25 GHz.

In LTE, all devices support the maximum carrier bandwidth of 20 MHz. However, given the very wide bandwidths possible in NR, it is not reasonable to require all devices to support the maximum carrier bandwidth. This has implications on several areas and requires a design different from LTE, for example the design of control channels as discussed later. Furthermore, NR allows for device-side *receiver-bandwidth adaptation* as a means to reduce the device energy consumption. Bandwidth adaptation refers to the use of a relatively modest bandwidth for monitoring control channels and receiving medium data rates, and to dynamically open up a wideband receiver only when needed to support very high data rates.

To handle these two aspects NR defines *bandwidth parts* that indicate the bandwidth over which a device is currently assumed to receive transmissions of a certain numerology. If a device is capable of simultaneous reception of multiple bandwidths parts, it would in principle be possible to, on a single carrier, mix transmissions of different numerologies for a single device although NR currently supports a single active bandwidth part at a time.

The NR time-domain structure is illustrated in Fig. 5.3 with a 10 ms radio frame is divided into ten 1 ms subframes. A subframe is in turn divided into slots consisting of 14 OFDM symbols each, that is, the duration of a slot in milliseconds depends on the numerology. For the 15 kHz subcarrier spacing, an NR slot has structure that is identical to the structure of an LTE subframe which is beneficial from a coexistence perspective. Since a slot is defined as a fixed number of OFDM symbols, a higher subcarrier spacing leads to a shorter slot duration. In principle this could be used to support lower-latency transmission, but as the cyclic prefix also shrinks when increasing the subcarrier spacing, it is not a feasible approach in all deployments. The subcarrier spacing (and cyclic prefix) is primarily selected based on the deployment scenario. To obtain low latency, NR supports a flexible approach by allowing for transmissions over a fraction of a slot, sometimes referred to as "mini-slot" transmission. Such transmissions can also preempt an already ongoing slot-based transmission to another device, allowing for immediate transmission of data requiring very low latency.

---

[b] Originally, FR1 stopped at 6 GHz but was later extended to 7.125 GHz to accommodate the 6 GHz unlicensed band.

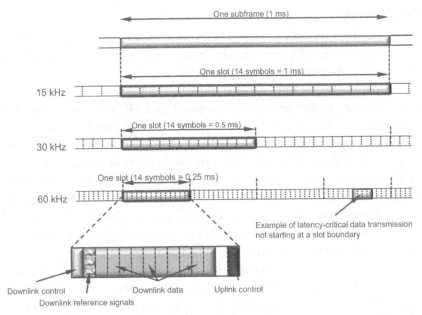

Fig. 5.3 Frame structure (TDD assumed in this example, not all subcarrier spacings shown).

Having the flexibility of starting a data transmission not only at the slot boundaries is also useful when operating in unlicensed spectra. In unlicensed spectra the transmitter is typically required to ensure that the radio channel is not occupied by other transmissions prior to starting a transmission, a procedure commonly known as "listen-before-talk". Clearly, once the channel is found to be available it is beneficial to start the transmission immediately, rather than wait until the start of the slot, in order to avoid some other transmitter initiating a transmission on the channel.

Operation in the mm-wave domain is another example of the usefulness of "mini-slot" transmissions as the available bandwidth in such deployments is often very large and even a few OFDM symbols can be sufficient to carry the available payload. This is of particular use in conjunction with *analog beamforming*, discussed later, where transmissions to multiple devices in different beams cannot be multiplexed in the frequency domain but only in the time domain.

Channel estimation at the receiver relies on user-specific demodulation reference signals (DM-RS). Not only does this enable efficient beamforming and multi-antenna operation as discussed later, it is also in line with the ultra-lean design principle described earlier. In contrast to cell-specific reference signals, demodulation reference signals are not transmitted unless there are data to transmit, thereby improving network energy performance and reducing interference.

The overall NR time/frequency structure, including bandwidth parts, is the topic of Chapter 7.

5.1.5 Duplex schemes

The duplex scheme to use is typically given by the spectrum allocation at hand. For lower frequency bands, allocations are often paired, implying frequency-division duplex (FDD) as illustrated in Fig. 5.4. At higher frequency bands, unpaired spectrum allocations are increasingly common, calling for time-division duplex (TDD). Given the significantly higher carrier frequencies supported by NR compared to LTE, efficient support for unpaired spectra is an even more critical component of NR, compared to LTE.

NR can operate in both paired and unpaired spectra using a *single* frame structure unlike LTE where two different frame structures are used (and later expanded to three when support for unlicensed spectra was introduced in release 13). The basic NR frame structure is designed such that it can support both half-duplex and full-duplex operation. In half duplex, the device cannot transmit and receive at the same time. Examples hereof are TDD and half-duplex FDD. In full-duplex operation, on the other hand, simultaneous transmission and reception is possible with FDD as a typical example.

As already mentioned, TDD increases in importance when moving to higher frequency bands where unpaired spectrum allocations are more common. These frequency bands are less useful for wide-area coverage with very large cells due to their propagation conditions but are highly relevant for local-area coverage with smaller cell sizes. Furthermore, some of the problematic interference scenarios in wide-area TDD networks are less pronounced in local area deployments with lower transmission power and below-rooftop antenna installations. In such denser deployments with smaller cell sizes, the per-cell traffic variations are more rapid compared to large-cell deployments with a large number of active devices per cell. To address such scenarios, *dynamic TDD,* that is, the possibility for dynamic assignment and re-assignment of time-domain resources between the downlink and uplink transmission directions, is a key NR technology component. This is in contrast to LTE where the uplink-downlink allocation does not change over time.[c] Dynamic TDD enables following rapid traffic variations which are particularly pronounced in dense deployments with a relatively small number of users per cell. For example, if a user is (almost) alone in a cell and needs to download a large object, most of the resources should be utilized in the downlink direction and only a small

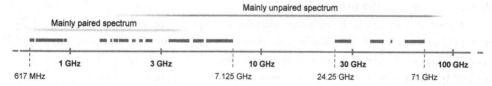

Fig. 5.4 Spectrum and duplex schemes.

---

[c] In later LTE releases, the eIMTA features allows some dynamics in the uplink-downlink allocation.

fraction in the uplink direction. At a later point in time, the situation may be different and most of the capacity is needed in the uplink direction.

The basic approach to dynamic TDD is for the device to monitor for downlink control signaling and follow the scheduling decisions. If the device is instructed to transmit, it transmits in the uplink, otherwise it will attempt to receive any downlink transmissions. The uplink-downlink allocation is then completely under the control of the scheduler and any traffic variations can be dynamically tracked. There are deployment scenarios where dynamic TDD may not be useful, but it is much simpler to restrict the dynamics of a dynamic scheme in those scenarios when needed rather than trying to add dynamics to a fundamentally semi-static design as LTE. For example, in a wide-area network with above-rooftop antennas, the inter-cell interference situation requires coordination of the uplink-downlink allocation between the cells. In such situations, a semi-static allocation is appropriate with operation along the lines of LTE. This can be obtained by the appropriate scheduling implementation. There is also the possibility to semi-statically configure the transmission direction of some or all of the slots, a feature that can allow for reduced device energy consumption as it is not necessary to monitor for downlink control channels in slots that are a priori known to be reserved for uplink usage.

5.1.6 Low-latency support

The possibility for very low latency is an important characteristic of NR and has impacted many of the NR design details. One example is the use of "front-loaded" reference signals and control signaling as illustrated in Fig. 5.3. By locating the reference signals and downlink control signaling carrying scheduling information at the beginning of the transmission and not using time-domain interleaving across OFDM symbols, a device can start processing the received data immediately without prior buffering, thereby minimizing the decoding delay. The possibility for transmission over a fraction of a slot, sometimes referred to as "mini-slot" transmission, is another example of a low-latency feature.

The requirements on the device (and network) processing times are tightened significantly in NR compared to LTE. As an example, a device has to respond with a hybrid-ARQ acknowledgement in the uplink approximately one slot (or even less depending on device capabilities) after receiving the downlink data transmission. Similarly, the time from grant reception to uplink data transfer is in the same range.

The higher-layer protocols MAC and RLC have also been designed with low latency in mind with header structures chosen to enable processing without knowing the amount of data to transmit, see Chapter 6. This is especially important in the uplink direction as the device may only have a few OFDM symbols for processing after receiving the uplink grant until the transmission should take place. In contrast, the LTE protocol design requires the MAC and RLC protocol layers to know the amount of data to transmit before any processing can take place, which makes support for a very low latency more challenging.

5.1.7 Scheduling and data transmission

One key characteristic of mobile radio communication is the large and typically rapid variations in the instantaneous channel conditions stemming from frequency-selective fading, distance-dependent path loss, and random interference variations due to transmissions in other cells and by other devices. Instead of trying to combat these variations, they can be exploited through *channel-dependent scheduling* where the time-frequency resources are dynamically shared between users (see Chapter 14 for details). Dynamic scheduling is used in LTE as well and on a high level, the NR scheduling framework is similar to the one in LTE. The scheduler, residing in the base station, takes scheduling decisions based on channel-quality reports obtained from the devices. It also takes different traffic priorities and quality-of-service requirements into account when forming the scheduling decisions sent to the scheduled devices.

Each device monitors several *physical downlink control channels* (PDCCHs) for *downlink control information* (DCI), typically once per slot although it is possible to configure more frequent monitoring to support traffic requiring very low latency. Upon detection of a valid PDCCH, the device follows the scheduling decision and receives (or transmits) one unit of data known as a transport block in NR.

In the case of downlink data transmission, the device attempts to decode the downlink transmission. Given the very high data rates supported by NR, channel-coding data transmission is based on low-density parity-check (LDPC) codes [64]. LDPC codes are attractive from an implementation perspective, especially at higher code rates where they can offer a lower complexity than the Turbo codes used in LTE.

Hybrid automatic repeat-request (ARQ) retransmission using incremental redundancy is used where the device reports the outcome of the decoding operation to the base station (see Chapter 13 for details). In the case of erroneously received data, the network can retransmit the data and the device combines the soft information from multiple transmission attempts. However, retransmitting the whole transport block could in this case become inefficient. NR therefore supports retransmissions on a finer granularity known as *code-block groups* (CBGs). This can also be useful when handling *preemption*. An urgent transmission to a second device may use only one or a few OFDM symbols and therefore cause high interference to the first device in some OFDM symbols only. In this case it may be sufficient to retransmit the interfered CBGs only and not the whole data block. Handling of preempted transmission can be further assisted by the possibility to indicate to the first device the impacted time-frequency resources such that it can take this information into account in the reception process.

Although dynamic scheduling is the basic operation of NR, operation without a dynamic grant can be configured. In this case, the device is configured in advance with resources that can be (periodically) used for uplink data transmission (or downlink data reception). Once a device has data available it can immediately commence uplink transmission without going through the scheduling request-grant cycle, thereby enabling lower latency.

5.1.8 Control channels

Operation of NR requires a set of physical-layer control channels to carry downlink control information (DCI), for example scheduling decisions, and uplink control information (UCI) to provide feedback information in the uplink. A detailed description of the structure of these control channels is provided in Chapter 10.

Downlink control channels are known as PDCCHs (physical downlink control channels). One major difference compared to LTE is the more flexible time-frequency structure of downlink control channels where PDCCHs are transmitted in one or more *control resource sets* (CORESETs) which, unlike LTE where the full carrier bandwidth is used, can be configured to occupy only part of the carrier bandwidth. This is needed in order to handle devices with different bandwidth capabilities and is also in line with the principles for forward compatibility as discussed earlier. Another major difference compared to LTE is the support for beamforming of the control channels, which has required a different reference signal design with each control channel having its own dedicated reference signal.

Uplink control information such as hybrid-ARQ acknowledgements, channel-state feedback for multi-antenna operation, and scheduling request for uplink data awaiting transmission, is transmitted using the *physical uplink control channel* (PUCCH). There are several different PUCCH formats, depending on the amount of information and the duration of the PUCCH transmission. A short duration PUCCH can be transmitted in the last one or two symbols of a slot and thus support very fast feedback of hybrid-ARQ acknowledgements in order to realize so-called self-contained slots where the delay from the end of the data transmission to the reception of the acknowledgement from the device is in the order of an OFDM symbol, corresponding to a few tens of microseconds depending on the numerology used. This can be compared to almost 3 ms in LTE and is yet another example on how the focus on low latency has impacted the NR design. For situations when the duration of the short PUCCH is too short to provide sufficient coverage, there are also possibilities for longer PUCCH durations.

For coding of the physical-layer control channels, for which the information blocks are small compared to data transmission and hybrid-ARQ is not used, polar codes [17] and Reed-Muller codes have been selected.

5.1.9 Beam-centric design and multi-antenna transmission

Support for (a large number of) steerable antenna elements for both transmission and reception is a key feature of NR. At higher frequency bands, the large number of antennas elements are primarily used for beamforming to extend coverage, while at lower frequency bands they enable full-dimensional MIMO, sometimes referred to as massive MIMO, and interference avoidance by spatial separation.

NR channels and signals, including those used for control and synchronization, have all been designed to support beamforming (Fig. 5.5). Channel-state information (CSI) for

Fig. 5.5 Beamforming in NR.

operation of massive multi-antenna schemes can be obtained by feedback of CSI reports based on transmission of CSI reference signals in the downlink, as well as using uplink measurements exploiting channel reciprocity.

To provide implementation flexibility, NR is deliberately including functionality to support analog beam-forming as well as digital precoding/beam-forming, see Chapter 11. At high frequencies, analog beamforming, where the beam is shaped after digital-to-analog conversion, may be necessary from an implementation perspective, at least initially. Analog beamforming results in the constraint that a receive or transmit beam can only be formed in one direction at a given time instant and requires beam-sweeping where the same signal is repeated in multiple OFDM symbols but in different transmit beams. By having beam-sweeping possibility, it is ensured that any signal can be transmitted with a high gain, narrow beamformed transmission to reach the entire intended coverage area.

Signaling to support beam-management procedures is specified, such as an indication to the device to assist selection of a receive beam (in the case of analog receive beamforming) to be used for data and control reception. For a large number of antennas, beams are narrow and beam tracking can fail, therefore beam-recovery procedures have also been defined, which can be triggered by a device. Moreover, a cell may have multiple transmission points, each with multiple beams, and the beam-management procedures allow for device-transparent mobility and seamless handover between the beams of different points. Additionally, uplink-centric and reciprocity-based beam management is possible by utilizing uplink signals.

With the use of a massive number of antenna elements for lower frequency bands, the possibility to separate users spatially increases both in uplink and downlink but requires that the transmitter has channel knowledge. For NR, extended support for such multi-user spatial multiplexing is introduced, either by using a high-resolution channel-state-information feedback using a linear combination of DFT vectors, or uplink sounding reference signals targeting the utilization of channel reciprocity.

Twelve (increased to 24 in release 17) orthogonal demodulation reference signals are specified for multi-user MIMO transmission purposes while an NR device can maximally receive eight MIMO layers in the downlink and transmit up to four layers in the uplink (increased to eight layers in release 17). Moreover, additional configuration of a phase tracking reference signal is introduced in NR since the increased phase noise power at high carrier frequency bands otherwise will degrade demodulation performance for larger modulation constellations, for example 64 QAM.

Distributed MIMO implies that the device can receive multiple independent physical data shared channels (PDSCHs) per slot to enable simultaneous data transmission from multiple transmission points to the same user. In essence, some MIMO layers are transmitted from one site while other layers are transmitted from another site. This can be handled through proper network implementation in release 15 although the multi-transmission point (multi-TRP) support in release 16 provides further enhancements.

Multi-antenna transmission in general, as well as a more detailed discussion on NR multi-antenna precoding, is described in Chapter 11 with beam management being the topic of Chapter 12.

5.1.10 Initial access

Initial access refers to the procedures allowing a device to find a cell to camp on, receive the necessary system information, and request a connection through random access. The basic structure of NR initial access, described in Chapters 16 and 17, is similar to the corresponding functionality of LTE [26]:

There is a pair of downlink signals, the *Primary Synchronization Signal* (PSS) and *Secondary Synchronization Signal* (SSS), that is used by UEs to find, synchronize to, and identify a network

There is a downlink *Physical Broadcast Channel* (PBCH) transmitted together with the PSS/SSS. The PBCH carries a minimum amount of system information including indication where the remaining broadcast system information is transmitted. In the context of NR, the PSS, SSS, and PBCH are jointly referred to as a *Synchronization Signal Block* (SSB).

There is a four-stage random-access procedure, commencing with the uplink transmission of a *random-access preamble*. From release 16 it is also possible ot use a two-sate random-access procedure.

However, there are some important differences between LTE and NR in terms of initial access. These differences come mainly from the ultra-lean principle and the beam-centric design, both of which impact the initial access procedures and partly lead to different solutions compared to LTE.

In LTE, the PSS, SSS, and PBCH are located at the center of the carrier and are transmitted once every 5 ms. Thus, by dwelling on each possible carrier frequency during at least 5 ms, a device is guaranteed to receive at least one PSS/SSS/PBCH transmission if a carrier exists at the specific frequency. Without any a priori knowledge a device must search all possible carrier frequencies over a carrier raster of 100 kHz.

To enable higher NR network energy performance in line with the ultra-lean principle, the SSB is, by default, transmitted once every 20 ms. Due to the longer period between consecutive SSBs, compared to the corresponding signals/channels in LTE, a device searching for NR carriers must dwell on each possible frequency for a longer time. To reduce the overall search time while keeping the device complexity comparable to LTE, NR supports a *sparse frequency raster* for the SSB. This implies that the possible frequency-domain positions of the SSB could be significantly sparser, compared to the possible positions of an NR carrier (the *carrier raster*). As a consequence, the SSB will typically not be located at the center of the NR carrier, which has impacted the NR design.

The sparse SSB raster in the frequency domain enables a reasonable time for initial cell search, at the same time as the network energy performance can be significantly improved due to the longer SSB period.

Network-side beam-sweeping is supported for both downlink SSB transmission and uplink random-access reception as a means to improve coverage, especially in the case of operation at higher frequencies. It is important to realize that beam sweeping is a *possibility* enabled by the NR design. It does not imply that it must be used. Especially at lower carrier frequencies, beam sweeping may not be needed.

5.1.11 Interworking and LTE coexistence

As it is difficult to provide full coverage at higher frequencies, interworking with systems operating at lower frequencies is important. In particular, a coverage imbalance between uplink and downlink is a common scenario, especially if uplink and downlink are in different frequency bands. The higher transmit power for the base station compared to the mobile device results in the downlink achievable data rates often being bandwidth limited, making it more relevant to operate the downlink in higher spectrum where wider bandwidth may be available. In contrast, the uplink is more often power limited, reducing the need for wider bandwidth. Instead, higher data rates may be achieved on lower-frequency spectra, despite there being less available bandwidth, due to less radio-channel attenuation.

Through interworking, a high-frequency NR system can complement a low-frequency system (see Chapter 18 for details). The lower frequency system can be either NR or LTE, and NR supports interworking with either of these. The interworking can be realized at different levels, including intra-NR carrier aggregation, dual connectivity[d] with a common packet data convergence protocol (PDCP) layer, and handover.

---

[d] In the December 2017 version of release 15, dual connectivity is only supported between NR and LTE. Dual connectivity between NR and NR is part of the final June 2018 release 15.

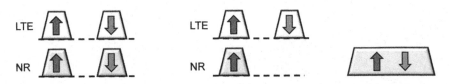

Fig. 5.6 Example of NR-LTE coexistence.

However, the lower frequency bands are often already occupied by current technologies, primarily LTE. LTE/NR *spectrum co-existence*, that is, the possibility for an operator to deploy NR in the same spectrum as an already existing LTE deployment has therefore been identified as a way to enable early NR deployment in lower frequency spectra without reducing the amount of spectrum available to LTE.

Two co-existence scenarios were identified in 3GPP and guided the NR design:

In the first scenario, illustrated in the left part of Fig. 5.6, there is LTE/NR co-existence in both downlink and uplink. Note that this is relevant for both paired and unpaired spectra although a paired spectrum is used in the illustration.

In the second scenario, illustrated in the right part of Fig. 5.6, there is co-existence only in the uplink transmission direction, typically within the uplink part of a lower-frequency paired spectrum, with NR downlink transmission taking place in the spectrum dedicated to NR, typically at higher frequencies. This scenario attempts to address the uplink-downlink imbalance discussed above. Carrier aggregation or the *supplementary uplink* (SUL) can both be used to handle this scenario.

The possibility for an LTE-compatible NR numerology based on 15 kHz sub-carrier spacing, enabling identical time/frequency resource grids for NR and LTE, is one of the fundamental tools for such coexistence. The flexible NR scheduling with a scheduling granularity as small as one symbol can then be used to avoid scheduled NR transmissions to collide with key LTE signals such as cell-specific reference signals, CSI-RS, and the signals/channels used for LTE initial access. Reserved resources, introduced for forward compatibility (see Section 5.1.3), can also be used to further enhance NR-LTE co-existence. It is possible to configure reserved resources matching the cell-specific reference signals in LTE, thereby enabling an enhanced NR-LTE overlay in the downlink.

5.2 NR evolution and 5G advanced

Release 15 forms the foundation for NR, providing a highly capable 5G standard. To further improve the performance and to meet new use cases and deployment scenarios, release 16 provide additional enhancements and mark the start of the NR evolution, an

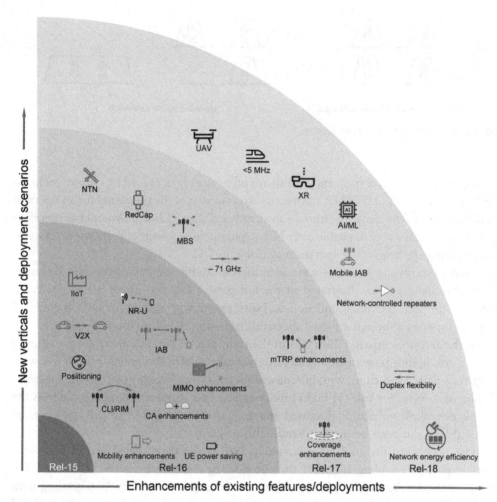

Fig. 5.7 Illustration of the NR evolution.

evolution that continues in the following releases. From release 18 and onwards, the name 5G Advanced will be used to highlight the significant enhancements compared to the first release of NR.

On a high level, the enhancements can be grouped into two categories, see Fig. 5.7:

- improvements of already existing features such as multi-antenna enhancements, carrier aggregation enhancements, mobility enhancements, and power-saving improvements; and
- new features addressing new deployment scenarios and verticals, for example integrated access and backhaul, support for unlicensed spectra, intelligent transportation systems, industrial IoT, and non-terrestrial networks (NTN).

Clearly, there is not a sharp and well-defined border between these two categories and the grouping can be done in many different ways. Fig. 5.7 should merely be seen as an illustration giving some structure to the discussion. In the following, a brief overview of these enhancements is given.

5.2.1 Multi-antenna enhancements

The multi-antenna enhancements in release 16 covers several aspects as detailed in Chapters 11 and 12. Enhancements to the CSI reporting for MU–MIMO are provided by defining a new codebook, providing increased throughput and/or reduced overhead. The beam-recovery procedures are also improved, reducing the impact from a beam failure. Finally, support for transmissions to a single device from multiple transmission points, often referred to as multi-TRP, is also added, including the necessary control signaling enhancements. Multi-TRP can provide additional robustness towards blocking of the signal from the base station to the device, something which is particularly important for URLLC scenarios.

Building on release 16, release 17 primarily addresses beam management, multi-TRP, and reciprocity-based operation as illustrated in Fig. 5.8.

Beam management has been part of NR from the start, based on the *transmission configuration indicator* (TCI) framework. In release 17, the TCI framework has been revised to reduce overhead by focusing on the common scenario where the device is using the same beam for both reception and transmission. DCI-based beam switching, which is possible for PDSCH only in released 16, is extended to cover all channels.

Multi-TRP is introduced in release 16 covering PDSCH only and extended to cover also PUSCH, PDCCH, and PUCCH in release 17. To better support non-coherent joint transmission, the CSI reporting scheme is extended to cater for this scenario. Enhancements for high-speed trains, where a device between two transmission points

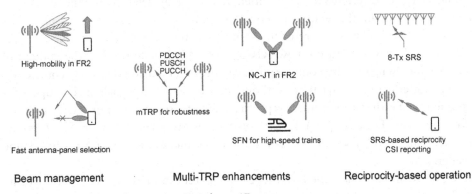

Fig. 5.8 Multi-antenna enhancements in release 17.

may experience high Doppler but with opposite signs from the two transmission points, are also introduced.

Release 17 also introduces an additional CSI codebook that allows for reduced reporting overhead by utilizing partial channel reciprocity which may be available even for FDD-based network deployments. To aid this, support for more than four antenna ports for SRS, together with a more flexible SRS triggering mechanism, is also introduced.

Release 18 continues the multi-antenna evolution and provides enhancements in the areas of enhanced CSI reporting for coherent joint transmission and increased multi-user MIMO capacity.

5.2.2 Carrier aggregation and dual connectivity enhancements

Dual connectivity and carrier aggregation are both part of NR from the first release. One use case is to improve the overall data rates. Given the bursty nature of most data traffic, rapid setup and activation of additional carriers is important in order to benefit from the high data rates resulting from carrier aggregation. If the additional carriers are not rapidly activated, the data transaction might be over before the extra carriers are active. Having all carriers in the device permanently activated, which would address the latency aspect, is not realistic from a power consumption perspective. Therefore, release 16 provides functionality for early reporting of measurements on serving and neighboring cells, as well as mechanisms to reduce signaling overhead and latency for activating additional cells. Having early knowledge of various measurements enables the network to quickly select, for example, an appropriate MIMO scheme. Without early reporting, the network needs to rely on less efficient single-layer transmission until the necessary channel-state information is available.

In earlier technologies, for example LTE, the absence of encryption for early RRC signaling has typically resulted in measurement reports being delayed until the (extensive) signaling for setting up the security protocols is complete. Thus, it will take some time before the network is fully aware of the situation at the device and can schedule data accordingly. However, with the new RRC_INACTIVE state in NR, the context of the device, including security configuration, can be preserved and the RRC connection be resumed after periods of inactivity without need for extensive signaling. This opens for the possibility of earlier measurement reporting and faster setup of carrier aggregation and dual connectivity. For example, release 16 enables measurement configuration upon the device entering RRC_INACTIVE state and measurement reporting during the resume procedure.

Release 16 also enhances the coexistence of different numerologies on different carriers by supporting cross-carrier scheduling with different numerologies on the scheduling and scheduled carriers. This was originally planned to be part of release 15 but was postponed to release 16 due to time limitations.

One relatively common scenario is to use carrier aggregation with a primary carrier on a lower frequency band and secondary carriers on higher frequency bands. By doing so, the more robust propagation conditions of a low frequency band are combined with the significantly wider transmission bandwidths available in the higher frequency bands. However, prior to release 17, it is not possible to use cross-carrier scheduling from a higher frequency band to schedule transmissions on the lower frequency band in this setup. This may lead to the PDCCH capacity being a bottleneck with a large part of the scheduling information being transmitted on a relatively narrow-band low-frequency carrier. To address this and increase the PDCCH capacity, cross-carrier scheduling can be done also from the higher frequency bands in release 17.

The dual connectivity framework also received some updated in release 17 with faster activation of secondary carriers.[e] This allows the secondary carriers(s) to be kept in a low power state but be rapidly activated to follow bursty traffic.

The carrier aggregation evolution continues in release 18 where among other features, a scheme for scheduling multiple carriers with a single DCI is introduced to reduce the overhead (in releases prior to release 18, there is one DCI per carrier scheduled).

5.2.3 Mobility enhancements

Mobility is essential for any cellular system and NR already from the start has extensive functionality in this area. Nevertheless, enhancements in terms of latency and robustness are relevant to further improve the performance.

Latency, that is the time it takes to perform a handover, needs to be sufficiently small. At high frequency ranges, extensive use of beamforming is necessary. Due to the beam sweeping used, the handover interruption time can be larger than at lower frequencies. Release 16 therefore introduces enhancements such as *dual active protocol stack* (DAPS), which in essence is a make-before-break solution to significantly reduce the interruption time.

The basic way for handling mobility and handover between cells is to use measurements reports from the device, for example reports on the received power from other neighboring cells. When the network, based on these reports, determines a handover is desirable, it will instruct the device to establish a connection to the new cell. The device follows these instructions and, once the connection with the new cell is established, responds with an acknowledgment message. This procedure in most cases work well, but in scenarios where the device experiences a very sudden drop in signal quality from the serving base station, it may not be able to receive the handover command before the

[e] In the specifications, the primary carrier is referred to as the PCell and the secondary carriers as SCells, see Chapter 7.

connection is lost. To mitigate this, release 16 provides for conditional handovers, where the device is informed in advance about candidate cells to handover to, as well as a set of conditions when to execute handover to that particular cell. This way, the device can by itself conclude when to perform a handover and thereby maintain the connection even if the link from the serving cell experiences a very sudden drop in quality. In release 17, the conditional handover is extended to also cover the secondary cell group in a dual connectivity scenario (in release 16 it is limited to the master cell group only).

Release 18 introduces additional mobility enhancements, for example, L1/L2-triggered mobility (LTM) to facilitate faster inter-cell mobility than the traditional mobility mechanism based on RRC signaling. In LTM, multiple candidate cells are configured through RRC and L1/L2 signaling is used for fast switching between the preconfigured candidate cells. This can be seen as an extension of the L1/L2-based beam management present in earlier NR releases.

5.2.4 Device power saving enhancements

From an end-user perspective, device power consumption is a very important aspect. There are mechanisms already in the first release of NR to help reducing the device power consumption, most notably discontinuous reception and bandwidth adaptation. These enhancements can, on a high level, be seen as mechanisms to rapidly turn on/off features that are needed when actively transferring data only. For example, high data rates and low latency are important when transferring data and NR therefore specifies features for short delays between control signaling and the associated data, as well as a flexible MIMO scheme supporting a large number of layers. However, when not actively transferring data, some of these aspects are less relevant and relaxing the latency budget and reducing the number of MIMO layers can help reducing the power consumption in the device. Release 16 therefore enhances cross-slot scheduling, such that the device does not have to be prepared to receive data in the same slot as the associated PDCCH, adds the possibility for a PDCCH-based wake-up signal, and introduces fast (de)activation of cell dormancy, all of which are discussed in Chapter 14. Signaling where the device can indicate a preference to transfer from connected state to idle/inactive state, and assistance information from the device on a recommended set of parameters to maximize the power saving gains are other examples of enhancements in release 16.

Further enhancements in the area of device power savings are introduced in release 17. For example, in connected state the radio-link monitoring and beam-failure detection can be relaxed when the device operated in a low-mobility scenario to conserve energy. It is also possible to adapt the PDCCH monitoring by dynamically switch between search spaces or by skipping PDCCH monitoring directly after a data burst. In idle state, a *paging early indicator* (PEI), based on a new DCI format, can be used to indicate to the device whether paging will occur in an upcoming paging occasion or not. This allows the device to save energy otherwise needed to prepare for reception of potential paging messages on the PDSCH, see Chapter 14.

5.2.5 Cross-link interference mitigation and remote interference management

Cross-link interference (CLI) handling and remote interference management (RIM) refers to two enhancements added in release 16 to handle interference scenarios in TDD systems, see Fig. 5.9. Both enhancements are discussed in further detail in Chapter 19 and summarized below.

CLI primarily targets small-cell deployments using dynamic TDD and introduces new measurements to detect crosslink interference between downlink and uplink where downlink transmissions in one cell interferes with uplink reception in a neighboring cell (and vice versa). Based on the measurements, both by devices and neighboring cells, the scheduler can improve the scheduling strategy to reduce the impact from cross-link interference.

RIM targets wide-area TDD deployments where certain weather conditions can create atmospheric ducts leading to interference from very distant base stations. Ducting is a rare event, but when it occurs downlink transmission from a base stations hundreds of kilometers away may cause very strong interference to uplink reception at thousands of base stations. The impact on the network performance is significant. With RIM, problematic interference scenarios can be managed in an automated way in contrast to manual intervention approaches.

5.2.6 Integrated access and backhaul/network-controlled repeaters

Integrated access and backhaul (IAB) extends NR to also support wireless backhaul, as an alternative to, for example, fiber backhaul. This enables the use of NR for a wireless link from central locations to distributed cell sites and between cell sites. For example, deployments of small cells in dense urban networks can be simplified, or deployment of temporary sites used for special events can be enabled.

IAB can be used in any frequency band in which NR can operate. However, it is anticipated that mm-wave spectra will be the most relevant spectrum for IAB due to the amount of spectrum available. As higher-frequency spectra typically are unpaired, this also means that IAB can be expected to primarily operate in TDD mode on the backhaul link.

Fig. 5.10 illustrates the basic structure of a network utilizing IAB. An *IAB node* connects to the network via an (IAB) *donor node* which, essentially, is a normal base station utilizing conventional (non-IAB) backhaul. The IAB node creates cells of its own and appears as a normal base station to devices connecting to it. Thus, there is no specific device impact from IAB, which is solely a network feature. This is important as it allows

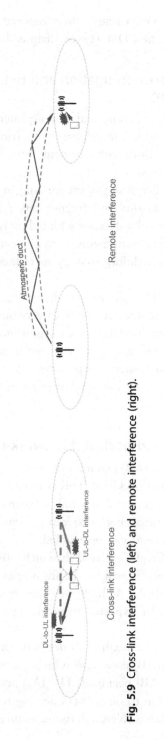

Fig. 5.9 Cross-link interference (left) and remote interference (right).

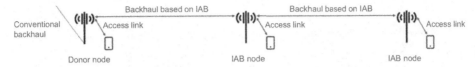

Fig. 5.10 Integrated access and backhaul.

also for legacy (release 15) devices to access the network via an IAB node. Additional IAB nodes can connect to the network via the cells created by an IAB node, thereby enabling multi-hop wireless backhauling.

In release 18, IAB is extended with enhanced for support for mobile IAB nodes, including full inter-donor mobility. This allows for IAB nodes to be placed on, for example, buses.

As an alternative to IAB nodes, repeaters can be used. A repeater essentially receives, amplifies, and retransmits the received analog signal. Repeaters have been part of cellular systems for many years but release 18 introduces support for *network-controlled* repeaters. The network control, for example, allows for scheduler-controlled beam-forming on the repeater-device link, which may improve the repeater coverage and reduce the interference caused by the repeater,

Further details on IAB and network-controlled repeaters are found in Chapter 24.

5.2.7 NR in unlicensed spectra

Spectrum availability is essential to wireless communication, and the large amount of spectra available in unlicensed bands is attractive for increasing data rates and capacity for 3GPP systems. In release 16, NR is enhanced to enable operation in unlicensed spectra. NR supports a similar setup as LTE – license-assisted access – where the device is attached to the network using a licensed carrier, and one or more unlicensed carriers are used to boost data rate using the carrier aggregation framework. However, unlike LTE, NR also supports standalone unlicensed operation, without the support from a carrier in licensed spectra, see Fig. 5.11. This will greatly add to the deployment flexibility of NR in unlicensed spectra compared to LTE-LAA.

Release 16 provides a global framework which allows operation not only in the existing 5 GHz unlicensed bands (5150–5925 MHz), but also in new bands such as the 6 GHz band (5925–7125 MHz) when they become available. In release 17, the spectrum range for NR is extended up to 71 GHz, allowing the exploitation of the 60 GHz unlicensed band in addition.

Several key principles important for operation in unlicensed spectra are already part of NR in release 15, for example ultra-lean transmission and the flexible frame structure, but there are also new mechanisms added in release 16, most notable the channel-access procedure used to support listen-before-talk (LBT), as described in Chapter 20. NR largely reuses the same channel-access mechanisms as LAA with some enhancements. In fact, the

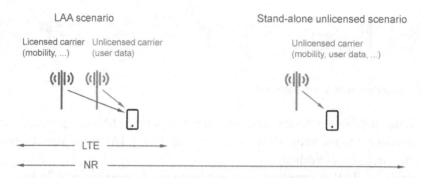

Fig. 5.11 NR in unlicensed spectra; license-assisted operation (left) and fully stand-alone (right).

same multi-standard specification, [88], is used for both LTE and NR. Reusing the mechanism developed for LTE-LAA (which to a large extent is used by Wi-Fi as well) is highly beneficial from a coexistence perspective. During the studies in 3GPP, it was demonstrated that replacement of one Wi-Fi network with an NR network in unlicensed spectra can lead to improved performance, not only for the network migrated to NR but also for the remaining Wi-Fi network.

5.2.8 Extension beyond 52.5 GHz

One major enhancement in release 17 is the extension of the upper limit of FR2 from 52.6 GHz to 71 GHz. The range 52.6–71 GHz is known as FR2-2[f] and is equally applicable to licensed and unlicensed spectra, but one reason for the extension is the unlicensed 60 GHz band. Support of frequencies above 52.6 GHz impacts NR in several ways, among them phase noise. Phase noise can be handled by more advanced receiver algorithms and does not necessarily require a larger subcarrier spacing. However, to avoid the FFT size to unnecessary limit the carrier bandwidth, NR is extended with subcarrier spacings of 480 and 960 kHz to be able to exploit carrier bandwidths up to 2 GHz.

5.2.9 Intelligent transportation systems, vehicle-to-anything, and sidelinks

Intelligent transportation systems (ITS) is one example of a new vertical in focus for NR release 16. ITS provide a range of transport- and traffic-management services that together will improve traffic safety, and reduce traffic congestion, fuel consumption and environmental impacts. Examples hereof are vehicle platooning, collision avoidance, cooperative lane change, and remote driving. To facilitate ITS, communication is required not only between vehicles and the fixed infrastructure but also between vehicles.

---

[f] The range 24.5–52.6 GHz is known as FR2-1. Together, FR2-1 and FR2-2 forms FR2.

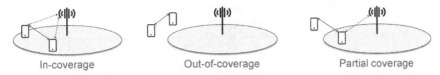

In-coverage   Out-of-coverage   Partial coverage

Fig. 5.12 Device-to-device communication under network control when in coverage.

Communication with the fixed infrastructure is obviously already catered for by using the uplink and downlink. To handle the case of direct communication between vehicles (vehicle-to-vehicle, V2V, communication), release 16 introduces the sidelink described in Chapter 26. The sidelink is also the basis for other, non–ITS-related, enhancements in release 17 and could therefore be seen as a general addition not tied to any particular vertical. Besides the sidelink, many of the enhancements introduced for the cellular uplink/downlink interface are also relevant for supporting ITS services. In particular, the ultra-reliable low-latency communication (URLLC) is instrumental in enabling remote-driving services.

The sidelink supports not only physical-layer unicast transmissions but also groupcast and broadcast transmissions. Unicast transmission, where two devices communicate directly, support advanced multi-antenna schemes relying on CSI feedback, hybrid ARQ, and link adaptation. Broadcast and multicast modes are relevant when transmitting information relevant for multiple devices in the neighborhood, for example safety messages. In this case feedback-based transmissions schemes are less suitable although hybrid-ARQ is possible.

The sidelink can operate in in-coverage, out-of-coverage and partial coverage scenarios (see Fig. 5.12) and can make use of all NR frequency bands. When there is cellular coverage, the base station may also take the role of scheduling al the sidelink transmissions.

In release 17, the sidelink design is generalized to also match use cases other than V2X, for example wearables. The resource allocation mechanism has been updated to cater for the limited battery energy in many devices, something that was less of an issue for V2X applications where the device could have access to the vehicle battery. Support for sidelink relaying, that is, using intermediate devices as relay nodes to extend network coverage, is also part of release 17.

Further enhancements take place in release 18, supporting sidelink carrier aggregation, operation in unlicensed spectrum, and potential extension to FR2.

5.2.10 Machine-type communication and internet-of-things

Machine-type communication and internet-of-things are very wide terms, covering a wide range of use cases and scenarios, including

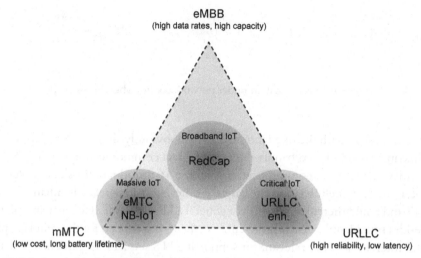

Fig. 5.13 Relative positioning of different IoT technologies.

- low-power wide-area (LPWA) networks, also known as massive IoT and character-ized by a large number of very low-cost devices, each with modest data-rate require-ments and a long battery lifetime;
- ultra-reliable low-latency communication (URLLC), sometimes referred to as critical IoT and where reliability and latency are the most important aspects;
- non-latency-critical industrial applications, also referred to as broadband IoT, for example surveillance cameras, requiring data rates of a few ten Mbit/s at most at rea-sonable cost and relaxed latency requirements.

Given the vastly different requirements of these three areas, different set of features and technologies are used as illustrated in Fig. 5.13.

LPWA is, as already mentioned, well served by the LTE-derived NB-IoT and eMTC technologies [26]. Although both NB-IoT and eMTC existed before NR was standard-ized, the spectrum-sharing mechanisms in NR enable a tight integration of these tech-nologies into an NR carrier, forming an integrated solution.

URLLC support is part of NR from the beginning. However, to widen the set of industrial IoT use cases addressed and to better support new use cases, such as factory automation, electrical power distribution, and transport industry (including the remote driving use case), release 16 adds enhancements to increase the reliability and further reduce the already low latency in release 15. Time-sensitive networking (TSN), where latency variations and accurate clock distribution are as important as low latency in itself, is another area targeted by the enhancements. Another example is a mechanism to prioritize traffic flows within and between devices. For example, uplink preemption where an ongoing low-priority uplink transmission can be cancelled as well as enhanced power control to increase the transmission power of a high-priority uplink transmission, are introduced. In general, many of the additions, described in Chapter 21, can be viewed as a collection of smaller improvements that together significantly enhance NR in the area of URLLC.

Further enhancements to industrial IoT and URLLC are introduced in release 17, aiming at increased spectral efficiency and system capacity, and improved support for time-sensitive communication. The enhancements related to PUCCH repetition, CSI reporting, hybrid-ARQ-related signaling, intra-device multiplexing, and TSN assistance information. Release 17 also improved the support of URLLC in unlicensed spectra when operating in a controlled environment. In such environments the interference variations are often smaller and channel access can be simplified, avoiding the random back-off typically encountered otherwise.

Non-latency-critical applications have in the past often been handled by simpler LTE devices, for example LTE cat 1–4. These categories are more capable than NB-IoT and eMTC, but less capable than NR. To provide an NR-based solution for use cases having higher requirements than what eMTC/NB-IoT provides but lower than what can be provided by the full set of NR features, release 17 introduces *reduced capability* (RedCap) devices. These devices are NR devices but with lower complexity than regular NR by limiting some of the features, see Fig. 5.14. RedCap is further evolved in release 18 with enhancements to the DRX handling and further data rate reduction. A more detailed discussion of RedCap is found in Chapter 22.

Independent of RedCap, release 17 also introduced enhancements for small data transmission, a way to reduce the overhead for the small and infrequent payloads often encountered in IoT applications. This is achieved by allowing uplink data transmission in

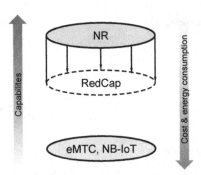

Fig. 5.14 Illustration of the relation between NR, RedCap, and eMTC/NB-IoT.

inactive state and not enforcing a transition to connected state prior to transmitting a small data packet.

5.2.11 Positioning

There is a range of applications, for example logistics and manufacturing, that require accurate positioning, not only outdoors but also indoors. NR is therefore extended in release 16 to provide better positioning support. *Global navigation satellite systems* (GNSS), assisted by cellular networks, have for many years been used for positioning. This provides accurate positioning but is typically limited to outdoor areas with satellite visibility and additional positioning methods are therefore important.

Architecture-wise, NR positioning is based on the use of a location server, similar to LTE. The location server collects and distributes information related to positioning (device capabilities, assistance data, measurements, position estimates, and so forth) to the other entities involved in the positioning procedures. A range of positioning methods, both downlink-based and uplink-based, are used separately or in combination to meet the accuracy requirements for different scenarios, see Fig. 5.15.

Downlink-based positioning is supported by providing a new reference signal, the *positioning reference signal* (PRS). Compared to LTE, the PRS has a more regular structure and a much larger bandwidth, which allows for a cleaner and more precise time–of–arrival estimation. The device can measure and report the time-of-arrival difference for PRSs received from multiple distinct base stations, and the reports are used by the location server to determine the position of the device. If different PRSs are transmitted in different beams, the reports will indirectly give information in which direction from a base station the device is located.

Uplink-based positioning is based on sounding reference signals (SRSs) transmitted from the devices, which are extended to improve the accuracy. Using these (extended) SRSs, the base stations can measure and report the arrival time, the received power, the

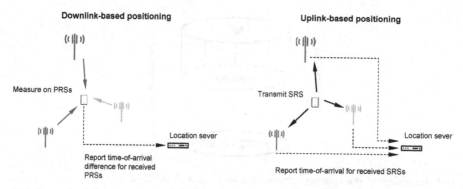

Fig. 5.15 Examples of downlink-based (left) and uplink-based (right) positioning.

angle-of-arrival (if receiver beamforming is used), and the difference between downlink transmission time and uplink SRS reception time. All these measurements are collected from the base station and fed to the location server to determine the position.

Positioning requires accurate timing. If different signals pass through different RF chains, the timing may differ. Release 17 added support for grouping signals known to passing the same RF processing together, thereby improving the timing knowledge to obtain better positioning estimates. Other enhancements in release 17 include angle-based positioning, positioning also in the RRC_INACTIVE state, and various tools to reduce the overall positioning latency. In release 18, support for positioning of RedCap devices is added, as is bandwidth aggregation and carrier-phase positioning for increased accuracy.

Positioning and the associated reference signals are the topics of Chapter 27.

5.2.12 Non-terrestrial networks

Terrestrial networks can provide very high capacity and good coverage and is the typical way cellular networks are deployed. Nevertheless, there are areas where terrestrial networks are costly or even infeasible. Coverage over oceans is an obvious example of this, but also in very remote areas it might be too costly to deploy a terrestrial network given the very small number of users. Given the reduced cost of launching satellites in recent years, non-terrestrial networks (NTN) can serve as a complement to terrestrial networks to provide coverage is remote areas where capacity is not the main concern

Support for non-terrestrial networks (NTN) has been added to NR in release 17, including low-earth orbit (LEO) as well as geostationary (GEO) satellite constellations. Handling of very high Doppler shifts and very long roundtrip times are two of the challenges addressed by release 17 enhancements, described in more detail in Chapter 25.

5.2.13 Broadcast and multicast

Data transmission in the first releases of NR focuses of data intended for a single user only and transmission protocols and radio-resource management procedures for user data are all designed with this in mind. However, there are use cases where *multiple* users are interested in the *same* data. Radio and TV broadcasting networks are two well-known examples of networks focusing on covering very large areas with the same content, with no or limited possibilities for transmission of data intended for a single user. Other examples, potentially more relevant to cellular systems, when delivery the same data to multiple devices is of interest are first responders, where a dispatcher need to communicate with a group of responders, and software upgrades of multiple devices.

To better address these types of scenarios, release 17 introduces broadcast and multicast functionality to NR as described in Chapter 23. The provision of broadcast/multicast services in a mobile-communication system implies that the same information is simultaneously provided to multiple devices, which in many cases reduces the amount of network resources required compared to individual transmissions to the devices.

5.2.14 Coverage enhancements

Coverage is one of the most important requirements on a cellular network. To a large extent, coverage for a certain data rate is determined by deployment aspects, for example transmission power and inter-site distance.

In release 17, several smaller enhancements are introduced to improve the uplink coverage which often is limited by the PUSCH performance. Channel estimation can be made jointly across both PUSCH and PUCCH, and across multiple slots. This allows the gNB to improve channel estimation by filtering the channel estimates across multiple slots transmitted back-to-back. A single transport block can be transmitted across multiple slots, which together with channel estimation across multiple slots improves the performance. There are also enhancements to dynamically control PUCCH repetitions.

In release 18, further enhancements are introduced, for example dynamic switching between OFDM and DFTS-OFDM in the uplink.

5.2.15 NR in less than 5 MHz of spectrum

The NR design is inherently flexible and a wide range of carrier bandwidths can be supported. The smallest carrier bandwidth is 5 MHz, which is small enough for typical cellular operators and is sufficient to support the 20 resource blocks required for the SSB. However, for some used cases, in particular railroad communication and some public safety applications, the available spectrum allocations can be smaller than 5MHz. To address thee use cases, release 18 reduces the smallest carrier bandwidth to approximately 3 MHz without redesigning the SSB structure. Thus, the changes are primarily related to RAN4 aspects such as the channel raster [113].

5.2.16 Extended reality (XR)

Extended reality (XR) is an umbrella term covering virtual reality (VR), augmented reality (AR) where, and mixed reality (MR). In virtual reality, computer-generated three-dimensional scenery is presented to the user, typically through VR glasses with one display per eye. By tracking the movements of the head, the virtual scenery is updated to follow user movements. In augmented reality, computer-generated scenes are overlaid on to of the real-world. Also in this case movements are tracked and the scenery is updated as the user moves. Mixed reality refers to use cases that are somewhere between AR and VR.

The development in the area of XR is rapid and XR glasses with reasonable size and weight start to appear commercially. This has triggered a huge interest in supporting these types of applications also in cellular networks and 3GPP therefore investigates potential enhancements to NR for better support of XR services [114]. XR is a demanding service in several respects. First, relatively high data rates are needed for the video information, both in the downlink to the XR glasses but also in the uplink from head-mounted cameras. Secondly, low latency is critical as the user may suffer motion sickness if the computer-generated picture is not updated fat enough. Sometimes the term *motion-to-photon* latency is used, which includes not only the data transfer part but also any video coding and video processing performed. Finally, the capacity required in a network with multiple XR users can be very large.

Although XR can be provided using NR with no specific enhancements, release 18 includes some enhancements to increase the capacity for this type of services [115]. The enhancements are primarily related to more efficient scheduling, for example improvements to configured grants and enhanced buffer-status reports. DRX mechanisms matching typical XR frame rates is another enhancement, aiming at extending the battery lifetime when using these types of services.

5.2.17 Unmanned aerial vehicles and drones

Unmanned aerial vehicles (UAVs), also known as drones, are aircrafts without any human pilot or crew. Although originally developed for military applications, they have quickly found multiple non-military uses as well, for example aerial photo, surveillance, and cargo delivery. Using cellular networks to control these drones is an attractive option and release 18 introduces a number of enhancements in this area [116]. NR was originally designed for terrestrial usage with base stations and devices on the ground, but the flexible NR framework facilitates straightforward support of drones flying in the air. One major difference compared to a fully terrestrial deployment is that a drone may see multiple cells at the same time, motivating the introduction of neighboring-cell measurement reports based on multiple cells. Other enhancements introduced include altitude-triggered device measurement reports and flight path reporting.

5.2.18 Duplex flexibility

NR is designed to support a wide range of duplex technologies – dynamic TDD, half-duplex FDD, and full-duplex FDD. All these duplex technologies assume that base station (or device) transmission and reception is separated at least in time or frequency. This separation greatly simplifies handling of self-interference. True full-duplex operation, that is, simultaneous transmission and reception on the same time/frequency resources, would suffer from significant interference that needs to be handled but can potentially offer benefits as well. For example, latency in a TDD network is strongly impacted by the uplink-downlink split, which often is semi-statically set in a wide area network, and data cannot be transmitted until the radio channel is the "right direction" (uplink for uplink data, downlink for downlink data). If data could be transmitted at any point

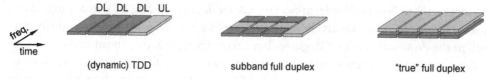

Fig. 5.16 TDD, subband full duplex, and "true" full duplex.

in time, regardless of a static uplink–downlink split, a reduced latency could be achieved. There may also be coverage and capacity benefits in some cases as data can be (continuously) transmitted in both directions at the same time. Proper handling of the self--interference and, even more important, inter-cell interference is crucial – and extremely challenging – if these benefits are to be harvested [117,118].

To investigate the feasibility and potential benefits for full duplex, a study is carried out in release 18 [119]. The focus is on subband full duplex at the gNB and conventional TDD operation at the device. In subband full duplex, uplink transmissions can occur in parts of the downlink slots, see Fig. 5.16. Self-interference is suppressed by the appropriate antenna design, linearization, and digital interference cancellation.

5.2.19 Network energy efficiency enhancements

Network energy efficiency is a very important aspect for an operator, both from a cost perspective and from a environmental angle (reduction of CO_2 emissions). This was identified early in the 5G design and was one of the major reasons behind the ultra-lean design (see Section 5.1.2). Implementations of 5G radio networks exploits the benefits from the ultra-lean design and modern 5G radio equipment consume considerably less energy than corresponding 4G equipment. To further improve network energy efficient, release 18 includes a study item [120] on network energy savings for NR. A range of techniques are studied, for example SSB-less cells and DRX/DTX also in the base station. Some of these techniques may become part of release 19, but is also valuable input for the future design of 6G.

5.2.20 Artificial intelligence and machine learning

In recent years, artificial intelligence (AI) and machine learning (ML) has been successfully applied to various problems such as image recognition and natural language interaction. To no surprise, AI/ML is being applied to many other areas as well and radio communication networks is no exception with a huge number of research papers studying the applicability of AI/ML to radio resource management as well as physical layer problems. The set of problems benefitting from AI/ML is still under discussion, but in general complex non-linear problems with access to vast amounts of data for training may benefit more than confined, mathematically well-defined (linear) problems. Already today, radio resource management, for example secondary carrier activation and hand-over decisions, can be implemented using AI/ML without explicit specification support.

To investigate the applicability of AI/ML to cellular communication and what specification support that might be needed, 3GPP has started a study in release 18 [121], focusing on three areas:

- CSI feedback, for example reduced overhead and/or improved accuracy;
- Beam management, for example beam prediction to reduce overhead and more accurate beam selection; and
- Positioning, for example ways to improve the accuracy in different (non-line-of-sight) scenarios.

The study will also familiarize 3GPP with AI/MML techniques and how such enhancements can be captured in the specifications, something which is very useful not only from an NR evolution perspective but also for a future 6G standard.

CHAPTER 6

Radio-interface architecture

This chapter contains a brief overview of the overall architecture of an NR radio-access network and the associated core network, followed by descriptions of the radio-access network user-plane and control-plane protocols.

6.1 Overall system architecture

In parallel to the work on the NR (New Radio) radio-access technology in 3GPP, the overall system architectures of both the *Radio-Access Network* (RAN) and the *Core Network* (CN) were revisited, including the split of functionality between the two networks.

The RAN is responsible for all radio-related functionality of the overall network including, for example, scheduling, radio-resource handling, retransmission protocols, coding, and various multi-antenna schemes. These functions will be discussed in detail in the subsequent chapters.

The 5G core network is responsible for functions not related to the radio access but needed for providing a complete network. This includes, for example, authentication, security, charging functionality, and setup of end-to-end connections. Handling these functions separately, instead of integrating them into the RAN, is beneficial as it allows for several radio-access technologies to be served by the same core network.

In addition to the 6G core, it is possible to connect the NR radio-access network also to the legacy LTE (Long-Term Evolution) core network known as the *Evolved Packet Core* (EPC). In fact, this is the case when operating NR in non-stand-alone mode where LTE and EPC handle functionality like connection set-up and mobility. Thus, the LTE and NR radio-access schemes and their corresponding core networks are closely related, unlike the transition from 3G to 4G where the 4G LTE radio-access technology cannot connect to a 3G core network.

Although this book focuses on the NR radio access, a brief overview of the 5G core network, as well as how it connects to the RAN, is useful as a background. For a detailed description of the 5G core network see [84].

6.1.1 5G core network

The 5G core network builds upon the EPC with three new areas of enhancement compared to EPC: service-based architecture, support for network slicing, and control-plane/user-plane split.

5G/5G-Advanced
https://doi.org/10.1016/B978-0-443-13173-8.00003-7
95

A service-based architecture is the basis for the 5G core. This means that the specification focuses on the services and functionalities provided by the core network, rather than nodes as such. This is natural as today's core network often is highly virtualized with the core network functionality running on generic computer hardware.

Network slicing is a term commonly seen in the context of 5G. A network slice is a logical network serving a certain business or customer need and consists of the necessary functions from the service-based architecture configured together. For example, one network slice can be set up to support mobile broadband applications with full mobility support and another slice can be set up to support a specific non-mobile latency-critical industry-automation application. These slices will all run on the same underlying physical core and radio networks, but, from the end-user application perspective, they appear as independent networks, each with its unique capabilities. In many aspects it is similar to configuring multiple virtual computers on the same physical computer. Edge computing, where parts of the end-user application run close to the core network edge, can also be part of such a network slice.

Control-plane/user-plane split is emphasized in the 5G core network architecture, including independent scaling of the capacity of the two. For example, if more control plane capacity is need, it is straight forward to add it without affecting the user-plane of the network.

On a high level, the 5G core can be illustrated as shown in Fig. 6.1. The figure uses a service-based representation, where the services and functionalities are in focus. In the specifications there is also an alternative, reference-point description, focusing on the point-to-point interaction between the functions, but that description is not captured in the figure.

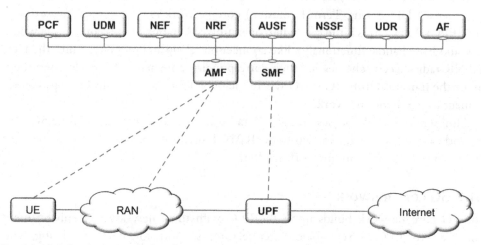

Fig. 6.1 High-level core network architecture (service-based description).

The user plane consists of the *User Plane Function* (UPF) which is a gateway between the RAN and external networks such as the Internet. Its responsibilities include packet routing and forwarding, packet inspection, quality-of-service handling and packet filtering, and traffic measurements. It also serves as an anchor point for (inter-RAT) mobility when necessary.

The control plane consist of several functions. The *Session Management Function* (SMF) handles, among other functions, IP address allocation for the device (also known as *User Equipment,* UE), control of policy enforcement, and general session-management functions. The *Access and Mobility Management Function* (AMF) is in charge of control signaling between the core network and the device, security for user data, idle-state mobility, and authentication. The functionality operating between the core network, more specifically the AMF, and the device is sometimes referred to as the *Non-Access Stratum* (NAS), to separate it from the *Access Stratum* (AS) which handles functionality operating between the device and the radio-access network.

In addition, the core network can also handle other types of functions, for example the *Policy Control Function* (PCF) responsible for policy rules, the *Unified Data Management* (UDM) responsible for authentication credentials and access authorization, the *Network Exposure Function* (NEF), the *NR Repository Function* (NRF), the *Authentication Server Function* (AUSF) handling authentication functionality, the *Network Slice Selection Function* (NSSF) handling the slide selected to serve the device, the *Unified Data Repository* (UDR), and the *Application Function* (AF) influencing traffic routing and interaction with policy control. These functions are not discussed further in this book and the reader is referred to [13,84] for further details.

It should be noted that the core network functions can be implemented in many ways. For example, all the functions can be implemented in a single physical node, distributed across multiple nodes, or executed on a cloud platform.

The description above focused on the 5G core network, developed in parallel to the NR radio access and capable of handling both NR and LTE radio accesses. However, to allow for an early introduction of NR in existing networks, it is also possible to connect NR to EPC, the LTE core network. This is illustrated as "option 3" in Fig. 6.2 and is also known as "non-standalone operation" as LTE is used for control-plane functionality such as initial access, paging, and mobility. The nodes denoted eNB and gNB will be discussed

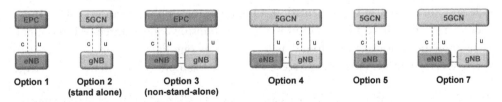

Fig. 6.2 Different combinations of core networks and radio-access technologies.

in more detail in the next section; for the time being eNB and gNB can be thought of as base stations for LTE and NR, respectively.

In option 3, the EPC core network is connected to the eNB. All control-plane functions are handled by LTE, and NR is used only for the user-plane data. The gNB is connected to the eNB and user-plane data from the EPC can be forwarded from the eNB to the gNB. There are also variants of this: option 3a and option 3x. In option 3a, the user plane part of both the eNB and gNB are directly connected to the EPC. In option 3x, only the gNB user plane is connected to the EPC and user-plane data to the eNB is routed via the gNB.

For stand-alone operation, the gNB is connected directly to the 5G core as shown in option 2. Both user-plane and control-plane functions are handled by the gNB. Option 1 is LTE and otions 4, 5, and 7 show various possibilities for connecting an LTE eNB to the 5GCN.

6.1.2 Radio-access network

The radio-access network can have two types of nodes connected to the 5G core network:

- a gNB, serving NR devices using the NR user-plane and control-plane protocols; or
- an ng-eNB, serving LTE devices using the LTE user-plane and control-plane protocols[a].

A radio-access network consisting of both ng-eNBs for LTE radio access and gNBs for NR radio access is known as an NG-RAN, although the term RAN will be used in the following for simplicity. Furthermore, it will be assumed that the RAN is connected to the 5G core and hence 5G terminology such as gNB will be used. In other words, the description will assume a 5G core network and an NR-based RAN as shown in option 2 in Fig. 6.2. However, as already mentioned, the first version of NR operates in non-standalone mode where NR is connected to the EPC using option 3. The principles are in this case similar although the naming of the nodes and interfaces differs slightly.

The gNB (or ng-eNB) is responsible for all radio-related functions in one or several cells, for example radio resource management, admission control, connection establishment, routing of user-plane data to the UPF and control-plane information to the AMF, and *quality-of-service* (QoS) flow management. It is important to note that a gNB is a *logical* node and not a physical implementation. One common implementation of a gNB is a three-sector site, where a base station is handling transmissions in three cells, although other implementations can be found as well, such as one baseband processing unit to which several remote radio heads are connected. Examples of the latter are a large

---

[a] Fig. 6.2 is simplified as it does not make a distinction between eNB connected to the EPC and ng-eNB connected to the 5GCN.

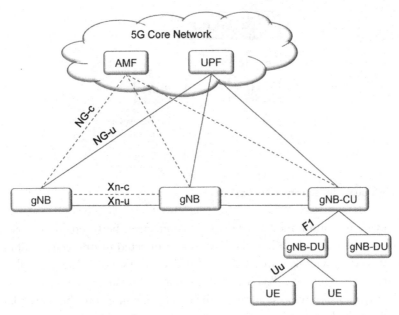

Fig. 6.3 Radio-access network interfaces.

number of indoor cells, or several cells along a highway, belonging to the same gNB. Thus, a base station is a *possible* implementation of, but *not the same* as, a gNB.

As can be seen in Fig. 6.3, the gNB is connected to the 5G core network by means of the *NG interface*, more specifically to the UPF by means of the *NG user-plane part* (NG-u) and to the AMF by means of the *NG control-plane part* (NG-c). One gNB can be connected to multiple UPFs/AMFs for the purpose of load sharing and redundancy.

The *Xn interface*, connecting gNBs to each other, is mainly used to support dual connectivity and lossless active-state mobility between cells by means of packet forwarding. It may also be used for multi-cell *Radio Resource Management* (RRM) functions.

There is also a standardized way to split the gNB into two parts, a central unit (gNB-CU) and one or more distributed units (gNB-DU) using the *F1 interface*. In case of a split gNB, the RRC, PDCP, and SDAP protocols, described in more detail later, reside in the gNB-CU and the remaining protocol entities (RLC, MAC, PHY) in the gNB-DU. However, in practice, radio resource management is split across the CU and DU as some quantities are known in the CU and others in the DU. The DU therefore decides on most of the RRC configuration, transmits this information to the CU, which forwards it to the UE. This adds complexity to the CU–DU split and in practice CU and DU are typically colocated and implemented in the same node.

The interface between the gNB (or the gNB-DU) and the device is known as the *Uu interface*.

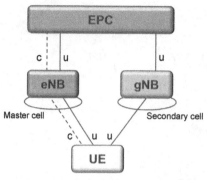

Fig. 6.4 LTE-NR dual connectivity using option 3.

For a device to communicate at least one connection between the device and the network is required. As a baseline, the device is connected to one cell handling all the uplink as well as downlink transmissions. All data flows, user data as well as RRC signaling, are handled by this cell. This is a simple and robust approach, suitable for a wide range of deployments. However, allowing the device to connect to the network through multiple cells can be beneficial in some scenarios. One example is user-plane aggregation where flows from multiple cells are aggregated in order to increase the data rate. Another example is control-plane/user-plane separation where the control plane communication is handled by one node and the user plane by another. The scenario of a device connected to *two* cells[b] is known as *dual connectivity*.

Dual connectivity between LTE and NR is of particular importance as it is the basis for non-stand-alone operation using option 3 as illustrated in Fig. 6.4. The LTE-based master cell handles control-plane and (potentially) user-plane signaling, and the NR-based secondary cell handles user-plane only, in essence boosting the data rates.

Dual connectivity between NR and NR is not part of the December 2017 version of release 15 but is possible in the final June 2018 version of release 15.

6.2 Radio protocol architecture

With the overall network architecture in mind, the RAN protocol architecture for the user and control planes can be discussed. Fig. 6.5 illustrates the RAN protocol architecture. The AMF is, as discussed in the previous section, not part of the RAN but is included in the figure for completeness. Furthermore, user data from the internet enters the SDAP via the UPF in the core network although this is not shown in the figure.

---

[b] Actually, two cell *groups*, the master cell group (MCG) and the secondary cell group (SCG) in case of carrier aggregation as carrier aggregation implies multiple cells in each of the two cell groups.

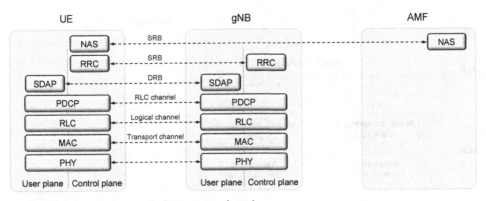

Fig. 6.5 User plane and control plane protocol stack.

A general overview of the NR protocol architecture for the downlink is illustrated in Fig. 6.6 (the names of the protocol layers will be described later). Many of the protocol layers are similar to those in LTE, although there are some differences as well. One of the differences is the QoS handling in NR when connected to a 5G core network, where the user-plane SDAP protocol layer accepts one or more QoS flows carrying IP packets according to their QoS requirements. In case of NR connected to the EPC, the SDAP is not used.

As will become clear in the subsequent discussion, not all the entities illustrated in Fig. 6.6 are applicable in all situations. For example, ciphering is not used for broadcasting of the basic system information, and SDAP is not used for control plane messages. The uplink protocol structure is similar to the downlink structure in Fig. 6.6, although there are some differences with respect to, for example, transport-format selection and the control of logical-channel multiplexing.

For each connected device, there is one or more *PDU sessions*, each with one or more *QoS flows* and *data radio bearers* (DRBs) in the user plane carrying the user data. Multiple DRBs can be configured, for example to handle different traffic priorities. In a similar manner, there are *signaling radio bearers* (SRBs) to handle control-plane messages. Up to four different SRB types can be used: SRB0 for RRC messages using common channels, SRB1 for RRC messages (and piggybacked NAS messages) using dedicated channels, SRB2 for NAS messages after security being configured, and SRB3 for certain RRC messages in case of non-standalone operation.

The different protocol entities of the radio-access network are summarized below and described in more detail in the following sections:

• *Service Data Application Protocol* (SDAP) is responsible for mapping QoS bearers to radio bearers according to their quality-of-service requirements. This protocol layer is not

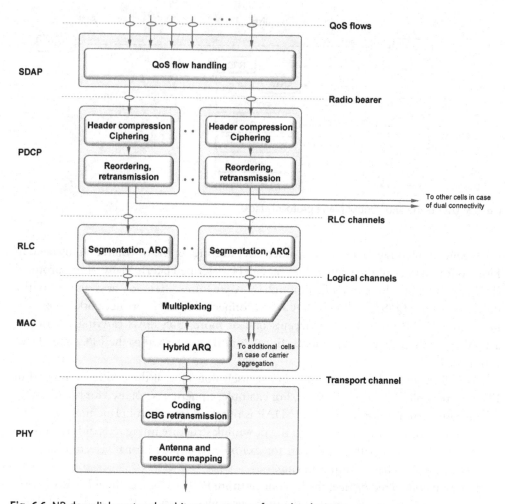

Fig. 6.6 NR downlink protocol architecture as seen from the device.

present in LTE but introduced in NR when connecting the user plane to the 5G core network due to the new quality-of-service handling.

- *Packet Data Convergence Protocol* (PDCP) performs IP header compression, ciphering, and integrity protection. It also handles retransmissions, in-sequence delivery, and duplicate removal[c] in the case of handover. For dual connectivity with split bearers, PDCP can provide routing and duplication. Duplication and transmission from different cells can be used to provide diversity for services requiring very high reliability. There is one PDCP entity per radio bearer configured for a device.

---

[c] Duplicate detection is part of the June 2018 release and not present in the December 2017 release of NR.

- *Radio-Link Control* (RLC) is responsible for segmentation and retransmission handling. The RLC provides services to the PDCP in the form of *RLC channels*. There is one RLC entity per RLC channel (and hence per radio bearer) configured for a device. Unlike LTE, the NR RLC does not support in-sequence delivery of data to higher protocol layers, a change motivated by the reduced delays as discussed below.
- *Medium-Access Control* (MAC) handles multiplexing of logical channels, hybrid-ARQ retransmissions, and scheduling and scheduling-related functions. The scheduling functionality is located in the gNB for both uplink and downlink. The MAC provides services to the RLC in the form of *logical channels*. The header structure in the MAC layer has been designed to allow for efficient support of low-latency processing.
- *Physical Layer* (PHY) handles coding/decoding, modulation/demodulation, multi-antenna mapping, and other typical physical-layer functions. The physical layer offers services to the MAC layer in the form of *transport channels*.

To summarize the flow of downlink user data through all the protocol layers, an example illustration with three IP packets, two on one radio bearer and one on another radio bearer, is given in Fig. 6.7. In this example, there are two radio bearers and one RLC SDU is segmented and transmitted in two different transport blocks. The data flow in the case of uplink transmission is similar.

The SDAP protocol maps the IP packets to the different radio bearers; in this example IP packets *n* and *n+1* are mapped to radio bearer *x* and IP packet *m* is mapped to radio bearer *y*. In general, the data entity from/to a higher protocol layer is known as a *Service Data Unit* (SDU) and the corresponding entity to/from a lower protocol layer entity is called a *Protocol Data Unit* (PDU). Hence, the output from the SDAP is an SDAP PDU, which equals a PDCP SDU.

The PDCP protocol performs (optional) IP-header compression, followed by ciphering, for each radio bearer. A PDCP header is added, carrying information required for deciphering in the device as well as a sequence number used for retransmission

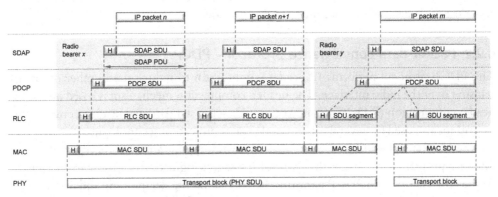

Fig. 6.7 Example of user-plane data flow.

and in-sequence delivery, if configured. The output from the PDCP is forwarded to the RLC.

The RLC protocol performs segmentation of the PDCP PDUs if necessary and adds an RLC header containing a sequence number used for handling retransmissions. Unlike LTE, the NR RLC is not providing in-sequence delivery of data to higher layers. The reason is additional delay incurred by the reordering mechanism, a delay that might be detrimental for services requiring very low latency. If needed, in-sequence delivery can be provided by the PDCP layer instead.

The RLC PDUs are forwarded to the MAC layer, which multiplexes a number of RLC PDUs and attaches a MAC header to form a transport block. Note that the MAC headers are distributed across the MAC PDU such that the MAC header related to a certain RLC PDU is located immediately prior to the RLC PDU. This is different compared to LTE, which has all the header information at the beginning of the MAC PDU, and is motivated by efficient low-latency processing. With the structure in NR, the MAC PDU can be assembled "on the fly" as there is no need to assemble the full MAC PDU before the header fields can be computed. This reduces the processing time and hence the overall latency.

The remainder of this section contains an overview of the SDAP, RLC, MAC, and physical layers.

6.2.1 Service data adaptation protocol – SDAP

The Service Data Adaptation Protocol (SDAP), relevant for the user plane only, is responsible for mapping between a quality-of-service flow from the 5G core network and a data radio bearer, as well as marking the *quality-of-service flow identifier* (QFI) in uplink and downlink packets. The reason for the introduction of SDAP in NR is the quality-of-service handling when connected to the 5G core as discussed in more detail in Section 6.4. In this case the SDAP is responsible for the mapping between QoS flows and radio bearers as described in Section 6.4. If the gNB is connected to the EPC, as is the case for non-standalone mode, the SDAP is not used.

6.2.2 Packet-data convergence protocol – PDCP

The PDCP protocol performs IP header compression to reduce the number of bits to transmit over the radio interface. The header-compression mechanism is based on robust header compression (ROHC) framework [36], a set of standardized header-compression algorithms also used for several other mobile-communication technologies. PDCP is also responsible for ciphering to protect against eavesdropping and, for the control plane, integrity protection to ensure that control messages originate from the correct source. At the receiver side, the PDCP performs the corresponding deciphering and decompression operations.

The PDCP is also responsible for duplicate removal and (optional) in-sequence delivery, functions useful for example in case of intra-gNB handover. Upon handover, undelivered downlink data packets will be forwarded by the PDCP from the old (source) gNB to the new (target) gNB. The PDCP entity in the device will also handle retransmission of all uplink packets not yet delivered to the gNB as the hybrid-ARQ buffers are flushed upon handover. In this case, some PDUs may be received in duplicate, both over the connection to the old gNB and the new gNB. The PDCP will in this case remove any duplicates. The PDCP can also be configured to perform reordering to ensure in-sequence delivery of SDUs to higher-layer protocols if desirable.

Duplication in PDCP can also be used for additional diversity. Packets can be duplicated and transmitted on multiple cells, increasing the likelihood of at least one copy being correctly received. This can be useful for services requiring very high reliability. At the receiving end, the PDCP duplicate removal functionality removes any duplicates. In essence, this results in selection diversity.

Dual connectivity is another area where PDCP plays an important role. In dual connectivity, a device is connected to two cells, or in general, two cell groups[d], the *Master Cell Group* (MCG) and the *Secondary Cell Group* (SCG). The two cell groups can be handled by different gNBs. A radio bearer is typically handled by one of the cell groups, but there is also the possibility for *split bearers* in which case one radio bearer is handled by both cell groups. In this case the PDCP is in charge of distributing the data between the MCG and the SCG as illustrated in Fig. 6.8.

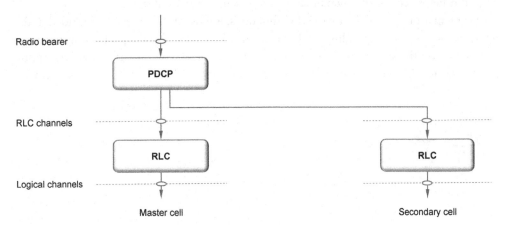

Fig. 6.8 Dual connectivity with split bearer.

---

[d] The reason for the term cell *group* is to cover also the case of carrier aggregation where there are multiple cells, one per aggregated carriers, in each cell group.

The June 2018 version of release 15, as well as later releases, support dual connectivity in general while the December 2017 version of release 15 is limited to dual connectivity between LTE and NR only. This is of particular importance as it is the basis for non-stand-alone operation using option 3 as illustrated in Fig. 6.4. The LTE-based master cell handles control-plane and (potentially) user-plane signaling, and the NR-based secondary cell handles user-plane only, in essence boosting the data rates.

6.2.3 Radio-link control

The RLC protocol is responsible for segmentation of RLC SDUs from the PDCP into suitably sized RLC PDUs. It also handles retransmission of erroneously received PDUs, as well as removal of duplicate PDUs. Depending on the type of service, the RLC can be configured in one of three modes – transparent mode, unacknowledged mode, and acknowledged mode – to perform some or all of these functions. Transparent mode is, as the name suggests, transparent and no headers are added. Unacknowledged mode supports segmentation and duplicate detection while acknowledged mode in addition supports retransmission of erroneous packets.

One major difference compared to LTE is that the RLC does not ensure in-sequence delivery of SDUs to upper layers. Removing in-sequence delivery from the RLC reduces the overall latency as later packets do not have to wait for retransmission of an earlier missing packet before being delivered to higher layers but can be forwarded immediately. Another difference is the removal of concatenation from the RLC protocol to allow RLC PDUs to be assembled in advance, prior to receiving the uplink scheduling grant. This also helps reduce the overall latency as discussed in Chapter 13.

Segmentation, one of the main RLC functions, is illustrated in Fig. 6.9. Included in the figure is also the corresponding LTE functionality, which also supports concatenation. Depending on the scheduler decision, a certain amount of data, that is, certain transport-block size, is selected. As part of the overall low-latency design of NR, the scheduling

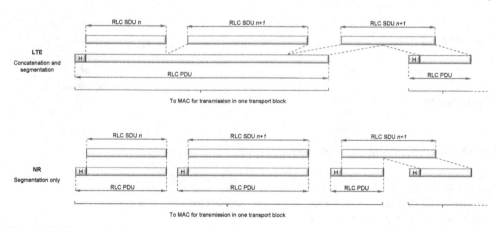

Fig. 6.9 RLC segmentation.

decision in case of an uplink transmission is known to the device just before transmission, in the order of a few OFDM symbols before. In the case of concatenation in LTE, the RLC PDU cannot be assembled until the scheduling decision is known, which results in an additional delay until the uplink transmission and cannot meet the low latency requirement of NR. By removing the concatenation from RLC, the RLC PDUs can be assembled in advance and upon receipt of the scheduling decision the device only has to forward a suitable number of RLC PDUs to the MAC layer, the number depending on the scheduled transport block size. To completely fill up the transport block size, the last RLC PDU may contain a segment of an SDU. The segmentation operation is simple. Upon receiving the scheduling grant, the device includes the amount of data needed to fill up the transport block and updates the header to indicate it is a segmented SDU.

The RLC retransmission mechanism is also responsible for providing error-free delivery of data to higher layers. To accomplish this, a retransmission protocol operates between the RLC entities in the receiver and transmitter. By monitoring the sequence numbers indicated in the headers of the incoming PDUs, the receiving RLC can identify missing PDUs (the RLC sequence number is independent of the PDCP sequence number). Status reports are fed back to the transmitting RLC entity, requesting retransmission of missing PDUs. Based on the received status report, the RLC entity at the transmitter can take the appropriate action and retransmit the missing PDUs if needed.

Although the RLC is capable of handling transmission errors due to noise, unpredictable channel variations, and so forth, error-free delivery is in most cases handled by the MAC-based hybrid-ARQ protocol. The use of a retransmission mechanism in the RLC may therefore seem superfluous at first. However, as will be discussed in Chapter 13, this is not the case and the use of both RLC- and MAC-based retransmission mechanisms is in fact well motivated by the differences in the feedback signaling.

The details of RLC are further described in Section 13.2.

6.2.4 Medium-access control

The MAC layer handles logical-channel multiplexing, hybrid-ARQ retransmissions, and scheduling and scheduling-related functions, including handling of different numerologies. It is also responsible for multiplexing/demultiplexing data across multiple component carriers when carrier aggregation is used.

6.2.4.1 Logical channels and transport channels

The MAC provides services to the RLC in the form of *logical channels*. A logical channel is defined by the *type* of information it carries and is generally classified as a *control channel*, used for transmission of control and configuration information necessary for operating an NR system, or as a *traffic channel*, used for the user data. The set of logical-channel types specified for NR includes:

- The *Broadcast Control Channel* (BCCH), used for downlink transmission of *system information* from the network to all devices in a cell. Prior to accessing the system, a device needs to acquire the system information to find out how the system is configured and, in general, how to behave properly within a cell. Note that, in the case of non-stand-alone operation, system information is provided by the LTE system and there is no BCCH.

- The *Paging Control Channel* (PCCH), used for paging of devices whose location on a cell level is not known to the network. The paging message therefore needs to be transmitted in multiple cells. Note that, in the case of non-stand-alone operation, paging is provided by the LTE system and there is no PCCH.
- The *Common Control Channel* (CCCH), used for transmission of control information in conjunction with random access when the device has no RRC connection to the network. SRB0 is carried on the CCCH.
- The *Dedicated Control Channel* (DCCH), used for transmission of control information to/from a device. This channel is used for individual configuration of devices such as setting various parameters in devices and carries SRBs 1, 2, and 3.
- The *Dedicated Traffic Channel* (DTCH), used for transmission of user data to/from a device. This is the logical channel type used for transmission of all unicast uplink and downlink user data on one or more DRBs.

Later releases of NR introduce support for multicast/broadcast transmission and sidelink communication, requiring additional logical-channel types:

- The *MBS Traffic Channel* (MTCH), used for transmission of multicast/broadcast user data, see Chapter 23.
- The *MBS Control Channel* (MCCH), used for MBS-related control information, see Chapter 23.
- The *Sidelink Broadcast Control Channel* (SBCCH), used for broadcast of sidelink-related control information, see Chapter 26.
- The *Sidelink Common Control Channel* (SCCH), used for sidelink control information, see Chapter 26.
- The *Sidelink Traffic Channel* (STCH), used for sidelink user data between devices.

From the physical layer, the MAC layer uses services in the form of *transport channels*. A transport channel is defined by *how* and *with what characteristics* the information is transmitted over the radio interface. Data on a transport channel are organized into *transport blocks*. Associated with each transport block is a *Transport Format* (TF), specifying *how* the transport block is to be transmitted over the radio interface. The transport format includes information about the transport-block size, the modulation-and-coding scheme, and the antenna mapping. By varying the transport format, the MAC layer can thus realize different data rates, a process known as *transport-format selection*.

The following transport-channel types are defined for NR:

- The *Broadcast Channel* (BCH) has a fixed transport format, provided by the specifications. It is used for transmission of parts of the BCCH system information in the downlink, more specifically the so-called *Master Information Block* (MIB), as described in Chapter 16.

- The *Paging Channel* (PCH) is used for downlink transmission of paging information from the PCCH logical channel. The PCH supports *discontinuous reception* (DRX) to allow the device to save battery power by waking up to receive the PCH only at predefined time instants.
- The *Downlink Shared Channel* (DL-SCH) is the main transport channel used for transmission of downlink data in NR. It supports key NR features such as dynamic rate adaptation and channel-dependent scheduling in the time and frequency domains, hybrid ARQ with soft combining, and spatial multiplexing. It also supports DRX to reduce device power consumption while still providing an always-on experience. The DL-SCH is also used for transmission of the parts of the BCCH system information not mapped to the BCH. Each device has a DL-SCH per cell it is connected to. At time instants when system information is received there is one additional DL-SCH from the device perspective.
- The *Uplink Shared Channel* (UL-SCH) is the uplink counterpart to the DL-SCH – that is, the uplink transport channel used for transmission of uplink data.
- The *Random-Access Channel* (RACH), used during random access, is also (for historical reasons) defined as a transport channel, although it does not carry transport blocks.

Additional transport-channel types are introduced in release 16 to support sidelink communication (see Chapter 26):

- The *Sidelink Broadcast Channel* (SL-BCH) to handle sidelink broadcast information.
- The *Sidelink Shared Channel* (SL-SCH) to carry, among other things, sidelink user data.

Part of the MAC functionality is multiplexing of different logical channels and mapping of the logical channels to the appropriate transport channels. The mapping between logical-channel types and transport-channel types is given in Fig. 6.10. This figure clearly indicates how DL-SCH and UL-SCH are the main downlink and uplink transport channels respectively. In the figures, the corresponding physical channels, described further later, are also included together with the mapping between transport channels and physical channels.

   To support priority handling, multiple logical channels, where each logical channel has its own RLC entity, can be multiplexed into one transport channel by the MAC layer. At the receiver, the MAC layer handles the corresponding demultiplexing and forwards the RLC PDUs to their respective RLC entity. To support the demultiplexing at the receiver, a MAC header is used. The placement of the MAC headers has been improved compared to LTE, again with low latency operation in mind. Instead of locating all the MAC header information at the beginning of a MAC PDU, which implies that assembly of the MAC PDU cannot start until the scheduling decision is available and the set of SDUs to include is known, the sub-header corresponding to a certain MAC SDU is placed immediately before that SDU as shown in Fig. 6.11. This allows the PDUs to be

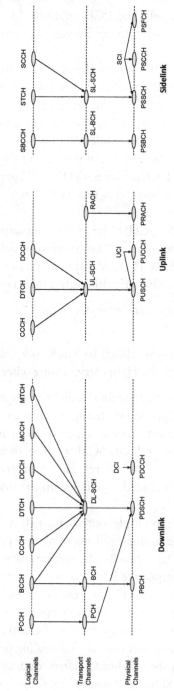

Fig. 6.10 Mapping between logical, transport, and physical channels.

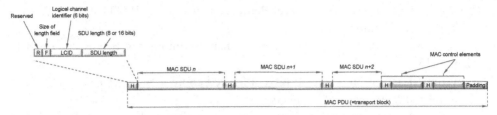

Fig. 6.11 MAC SDU multiplexing and header insertion (uplink case).

preprocessed before having received the scheduling decision. If necessary, padding can be appended to align the transport block size with those supported in NR.

The sub-header contains the identity of the logical channel (LCID) from which the RLC PDU originated and the length of the PDU in bytes.[e] There is also a flag indicating the size of the length indicator, as well as a bit reserved for future use. The LCID normally uses 6 bits, that is, up to 32 different logical channels and 32 reserved values, but it is possible to expand the LCID to 8 or 16 bits, referred to as eLCID. This can be useful in, for example, IAB where a large number of logical channels may need to be handled by the donor node.

In addition to multiplexing of different logical channels, the MAC layer can also insert *MAC control elements* into the transport blocks to be transmitted over the transport channels. A MAC control element is used for inband control signaling and is identified with reserved values in the LCID field, where the LCID value indicates the type of control information. Both fixed and variable length MAC control elements are supported, depending on their usage. For downlink transmissions, MAC control elements are located at the beginning of the MAC PDU, while for uplink transmissions the MAC control elements are located at the end, immediately before the padding (if present). Again, the placement is chosen in order to facilitate low-latency operation in the device.

MAC control elements are, as mentioned earlier, used for inband control signaling. It provides a faster way to send control signaling than RLC without having to resort to the restrictions in terms of payload sizes and reliability offered by physical-layer L1/L2 control signaling (PDCCH or PUCCH). There are multiple MAC control elements, used for various purposes, for example:

- Scheduling-related MAC control elements, such as buffer status reports and power headroom reports used to assist uplink scheduling as described in Chapter 14 and the configured grant confirmation MAC control element used when configuring semi-persistent scheduling;

---

[e] LCID 0 indicates the CCCH, 1–32 different DCCHs/DTCHs/MTCHs, and the remaining values are reserved or used for MAC control elements.

- Random-access related MAC control elements such as the C-RNTI and contention-resolution MAC control elements;
- Timing-advance MAC control element to handle timing advance as described in Chapter 15;
- Activation/deactivation of previously configured component carriers;
- DRX-related MAC control elements;
- Activation/deactivation of PDCP duplication detection; and
- Activation/deactivation of CSI reporting and SRS transmission (see Chapter 8).

The MAC entity is also responsible for distributing data from each flow across the different component carriers, or cells, in the case of carrier aggregation. The basic principle for carrier aggregation is independent user-plane processing of the component carriers in the physical layer, including control signaling and hybrid-ARQ retransmissions, while carrier aggregation is invisible above the MAC layer. Carrier aggregation is therefore mainly seen in the MAC layer, as illustrated in Fig. 6.12, where logical channels, including any MAC control elements, are multiplexed to form transport blocks per component carrier with each component carrier having its own hybrid-ARQ entity.

Both carrier aggregation and dual connectivity result in the device being connected to more than one cell. Despite this similarity, there are fundamental differences, primarily related to how tightly the different cells are coordinated and whether they reside in the same or in different gNBs.

Carrier aggregation implies a very tight coordination with all the cells belonging to the same gNB. Scheduling decisions are taken jointly for all the cells the device is connected to by one joint scheduler.

Dual connectivity, on the other hand, allows for a much looser coordination between the cells. The cells can belong to different gNBs, and they may even belong to different radio access technologies as is the case for NR-LTE dual connectivity in case of non-

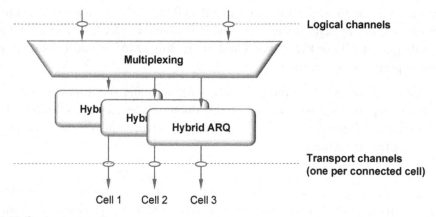

Fig. 6.12 Carrier aggregation.

standalone operation. One drawback with dual connectivity is that it will, unike carrier aggregation, typically result in approximately the same data rate as the best link, given the loose coordination between the cells and the associated delays.

Carrier aggregation and dual connectivity can also be combined. This is the reason for the terms master cell group and secondary cell group. Within each of the cell groups, carrier aggregation can be used.

6.2.4.2 Hybrid ARQ with soft combining

Hybrid ARQ with soft combining provides robustness against transmission errors. As hybrid-ARQ retransmissions are fast, many services allow for one or multiple retransmissions, and the hybrid-ARQ mechanism therefore forms an implicit (closed loop) rate-control mechanism. The hybrid-ARQ protocol is part of the MAC layer, while the physical layer handles the actual soft combining.[f]

Hybrid ARQ is not applicable for all types of traffic. For example, broadcast transmissions, where the same information is intended for multiple devices, typically do not rely on hybrid ARQ. Hence, hybrid ARQ is only supported for the DL-SCH and the UL-SCH, although its usage is up to the gNB implementation.[g]

The hybrid-ARQ protocol uses multiple parallel stop-and-wait processes, see Fig. 6.13. Upon receipt of a transport block, the receiver tries to decode the transport block and informs the transmitter about the outcome of the decoding operation through a single acknowledgment bit indicating whether the decoding was successful or if a retransmission of the transport block is required. Note that the use of multiple parallel

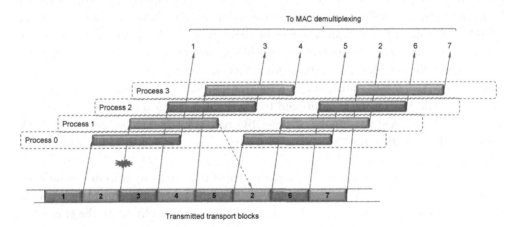

Fig. 6.13 Multiple parallel hybrid-ARQ processes.

---

[f] The soft combining is done before or as part of the channel decoding, which clearly is a physical-layer functionality. Also, the per-CBG retransmission handling is formally part pf the physical layer.

[g] The SL-SCH also supports hybrid-ARQ, see Chapter 26.

hybrid-ARQ processes can result in data being delivered from the hybrid-ARQ mechanism out of sequence. For example, transport block 3 in the figure was successfully decoded before transport block 2 which required retransmissions. For many applications this is acceptable and, if not, in-sequence delivery can be provided through the PDCP protocol. In sequence may result in longer latency; in the example in Fig. 6.13 packets numbers 3, 4, and 5 need to be delayed until packet number 2 is correctly received before delivering them to higher layers, while without in-sequence delivery each packet can be forwarded as soon as it is correctly received.

An asynchronous hybrid-ARQ protocol is used for both downlink and uplink – that is, an explicit hybrid-ARQ process number is used to indicate which process is being addressed for the transmission. In an asynchronous hybrid-ARQ protocol, the retransmissions are in principle scheduled similar to the initial transmissions. The use of an asynchronous uplink protocol, instead of a synchronous one as in LTE, is necessary to support dynamic TDD where there is no fixed uplink/downlink allocation. It also offers better flexibility in terms of prioritization between data flows and devices and is beneficial for the extension to unlicensed spectrum operation in release 16.

One additional feature of the hybrid-ARQ mechanism is the possibility for retransmission of *codeblock groups*, a feature that can be beneficial for very large transport blocks or when a transport block is partially interfered by another preempting transmission. As part of the channel-coding operation in the physical layer, a transport block is split into one or more codeblocks with error-correcting coding applied to each of the codeblocks. Even for modest data rates there can be multiple code blocks per transport block and at Gbps data rates there can be hundreds of code blocks per transport block. In many cases, especially if the interference is bursty and affects a small number of OFDM symbols in the slot, only a few of these code blocks in the transport block may be corrupted while the majority of code blocks are correctly received. To correctly receive the transport block, it is therefore sufficient to retransmit the erroneous code blocks only. At the same time, the control signaling overhead would be too large if individual code blocks can be addressed by the hybrid-ARQ mechanism. Therefore, *codeblock groups* (CBGs) are defined. If per-CBG retransmission is configured, feedback is provided per CBG and only the erroneously received code block groups are retransmitted (Fig. 6.14).

The hybrid-ARQ mechanism will rapidly correct transmission errors due to noise or unpredictable channel variations. As discussed earlier, the RLC is also capable of requesting retransmissions, which at first sight may seem unnecessary. However, the reason for having two retransmission mechanisms on top of each other can be seen in the feedback signaling – hybrid ARQ provides fast retransmissions but due to errors in the feedback the residual error rate is typically too high for, for example, good TCP performance, while RLC ensures (almost) error-free data delivery but slower retransmissions than the

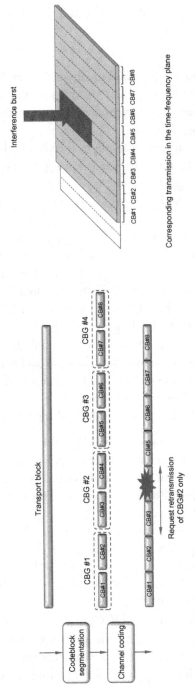

Fig. 6.14 Codeblock group retransmission.

hybrid-ARQ protocol. Hence, the combination of hybrid ARQ and RLC provides an attractive combination of small round-trip time and reliable data delivery.

6.2.5 Physical layer

The physical layer is responsible for coding, physical-layer hybrid-ARQ processing, modulation, multi-antenna processing, and mapping of the signal to the appropriate physical time–frequency resources. It also handles mapping of transport channels to physical channels, as shown in Fig. 6.10.

As mentioned in the introduction, the physical layer provides services to the MAC layer in the form of transport channels. Data transmission in downlink and uplink uses the DL-SCH and UL-SCH transport-channel types, respectively. In the case of carrier aggregation, there is one DL-SCH (or UL-SCH) per component carrier seen by the device.

A *physical channel* corresponds to the set of time–frequency resources used for transmission of a particular transport channel and each transport channel is mapped to a corresponding physical channel, as shown in Fig. 6.10. In addition to the physical channels with a corresponding transport channel, there are also physical channels without a corresponding transport channel. These channels, known as L1/L2 control channels, are used for *downlink control information* (DCI), providing the device with the necessary information for proper reception and decoding of the downlink data transmission, *uplink control information* (UCI) used for providing the scheduler and the hybrid-ARQ protocol with information about the situation at the device, and *sidelink control information* (SCI) used in conjunction with sidelink transmissions.

The following physical-channel types are defined for NR:

- The *Physical Downlink Shared Channel* (PDSCH) is the main physical channel used for unicast data transmission, but also for transmission of, for example, paging information, random-access response messages, and delivery of parts of the system information.
- The *Physical Broadcast Channel* (PBCH) carries part of the system information, required by the device to access the network.
- The *Physical Downlink Control Channel* (PDCCH) is used for downlink control information, mainly scheduling decisions, required for reception of PDSCH, and for scheduling grants enabling transmission on the PUSCH.
- The *Physical Uplink Shared Channel* (PUSCH) is the uplink counterpart to the PDSCH. There is at most one PUSCH per uplink component carrier per device.
- The *Physical Uplink Control Channel* (PUCCH) is used by the device to send hybrid-ARQ acknowledgments, indicating to the gNB whether the downlink transport block(s) was successfully received or not, to send channel-state reports aiding downlink channel-dependent scheduling, and for requesting resources to transmit uplink data upon.
- The *Physical Random-Access Channel* (PRACH) is used for random access.

Note that some of the physical channels, more specifically the channels used for downlink and uplink control information (PDCCH and PUCCH) do not have a corresponding transport channel mapped to them. With the introduction of sidelink in release 16, additional physical channel types are defined, the *Physical Sidelink Control Channel* (PSCCH), the *Physical Sidelink Shared Channel* (PSSCH), the *Physical Sidelink Feedback Channel* (PSFCH), and the *Physical Sidelink Broadcast Channel* (PSBCH), see Chapter 26 for details.

6.3 Scheduling

One of the basic principles of NR radio access is shared-channel transmission – that is, time–frequency resources are dynamically shared between users. The *scheduler* is part of the MAC layer and controls the assignment of uplink and downlink resources in terms of so-called *resource blocks* in the frequency domain and OFDM symbols and slots in the time domain. Sidelink scheduling is treated in Chapter 26 and not discussed further in this section.

The basic operation of the scheduler is *dynamic* scheduling, where the gNB takes a scheduling decision, typically once per slot, and sends scheduling information to the selected set of devices. Although per-slot scheduling is a common case, neither the scheduling decisions, nor the actual data transmission is restricted to start or end at the slot boundaries. This is useful to support low-latency operation as well as unlicensed spectrum operation as mentioned in Chapter 7.

Uplink and downlink scheduling are separated in NR, and uplink and downlink scheduling decisions can be taken independent of each other (within the limits set by the duplex scheme in case of half-duplex operation).

The downlink scheduler is responsible for (dynamically) controlling which devices to transmit to and, for each of these devices, the set of resource blocks upon which the device's DL-SCH should be transmitted. Transport-format selection (selection of transport-block size, modulation scheme, and antenna mapping) and logical-channel multiplexing for downlink transmissions are controlled by the gNB, as illustrated in the left part of Fig. 6.15.

The uplink scheduler serves a similar purpose, namely to (dynamically) control which devices are to transmit on their respective UL-SCH and on which uplink time–frequency resources (including component carrier). Even though the gNB scheduler determines the transport format for the device, it is important to point out that the uplink scheduling decision does not explicitly schedule a certain logical channel but rather the device as such. Thus, although the gNB scheduler controls the payload of a scheduled device, the device is responsible for selecting *from which radio bearer(s)* the data are taken according to a set of rules, the parameters of which can be configured by the gNB. This is illustrated in the right part of Fig. 6.15, where the gNB scheduler controls the transport format and the device controls the logical-channel multiplexing.

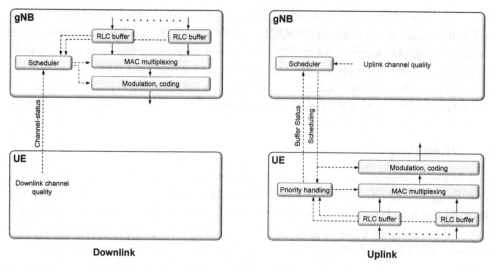

Fig. 6.15 Transport-format selection in (a) downlink and (b) uplink.

Although the scheduling strategy is implementation specific and not specified by 3GPP, the overall goal of most schedulers is to take advantage of the channel variations between devices and preferably schedule transmissions to a device on resources with advantageous channel conditions in both the time and frequency domain, often referred to as *channel-dependent scheduling*.

Downlink channel-dependent scheduling is supported through *channel-state information* (CSI), reported by the device to the gNB and reflecting the instantaneous downlink channel quality in the time and frequency domains, as well as information necessary to determine the appropriate antenna processing in the case of spatial multiplexing. In the uplink, the channel-state information necessary for uplink channel-dependent scheduling can be based on a *sounding reference signal* transmitted from each device for which the gNB wants to estimate the uplink channel quality. To aid the uplink scheduler in its decisions, the device can transmit buffer-status and power-headroom information to the gNB using MAC control elements. This information can only be transmitted if the device has been given a valid scheduling grant. For situations when this is not the case, a scheduling request indicating that the device needs uplink resources is provided as part of the uplink L1/L2 control-signaling structure (see Chapter 10).

Although dynamic scheduling is the baseline mode-of-operation, there are also possibilities for data transmission/reception without a dynamic grant. Configured grants and semi-persistent scheduling can be used to reduce the control-signaling overhead. Release 17 also introduces small data transmission to allow small amounts of data to be transmitted as part of the random-access procedure without explicit dynamic scheduling.

Downlink semi-persistent scheduling can be used in the downlink. A semi-static scheduling pattern is signaled in advance to the device. Upon activation by L1/L2 control signaling, which also includes parameters such as the time-frequency resources and coding-and-modulation scheme to use, the device receives downlink data transmissions according to the preconfigured pattern.

Uplink configured grants can be either of type 1 or type 2. The two alternatives are fairly similar, but differs in how to activate the scheme. In type 1, RRC configures all parameters, including the time-frequency resources and the modulation-and-coding scheme to use, and also activates the uplink transmission according to the parameters. Type 2, on the other hand, is similar to semi-persistent scheduling where RRC configures the scheduling pattern in time. Activation is done using L1/L2 signaling which includes the necessary transmission parameters (except the periodicity which is provides through RRC signaling). In both type 1 and type 2, the device does not transmit in the uplink unless there are data to convey.

When communicating the scheduling decisions to the devices, each device need to have an identity, unique within the cell. For this purpose, the *cell radio network temporary identifier* (C-RNTI) is used. There are also multiple other identities defined, for example to transmit paging messages, system information, and various control messages as outlined in Table 6.1. Many of these identities will be discussed in the upcoming chapters.

6.4 Quality-of-service handling

QoS handling is essential for the realization of network slicing. For each connected device, there is, as described earlier, one or more PDU sessions, each with one or more QoS flows and data radio bearers. The IP packets are mapped to the QoS flows according to the QoS requirements, for example in terms of delay or required data rate, as part of the UDF functionality in the core network. Each packet can be marked with a *QoS Flow Identifier* (QFI) to assist uplink QoS handling. The second step, mapping of QoS flows to data radio bearers, is done by the SDAP in the radio-access network. Thus, the core network is aware of the service requirements while the radio-access network only maps the QoS flows to radio bearers. The QoS-flow-to-radio-bearer mapping is not necessarily a one-to-one mapping; multiple QoS flows can be mapped to the same data radio bearer (Fig. 6.16).

There are two ways of controlling the mapping from quality-of-service flows to data radio bearers in the uplink: reflective mapping and explicit configuration.

In the case of reflective mapping, the device observes the QFI in the downlink packets for the PDU session. This provides the device with knowledge about which IP flows are mapped to which QoS flow and radio bearer. The device then uses the same mapping for the uplink traffic.

In the case of explicit mapping, the quality-of-service flow to data radio bearer mapping is configured in the device using RRC signaling.

Table 6.1 Different RNTIs and their pupose.

RNTI	Purpose
C-RNTI	Device identity, unique within the cell and used when dynamically scheduling data.
MCS-C-RNTI	Similar to C-RNTI but with a different modulation and coding scheme
CS-RNTI	Device identity used for semi-persistent/configured-grant scheduling.
P-RNTI	Used for paging.
SI-RNTI	Used for scheduling system information.
RA-RNTI	Used for the random-access response.
TC-RNTI	Temporary C-RNTI, used in conjunction with random access.
MSGB-RNTI	Used for transmission of MsgB in the two-step random-access procedure.
TPC-PUSCH-RNTI	Used for transmission of PUSCH power control commands
TPC-PUCCH-RNTI	Used for transmission of PUCCH power control commands
TPC-SRS-RNTI	Used for transmission of SRS power control commands
INT-RNTI	Used for transmission of the preemption indicator
SFI-RNTI	Used for transmission of the slot-format indicator
SP-CSI-RNTI	Activation of Semi-persistent CSI reporting on PUSCH
CI-RNTI	Cancellation indicator used for cancelling uplink transmissions
PS-RNTI	Used to control DRX behavior
SL-RNTI	Dynamically scheduled sidelink transmission
SLCS-RNTI	Configured sidelink transmission
SL Semi-Persistent Scheduling V-RNTI	Used for V2X transmissions
AI-RNTI	Used when transmitting the availability indicator
G-RNTI	Dynamically scheduled MBS PTM transmission
G-CS-RNTI	Configured scheduled multicast transmission
MCCH-RNTI	Dynamically scheduled MCCH transmissions (for MBS)
PEI-RNTI	Paging Early Indicator
CG-SDT-CS-RNTI	Used for retransmissions when SDT is used with configured grants
NCR-RNTI	Used for controlling network-controlled repeaters

6.5 Radio resource control

The control plane consists of NAS and RRC and is, among other things, responsible for connection setup, mobility, and security. Establishing a connection between the device and the network is an intricate process with many steps, including signaling between the device and the network but also between network nodes and functionalities (see Fig. 6.17 for a simplified example).

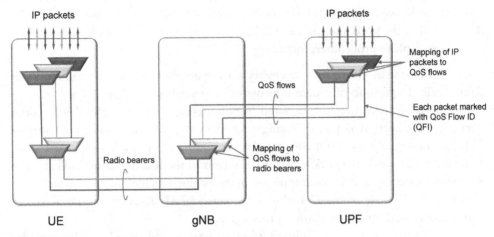

Fig. 6.16 QoS flows and radio bearers during a PDU session.

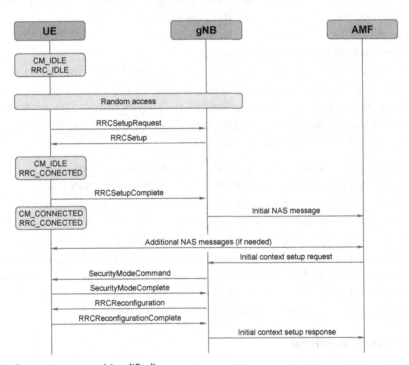

Fig. 6.17 Connection setup (simplified).

The NAS control–plane functionality operates between the AMF in the core network and the device. It includes authentication, security, registration management, and mobility management (of which paging, described later, is a subfunction). It is also responsible for assigning an IP address to a device as part of the session management functionality.

The *Radio Resource Control* (RRC) control-plane functionality operates between the RRC located in the gNB and the device. RRC is responsible for handling the RAN-related control-plane procedures, including:

- Broadcast of system information necessary for the device to be able to communicate with a cell. Acquisition of system information is described in Chapter 16.
- Transmission of core-network initiated paging messages to notify the device about incoming connection requests. Paging is used in the idle state (described further below) when the device is not connected to a cell. It is also possible to initiate paging from the radio-access network when the device is in the inactive state. Indication of system-information updates is another use of the paging mechanism, as are *Earthquake and Tsunami Warning System* (ETWS) and *Commercial Mobile Alert Systems* (CMAS), two systems used for public warning messages.
- Connection management, including setting up bearers and mobility. This includes establishing an RRC context – that is, configuring the parameters necessary for communication between the device and the radio-access network.
- Mobility functions such as cell (re)selection, see below.
- Measurement configuration and reporting.
- Handling of device capabilities; when connection is established the device will upon request from the network announce its capabilities as not all devices are capable of supporting all the functionality described in the specifications. The network needs to take these capabilities, for example the maximum number of MIMO layers a device supports, into account when transmitting data.

RRC messages are transmitted to the device using signaling radio bearers (SRBs), using the same set of protocol layers (PDCP, RLC, MAC and PHY) as described in Section 6.1. The SRB, more specifically SRB0, is mapped to the common control channel (CCCH) during establishment of connection and, once a connection is established, SRBs 1, 2, and 3 are mapped to the dedicated control channel (DCCH). Control-plane and user-plane data can be multiplexed in the MAC layer and transmitted to the device in the same transport block. The aforementioned MAC control elements can also be used for control of radio resources in some specific cases where low latency is more important than ciphering, integrity protection, and reliable transfer.

6.5.1 RRC state machine

In most wireless communication systems, the device can be in different states depending on the traffic activity. This is true also for NR and an NR device can be in one of three RRC states, RRC_IDLE, RRC_CONNECTED, and RRC_INACTIVE (see Fig. 6.18). The two first two RRC states, RRC_IDLE and RRC_CONNECTED, are similar to the counterparts in LTE while RRC_INACTIVE is a new state introduced in NR and not present in the original LTE design. There are also core network states not

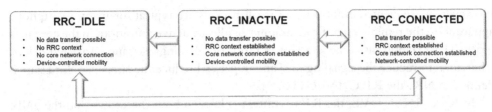

Fig. 6.18 RRC states.

discussed further herein, CM_CONNECTED and CM_IDLE, depending on whether the device has established a connection with the core network or not.

In RRC_IDLE, there is no RRC context – that is, the parameters necessary for communication between the device and the network – in the radio-access network and the device does not belong to a specific cell. From a core network perspective, the device is in the CM_IDLE state. No data transfer may take place as the device sleeps most of the time to reduce battery consumption. In the downlink, devices in idle state periodically wake up to receive paging messages, if any, from the network. Mobility is handled by the device through cell reselection, see Section 6.6.5, and paging is initiated by the core network. Uplink synchronization is not maintained and hence the only uplink transmission activity that may take place is random access, discussed in Chapter 17, to move the device to connected state. As part of moving to a connected state, the RRC context is established in both the device and the network.

In RRC_CONNECTED, the RRC context is established and all parameters necessary for communication between the device and the radio-access network are known to both entities. From a core network perspective, the device is in the CM_CON-NECTED state. The cell to which the device belongs is known and the an C-RNTI has been assigned to the device. The connected state is intended for data transfer to/from the device, but *discontinuous reception* (DRX) can be configured to reduce device power consumption (DRX is described in further detail in Section 14.5). Since there is an RRC context established in the gNB in the connected state, leaving DRX and starting to receive/transmit data is relatively fast as no connection setup with its associated signaling is needed. Mobility is managed by the radio-access network as described in Section 6.6.1, that is, the device provides neighboring-cell measurements to the network which instructs the device to perform a handover when needed. Uplink time alignment may or may not exist but needs to be established using random access and maintained as described in Chapter 17 for data transmission to take place.

In LTE, only idle and connected states are supported.[h] A common case in practice is to use the idle state as the primary sleep state to reduce the device power consumption.

---

[h] This holds for LTE connected to EPC. Starting from release 15, it is also possible to connect LTE to the 5G core network in which case there is LTE support for the inactive state.

However, as frequent transmissions of small packets are typical for many smartphone applications, the result is a significant amount of idle-to-active transitions in the core network. These transitions come at a cost in terms of signaling load and associated delays. Therefore, to reduce the signaling load and in general reduce the latency, a third state is defined in NR, the RRC_INACTIVE state.

In RRC_INACTIVE, the RRC context is kept in both the device and the gNB. The core network connection is also kept, that is, the device is in CM_CONNECTED from a core network perspective. Hence, transition to connected state for data transfer is fast as no core network signaling is needed. The RRC context is already in place in the network and inactive-to-active transitions can be handled in the radio-access network. At the same time, the device is allowed to sleep in a similar way as in the idle state and mobility is handled through cell reselection, that is, without involvement of the network. Except for the small-data transmission enhancements introduced in release 17, no data transmission can take place in RRC_INACTIVE. Thus, RRC_INACTIVE can be seen as a mix of the idle and connected states.[i]

6.5.2 Radio link monitoring

Once a link has been established in connected state, the device performs *radio-link monitoring* (RLM), that is, the estimated quality in terms of block-error rate of a hypothetical PDCCH transmission is monitored. Based the signal-to-interference-and-noise ratio on an SSB or CSI-RS, the device computes the corresponding block-error rate. If the block-error rate exceeds 10%, out-of-sync is declared. Unless the block-error rate falls below 2% (the threshold for in-sync) during a certain time, *radio-link failure* (RLF) is declared. Upon declaring radio-link failure, the device attempts to reestablish the connection by selecting a suitable cell and performing random access, similar to how a connection is initially set up. For some handover procedures discussed in the next section, more specifically DAPS and CHO, upon detecting radio-link failure the device attempts to perform a handover to one of the preconfigured target cells. If connection reestablishment is not successful within a certain time, the device enters idle state and needs to go through the full connection establishment procedure.

6.6 Mobility

Efficient mobility handling is a key part of any mobile communication system and NR is no exception. Mobility is a large and fairly complex area, including not only the measurement reports from the device but also the proprietary algorithms implemented in

---

[i] In LTE release 13, the RRC suspend/resume mechanism was introduced to provide similar functionality as RRC_INACTIVE in NR. However, the connection to the core network is not maintained in RRC suspend/resume.

the radio-access network, involvement from the core network, and how to update the routing of the data in the transport network. Mobility can also occur between different radio-access technologies, for example between NR and LTE. Covering the whole mobility area in detail would fill a book on its own and, in the following, only a brief summary is provided, focusing on the radio-network aspects.

Depending on the state of the device – idle, inactive, or connected – different mobility principles are used. For the connected state, network-controlled mobility is used where the device measures and reports the signal quality for neighboring candidate cells and the network decides when to hand over to another cell. For the inactive and idle states, the device is handling the mobility autonomously using cell reselection. The two mechanisms are briefly described in the following.

6.6.1 Network-controlled mobility

In the connected state the device has a connection established to the network. The aim of mobility in this case is to ensure that this connectivity is retained without any interruption or noticeable degradation as the device moves within the network. The basis for this is network-controlled mobility which comes in two flavors: beam-level mobility and cell-level mobility, both based on filtered measurements as illustrated in Fig. 6.19.

Beam-level mobility is handled in lower layers, MAC and the physical layer, and is essentially identical to beam management as discussed in Chapter 12. The device remains in the same cell.

Cell-level mobility, on the other hand, implies changing the serving cell. The device is configured with measurements to perform on candidate cells, filtering of the measurements, and event-triggered reporting to the network. Since the network is in charge of determining when the device should be moved to a different cell, the device location is known to the network on a cell level (or possibly with even finer granularity but that is not relevant for this discussion). Changing the serving cell typically involves RRC signaling and is therefore slower than beam-level mobility, but in later releases enhancements such as conditional handover and L1/L2 triggered mobility have been introduced to speed up the cell change.

The first step in cell-level mobility, which is continuously executed when in connected state, is to search for candidate cells using the cell search mechanism described in Chapter 16. The search can be on the same frequency as the currently connected cell, but it is also possible to search on other frequencies (or even other radio-access technologies). In the latter case, the device may need to use network-configured measurements gaps during which data reception is stopped and the receiver is temporarily retuned to another frequency for measurement purposes.

Once a candidate cell has been found, the device measures the *reference signal received power* (RSRP, a power measurement) or *reference signal received quality* (RSRQ, essentially

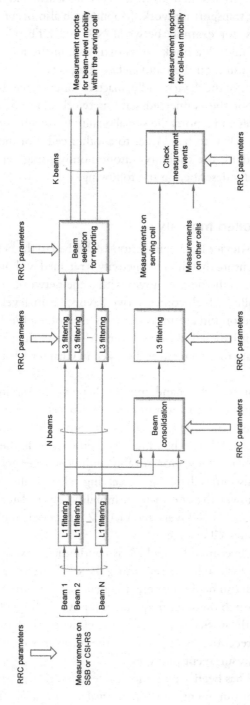

Fig. 6.19 Illustration of measurements and filtering for cell-level and beam-level mobility.

a signal-to-noise-and-interference-ratio measurement). In NR, the measurements are typically performed on the SSB, but it is also possible to measure on the CSI-RS. Filtering, for example averaging over some hundred milliseconds, can be configured and is normally used. Without averaging the measurement reports might fluctuate and result in incorrect handover decisions and possibly ping-pong effects with the device repeatedly moved back and forth between two cells. For stable operation, handover decisions should be taken on the average signals strength and not account for the rapid fading phenomenon on millisecond time scale.

A measurement event is a condition which should be fulfilled before the measured value is reported to the network. In NR, six different intra-RAT triggering conditions, or events, can be configured:

- A1 (*serving becomes better than threshold*). One usage is to cancel an ongoing handover if the device suddenly moves back into good coverage before the handover is completed.
- A2 (*serving becomes worse than threshold*). This event does not involve any measurements on cells other than the serving cell and can thus be used to trigger a handover when the device moves close to the cell border.
- A3 (*neighbor becomes offset better than SpCell*), which compares the serving cell to a candidate cell. It can thus be used to trigger handover when the device moves towards the cell edge and another cell becomes better than the current one.
- A4 (*neighbor becomes better than threshold*), which only considers the signal strength of candidate cells. One possible usage is to trigger handover for load balancing reasons.
- A5 (*SpCell becomes worse than threshold1 and neighbor becomes better than threshold2*) which can be used in a similar way as avent A3.
- A6 (*neighbor becomes offset better than SCell*), which can be used to determine which SCells to use in a carrier aggregation scenario.

Take event A3 (*neighbor becomes offset better than SpCell*) as an example, illustrated in Fig. 6.20. This event compares the filtered measurement (RSRP or RSRQ) on the candidate cell with the same quantity for the cell currently serving the device. A configurable cell-specific offset can be included for both the current and the candidate cell to handle any different in transmission power between the two. There is also a configurable threshold included to avoid triggering the event if the difference between the two cells is very small. If the measured value for the candidate cell is better than the current cell, including the offsets and the threshold, the event is triggered, and, after a configurable delay, a measurement report is transmitted to the network. The measurement report can be configured to be repeated multiple times to ensure it reaches the gNB. It is also possible to configure reporting when leaving the event.

Although the description above assumed event A3, other events can be configured following the same framework. The configuration and choice of events to use depend

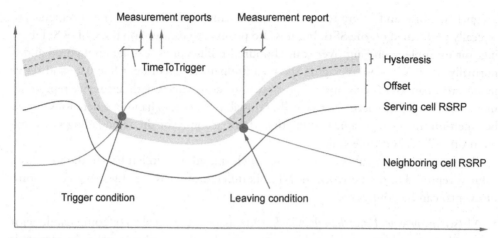

Fig. 6.20 Example of event A3.

on the mobility strategy implemented in the network for a particular deployment. In addition to the intra-RAT events A1–A6, there are also events related to inter-RAT mobility (B1, B2), interference levels (I1), measurements to support relays (X1, X2, Y1, Y2) and sidelinks (C1, C2). Periodic reporting can also be specified.

The purpose of using configurable events is to avoid unnecessary measurement reports to the network. An alternative to event-driven reporting is periodic reporting, but this would in most cases result in a significantly higher overhead as the reports need to be frequent enough to account for the device moving around in the network. Infrequent periodic reports would be preferable from an uplink overhead perspective, but would also result in an increased risk of missing an important handover occasion. By configuring the events of interest, reports are only transmitted when the situation has changed which is a much better choice [72].

Upon reception of a measurement report, the network can decide whether to perform a handover or not. Additional information other than the measurement report can be taken into account, such as whether there is sufficient capacity available in the candidate target cell for the handover. The network may also decide to handover a device to another cell even if no measurement report has been received, for example for load balancing purposes. If the network decides to handover the device to another cell, independent of the reason, a series of messages are exchanged, see Fig. 6.21 for a simplified view.

The source gNB sends a handover request to the target gNB (if the source and target cell belong to the same gNB there is no need for this message as the situation in the target cell is already known to the gNB). If the target gNB accepts (it may reject the request, for example if the load in that cell is too high), the source gNB instructs the device to switch to the target cell. This is done by sending an RRC reconfiguration message to the device,

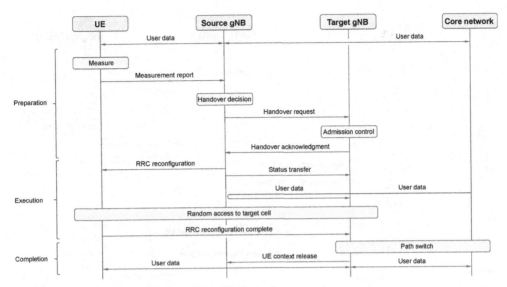

Fig. 6.21 Connected state handover (simplified view).

containing the necessary information for the device to access the target cell. To be able to connect to the new cell, uplink synchronization is required. Therefore, the device is instructed to perform random access toward the target cell. Once synchronization is established, a handover complete message is sent to the target cell, indicating that the device successfully has connected to the new cell. In parallel, any data buffered in the source gNB are moved to the target gNB and new incoming downlink data is rerouted to the target cell (which now is the serving cell). The handover completion part in Fig. 6.21 also includes network–internal signaling such as switching the data routing path to the target cell and to release the device context in the source cell.

6.6.2 Conditional handover (CHO)

The conventional, network–controlled handover procedure described above is a good baseline solution, applicable to most cases. However, it requires the device to maintain the connection to the source cell during the handover procedure until the handover is completed and the device is connected to the target cell. In scenarios where sudden changes in the received channel quality may occur, there is a risk that the radio channel quality between the device and the source cell deteriorates to, in the worst case, a level where the device is not able to receive the handover command from the source cell. This will result in a radio link failure, forcing the device to the idle state and to reestablish the connection, which clearly causes a non-negligible interruption in the connection.

To provide additional robustness in such scenarios, release 16 introduces a conditional handover mechanism as a complement. In conditional handover, the network configures

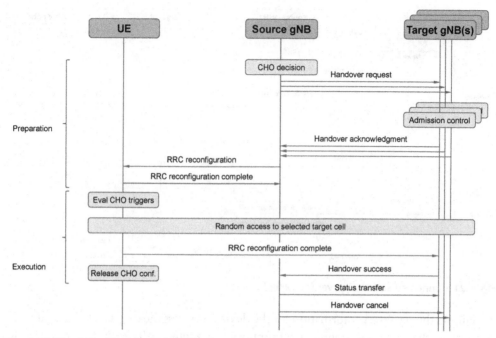

Fig. 6.22 Conditional handover (simplified).

the device with a list of candidate cells and, for each candidate cell, a condition when the handover should be triggered. For example, the trigger could be that the RSRP of a candidate cell is better than the current serving cell. Prior to configuring the device, the source cell ensures that the candidate target cells are capable of receiving a potential handover. When the triggering condition is fulfilled, the device autonomously initiates the handover, including performing a random access to the target cell, and sends a *RRCReconfigurationComplete* message to the target cell to indicate that the handover has completed, see Fig. 6.22. This way, there is no risk of losing a handover command despite sudden drops in channel quality towards the source cell. Once the conditional handover is completed, the device releases the conditional handover configuration. There are also some network-internal messages not shown in Fig. 6.22 to handle data routing between the core network and the target gNB, similar to the case of network-controlled handover. Note that L3 handover commands have precedence over conditional handover.

Conditional handovers also play an important role when extending NR to support non-terrestrial networks as described in Chapter 25.

6.6.3 Dual active protocol stacks (DAPS)

Network-controlled mobility uses RRC signaling as described above when switching from the source to the target cell. Processing of the RRC signaling and establishing the connection to the target cell takes some time, resulting in a brief interrupt in the data transfer during the handover procedure. In most cases this is not an issue, but it is still preferable to minimize this interruption time. To address this, release 16 supports *dual active protocol stacks* (DAPS) to realize a make-before-break type of handover to reduce handover interruption time. In DAPS, the data transfer between the source cell and the device continues while the device is establishing a connection to the target cell and only when the new connection is established is the device disconnected from the source cell. DAPS is beneficial from a handover interruption time point-of-view but requires the device to be able to maintain two connections at the same time. Although DAPS transmits all uplink data to the source cell until the random-access procedure to the target cell has completed, uplink transmission of control information needs to be maintained to both the cells during the procedure to support downlink data delivery. This adds complexity to the device.

6.6.4 L1/L2-triggered mobility (LTM)

One drawback with DAPS is, as noted above, the need to maintain uplink transmission to multiple cells. Another way to reduce the handover interruption time without the complexity of maintaining dual protocol stacks is the L1/L2-triggered mobility (LTM) introduced in release 18. With LTM, the device is provided with a set of candidate cells in advance, similar to what is done with conditional handovers. Communication takes place with the source cell and the device measures to on the candidate cells, that is, a similar approach as conventional network-controlled mobility. However, the measurements are L1 measurements and not L3 measurements with a fairly long time averaging as used for conventional mobility reports. Furthermore, the measurement reports are sent to the gNB using L1/L2 signaling, not RRC signaling. This significantly speeds up the measuring and reporting.

Based on the L1/L2 report, the gNB may take a handover decision to one of the pre-configured cells and informs the device using a MAC control element. The device applies the stored configuration, connects to the target cell, and indicates successful completion of the LTM procedure, see Fig. 6.23.

One of the bigger contributors to handover interruption time is acquiring downlink synchronization. To reduce this, it is possible to obtain downlink synchronization to the candidate target cells prior to sending measurement reports and executing the handover command. This speeds up the handover process further.

LTM supports both inter and intra-frequency handovers, but all the cells need to belong to the same gNB, that is, inter-gNB handovers are not possible with LTM.

6.6.5 Idle-state mobility – cell reselection

Cell reselection is the mechanism used for device mobility in idle and inactive states. The device by itself finds and selects the best cell to camp on and the network is not

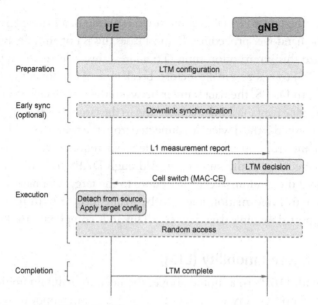

Fig. 6.23 L1/L2-triggered mobility (simplified).

directly involved in the mobility events (other than providing the configuration to the device). One reason for using a different scheme than in connected state is that the requirements are different. For example, there is no need to worry about any handover interruption as there is no ongoing data transmission. Low power consumption, on the other hand, is an important aspect as a device typically spends the vast majority of their time in idle state.

In order find the best cell to camp upon, the device searches for, and measures on, SSBs similar to the initial cell search as described in Chapter 16. Once the device discovers an SSB with a received power that exceeds the received power of its current SSB by a certain threshold it reads the system information (SIB1) of the new cell to determine (among other things) if it is allowed to camp on this particular cell.

From the perspective of device-initiated data transaction there is no need to update the network with information about the location of the device and the idle-state mobility procedure described so far would be sufficient. If there are data to be transmitted in the uplink, the device can initiate the transition from idle (or inactive) state to connected state by using random access. However, there may also be data coming to the network which needs to be transmitted to the device and hence there is a need for a mechanism to ensure that the device is reachable by the network. This mechanism is known as paging, where the network notifies the device by means of a paging message. Before describing the transmission of a paging message in Section 6.6.7, the area over which such a paging message is transmitted, a key aspect of paging, will be discussed.

6.6.6 Tracking the device

In principle, the network could transmit the page the device over the entire coverage of the network, by broadcasting the paging message from every cell. However, that would obviously imply a very high overhead in terms of paging-message transmissions as the vast majority of the paging transmissions would take place in cells where the target device is not located. On the other hand, if the paging message is only to be transmitted in the cell in which the device is located, there is a need to track the device on a cell level. This would imply that the device would have to inform the network every time it moves out of the coverage of one cell and into the coverage of another cell. This would also lead to very high overhead, in this case in terms of the signaling needed to inform the network about the updated device location. For this reason, a compromise between these two extremes is typically used, where devices are only tracked on a cell-group level:

- The network only receives new information about the device location if the device moves into a cell outside of the current cell group;
- When paging the device, the paging message is broadcast over all cells within the cell group.

For NR, the basic principle for such tracking is the same for idle state and inactive state although the grouping is somewhat different in the two cases.

As illustrated in Fig. 6.24, NR cells are grouped into *RAN areas*, where each RAN area is identified by an *RAN Area Identifier* (RAI). The RAN areas, in turn, are grouped into even larger *tracking areas*, with each tracking area being identified by a *Tracking Area Identifier* (TAI). Thus, each cell belongs to one RAN area and one tracking area, the identities of which are provided as part of the cell system information.

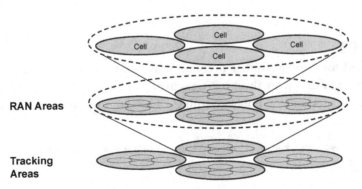

Fig. 6.24 RAN areas and tracking areas.

The Tracking areas are the basis for device tracking on core-network level. Each device is assigned a *UE registration area* by the core network, consisting of a list of tracking area identifiers. When a device enters a cell that belongs to a tracking area not included in the assigned UE registration area it accesses the network, including the core network, and performs a *NAS registration update*. The core network registers the device location and updates the device UE registration area, in practice providing the device with a new TAI list that includes the new TAI.

The reason the device is assigned *a set of* TAIs, that is, a set of tracking areas is to avoid repeated NAS registration updates if a device moves back and forth over the border of two neighbor tracking areas. By keeping the old TAI within the updated UE registration area no new update is needed if the device moves back into the old TA.

The RAN area is the basis for device tracking on radio-access-network level. Devices in inactive state can be assigned a *RAN notification area* that consists of either of the following:

- A list of cell identities;
- A list of RAIs, in practice a list of RAN areas; or
- A list of TAIs, in practice a list of tracking areas.

Note the first case is essentially the same has having each RAN area consist of a single cell while the last case is essentially the same as having the RAN areas coincide with the tracking areas.

The procedure for RAN notification area updates is similar to updates of the UE registration area. When a device enters a cell that is not directly or indirectly (via a RAN/tracking area) included in the RAN notification area, the device accesses the network and makes an *RRC RAN notification area update*. The radio network registers the device location and updates the device RAN notification area. As a change of tracking area always implies a change also of the device RAN area, an RRC RAN notification area update is done implicitly every time a device makes a UE registration update.

6.6.7 Paging

Paging is used for network-initiated connection setup when the device is in idle or inactive states, and to convey short messages in any of the states for indication of system information updates and/or public warning[j]. The same mechanism as for "normal"

---

[j] These warning messages are known as earthquake and tsunami warning system (ETWS) and commercial mobile alert service (CMAS).

downlink data transmission on the DL-SCH is used and the mobile device monitors the L1/L2 control signaling for downlink scheduling assignments using a special RNTI for paging purposes, the P-RNTI. Since the location of the device is not known on a cell level (unless the device is in connected state), the paging message is typically transmitted across multiple cells in the tracking area (for CN-initiated paging) or in the RAN notification are (for RAN-initiated paging).

Upon detection of a PDCCH with the P-RNTI, the device checks the PDCCH content. Two of the bits in the PDCCH indicate one of a short message on the PDCCH, paging information carried on the PDSCH, or both.

Short messages are relevant for all devices, irrespective of whether they are in idle, inactive, or connected states, and the PDCCH contains (among other fields) eight bits for the short messages. One of the eight bits indicates whether (parts of) the system information – more specifically SIBs other than SIB6, 7, or 8 – has been updated. If so, the device reacquires the updated SIBs as described in Chapter 16. Similarly, another of the bits is used to indicate reception of a public warning message, for example about an ongoing earthquake.

Paging messages are transmitted on the PDSCH, similar to user data, and the PDCCH contains the scheduling information necessary to receive this transmission. Only devices in idle or inactive states are concerned about paging messages as devices in active state can be contacted by the network through other means. One paging message on the DL-SCH can contain pages for multiple devices. A device in idle or inactive states receiving a paging message checks if it contains the identity of the device in question. If so, the device initiates a random-access procedure to move from idle/inactive state to connected state. As the uplink timing is unknown during in idle and inactive states, no hybrid-ARQ acknowledgements can be transmitted and consequently hybrid ARQ with soft combining is not used for paging messages.

An efficient paging procedure should, in addition to deliver paging messages, preserve power. Discontinuous reception is therefore used, allowing devices in idle or inactive states to sleep with no receiver processing most of the time and to briefly wake up at predefined time intervals, known as paging occasions. In a paging occasion, which can be one or more consecutive slots, the device monitors for a PDCCH with P-RNTI. If the device detects a PDCCH transmitted with the P-RNTI in a paging occasion, is processes the paging message as described earlier, otherwise it sleeps according to the paging cycle until the next paging occasion. Release 17 provides additional enhancements for paging to further reduce the device power consumption through an early indication of paging occasions, see Section 14.5.7 for details.

The paging occasions are determined from the system frame number, the device identity, and parameters such as the paging periodicity configured by the network. The identity used is the so-called 5G-S-TMSI, an identity coupled to the subscription as an idle state device does not have a C–RNTI allocated. For RAN-initiated paging there is also the possibility to use an identity previously configured in the device. Different paging periodicities can be configured for paging initiated by the radio-access network (once every 32, 64, 128, or 256 frames) and the core network.

Since different devices have different 5G-S-TMSI, they will compute different paging instances. Hence, from a network perspective, paging may be transmitted more often than once per 32 frames, although not all devices can be paged at all paging occasions as they are distributed across the possible paging instances as shown in Fig. 6.25. Furthermore, the cost of a short paging cycle is minimal from a network perspective as resources not used for paging can be used for normal data transmission and are not wasted. However, from a device perspective, a short paging cycle increases the power consumption as the device needs to wake up frequently to monitor the paging instants. The configuration is therefore a balance between fast paging and low device power consumption.

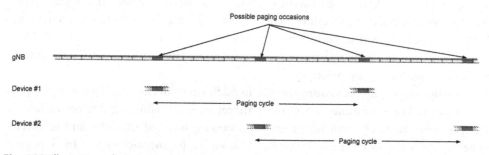

Fig. 6.25 Illustration of paging cycles.

CHAPTER 7

Overall transmission structure

Prior to discussing the detailed NR downlink and uplink transmission schemes, a description of the basic time-frequency transmission resource of NR will be provided in this chapter, including bandwidth parts, supplementary uplink, carrier aggregation, duplex schemes, antenna ports, and quasi-colocation.

7.1 Transmission scheme

OFDM was found to be a suitable waveform for NR due to its robustness to time dispersion and ease of exploiting both the time and frequency domains when defining the structure for different channels and signals. It is therefore the basic transmission scheme for both the downlink and uplink transmission directions in NR. However, unlike LTE where DFT-precoded OFDM is the sole transmission scheme in the uplink, NR uses OFDM as the baseline uplink transmission scheme with the possibility for complementary DFT-precoded OFDM. The reasons for DFT-precoded OFDM in the uplink are the same as in LTE, namely to reduce the cubic metric and obtain a higher power-amplifier efficiency, but the use of DFT-precoding also has several drawbacks including:

- Spatial multiplexing ("MIMO") receivers become more complex. This was not an issue when DFT-precoding was agreed in the first LTE release as it did not support uplink spatial multiplexing but becomes important when supporting uplink spatial multiplexing.
- Maintaining symmetry between uplink and downlink transmission schemes is in many cases beneficial, something which is lost with an DFT-precoded uplink. One example of the benefits with symmetric schemes is sidelink transmission, that is, direct transmissions between devices. When sidelinks were introduced in LTE, it was agreed to keep the uplink transmission scheme which requires the devices to implement a receiver for DFT-precoded OFDM in addition to the OFDM receiver already present for downlink transmissions. The introduction of sidelink support in NR release 16, see Chapter 26, was thus more straightforward as the device already had support for OFDM transmission and reception.

Hence, NR has adopted OFDM in the uplink with *complementary* support for DFT-precoding for data transmission. When DFT-precoding is used, uplink transmissions are restricted to a single layer only while uplink transmissions of up to four layers are

5G/5G-Advanced
https://doi.org/10.1016/B978-0-443-13173-8.00006-2

137

possible with OFDM (up to eight layers in release 18). Support for DFT-precoding is mandatory in the device and the network can therefore configure DFT-precoding for a particular device if/when needed.[a]

One important aspect of OFDM is the selection of the numerology, in particular the subcarrier spacing and the cyclic prefix length. A large subcarrier spacing is beneficial from a frequency-error perspective as it reduces the impact from frequency errors and phase noise. It also allows implementations covering a large bandwidth with a modest FFT size. However, for a given cyclic prefix length in microseconds, the relative overhead increases the larger the subcarrier spacing is and from this perspective a smaller subcarrier spacing would be preferable. The selection of the subcarrier spacing therefore needs to carefully balance overhead from the cyclic prefix against sensitivity to Doppler spread/shift and phase noise.

For LTE, a choice of 15 kHz subcarrier spacing and a cyclic prefix of approximately 4.7 μs was found to offer a good balance between these different constraints for scenarios for which LTE was originally designed – outdoor cellular deployments up to approximately 3 GHz carrier frequency.

NR, on the other hand, is designed to support a wide range of deployment scenarios, from large cells with sub-1 GHz carrier frequency up to mm-wave deployments with very wide spectrum allocations. Having a single numerology for all these scenarios is not efficient or even possible. For the lower range of carrier frequencies, from below 1 GHz up to a few GHz, the cell sizes can be relatively large and a cyclic prefix capable of handling the delay spread expected in these type of deployments, a couple of microseconds, is necessary. Consequently, a subcarrier spacing in the LTE range or somewhat higher, in the range of 15–30 kHz, is needed. For higher carrier frequencies approaching the mm-wave range, implementation limitations such as phase noise become more critical and the available channel bandwidths are larger, calling for higher subcarrier spacings. At the same time, the expected cell sizes are smaller at higher frequencies as a consequence of the more challenging propagation conditions. The extensive use of beamforming at high frequencies also helps reduce the expected delay spread. Hence, for these types of deployments a higher subcarrier spacing and a shorter cyclic prefix are suitable.

From this discussion, it is seen that a flexible numerology is needed. NR therefore supports a scalable numerology with a range of subcarrier spacings, based on scaling a baseline subcarrier spacing of 15 kHz. The reason for the choice of 15 kHz is coexistence with LTE and the LTE-derived NB-IoT and eMTC on the same carrier. This is an important requirement, for example for an operator which has deployed NB-IoT or eMTC to support machine-type communication. Unlike smartphones, such MTC devices can have a

[a] The waveform to use for the uplink random-access messages is configured as part of the system information.

relatively long replacement cycle, 10 years or longer. Without provisioning for coexistence, the operator would not be able to migrate the carrier to NR until all the MTC devices have been replaced. Another example is gradual migration where the limited spectrum availability may force an operator to share a single carrier between LTE and NR in the time domain. LTE coexistence is further discussed in Chapter 18.

Consequently, 15 kHz subcarrier spacing was selected as the baseline for NR. From the baseline subcarrier spacing, subcarrier spacings ranging from 15 kHz up to 240 kHz with a proportional change in cyclic prefix duration as shown in Table 7.1 are derived. Note that 240 kHz is supported for the SSB only (see Section 16.1) and not for regular data transmission. In release 17, two additional subcarrier spacings – 480 kHz and 960 kHz – are added as the upper frequency limit is increased from 52.6 GHz to 71 GHz. Although phase noise increases as the carrier frequency increases, the two largest subcarrier spacings are mainly motivated by the desire to support a wide carrier bandwidth without increasing the FFT size. From a pure phase noise perspective, 120 kHz is feasible with the appropriate receiver-side phase noise compensation.

Although the NR physical-layer specification is band-agnostic, not all supported numerologies are relevant for all frequency bands. For each frequency band, radio requirements are therefore defined for a subset of the supported numerologies as discussed in Chapter 28.

The useful symbol time T_u depends on the subcarrier spacing as shown in Table 7.1 with the overall OFDM symbol time being the sum of the useful symbol time and the cyclic-prefix length T_{CP}. In LTE, two different cyclic prefixes are defined, normal cyclic prefix and extended cyclic prefix. The extended cyclic prefix, although less efficient from a cyclic-prefix-overhead point of view, was intended for specific environments with excessive delay spread where performance was limited by time dispersion. However, extended cyclic prefix was not used in practical deployments (except for MBSFN transmission), rendering it an unnecessary feature in LTE for unicast transmission. With this

Table 7.1 Subcarrier spacings supported by NR.

Subcarrier spacing	Useful symbol time, T_u	Cyclic prefix, T_{CP}		Comment
		Normal	Extended	
15 kHz	66.7 μs	4.7 μs	–	Mandatory in FR1
30 kHz	33.3 μs	2.3 μs	–	Mandatory in FR1
60 kHz	16.7 μs	1.2 μs	4.2 μs	Optional
120 kHz	8.33 μs	0.59 μs	–	Mandatory in FR2
240 kHz	4.17 μs	0.29 μs	–	Mandatory in FR2
480 kHz	2.08 μs	0.15 μs	–	Introduced in release 17
960 kHz	1.04 μs	0.073 μs	–	Introduced in release 17

in mind, NR defines a normal cyclic prefix only, with the exception of 60 kHz subcarrier spacing[b] where both normal and extended cyclic prefix are defined for reasons discussed below.

To provide consistent and exact timing definitions, different time intervals within the NR specifications are defined as multiples of a basic time unit $T_c = 1/(480000 \cdot 4096)$.[c] The basic time unit T_c can thus be seen as the sampling time of an FFT-based transmitter/receiver implementation for a subcarrier spacing of 480 kHz with an FFT size equal to 4096. This is similar to the approach taken in LTE, which uses a basic time unit $T_s = 64 T_c$.

7.2 Time-domain structure

In the time domain, NR transmissions are organized into *frames* of length 10 ms, each of which is divided into 10 equally sized *subframes* of length 1 ms. A subframe is in turn divided into slots consisting of 14 OFDM symbols each. On a higher level, each frame is identified by a *System Frame Number* (SFN). The SFN is used to define different transmission cycles that have a period longer than one frame, for example paging sleep-mode cycles. The SFN period equals 1024, thus the SFN repeats itself after 1024 frames or 10.24 seconds. In release 17, a *Hyper Frame Number* (HFN) is defined to handle even longer time periods of up to almost 3 hours. The HFN is incremented whenever the SFN wraps around.

For the 15 kHz subcarrier spacing, an NR slot has the same structure as an LTE subframe with normal cyclic prefix. This is beneficial from an NR-LTE coexistence perspective and is, as mentioned above, the reason for choosing 15 kHz as the basic subcarrier spacing. However, it also means that the cyclic prefix for the first and eighth symbols in a 15 kHz slot are slightly larger than for the other symbols.

The time-domain structure for higher subcarrier spacings in NR is derived by scaling the baseline 15 kHz structure by powers of two. In essence, an OFDM symbol is split into two OFDM symbols of the next higher numerology, see Fig. 7.1, and 14 consecutive symbols form a slot. Scaling by powers of two is beneficial as it maintains the symbol boundaries across numerologies which simplifies mixing different numerologies on the same carrier and this is the motivation for the higher subcarrier spacings being expressed as $2^\mu \cdot 15$ kHz with quantity μ being known as the *subcarrier spacing configuration*. For the OFDM symbols with a somewhat larger cyclic prefix, the excess samples are allocated to the first of the two symbols obtained when splitting one symbol as illustrated in Fig. 7.1.[d]

---

[b] 60 kHz subcarrier spacing is an optional feature and not supported by all devices.

[c] Fractions of T_c are used for very short time durations.

[d] This also implies that the slot length for subcarrier spacings of 60/120/240/480/960 kHz is not exactly 0.25/0.125/0.0625/0.03125/0.015625 ms as some slots have the excess samples while others do not.

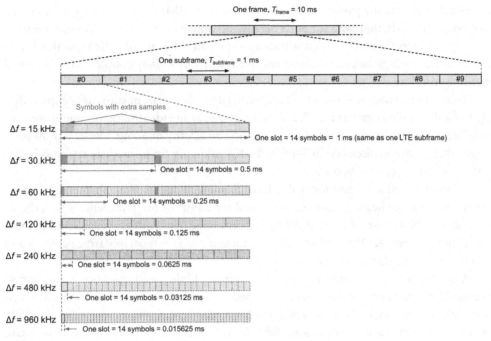

Fig. 7.1 Frames, subframes, and slots in NR.

Regardless of the numerology, a subframe has a duration of 1 ms long and consists of 2^{μ} slots. It thus serves as a numerology-independent time reference, which is useful especially in the case of multiple numerologies being mixed on the same carrier, while a slot is the typical dynamic scheduling unit. In contrast, LTE with its single subcarrier spacing uses the term subframe for both these purposes.

Since a slot is defined as a fixed number of OFDM symbols, a higher subcarrier spacing leads to a shorter slot duration. In principle this can be used to support lower-latency transmission, but as the cyclic prefix also shrinks when increasing the subcarrier spacing, it is not a feasible approach in all deployments. Therefore, to facilitate a fourfold reduction in the slot duration and the associated delay while maintaining a cyclic prefix similar to the 15 kHz case, an extended cyclic prefix is defined for 60 kHz subcarrier spacing. However, it comes at the cost of increased overhead in terms of cyclic prefix and is a less efficient way of providing low latency. The subcarrier spacing is therefore primarily selected to meet the deployment scenario in terms of, for example, carrier frequency, expected delay spread in the radio channel, and any coexistence requirements with LTE-based systems on the same carrier.

An alternative and more efficient way to support low latency is to decouple the transmission duration from the slot duration. Instead of changing subcarrier spacing and/or slot duration, the latency-critical transmission uses whatever number of OFDM symbols necessary to deliver the payload and can start at any OFDM symbol without waiting for a slot boundary. NR therefore supports occupying only part of a slot for the transmission, sometimes referred to as "mini-slot transmission". In other words, the term slot is primarily a numerology-dependent time reference and only loosely coupled with the actual transmission duration.

There are multiple reasons why it is beneficial to allow transmission to occupy only a part of a slot as illustrated in Fig. 7.2. One reason is, as already discussed, support of very low latency. Such transmissions can also preempt an already ongoing, longer transmission to another device as discussed in Section 14.1.2, allowing for immediate transmission of data requiring very low latency.

Another reason is support for analog beamforming as discussed in Chapters 11 and 12, where at most one beam at a time can be used for transmission. Different devices therefore need to be time-multiplexed. With the very large bandwidths available in the mm-wave range, a few OFDM symbols can be sufficient even for relatively large payloads and using a complete slot would be excessive.

A third reason is operation in unlicensed spectra. Unlicensed operation is not part of release 15 but the extension to operation in unlicensed spectra in release 16 was foreseen already at an early stage of NR design. In unlicensed spectra, listen-before-talk is typically used to ensure the radio channel is available for transmission. Once the listen-before-talk operation has declared the channel available, it is beneficial to start transmission immediately to avoid another device occupying the channel, something which is possible with decoupling the actual data transmission from the slot boundaries. If data transmission would have to wait until the start of a slot boundary, there could be a risk of another device grabbing the channel before the data transmission starts.

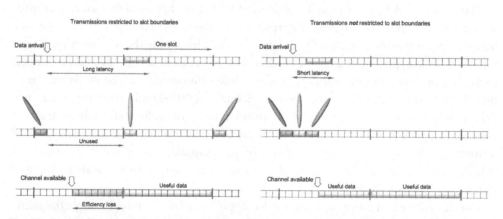

Fig. 7.2 Decoupling transmissions from slot boundaries to achieve low latency (top), more efficient beam sweeping (middle), and better support for unlicensed spectrum (bottom).

7.3 Frequency-domain structure

When the first release of LTE was designed, it was decided that all devices should be capable of the maximum carrier bandwidth of 20 MHz which was a reasonable assumption at the time, given the relatively modest bandwidth compared to NR. On the other hand, NR is designed to support very wide bandwidths, up to 400 MHz for a single carrier in release 15 and up to 2 GHz in release 17. Mandating all devices to handle such wide carriers is not reasonable from a cost perspective. Hence, an NR device may see only a part of the carrier and, for efficient utilization of the carrier, the part of the carrier received by the device may not be centered around the carrier frequency. This has implications for, among other things, the handling of the DC subcarrier.

In LTE, the DC subcarrier is not used as it may be subject to disproportionally high interference due to, for example, local-oscillator leakage. Since all LTE devices can receive the full carrier bandwidth and are centered around the carrier frequency, this was straightforward[e]. NR devices, on the other hand, may not be centered around the carrier frequency and each NR device may have its DC located at different locations in the carrier, unlike LTE where all devices typically have the DC coinciding with the center of the carrier. Therefore, having special handling of the DC subcarrier would be cumbersome in NR and instead it was decided to exploit also the DC subcarrier for data as illustrated in Fig. 7.3, accepting that the quality of this subcarrier may be degraded in some situations.

A *resource element*, consisting of one subcarrier during one OFDM symbol, is the smallest physical resource in NR. Furthermore, as illustrated in Fig. 7.4, 12 consecutive subcarriers in the frequency domain are called a *resource block*.

Note that the NR definition of a resource block differs from the LTE definition. An NR resource block is a one-dimensional measure spanning the frequency domain only while LTE uses two-dimensional resource blocks of 12 subcarriers in the frequency domain and one LTE slot in the time domain. One reason for defining resource blocks in the frequency domain only in NR is the flexibility in time duration for different transmissions whereas in LTE, at least in the original release, transmissions occupied a complete slot.[f]

NR supports multiple numerologies on the same carrier. Since a resource block is 12 subcarriers, the frequency span measured in Hz is different. The resource block boundaries are aligned across numerologies such that two resource blocks at a subcarrier spacing of Δf occupy the same frequency range as one resource block at a subcarrier spacing of $2\Delta f$. In the NR specifications, the alignment across numerologies in terms of resource

---

[e] In case of carrier aggregation, multiple carriers may use the same power amplifier in which case the DC subcarrier of the transmission does not necessarily coincide with the unused DC subcarrier in the LTE grid.

[f] There are some situations in LTE, for example the DwPTS in LTE/TDD, where a transmission does not occupy a full slot.

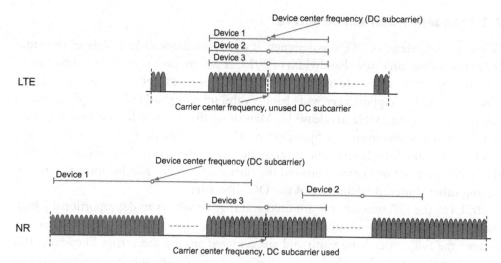

Fig. 7.3 Handling of the DC subcarrier in LTE and NR.

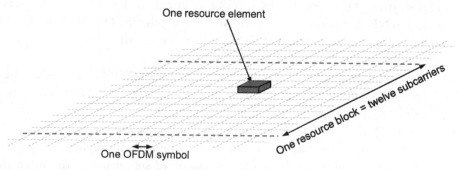

Fig. 7.4 Resource element and resource block.

block boundaries, as well as symbol boundaries, is described through multiple *resource grids* where there is one resource grid per subcarrier spacing and antenna port (see Section 7.9 for a discussion of antenna ports), covering the full carrier bandwidth in the frequency domain and one subframe in the time domain (Fig. 7.5).

The resource grid models the transmitted signal as seen by the device for a given subcarrier spacing. However, the device needs to know where in the carrier the resource blocks are located. In LTE, where there is a single numerology and all devices support the full carrier bandwidth, this is straightforward. NR, on the other hand, supports multiple numerologies and, as discussed further later in conjunction with bandwidth parts, not all devices may support the full carrier bandwidth. Therefore, a common reference point, known as *point A*, together with the notion of two types of resource blocks,

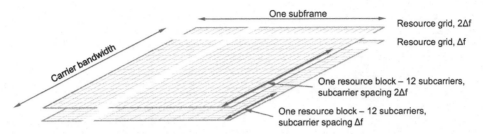

Fig. 7.5 Resource grids for two different subcarrier spacings.

common resource blocks and *physical resource blocks*, are used.[g] Reference point A coincides with subcarrier 0 of common resource block 0 for all subcarrier spacings. This point serves as a reference from which the frequency structure can be described and point A may be located outside the actual carrier. Upon detecting an SSB as part of the initial access (see Chapter 16), the device is signaled the location of point A as part of the broadcast system information (SIB1).

The physical resource blocks, which are used to describe the actual transmitted signal, are then located relative to this reference point as illustrated in Fig. 7.6. For example, physical resource block 0 for subcarrier spacing Δf is located m resource blocks from reference point A or, expressed differently, corresponds to common resource block m. Similarly, physical resource block 0 for subcarrier spacing $2\Delta f$ corresponds to common resource block n. The starting points for the physical resource blocks are signaled independently for each numerology (m and n in the example in Fig. 7.6), a feature that is useful for implementing the filters necessary to meet the out-of-band emission requirements (see Chapter 28). The guard in Hz needed between the edge of the carrier and the first used subcarrier is larger the larger the subcarrier spacing is, which can be accounted for by independently setting the offset between the first used resource block and reference point A. In the example in Fig. 7.6, the first used resource block for subcarrier spacing $2\Delta f$ is located further from the carrier edge than for subcarrier spacing Δf to avoid excessively steep filtering requirements for the higher numerology or, expressed differently, to allow a larger fraction of the spectrum to be used for the lower subcarrier spacing.

The location of the first usable resource block, which is the same as the start of the resource grid in the frequency domain, is signaled to the device. Note that this may or may not be the same as the first resource block of a bandwidth part (bandwidth parts are described in Section 7.4).

An NR carrier can at most be 275 resource blocks wide, which corresponds to $275 \cdot 12 = 3300$ used subcarriers. This also defines the largest possible carrier bandwidth

---

[g] There is a third type of resource block, *virtual resource blocks*, which are mapped to physical resource blocks when describing the mapping of the PDSCH/PUSCH, see Chapter 9. Furthermore, in release 16, *interleaved resource blocks* are defined to support unlicensed spectra, see Chapter 20.

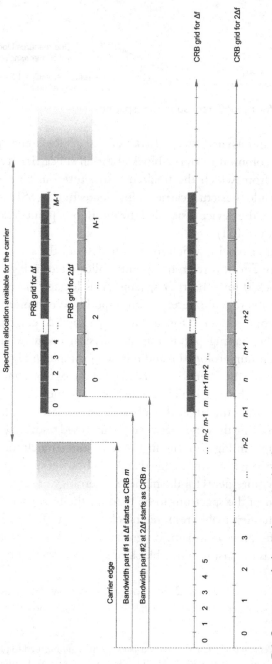

Fig. 7.6 Common and physical resource blocks.

in NR for each numerology. However, there is also an agreement to limit the per-carrier bandwidth to 400 MHz in release 15 and 2 GHz in release 17, resulting in the maximum carrier bandwidths of 50/100/200/400/1600/2000 MHz for subcarrier spacings of 15/30/60/120/480/960 kHz, respectively as mentioned in Chapter 5.[h] The smallest possible carrier bandwidth of 11 resource blocks is given by the RF requirements on spectrum utilization (see Chapter 28). However, for the numerology used for the SSB (see Chapter 16) at least 20 resource blocks are required and is thus the smallest practical carrier bandwidth.

7.4 Bandwidth parts

As discussed earlier, LTE is designed under the assumption that all devices are capable of the maximum carrier bandwidth of 20 MHz. This avoided several complications, for example around the handling of the DC subcarrier as already discussed, while having a negligible impact on the device cost. It also allowed control channels to span the full carrier bandwidth to maximize frequency diversity.

The same assumption – all devices being able to receive the full carrier bandwidth – is not reasonable for NR, given the very wide carrier bandwidth supported. Consequently, means for handling different device capabilities in terms of bandwidth support must be included in the design. Furthermore, reception of a very wide bandwidth can be costly in terms of device energy consumption compared to receiving a narrower bandwidth. Using the same approach as in LTE where the downlink control channels would occupy the full carrier bandwidth would therefore significantly increase the power consumption of the device. A better approach is, as done in NR, to use *receiver-bandwidth adaptation* such that the device can use a narrower bandwidth for monitoring control channels and to receive small-to-medium sized data transmissions and to open the full bandwidth when a large amount of data is scheduled.

To handle these two aspects – support for devices not capable of receiving the full carrier bandwidth and receiver-side bandwidth adaptation – NR defines *bandwidth parts* (BWPs), see Fig. 7.7. A bandwidth part is characterized by a numerology (subcarrier spacing and cyclic prefix) and a set of consecutive resource blocks in the numerology of the BWP, starting at a certain common resource block.

When a device enters the connected state it has obtained information from the PBCH about the *control resource set* (CORESET; see Section 10.1.2) where it can find the control channel used to schedule the remaining system information (see Chapter 16 for details). In addition, the PBCH information defining the CORESET configuration also defines and activates the *initial* bandwidth part in the downlink. The initial downlink bandwidth part covers the same set of resource blocks as CORESET#0, unless a different value is provided

---

[h] For 960 kHz, not all 3300 subcarriers are used (if they were, 3.2 GHz would be possible).

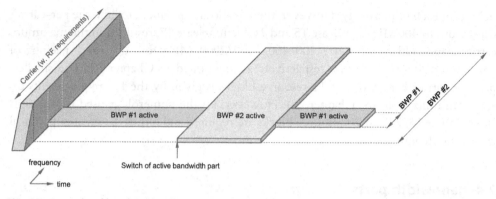

Fig. 7.7 Example of bandwidth adaptation using bandwidth parts.

as part of the remaining system information in SIB1. The initial active uplink bandwidth part is obtained from the system information scheduled using the downlink PDCCH. In case of non-standalone NR operation, the initial uplink and downlink bandwidth parts are obtained from the configuration information provided over the LTE carrier.

Once connected, a device can be configured with up to four downlink bandwidth parts and up to four uplink bandwidth parts for each serving cell. In the case of SUL operation (see Section 7.7), there can be up to four additional uplink bandwidth parts on the supplementary uplink carrier.

On each serving cell, at a given time instant one of the configured downlink bandwidth parts is referred to as the *active downlink bandwidth part* for the serving cell and one of the configured uplink bandwidth parts is referred to as the *active uplink bandwidth part* for the serving cell. For unpaired spectra a device may assume that the active downlink bandwidth part and the active uplink bandwidth part of a serving cell have the same center frequency. This simplifies the implementation as a single oscillator can be used for both directions. The gNB can activate and deactivate bandwidth parts using the same downlink control signaling as for scheduling information, see Chapter 10, thereby achieving rapid switching between different bandwidth parts.

In the downlink, a device is not assumed to be able to receive downlink data transmissions, more specifically the PDCCH or PDSCH, outside the active bandwidth part. Furthermore, the numerology of the PDCCH and PDSCH is restricted to the numerology configured for the bandwidth part. Thus, on a given carrier, a device can only receive one numerology at a time as multiple bandwidth parts cannot be simultaneously active.[i] Mobility measurements can still be done outside an active bandwidth part but

---

[i] The NR structure in principle allows multiple active bandwidths parts, but no need has been identified so far and consequently devices are only required to support a single active bandwidth part per carrier and "transmission direction" (uplink or downlink).

require a measurement gap similar to inter-frequency measurements. Hence, a device is not expected to monitor downlink control channels while doing measurements outside the active bandwidth part.

In the uplink, a device transmits PUSCH and PUCCH in the active uplink bandwidth part only.

Given the discussion, a relevant question is why two mechanisms, carrier aggregation and bandwidth parts, are defined instead of using the carrier-aggregation framework only. To some extent carrier aggregation could have been used to handle devices with different bandwidth capabilities as well as bandwidth adaptation. However, from an RF perspective there is a significant difference. A component carrier is associated with various RF requirements such as out-of-band emission requirements as discussed in Chapter 28, but for a bandwidth part inside a carrier there is no such requirement – it is all handled by the requirements set on the carrier as such. Furthermore, from a MAC perspective there are also some differences in the handling of, for example, hybrid ARQ retransmissions which cannot move between component carriers but can move between bandwidth parts on the same carrier.

7.5 Frequency-domain location of NR carriers

In principle, an NR carrier could be positioned anywhere within the spectrum and, similar to LTE, the basic NR physical-layer specification does not say anything about the exact frequency location of an NR carrier, including the frequency band. However, in practice, there is a need for restrictions on where an NR carrier can be positioned in the frequency domain to simplify RF implementation and to provide some structure to carrier assignments in a frequency band between different operators. In LTE, a 100 kHz carrier raster serves this purpose and a similar approach has been taken in NR. However, the NR raster has a much finer granularity of 5 kHz up to 3 GHz carrier frequency, 15 kHz for 3 to 24.25 GHz, and 60 kHz above 24.25 GHz. This raster has the benefit of being a factor in the subcarrier spacings relevant for each frequency range, as well as being compatible with the 100 kHz LTE raster in bands where LTE is deployed (below 3 GHz).

In LTE, this carrier raster also determines the frequency locations a device must search for as part of the initial access procedure. However, given the much wider carriers possible in NR and the larger number of bands in which NR can be deployed, as well as the finer raster granularity, performing initial cell search on all possible raster positions would be too time consuming. Instead, to reduce the overall complexity and not spend an unreasonable time on cell search, NR also defines a sparser *synchronization raster* which is used when an NR device searches for an SSB upon initial access. A consequence of having a sparser synchronization raster than carrier raster is that, unlike LTE, the synchronization signals may not be centered in the carrier (see Fig. 7.8 and Chapter 16 for further details).

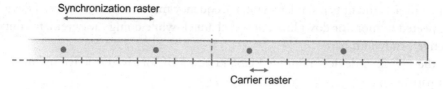

Fig. 7.8 NR carrier raster.

7.6 Carrier aggregation

The possibility for *carrier aggregation* is part of NR from the first release. Similar to LTE, multiple NR carriers can be aggregated and transmitted in parallel to/from the same device, thereby allowing for an overall wider bandwidth and correspondingly higher per-link data rates. The carriers do not have to be contiguous in the frequency domain but can be dispersed, both in the same frequency band as well as in different bands, resulting in three different scenarios:

- Intra-band aggregation with frequency-contiguous component carriers
- Intra-band aggregation with non-contiguous component carriers
- Inter-band aggregation with non-contiguous component carriers.

Although the overall structure is the same for all three cases, the RF complexity can be vastly different.

Up to 16 carriers, possibly of different bandwidths and different duplex schemes, can be aggregated allowing for overall transmission bandwidths of up to $16 \cdot 400\,\text{MHz} = 6.4\,\text{GHz}$ which is far beyond typical spectrum allocations. Even higher values can in theory be obtained in release 17; aggregating 16 carriers of $2\,\text{GHz}$ results in $32\,\text{GHz}$ which is much larger than what is needed from a practical perspective, especially as the highest subcarrier spacing is not suitable for all frequency ranges.

A device capable of carrier aggregation may receive or transmit simultaneously on multiple component carriers while a device not capable of carrier aggregation can access one of the component carriers. Thus, in most respects and unless otherwise mentioned, the physical-layer description in the following chapters applies to each component carrier separately in the case of carrier aggregation. It is worth noting that in in the case of inter-band carrier aggregation of multiple half-duplex (TDD) carriers, the transmission direction on different carriers does not necessarily have to be the same. This implies that a carrier-aggregation capable TDD device may need a duplex filter, unlike the typical scenario for a non-carrier-aggregation capable half-duplex device.

In the specifications, carrier aggregation is described using the term cell, that is, a carrier-aggregation-capable device is able to receive and transmit from/to multiple cells. One of these cells is referred to as the primary cell (PCell). This is the cell which the device initially finds and connects to, after which one or more secondary cells (SCells) can be configured once the device is in connected mode. The secondary cells can be rapidly activated or deactivated to meet the variations in the traffic pattern. Different devices may have different cells as their primary cell—that is, the configuration of the primary cell is device-specific. Furthermore, the number of carriers (or cells) does not have to be the same in uplink and downlink. In fact, a typical case is to have more carriers aggregated in the downlink than in the uplink. There are several reasons for this. There is typically more traffic in the downlink than in the uplink. Furthermore, the RF complexity from multiple simultaneously active uplink carriers is typically larger than the corresponding complexity in the downlink.

Carrier aggregation can also be combined with dual connectivity, see Fig. 7.9. In this case, in addition to PCells and SCells in the master cell group, there is one primary cell in the secondary cell group, referred to as the PSCell, which is the cell used for initial access when establishing the secondary cell group. The secondary cell group may also contain one or more SCells. In many cases, signaling messages in each of the cell groups occur only at the PCell and the PSCell. The term SpCell is defined to simplify the description of such cases and refers to both the PCell and the PSCell.

Scheduling grants and scheduling assignments, or in general terms downlink control information, is as a baseline transmitted separately per carrier scheduled, that is, there is one PDCCH for each scheduling grant or assignment sent to the device. However, in release 18, there is also a possibility to schedule multiple carriers using a single DCI, something that can be useful to reduce control signaling overhead.

Control information for scheduling purposes can be transmitted on either the same cell as the corresponding data, known as self-scheduling, or on a different cell than the corresponding data, known as cross-carrier scheduling, as illustrated in Fig. 7.10. In most cases, self-scheduling is sufficient. Transmissions on the PCell always use self-scheduling prior to release 17 when the possibility for an SCell to schedule the PCell is introduced. The reason for allowing this is the, fairly typical, scenario with the PCell on lower frequencies and one or more SCells on higher frequency bands. Placing the PCell on the lower frequency band is advantageous from a coverage perspective, but at the same time the carrier bandwidth is significantly smaller than on higher frequency bands. Restricting the PCell to use self-scheduling may therefore result in a limitation in PDCCH resources, something that can be alleviated by allowing cross-carrier scheduling of the PCell from one of the SCells.

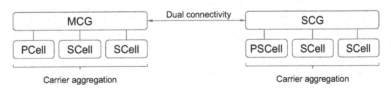

Fig. 7.9 PCell, SCell, and PSCell.

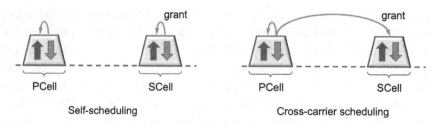

Fig. 7.10 Self scheduling and cross scheduling.

7.6.1 Uplink control signaling

Carrier aggregation uses L1/L2 control signaling for the same reason as when operating with a single carrier. The use of downlink control signaling for scheduling information was touched upon in the previous section. There is also a need for uplink control signaling, for example hybrid-ARQ acknowledgements to inform the gNB about the success or failure of downlink data reception. As baseline, all the feedback is transmitted on the PCell, motivated by the need to support asymmetric carrier aggregation with the number of downlink carriers supported by a device unrelated to the number of uplink carriers. For a large number of downlink component carriers, a single uplink carrier may thus carry a large number of acknowledgements. To avoid overloading a single carrier, it is possible to configure two *PUCCH groups* (see Fig. 7.11) where feedback relating to the first group of carriers is transmitted in the uplink of the PCell and feedback relating to the second group of carriers are transmitted on another cell known as the PUCCH-SCell.

When operating with TDD, uplink PUCCH transmissions may have to be deferred until an uplink slot occurs which would contribute to the overall latency. This is the case even if multiple uplink carriers are available as the PCell (or PUCCH-SCell) is semi-statically configured. To mitigate this, release 17 introduced *PUCCH cell switching* as discussed in Chapter 21. This allows the PUCCH to be switched to another cell, known as the PUCCH switching cell (PUCCH-sSCell). One PUCCH-sSCell can be configured in each of the two PUCCH groups. The switching can either be dynamically or semi-statically controlled. In the former case, the PUCCH cell indicator in the DCI selects which of two semi-statically configured cells that should be used for the PUCCH

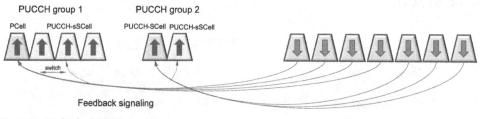

Fig. 7.11 Multiple PUCCH groups.

transmission. In the latter case, a semi-static pattern is configured, indicating for each of the PUCCH groups which of the two carriers to use for PUCCH transmission.

If carrier aggregation is used, the device may receive and transmit on multiple carriers, but reception on multiple carriers is typically only needed for the highest data rates. It is therefore beneficial to deactivate reception of carriers not used while keeping the configuration intact. Activation and deactivation of component carriers can be done through MAC signaling (more specifically, *MAC control elements*, discussed in Section 6.4.4.1) containing a bitmap where each bit indicates whether a configured SCell should be active or not.

7.7 Supplementary uplink

In addition to carrier aggregation, NR also supports so-called *supplementary uplink* (SUL). As illustrated in Fig. 7.12, SUL implies that a conventional downlink/uplink (DL/UL) carrier pair has an associated or *supplementary* uplink carrier with the SUL carrier typically operating in lower frequency bands. As an example, a downlink/uplink carrier pair operating in the 3.5 GHz band could be complemented with a supplementary uplink carrier in the 800 MHz band. Although Fig. 7.12 seems to indicate that the conventional DL/UL carrier pair operates on paired spectra with frequency separation between the downlink and uplink carriers, it should be understood that the conventional carrier pair could equally well operate in unpaired spectra with downlink/uplink separation by means of TDD. This would, for example, be the case in an SUL scenario where the conventional carrier pair operates in the unpaired 3.5 GHz band.

While the main aim of carrier aggregation is to enable higher peak data rates by increasing the bandwidth available for transmission to/from a device, the typical aim of SUL is to extend uplink coverage, that is, to provide higher uplink data rates in power-limited situations, by utilizing the lower path loss at lower frequencies. Furthermore, in an SUL scenario the non-SUL uplink carrier typically has significantly larger bandwidth compared to the SUL carrier. Thus, under good channel conditions such as the device being located relatively close to the cell site, the non-SUL carrier typically allows for substantially higher data rates compared to the SUL carrier. At the same time, under bad channel conditions, for example, at the cell edge, a lower-frequency SUL carrier typically allows for significantly higher data rates compared to the non-SUL carrier, due to the assumed lower path loss at lower frequencies. Hence, only in a relatively limited area do the two carriers provide similar data rates. As a consequence, *aggregating* the

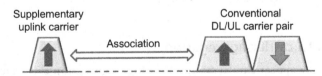

Fig. 7.12 Supplementary uplink carrier complementing a conventional DL/UL carrier pair.

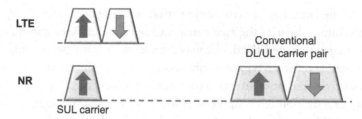

Fig. 7.13 SUL carrier co-existing with LTE uplink carrier.

throughput of the two carriers has in most cases limited benefits. At the same time, scheduling only a single uplink carrier at a time simplifies transmission protocols and in particular the RF implementation as various intermodulation interference issues are avoided. Note that for carrier aggregation the situation is different:

- The two (or more) carriers in a carrier–aggregation scenario are often of similar bandwidth and operating at similar carrier frequencies, making aggregation of the throughput of the two carriers more beneficial;
- Each uplink carrier in a carrier aggregation scenario is operating with its own downlink carrier, simplifying the support for simultaneous scheduling of multiple uplink transmissions in parallel.

Hence, only one of SUL and non-SUL is transmitting and simultaneous SUL and non-SUL transmission from a device is not possible.

One SUL scenario is when the SUL carrier is located in the uplink part of paired spectrum already used by LTE (see Fig. 7.13). In other words, the SUL carrier exists in an LTE/NR uplink co-existence scenario, see also Chapter 18. In many LTE deployments, the uplink traffic is significantly less than the corresponding downlink traffic. As a consequence, in many deployments, the uplink part of paired spectra is not fully utilized. Deploying an NR supplementary uplink carrier on top of the LTE uplink carrier in such a spectrum is a way to enhance the NR user experience with limited impact on the LTE network.

Finally, a supplementary uplink can also be used to reduce latency. In the case of TDD, the separation of uplink and downlink in the time domain may impose restrictions on when uplink data can be transmitted. By combining the TDD carrier with a supplementary carrier in paired spectra, latency-critical data can be transmitted on the supplementary uplink immediately without being restricted by the uplink-downlink partitioning on the normal carrier. The same benefit can be obtained with carrier aggregation, that is, it is not a SUL-specific benefit.

7.7.1 Relation to carrier aggregation

Although SUL may appear similar to uplink carrier aggregation there are some fundamental differences.

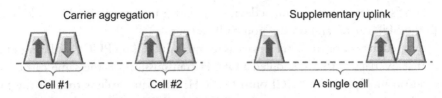

Fig. 7.14 Carrier aggregation vs supplementary uplink.

In the case of carrier aggregation, each uplink carrier has its own associated downlink carrier. Formally, each such downlink carrier corresponds to a cell of its own and thus different uplink carriers in a carrier-aggregation scenario correspond to different cells (see left part of Fig. 7.14).

In contrast, in case of SUL the supplementary uplink carrier does not have an associated downlink carrier of its own. Rather the supplementary carrier and the conventional uplink carrier share the same downlink carrier. As a consequence, the supplementary uplink carrier does not correspond to a cell of its own. Instead, in the SUL scenario there is a single cell with one downlink carrier and two uplink carriers (right part of Fig. 7.14).

It should be noted that in principle nothing prevents the combination of carrier aggregation with an additional supplementary uplink carrier, for example, a situation with carrier aggregation between two cells (two DL/UL carrier pairs) where one of the cells is an SUL cell. However, there are currently no band combinations defined for such carrier-aggregation/SUL combinations.

A relevant question is, if there is a *supplementary uplink*, is there such a thing as a *supplementary downlink*? The answer is yes – since the carrier aggregation framework allows for the number of downlink carriers to be larger than the number of uplink carriers, some of the downlink carriers can be seen as supplementary downlinks. One common scenario is to deploy an additional downlink carrier in unpaired spectra and aggregate it with a carrier in paired spectra to increase capacity and data rates in the downlink. No additional mechanisms beyond carrier aggregation are needed and hence the term supplementary downlink is mainly used from a spectrum point of view as discussed in Chapter 3.

7.7.2 Uplink control signaling

In the case of supplementary-uplink operation, a device is explicitly configured (by means of RRC signaling) to transmit PUCCH on either the SUL carrier or on the conventional (non-SUL) carrier.

In terms of PUSCH transmission, the device can be configured to transmit PUSCH on the same carrier as PUCCH. Alternatively, a device configured for SUL operation can be configured for dynamic selection between the SUL carrier or the non-SUL carrier. In the latter case, the uplink scheduling grant will include an *SUL/non-SUL indicator* that indicates on what carrier the scheduled PUSCH transmission should be carried. Thus, in the case of supplementary uplink, a device will never transmit PUSCH *simultaneously* on both the SUL carrier and on the non–SUL carrier.

As described in Section 10.2, if a device is to transmit UCI on PUCCH during a time interval that overlaps with a scheduled PUSCH transmission on the same carrier, the device instead multiplexes the UCI onto PUSCH. The same rule is true for the SUL scenario, that is, there is no simultaneous PUSCH and PUCCH transmission even on different carriers. Rather, if a device is to transmit UCI on PUCCH on a carrier (SUL or non–SUL) during a time interval that overlaps with a scheduled PUSCH transmission on either carrier (SUL or non–SUL), the device instead multiplexes the UCI onto the PUSCH.

An alternative to supplementary uplink would be to rely on dual connectivity with LTE on the lower frequency and NR on the higher frequency. Uplink data transmission would in this case be handled by the LTE carrier with, from a data rate perspective, benefits similar to supplementary uplink. However, in this case, the uplink control signaling related to NR downlink transmissions has to be handled by the high-frequency NR uplink carrier as each carrier pair has to be self-contained in terms of L1/L2 control signaling. Using a supplementary uplink avoids this drawback and allows L1/L2 control signaling to exploit the lower-frequency uplink. Another possibility would be to use carrier aggregation, but in this case a low-frequency downlink carrier has to be configured as well.

7.8 Duplex schemes

Spectrum flexibility is one of the key features of NR. In addition to the flexibility in transmission bandwidth, the basic NR structure also supports separation of uplink and downlink in time and/or frequency subject to either half duplex or full duplex operation, all using the same single frame structure. This provides a large degree of flexibility (Fig. 7.15):

- TDD – uplink and downlink transmissions use the same carrier frequency and are separated in time only;
- FDD – uplink and downlink transmissions use different frequencies but can occur simultaneously; and
- Half-duplex FDD – uplink and downlink transmissions are separated in frequency *and* time, suitable for simpler devices operating in paired spectra.

In principle, the same basic NR structure would also allow full duplex operation with uplink and downlink separated neither in time, nor in frequency, although this would result in a significant transmitter-to-receiver interference problem whose solution is still is in the research stage and left for the future. Some investigations on how to handle

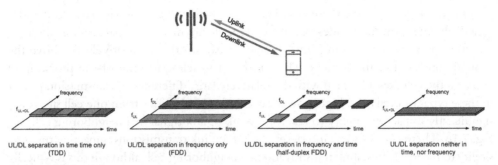

Fig. 7.15 Duplex schemes.

interference in a restricted form of full duplex, known as subband full duplex, has been carried out in release 18.

LTE also support both TDD and FDD, but unlike the *single* frame structure used in NR, LTE uses two *different* frame structures.[j] Furthermore, unlike LTE where the uplink-downlink allocation does not change over time,[k] the TDD operation for NR is designed with *dynamic* TDD as a key technology component.

7.8.1 TDD – time-division duplex

In the case of TDD operation, there is a single carrier frequency and uplink and downlink transmissions are separated in the time domain on a cell basis. Uplink and downlink transmissions are non-overlapping in time, both from a cell and a device perspective. TDD can therefore be classified as half-duplex operation.

In LTE, the split between uplink and downlink resources in the time domain is semi-statically configured and essentially remains constant over time. NR, on the other hand, uses *dynamic TDD* as the basis where (parts of) a slot can be dynamically allocated to either uplink or downlink as part of the scheduling decision. This enables following rapid traffic variations which are particularly pronounced in dense deployments with a relatively small number of users per base station. Dynamic TDD is particularly useful in small-cell and/or isolated cell deployments where the transmission power of the device and the base station is of the same order and the inter-site interference is reasonable. If needed, the scheduling decisions between the different sites can be coordinated. It is much simpler to restrict the dynamics in the uplink-downlink allocation *when needed* and thereby having a more static operation than trying to add dynamics to a fundamentally static scheme, which was done when introducing eIMTA for LTE in release 12.

---

[j] Originally, LTE supported frame structure type 1 for FDD and frame structure type 2 for TDD, but in later releases frame structure type 3 was added to handle operation in unlicensed spectrum.

[k] In LTE Rel-12 the eIMTA feature provides some support for time-varying uplink-downlink allocation.

One example when inter-site coordination is useful is a traditional macro-cell wide-area deployment. In such wide-area macro-type deployments, the base station antennas are often located above rooftop for coverage reasons – that is, relatively far above the ground compared to the devices. This can result in (close to) line-of-site propagation between the cell sites. Coupled with the relatively large difference in transmission power in these types of networks, high-power downlink transmissions from one cell site could significantly impact the ability to receive a weak uplink signal in a neighboring cell, see Fig. 7.16. There could also be interference from uplink transmissions impacting the possibility to receive a downlink transmission in a neighboring cell, although this is typically less of a problem as it impacts only a subset of the users in the cell.

The classical way of handling these interference problems is to (semi-)statically split the resources between uplink and downlink in the same way across all the cells in the network. In particular, uplink reception in one cell never overlaps in time with downlink transmission in a neighboring cell. The set of slots (or, in general, time-domain resources) allocated for a certain transmission direction, uplink or downlink, is identical across the whole networks and can be seen as a simple form of inter-cell coordination, albeit on a (semi-)static basis. Static or semi-static TDD operation is also necessary for handling coexistence with LTE, for example when an LTE carrier and an NR carrier are using the same sites and the same frequency band. Such restrictions in the uplink-downlink allocation can easily be achieved as part of the scheduling implementation by using a fixed pattern in each base station. There is also a possibility to semi-statically configure the transmission direction of some or all of the slots as discussed in Section 7.8.3, a feature that can allow for reduced device energy consumption as it is not necessary to monitor for downlink control channels in slots that are a priori known to be reserved for uplink usage.

An essential aspect of any TDD system, or half-duplex system in general, is the possibility to provide a sufficiently large *guard period* (or guard time), where neither downlink nor uplink transmissions occur. This guard period is necessary for switching from downlink to uplink transmission and vice versa and is obtained by using slot formats where the downlink ends sufficiently early prior to the start of the uplink. The required length of the guard period depends on several factors. First, it should be sufficiently large to provide the necessary time for the circuitry in base stations and the devices to switch from downlink to uplink. Switching is typically relatively fast, of the order of 20 μs or less, and in most deployments, does not significantly contribute to the required guard time.

Fig. 7.16 Interference scenarios is a TDD network.

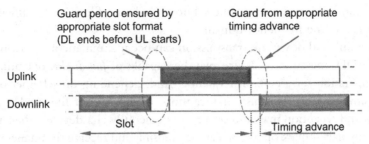

Fig. 7.17 Creation of guard time for TDD operation.

Second, the guard time should also ensure that uplink and downlink transmissions do not interfere at the base station. This is handled by advancing the uplink timing at the devices such that, at the base station, the last uplink subframe before the uplink-to-downlink switch ends before the start of the first downlink subframe. The uplink timing of each device can be controlled by the base station by using the timing advance mechanism, as will be elaborated upon in Chapter 15. Obviously, the guard period must be large enough to allow the device to receive the downlink transmission and switch from reception to transmission before it starts the (timing-advanced) uplink transmission (see Fig. 7.17). As the timing advance is proportional to the distance to the base station, a larger guard period is required when operating in large cells compared to small cells.

Finally, the selection of the guard period also needs to take interference between base stations into account. In a multi-cell network, inter-cell interference from downlink transmissions in neighboring cells must decay to a sufficiently low level before the base station can start to receive uplink transmissions. Hence, a larger guard period than motivated by the cell size itself may be required as the last part of the downlink transmissions from distant base stations otherwise may interfere with uplink reception. The amount of guard period depends on the propagation environments, but in some macro-cell deployments the inter-base-station interference is a non-negligible factor when determining the guard period. Depending on the guard period, some residual interference may remain at the beginning of the uplink period. Hence, it is beneficial to avoid placing interference-sensitive signals at the start of an uplink burst. In Chapter 19, some release 16 enhancements useful for improving the interference handling and setting of the guard period in TDD networks are discussed.

7.8.2 FDD – frequency-division duplex

In the case of FDD operation, uplink and downlink are carried on different carrier frequencies, denoted f_{UL} and f_{DL} in Fig. 7.15. During each frame, there is thus a full set of slots in both uplink and downlink, and uplink and downlink transmission can occur simultaneously within a cell. Isolation between downlink and uplink transmissions is achieved by transmission/reception filters, known as duplex filters, and a sufficiently large *duplex separation* in the frequency domain.

Even if uplink and downlink transmission can occur simultaneously within a cell in the case of FDD operation, a device may be capable of *full-duplex* operation or only *half-duplex* operation for a certain frequency band, depending on whether or not it is capable of simultaneous transmission/reception. In the case of full-duplex capability, transmission and reception may also occur simultaneously at a device, whereas a device capable of only half-duplex operation cannot transmit and receive simultaneously. Half-duplex operation allows for simplified device implementation due to relaxed or no duplex-filters. This can be used to reduce device cost, for example for low-end devices in cost-sensitive applications. Another example is operation in certain frequency bands with a very narrow duplex gap with correspondingly challenging design of the duplex filters. In this case, full duplex support can be *frequency-band dependent* such that a device may support only half-duplex operation in certain frequency bands while being capable of full-duplex operation in the remaining supported bands. It should be noted that full/half-duplex capability is a property of the *device*; the base station can operate in full duplex irrespective of the device capabilities. For example, the base station can transmit to one device while simultaneously receiving from another device.

From a network perspective, half-duplex operation has an impact on the sustained data rates that can be provided to/from a single device as it cannot transmit in all uplink subframes. The cell capacity is hardly affected as typically it is possible to schedule different devices in uplink and downlink in a given subframe. No provisioning for guard periods is required from a network perspective as the network is still operating in full duplex and therefore is capable of simultaneous transmission and reception. The relevant transmission structures and timing relations are identical between full-duplex and half-duplex FDD and a single cell may therefore simultaneously support a mixture of full-duplex and half-duplex FDD devices. Since a half-duplex device is not capable of simultaneous transmission and reception, the scheduling decisions must take this into account and half-duplex operation can be seen as a scheduling restriction.

7.8.3 Slot format and slot-format indication

Returning to the slot structure discussed in Section 7.2, it is important to point out that there is one set of slots in the uplink and another set of slots in the downlink, the reason being the time offset between the two as a function of timing advance. If both uplink and downlink transmission would be described using *the same* slot, which is often seen in various illustrations in the literature, it would not be possible to specify the necessary timing difference between the two.

Depending on whether the device is capable of full duplex, as is the case for FDD, or half duplex only, as is the case for TDD, a slot may not be fully used for uplink or downlink transmission. As an example, the downlink transmission in Fig. 7.17 had to stop prior to the end of the slot in order to allow for sufficient time for the device to switch to from downlink reception to uplink transmission. Since the necessary time between downlink and uplink depends on several factors, NR defines a wide range of *slot formats* defining which parts of a slot is used for uplink or downlink. Each slot format represents a combination of OFDM symbols denoted downlink, flexible, and uplink, respectively. The reason for having a third state, flexible, will be discussed further later, but one usage is to handle the necessary guard period in half-duplex schemes. A subset of the slot formats supported by NR are illustrated in Fig. 7.18. As seen in the figure, there are downlink-only and uplink-only slot formats which are useful for full-duplex operation (FDD), as well as partially filled uplink and downlink slots to handle the case of half-duplex operation (TDD).

The name slot format is somewhat misleading as there are separate slots for uplink and downlink transmissions, each filled with data in such a way that there is no simultaneous transmission and reception in the case of TDD. Hence, the slot format for a downlink slot should be understood as downlink transmissions can only occur in "downlink" or "flexible" symbols, and in an uplink slot, uplink transmissions can only occur in "uplink" or "flexible" symbols. Any guard period necessary for TDD operation is taken from the flexible symbols.

One of the key features of NR is, as already mentioned, the support for *dynamic TDD* where the scheduler dynamically determines the transmission direction. Since a

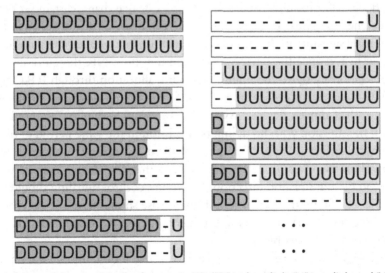

Fig. 7.18 A subset of the possible slot formats in NR. ("D" is downlink, "U" is uplink, and "-" is flexible).

half-duplex device cannot transmit and receive simultaneously, there is a need to split the resources between the two directions. In NR, three different signaling mechanisms provide information to the device on whether the resources are used for uplink or downlink transmission:

- Dynamic signaling for the scheduled device;
- Semi-static signaling using RRC; and
- Dynamic slot-format indication shared by a group of devices.

Some or all of these mechanisms are used in combination to determine the instantaneous transmission direction as will be discussed later. Although the following description uses the term dynamic TDD, the framework can in principle be applied to half-duplex operation in general, including half-duplex FDD.

The first mechanism and the basic principle is for the device to monitor for control signaling in the downlink and transmit/receive according to the received scheduling grants/assignments. In essence, a half-duplex device would view each OFDM symbol as a downlink symbol unless it has been instructed to transmit in the uplink. It is up to the scheduler to ensure that a half-duplex device is not requested to simultaneously receive and transmit and the term slot format may not make sense. For a full-duplex-capable device (FDD), there is obviously no such restriction, and the scheduler can independently schedule uplink and downlink.

The general principle above is simple and provides a flexible framework. However, if the network knows a priori that it will follow a certain uplink-downlink allocation, for example in order to provide coexistence with some other TDD technology or to fulfill some spectrum regulatory requirement, it can be advantageous to provide this information to the device. For example, if it is known to a device that a certain set of OFDM symbols is assigned to uplink transmissions, there is no need for the device to monitor for downlink control signaling in the part of the downlink slots overlapping with these symbols. This can help reduce the device power consumption. NR therefore provides the possibility to optionally signal the uplink-downlink allocation through RRC signaling.

The RRC-signaled pattern classifies OFDM symbols as "downlink," "flexible," or "uplink." For a half-duplex device, a symbol classified as "downlink" can only be used for downlink transmission with no uplink transmission in the same period of time. Similarly, a symbol classified as "uplink" means that the device should not expect any overlapping downlink transmission. "Flexible" means that the device cannot make any assumptions on the transmission direction. Downlink control signaling should be monitored and if a scheduling message is found, the devise should transmit/receive accordingly. Thus, the fully dynamic scheme outlined above is equivalent to semi-statically declaring all symbols as "flexible."

The RRC-signaled pattern is expressed as a concatenation of up to two sequences of downlink-flexible-uplink, together spanning a configurable period from 0.5 ms up to

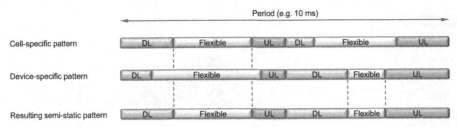

Fig. 7.19 Example of cell-specific and device-specific uplink-downlink patterns.

10 ms. Furthermore, *two* patterns can be configured, one cell-specific provided as part of system information and one signaled in a device-specific manner. The resulting pattern is obtained by combining these two where the dedicated pattern can further restrict the flexible symbols signaled in the cell-specific pattern to be either downlink or uplink. Only if both the cell-specific pattern *and* the device-specific pattern indicate flexible should the symbols be for flexible use (Fig. 7.19).

The third mechanism is to dynamically signal the current uplink-downlink allocation to a group of devices monitoring a special downlink control message known as the *slot-format indicator* (SFI). Similar to the previous mechanism, the slot format can indicate the number of OFDM symbols that are downlink, flexible, or uplink, and the message is valid for one or more slots.

The SFI message will be received by a group of one or more devices and can be viewed as a pointer into an RRC-configured table where each row in the table is constructed from a set of predefined downlink/flexible/uplink patterns one slot in duration. Upon receiving the SFI, the value is used as an index into the SFI table to obtain the uplink-downlink pattern for one or more slots as illustrated in Fig. 7.20. The set of predefined downlink/flexible/uplink patterns are listed in the NR specifications and cover a wide range of possibilities, some examples of which can be seen in Fig. 7.18 and in the left part of Fig. 7.20. The SFI can also indicate the uplink-downlink situations for other cells (cross-carrier indication).

Since a dynamically scheduled device will know whether the carrier is currently used for uplink transmission or downlink transmission from its scheduling assignment/grant, the group-common SFI signaling is primarily intended for *non-scheduled* devices. In particular, it offers the possibility for the network to overrule periodic transmissions of uplink *sounding reference signals* (SRS) or downlink measurements on *channel-state information reference signals* (CSI-RS). The SRS transmissions and CSI-RS measurements are used for assessing the channel quality as discussed in Chapter 8 and can be semi-statically configured. Overriding the periodic configuration can be useful in a network running with dynamic TDD (see Fig. 7.21 for an example illustration).

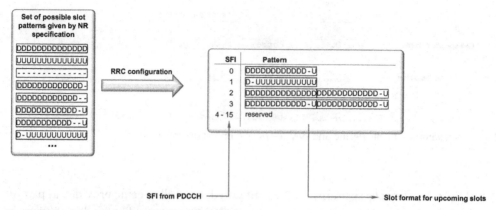

Fig. 7.20 Example of configuring the SFI table.

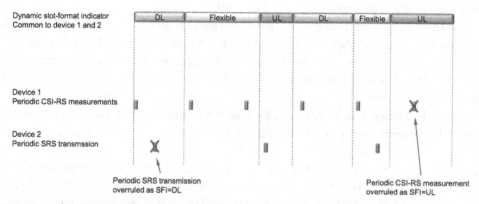

Fig. 7.21 Controlling periodic CSI-RS measurements and SRS transmsisisons by using the SFI.

The SFI cannot override a semi-statically configured uplink or downlink period, neither can it override a dynamically scheduled uplink or downlink transmission which take place regardless of the SFI. However, the SFI can override a symbol period semi-statically indicated as flexible by restricting it to be downlink or uplink. It can also be used to provide a reserved resource; if both the SFI and the semi-static signaling indicate a certain symbol to be flexible, then the symbol should be treated as reserved and not be used for transmission, nor should the device make any assumptions on the downlink transmission. This can be useful as a tool to reserve resource on an NR carrier, for example used for other radio-access technologies or for features added to future releases of the NR standard.

The description has focused on half-duplex devices in general and TDD in particular. However, the SFI can be useful also for full-duplex systems such as FDD as well, for example to override periodic SRS transmissions. Since there are two independent "carriers" in this case, one for uplink and one for downlink, two SFIs are needed, one for each carrier. This is solved by using the multi-slot support in the SFI; one slot is interpreted as the current SFI for the downlink and the other as the current SFI for the uplink.

7.9 Antenna ports

Downlink multi-antenna transmission is a key technology of NR. Signals transmitted from different antennas or signals subject to different and for the receiver unknown *multi-antenna precoders* (see Chapter 9) will experience different "radio channels" even if the set of antennas are located at the same site.[1]

In general, it is important for a device to understand what it can assume in terms of the relationship between the radio channels experienced by different downlink transmissions. This is, for example, important in order for the device to be able to understand what reference signal(s) should be used for channel estimation for a certain downlink transmission. It is also important in order for the device to be able to determine relevant channel-state information, for example for scheduling and link-adaptation purposes.

For this reason, the concept of *antenna port* is used in NR, following the same principles as in LTE. An antenna port is defined such that *the channel over which a symbol on the antenna port is conveyed can be inferred from the channel over which another symbol on the same antenna port is conveyed.* Expressed differently, each individual downlink transmission is carried out from a specific antenna port, the identity of which is known to the device. Furthermore, the device can assume that two transmitted signals have experienced the same radio channel *if and only if* they are transmitted from the same antenna port.[m]

In practice, each antenna port can, at least for the downlink, be seen as corresponding to a specific reference signal. A device receiver can then assume that this reference signal can be used to estimate the channel corresponding to the specific antenna port. The reference signals can also be used by the device to derive detailed channel-state information related to the antenna port.

The set of antenna ports defined in NR is outlined in Table 7.2. As seen in the table, there is a certain structure in the antenna port numbering such that antenna ports for different purposes have numbers in different ranges. For example, downlink antenna ports starting with 1000 are used for PDSCH. Different transmission layers for PDSCH can use

---

[1] An unknown transmitter-side precoder needs to be seen as part of the overall radio channel.

[m] For certain antenna ports, more specifically those that correspond to so-called demodulation reference signals, the assumption of same radio channel is only valid within a given scheduling occasion.

Table 7.2 Antenna ports in NR.[a]

Antenna port	Uplink	Downlink	Sidelink
0000-series	PUSCH and associated DM-RS	–	–
1000-series	SRS, precoded PUSCH	PDSCH	PSSCH
2000-series	PUCCH	PDCCH	PSCCH
3000-series	–	CSI-RS	CSI-RS
4000-series	PRACH	SSB	SSB
5000-series	–	PRS	PSFCH

[a]Sidelink and positioning reference signals (PRS) are introduced in release 16; see Chapters 26 and 27, respectively.

antenna ports in this series, for example 1000 and 1001 for a two-layer PDSCH transmission. The different antenna ports and their usage will be discussed in more detail in conjunction with the respective feature. Note that, even if they have the same antenna port number, an uplink antenna port is not the same as a downlink antenna port, that is, the numbering is separate for uplink, downlink, and sidelink.

It should be understood that an antenna port is an abstract concept that does not necessarily correspond to a specific physical antenna:

• Two different signals may be transmitted in the same way from multiple physical antennas. A device receiver will then see the two signals as propagating over a single channel corresponding to the "sum" of the channels of the different antennas and the overall transmission could be seen as a transmission from a single antenna port being the same for the two signals.

• Two signals may be transmitted from the same set of antennas but with different, for the receiver unknown, antenna transmitter-side precoders. A receiver will have to see the unknown antenna precoders as part of the overall channel implying that the two signals will appear as having been transmitted from two different antenna ports. It should be noted that if the antenna precoders of the two transmissions would have been known to be the same, the transmissions could have been seen as originating from the same antenna port. The same would have been true if the precoders would have been known to the receiver as, in that case, the precoders would not need to be seen as part of the radio channel.

The last of these two aspects motivates the introduction of QCL framework as discussed in the next section.

7.10 Quasi-colocation

Even if two signals have been transmitted from two different antennas, the channels experienced by the two signals may still have many *large-scale* properties in common.

As an example, the channels experienced by two signals transmitted from two different antenna ports corresponding to different physical antennas at the same site will, even if being different in the details, typically have the same or at least similar large-scale properties, for example, in terms of Doppler spread/shift, average delay spread, and average gain. It can also be expected that the channels will introduce similar average delay. Knowing that the radio channels corresponding to two different antenna ports have similar large-scale properties can be used by the device receiver, for example, in the setting of parameters for channel estimation.

In case of single-antenna transmission, this is straightforward. However, one integral part of NR is the extensive support for multi-antenna transmission, beamforming, and simultaneous transmission from multiple geographically separated sites. In these cases, the channels of different antenna ports relevant for a device may differ even in terms of large-scale properties.

For this reason, the concept of *quasi-colocation* (QCL) with respect to antenna ports is part of NR. A device receiver can assume that the radio channels corresponding to two different antenna ports have the same large-scale properties in terms of specific parameters such as average delay spread, Doppler spread/shift, average delay, and spatial Rx parameters *if and only if* the antenna ports are specified as being quasi-colocated. Whether or not two specific antenna ports can be assumed to be quasi-colocated with respect to a certain channel property is in some cases given by the NR specification. In other cases, the device may be explicitly informed by the network by means of signaling if two specific antenna ports can be assumed to be quasi-colocated or not.

The general principle of quasi-colocation is present already in the later releases of LTE when it comes to the temporal parameters. However, with the extensive support for beamforming in NR, the QCL framework has been extended to the spatial domain. *Spatial quasi-colocation* or, more formally, *QCL-TypeD* or *quasi-colocation with respect to RX parameters* is a key part of beam management. Although somewhat vague in its formal definition, in practice spatial QCL between two different signals implies that they are transmitted from the same place and in the same beam. As a consequence, if a device knows that a certain receiver beam direction is good for one of the signals, it can assume that the same beam direction is suitable also for reception of the other signal. In a typical situation, the NR specification states that certain transmissions, for example, PDSCH and PDCCH transmissions, are spatially quasi-colocated with specific reference signals, for example CSI-RS or SSB. The device may have decided on a specific receiver beam direction based on measurements on the reference signal in question and the device can then assume that the same beam direction is a good choice also for the PDSCH/PDCCH reception.

To summarize, in total there are four different QCL types defined in NR:

- QCL-TypeA – QCL with respect to Doppler shift, Doppler spread, average delay, and delay spread;

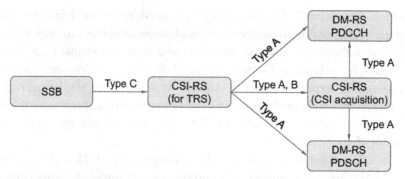

Fig. 7.22 Example of QCL relations between channels and signals.

- QCL–TypeB – QCL with respect to Doppler shift and Doppler spread;
- QCL–TypeC – QCL with respect to Doppler shift and average delay;
- QCL–TypeD – QCL with respect to spatial Rx parameters.

Thus, the QCL relations help the device to prepare for receiving a signal or channel where properties estimated from one signal can be used in the channel estimation for another channel or signal. In Fig. 7.22, the Doppler shift and the average delay obtained from the SSB can be used for improving the TRS-based estimate of time and frequency drift and, in the next step, for improving the DM–RS-based channel estimate for the PDCCH and PDSCH. The TRS will be discussed in the next chapter and a more extensive description of the usage of QCL relations in multi-antenna transmission is found on Chapter Additional QCL types can be added in future releases if necessary as the framework as such is generic.

CHAPTER 8

Channel measurements

Many transmission features in modern radio-access technologies are based on the availability of more or less detailed knowledge about different characteristics of the radio-channel. This may range from rough knowledge of the radio-channel path loss, for example, for mobility decisions to detailed knowledge of the channel amplitude and phase in the time, frequency, and/or spatial domain for selection of instantaneous transmission parameters. Many transmission features will also benefit from knowledge about the interference level experienced at the receiver side.

Such knowledge about different channel characteristics can be acquired in different ways. As an example, knowledge about downlink channel characteristics can be acquired by means of device measurements. The acquired information could then be reported to the network for the setting of different transmission parameters for subsequent downlink transmissions. Alternatively, if it can be assumed that the channel is reciprocal, that is, the channel characteristics of interest are the same in the downlink and uplink transmission directions, the network can, by itself, acquire knowledge about relevant downlink channel characteristics by estimating the same characteristics in the uplink direction.

The same alternatives exist when it comes to acquiring knowledge about uplink channel characteristics, that is, the network may determine the uplink characteristics of interest and either provide the information to the device or directly control subsequent uplink transmissions based on the acquired channel knowledge. Alternatively, assuming channel reciprocity, the device may, by itself, acquire knowledge about the relevant uplink channel characteristics by means of downlink measurements.

Regardless of the exact approach to acquire channel knowledge, there is typically a need for some *reference signal(s)* on which a receiver can measure/estimate the channel characteristics of interest. In this chapter we will describe the NR reference signals on which channel measurements are typically carried out, more specifically, the downlink *channel-state-information reference signals* (CSI-RS) and the uplink *sounding reference signals* (SRS). We will also provide an overview of the NR framework for physical-layer device measurements and corresponding device reporting to the network.

8.1 Channel-state-information reference signals – CSI-RS

The CSI-RS is the main tool for downlink channel measurements in NR. In addition to being used for detailed channel sounding for the selection of detailed downlink transmission parameters, CSI-RS can also be used, for example, for measurements related to beam management and mobility.

8.1.1 Basic CSI-RS structure

A configured CSI-RS resource may correspond to up to 32 CSI-RS antenna ports, that is, there may be up to 32 different reference signals multiplexed within the overall CSI-RS resource.

A CSI-RS resource is always configured on a per-device basis. It is important to understand though that configuration on a per-device basis does not mean that an actual CSI-RS transmission can only be used by a single device. Nothing prevents identical CSI-RS resources to be separately configured for multiple devices, in practice implying that the CSI-RS resource is shared between the devices.

As illustrated in Fig. 8.1, a single-port CSI-RS resource occupies a single resource element within a block corresponding to one resource block in the frequency domain and one slot in the time domain. In principle, the CSI-RS resource can be configured to occur anywhere within the resource block although in practice there are some restrictions to avoid collisions with other downlink physical channels and signals. Especially, a device can assume that a configured CSI-RS resource does not collide with

- Any CORESET (see Section 10.1.2) configured for the device;
- DM-RS (see Section 9.11) associated with PDSCH transmissions scheduled for the device;
- Any SSB (see Chapter 16) transmitted within the cell;

In case of a multi-port CSI-RS resource, that is, a CSI-RS resource corresponding to multiple antenna ports, the different antenna ports share the overall set of resource elements of the CSI-RS resource based on a combination of

- *Code-domain multiplexing* (CDM), implying that different per-antenna-port CSI-RS are transmitted on the same set of resource elements with separation achieved by modulating the different CSI-RS with different orthogonal patterns
- *Frequency-domain multiplexing* (FDM), implying that different per-antenna-port CSI-RS are transmitted on different sets of subcarriers within an OFDM symbol

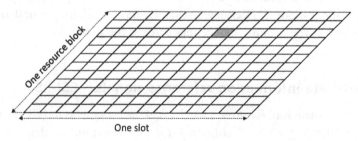

One resource block

One slot

Fig. 8.1 Single-port CSI-RS structure consisting of a single resource element within an RB/slot block.

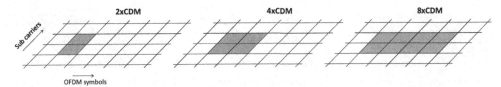

Fig. 8.2 Different CDM structures for multiplexing per-antenna-port CSI RS.

- *Time-domain mulitplexing* (TDM), implying that different per-antenna-port CSI-RS are transmitted in different OFDM symbols within a slot

Furthermore, as illustrated in Fig. 8.2, CDM between different per-antenna-port CSI-RS can be

- in the frequency domain with CDM over two adjacent subcarriers ("$2 \times$ CDM"), allowing for code-domain sharing between two per-antenna-port CSI-RS
- in the frequency and time domain with CDM over two adjacent subcarriers and two adjacent OFDM symbols ("$4 \times$ CDM"), allowing for code-domain sharing between up to four per-antenna-port CSI-RS
- in the frequency and time domain with CDM over two adjacent subcarriers and four adjacent OFDM symbols ("$8 \times$ CDM"), allowing for code-domain sharing between up to eight per-antenna-port CSI-RS

The different CDM alternatives of Fig. 8.2, in combination with FDM and/or TDM, can then be used to configure different structures for the overall multi-port CSI-RS resources where, in general, an N-port CSI-RS resource occupies a total of N resource elements within an RB/slot block.[a]

As a first example, Fig. 8.3 illustrates how a two-port CSI-RS resource consists of two adjacent resource elements in the frequency domain with CDM between the two antenna ports. In other words, the two-port CSI-RS resource has a structure identical to the basic $2 \times$ CDM structure in Fig. 8.2.

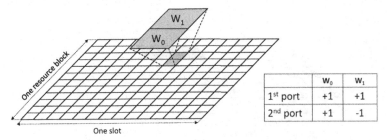

	W_0	W_1
1st port	+1	+1
2nd port	+1	-1

Fig. 8.3 Structure of two-port CSI-RS based on $2 \times$ CDM. The figure also illustrates the orthogonal pattern of each port.

---

[a] The so-called "density-3" CSI-RS used for TRS, see Section 8.1.7, is an exception to this rule.

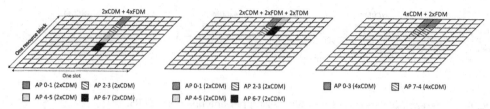

Fig. 8.4 Three different structures for eight-port CSI-RS.

In the case of a CSI-RS resource corresponding to more than two antenna ports there is some flexibility in the sense that, for a given number of ports, there are multiple different structures available for the CSI-RS resource based on different combinations of CDM, TDM and FDM

As an example, there are three different structures for an eight-port CSI-RS resource, see also Fig. 8.4.

- Frequency-domain CDM over two resource elements ($2 \times$ CDM) in combination with four times frequency-multiplexing (left part of Fig. 8.4). The overall CSI-RS resource thus consists of eight subcarriers within the same OFDM symbol.
- Frequency-domain CDM over two resource elements ($2 \times$ CDM) in combination with frequency and time multiplexing (middle part Fig. 8.4). The overall CSI-RS resource thus consists of four subcarriers within two OFDM symbols.
- Time/frequency-domain CDM over four resource elements ($4 \times$ CDM) in combination with two times frequency multiplexing. The overall CSI-RS resource thus once again consists of four subcarriers within two OFDM symbols.

Note that, in all these cases, the total number of resources blocks of the CSI-RS resource equals the number of antenna ports.

Finally, Fig. 8.5 illustrates one out of three possible structures for a 32-port CSI-RS resource based on a combination of $8 \times$ CDM and four times frequency multiplexing.

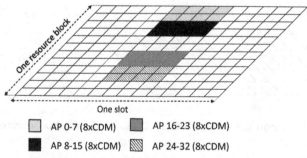

Fig. 8.5 One structure (out of three supported structures) for a 32-port CSI-RS.

Note that, once again, the total number of resources blocks of the CSI-RS resource equals the number of antenna ports. This example also illustrates that CSI-RS antenna ports separated in the frequency domain do not necessarily have to occupy consecutive subcarriers. Likewise, CSI-RS ports separated in the time domain do not necessarily have to occupy consecutive OFDM symbols.

In the case of a multi-port CSI-RS resource, the association between per-port CSI-RS and port number is done first in the CDM domain, then in the frequency domain, and finally in the time domain. This can, for example, be seen from the eight-port example of Fig. 8.4 where per-port CSI-RS separated by means of CDM correspond to consecutive port numbers. Furthermore, for the FDM+TDM case (center part of Fig. 8.4), port number zero to port number three are transmitted within the same OFDM symbol while port number four to port number seven are jointly transmitted within another OFDM symbol. Port number zero to three and port number four to seven are thus separated by means of TDM.

8.1.2 Frequency-domain structure of CSI-RS configurations

A CSI-RS resource is configured for a given downlink bandwidth part and is then assumed to be confined within, and use the numerology of, that bandwidth part.

The CSI-RS resource can be configured to cover the full bandwidth or just a fraction of the bandwidth of the bandwidth part. In the latter case, the bandwidth and frequency-domain starting position of the CSI-RS resource are provided as part of the CSI-RS configuration.

Within its configured bandwidth, a CSI-RS may be configured for transmission in every resource block in the frequency domain (*CSI-RS frequency density equal to one*). However, a CSI-RS may also be configured for transmission only in every second resource block (*frequency density equal to 1/2*). In the latter case, the configuration includes information about the set of resource blocks (odd resource blocks or even resource blocks) within which the CSI-RS will be transmitted. CSI-RS density equal to 1/2 is not supported for CSI-RS resources with 4, 8 and 12 antenna ports.

There is also a possibility to configure a *single-port* CSI-RS resource with a frequency density of 3 in which case the CSI-RS occupies three subcarriers within each resource block. This special CSI-RS structure is used as part of a so-called *Tracking Reference Signal* (TRS), see Section 8.1.7.

8.1.3 Time-domain property of CSI-RS configurations

In general, a CSI-RS resource can be configured for *periodic, semi-persistent,* or *aperiodic* transmission.

In the case of periodic transmission, a device can assume that CSI-RS transmissions occur every Nth slot, where N ranges from as low as four, that is, transmissions every

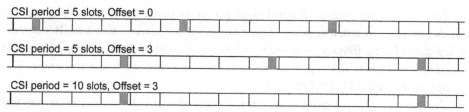

CSI period = 5 slots, Offset = 0

CSI period = 5 slots, Offset = 3

CSI period = 10 slots, Offset = 3

Fig. 8.6 Examples of periodic CSI-RS transmission.

fourth slot, to as high as 640, that is, transmission only every 640th slot. In addition to the periodicity, the device is also configured with a specific slot offset for the CSI-RS transmission, see Fig. 8.6.

In the case of semi-persistent CSI-RS transmission, a periodicity and corresponding slot offset are configured for the CSI-RS transmission in the same way for periodic CSI-RS transmission. However, actual CSI-RS transmission is activated and deactivated based on MAC-CE signaling (see Section 6.2.4.1). Once the CSI-RS transmission has been activated, the device can assume that the CSI-RS transmission will continue according to the configured periodicity/offset until it is explicitly deactivated. Similarly, once the CSI-RS transmission has been deactivated, the device cannot assume that there will be any further CSI-RS transmissions until the CSI-RS transmission is explicitly reactivated again.

In case of aperiodic CSI-RS, no periodicity is configured. Rather, each CSI-RS transmission is indicated by means of layer-1 signaling (DCI).

It should be mentioned that the property of periodic, semipersistent, or aperiodic is actually not a property of a CSI-RS resource but rather the property of a so-called CSI-RS *resource set*. As a consequence, activation/deactivation and triggering of semi-persistent and aperiodic CSI-RS transmissions respectively, are not done for a specific CSI-RS resources but for the set of CSI-RS resources within a CSI-RS resource set. CSI-RS resource sets are further discussed in Section 8.1.6.

8.1.4 CSI-IM – Resources for interference measurements

CSI-RS transmission on a configured CSI-RS resource can be used by a device to acquire information about the properties of the channels of the different CSI-RS antenna ports. It can also be used to estimate the interference level by subtracting the expected received signal from what is actually received on the CSI-RS resource.

However, the interference level can also be estimated from measurements on *CSI-Interference-Measurement* (CSI-IM) resources.

Fig. 8.7 illustrates the structure of a CSI-IM resource. As can be seen, there are two different CSI-IM structures, each consisting of four resource elements but with different time/frequency structures. Similar to a CSI-RS resource, the exact location of the CSI-IM resource within the RB/slot block is flexible and information about it is included as part of the CSI-IM configuration.

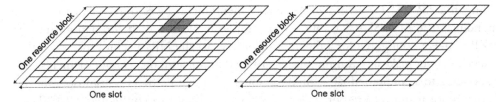

Fig. 8.7 Alternative structures for CSI-IM resource.

The time-domain property of a CSI-IM resource is the same as that of a CSI-RS resource, that is, a CSI-IM resource can be configured as periodic, semi-persistent (activation/deactivation by means of MAC CE), or aperiodic (indicated by DCI).[b]

In a typical case, a CSI-IM resource would correspond to resource elements where nothing is transmitted within the current cell while the activity within the CSI-IM resource in neighbor cells should correspond to normal activity of those cells. Thus, by measuring the receiver power within a CSI-IM resource, a device would get an estimate on the typical interference due to transmissions within other cells.

As there should be no transmissions on CSI-IM resources within the cell, devices should be configured with the corresponding resources as so-called ZP-CSI-RS resources (see below).

8.1.5 Zero-power CSI-RS

A CSI-RS resource as described above should more correctly be referred to as *non-zero-power* (NZP) CSI-RS resource to distinguish it from so-called *zero-power* (ZP) CSI-RS resources that can also be configured for a device.

If a device is scheduled for PDSCH reception on a resource that includes resource elements corresponding to a configured CSI-RS resource, the device can assume that the PDSCH rate matching and resource mapping avoid those resource elements. However, a device may also be scheduled for PDSCH reception on a resource that includes resource elements corresponding to a CSI-RS resource configured for a different device. The PDSCH must also in this case be rate matched around the resource elements of the CSI-RS. The configuration of a ZP-CSI-RS resource is a way to inform the device for which the PDSCH is scheduled about such rate matching.

A configured ZP-CSI-RS resource corresponds to a set of resource elements with the same structure as an NZP-CSI-RS resource. However, while a device can assume that an NZP-CSI-RS is actually transmitted and is something on which a device can carry out measurements, a configured ZP-CSI-RS only indicates a set of resource blocks to which the device should assume that PDSCH is not mapped.

It should be emphasized that, despite the name, a device cannot assume that there are no transmissions, that is, zero power within the resource elements corresponding to a

---

[b] Once again, this is strictly speaking a property of a *CSI-IM resource set.*

configured ZP-CSI-RS resource. As already mentioned, the resource elements corresponding to a ZP-CSI-RS resource may, for example, be used for transmission of NZP-CSI-RS configured for a different device. What the NR specification says is that a device cannot make *any* assumptions regarding transmissions on resource elements corresponding to a configured ZP-CSI-RS resource and that PDSCH transmission for the device is not mapped to resource elements corresponding to a configured ZP-CSI-RS resource.

8.1.6 CSI-RS resource sets

In addition to being configured with CSI-RS resources, a device will be configured with one or several *CSI-RS resource sets*, formally referred to as *NZP-CSI-RS-ResourceSets*. Each such resource set includes one or several configured CSI-RS resources or, more exactly, *pointers* to already configured CSI-RS resources. The resource set can then be used as part of *report configurations* describing measurements and corresponding reporting to be done by a device (see further details in Section 8.2). Alternatively, and despite the name, an *NZP-CSI-RS-ResourceSet* may include pointers to a set of SSBs. This reflects the fact that some device measurements, especially measurements related to beam management and mobility, may be based on either CSI-RS or SSB.

Above it was described how CSI-RS transmission could be periodic, semi-persistent, or aperiodic. As mentioned there, this categorization is strictly speaking not a property of a configured CSI-RS resource itself but a property of a configured CSI-RS resource set. Furthermore, a MAC CE command will jointly activate/deactivate transmission on all CSI-RS resources within a CSI-RS resource set configured for semi-persistent transmission. Likewise, DCI will trigger transmission on all CSI-RS resources within a CSI-RS resource set configured for aperiodic transmission.

Similarly, a device may be configured with *CSI-IM resource sets*, each including pointers to a set of configured CSI-IM resources that can be jointly activated/deactivated (semi-persistent CSI-IM resource set) or triggered (aperiodic CSI-IM resource set).

8.1.7 Tracking reference signal – TRS

Due to oscillator imperfections, the device must track and compensate for variations in time and frequency to successfully receive downlink transmissions. To assist the device in this task, a *tracking reference signal* (TRS) can be configured. Strictly speaking, the TRS is not a CSI-RS. Rather a TRS is a *resource set* consisting of *multiple* periodic CSI-RS. More specifically a TRS consists of four one-port, density-3 CSI-RS located within two consecutive slots (see Fig. 8.8). The CSI-RS within the resource set, and thus also the TRS in itself, can be configured with a periodicity of 10, 20, 40, or 80 ms. Note that the exact set of resource elements (subcarriers and OFDM symbols) used for the TRS CSI-RS may vary. There is always a four-symbol time-domain separation (three intermediate symbols)

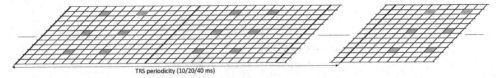

TRS periodicity (10/20/40 ms)

Fig. 8.8 TRS consisting of four one-port, density-3 CSI-RS located within two consecutive slots.

between the two CSI-RS within a slot though. This time domain separation sets the limit for the frequency error that can be tracked. Likewise, the frequency-domain separation (four subcarriers) sets the limit for the timing error that can be tracked.

There is also an alternative TRS structure with the same per-slot structure as the TRS structure of Fig. 8.8 but only consisting of two CSI-RS *within a single slot*, compared to two consecutive slots for the TRS structure in Fig. 8.8

8.1.8 Mapping to physical antennas

In Chapter 7, the concept of antenna ports and the relation to reference signals were discussed. A multi-port CSI-RS corresponds to a set of antenna ports and the CSI-RS can be used for sounding of the channels corresponding to those antenna ports. However, a CSI-RS port is often not mapped directly to a physical antenna, implying that the channel being sounded based on a CSI-RS is often not the actual physical radio channel. Rather, more or less any kind of (linear) transformation or *spatial filtering*, labeled F in Fig. 8.9, may be applied to the CSI-RS before mapping to the physical antennas. Furthermore, the number of physical antennas (N in Fig. 8.9) to which the CSI-RS is mapped may very well be larger than the number of CSI-RS ports.[c] When a device does channel sounding based on the CSI-RS, neither the spatial filter F nor the N physical antennas will be explicitly visible. What the device will see is just the M "channels" corresponding to the M CSI-RS ports.

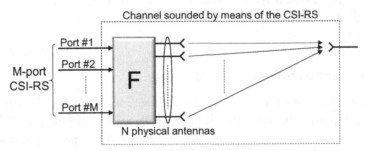

Fig. 8.9 CSI-RS applied to spatial filter (F) before mapping to physical antennas.

---

[c] Having N smaller than M does not make sense.

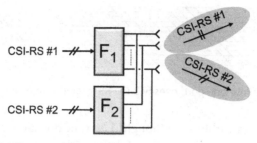

Fig. 8.10 Different spatial filters applied to different CSI-RS.

The spatial filter F may very well be different for different CSI-RS. The network could, for example, map two different configured CSI-RS such that they are beam-formed in different directions (see Fig. 8.10). To the device this will appear as two CSI-RS transmitted over two different channels, despite the fact that they are transmitted from the same set of physical antennas and are propagating via the same set of physical channels.

Although the spatial filter F is not explicitly visible to the device, the device still has to make certain assumptions regarding F. Especially, F has a strong relation to the concept of antenna ports discussed in Chapter 7. In essence one can say that two signals are transmitted from the same antenna port if they are mapped to the same set of physical antennas by means of the same transformation F.

As an example, in the case of downlink multi-antenna transmission (see Chapter 11), a device may measure on a CSI-RS and report a recommended *precoder matrix* to the network. The network may then use the recommended precoder matrix when mapping so called transmission layers to antenna ports. When selecting a suitable precoder matrix the device will assume that the network, if using the recommended matrix, will map the output of the precoding to the antenna ports of the CSI-RS on which the corresponding device measurements were carried out. In other words, the device will assume that the precoded signal will be mapped to the physical antennas by means of the same spatial filter F as applied to the CSI-RS.

8.2 Device measurements and reporting

An NR device can be configured to carry out different measurements, in most cases with corresponding reporting to the network. In general, such a configuration of measurements and corresponding reporting are done by means of a *report configuration*, in the 3GPP specifications [15] referred to as a *CSI-ReportConfig*.[d]

---

[d] Note that we are here talking about *physical-layer* measurements/reporting, to be distinguished from higher-layer reporting done by means of RRC signaling.

Each resource configuration describes/indicates:

- The specific quantity or set of quantities to be reported;
- The downlink resource(s) on which measurements should be carried out in order to derive the quantity or quantities to be reported;
- How the actual reporting is to be carried out, for example, when the reporting is to be done and what uplink physical channel to use for the reporting.

8.2.1 Report quantity

A report configuration indicates a quantity or set of quantities that the device is supposed to report. The report could, for example, include different combinations of *channel-quality indicator* (CQI), *rank indicator* (RI), and *precoder-matrix indicator* (PMI), jointly referred to as *channel-state information* (CSI). Alternatively, the report configuration may indicate reporting of received signal strength, more formally referred to as *reference-signal received power* (RSRP). RSRP has historically been a key quantity to measure and report as part of higher-layer *radio-resource management* (RRM) and is so also for NR. However, NR also supports layer-1 reporting of RSRP, for example, as part of the support for beam management (see Chapter 12). What is then reported is more specifically referred to as *L1-RSRP*, reflecting the fact that the reporting does not include the more long-term ("layer-3") filtering applied for the higher-layer RSRP reporting.

8.2.2 Measurement resource

In addition to describing what quantity to report, a report configuration also describes the set of downlink signals or, more generally, the set of downlink resources on which measurements should be carried out in order to derive the quantity or quantities to be reported. This is done by associating the report configuration with one or several resource sets as described in Section 8.1.6

A resource configuration is associated with at least one *NZP-CSI-RS-ResourceSet* to be used for measuring channel characteristics. As described in Section 8.1.6, a *NZP-CSI-RS-ResourceSet* may either contain a set of configured CSI-RS or a set of SSBs. Reporting of, for example, L1-RSRP for beam management can thus be based on measurements on either a set of SSBs or a set of CSI-RS.

Note that the resource configuration is associated with a resource *set*. Measurements and corresponding reporting are thus in the general case based on *a set* of CSI-RS resource or *a set* of SSBs.

In some cases, the set will only include a single reference-signal resource. An example of this is conventional feedback for link adaptation and multi-antenna precoding. In this case, the device would typically be configured with a resource set consisting of a single multi-port CSI-RS resource on which the device will carry out measurements to determine and report a combination of CQI, RI, and PMI.

On the other hand, in the case of beam management the resource set will typically consist of multiple CSI-RS resources, alternatively multiple SSBs, where, in practice, each CSI-RS resource or SSB is associated with a specific beam. The device measures on the set of signals within the resource set and, based on the measurements, provides a report to the network as input to the network beam-management functionality

There are also situations when a device needs to carry out measurements without any corresponding reporting to the network. One such case is when a device should carry out measurements for receiver-side downlink beam forming. As will be described in Chapter 12, in such a case a device may measure on downlink reference signals using different receiver beams. However, the result of the measurement is not reported to the network but only used internally within the device to select a suitable receiver beam. At the same time the device needs to be configured with the reference signals to measure on. Such a configuration is also covered by report configurations for which, in this case, the quantity to be reported is defined as "None."

8.2.3 Report types

In addition to the quantity to report and the set of resources to measure on, the report configuration also describes when and how the reporting should be carried out.

Similar to CSI-RS transmission, device reporting can be periodic, semi-persistent or aperiodic.

As the name suggests, periodic reporting is done with a certain configured periodicity. Periodic reporting is always done on the PUCCH physical channel. Thus, in the case of periodic reporting, the resource configuration also includes information about a periodically available PUCCH resource to be used for the reporting

In the case of semi-persistent reporting, a device is configured with periodically occurring reporting instances in the same way as for periodic reporting. However, actual reporting can be activated and deactivated by means of MAC signaling (MAC CE).

Similar to periodic reporting, semi-persistent reporting can be done on a periodically assigned PUCCH resource. Alternatively, semi-persistent reporting can be done on a semi-persistently allocated PUSCH. The latter is typically used for larger reporting payloads.

Aperiodic reporting is explicitly triggered by means of DCI signaling, more specifically within a CSI request field within the uplink scheduling grant (DCI formal 0-1). The DCI field may consist of up to 6 bits with each configured aperiodic report associated with a specific bit combination. Thus, up to 63 different aperiodic reports can be triggered.[e]

---

[e] The all-zero value indicates "no triggering."

Aperiodic reporting is always done on the scheduled PUSCH and thus requires an uplink scheduling grant. This is the reason why the triggering of aperiodic reporting is only included in the uplink scheduling grant and not in other DCI formats.

It should be noted that, in the case of aperiodic reporting, the report configuration could actually include multiple resource sets for channel measurements, each with its own set of reference signals (CSI-RS or SSB). Each resource set is associated with a specific value of the CSI-request field in the DCI. By means of the CSI request the network can, in this way, trigger the same type of reporting but based on different measurement resources. Note that the same could, in principle, have been done by configuring the device with multiple report configurations, where the different resource configurations would specify the same reporting configuration and report type but different measurement resources.

Periodic, semipersistent and aperiodic reporting should not be mixed up with periodic, semi-persistent, and aperiodic CSI-RS transmissions as described in Section 8.1.3 above. As an example, aperiodic reporting and semi-persistent reporting could very well be based on measurements on periodic CSI-RS resources. On the other hand, periodic reporting can only be based on measurements on periodic CSI-RS resources but not on aperiodic and semi-static CSI-RS resources. Table 8.1 summarizes the allowed combinations of reporting type (periodic, semi-persistent, and aperiodic) and resource type (periodic, semi-persistent, and aperiodic).

8.3 Sounding reference signals – SRS

To enable uplink channel measurements, a device can be configured for transmission of *sounding reference signals* (SRS). In some respects SRS can be seen as the uplink equivalence to the downlink CSI-RS in the sense that both CSI-RS and SRS are used for channel sounding, albeit in different transmission directions. Both CSI-RS and SRS can also serve as QCL references in the sense that other physical channels (PDCCH/PDSCH in the downlink, PUCCH/PUSCH in the uplink) can be configured as being quasi-co-located with CSI-RS and SRS respectively.

However, on a more detailed level, the structure of an SRS transmission is quite different from a CSI-RS transmission.

Table 8.1 Allowed combinations of report type and resource type.

Report type	Resource type		
	Periodic	Semi-persistent	Aperiodic
Periodic	Yes	—	—
Semi-persistent	Yes	Yes	—
Aperiodic	Yes	Yes	Yes

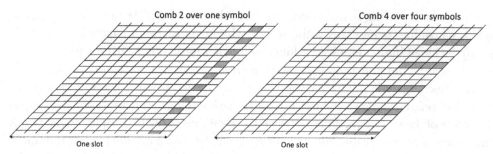

Fig. 8.11 Examples of SRS time/frequency structures.

- An SRS resource may correspond to up to four antenna ports while a CSI-RS resource may correspond to up to 32 antenna ports
- Being an uplink signal, SRS is designed for low cubic-metric [57], enabling higher device power-amplifier efficiency

The basic time/frequency structure of an SRS resource is exemplified in Fig. 8.11. In the first NR release (3GPP release 15), an SRS resource could span one, two, or four consecutive OFDM symbols and be located somewhere within the last six symbols of a slot. However, in later releases this has been extended so that an SRS resource can now span up to 14 consecutive symbols, that is, an entire slot. There is also higher flexibility in terms of the SRS location so that an SRS can now be located anywhere within a slot. Note though that an SRS transmission has to be confined to a slot, that is, it should not cross slot boundaries.

In the frequency domain, an SRS resource has a so-called "comb" structure, implying that an SRS is mapped to every Nth sub-carrier (illustrated in Fig. 8.11 for the case of "comb-2" and "comb-4"). The comb structure allows for the frequency multiplexing of SRS transmissions from different devices within the same bandwidth, see Fig. 8.12. In some cases, it can also be used to multiplex different antenna ports of the same SRS resource. Release 15 and 16 was limited to comb-2 and comb-4. However, in release 17 this was extended to also include a comb-8 structure, that is, SRS mapped to every 8th sub-carrier.

8.3.1 SRS sequences and Zadoff-Chu sequences

An SRS transmission consists of a set of resource elements in the frequency domain. The sequences applied to this set of resource elements are partly based on so-called *Zadoff-Chu* sequences (ZC). Due to their specific properties, Zadoff-Chu sequences are used at several places within the NR specifications, especially in the uplink transmission direction.

A Zadoff-Chu sequence of length M is given by the following expression:

$$z_i^u = e^{-j\frac{\pi u i(i+1)}{M}}; \qquad 0 \leq i < M \tag{8.1}$$

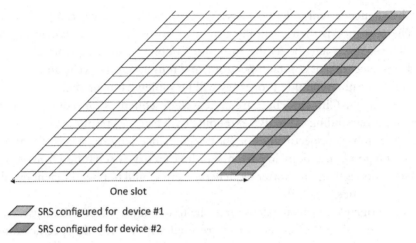

One slot

SRS configured for device #1
SRS configured for device #2

Fig. 8.12 Comb-based frequency multiplexing of SRS from two different devices assuming comb-2.

As can be seen from Eq. (8.1), a Zadoff-Chu sequence has a characterizing parameter u, referred to as the *root index* of the Zadoff-Chu sequence. For a given sequence length M, the number of root indices generating unique Zadoff-Chu sequences equals the number of integers that are relative prime to M. For this reason, Zadoff-Chu sequences of prime length are of special interest as they maximize the number of available Zadoff-Chu sequences. More specifically, assuming M being a prime number, there are M -1 unique Zadoff-Chu sequences of length M.

It can easily be shown that the discrete Fourier transform of a Zadoff-Chu sequence is also a Zadoff-Chu sequence.[f] From Eq. (8.1) it is obvious that a Zadoff-Chu sequence has constant amplitude. This is then true for both the time and frequency domain. Thus, if one applies a Zadoff-Chu sequence in the frequency domain the resulting signal will have a flat spectrum as well as a constant power in the time domain. The constant power in the time domain will make the signal good from a power-amplifier-efficiency point of-view. At the same time, the flat spectrum, which corresponds to zero cyclic autocorrelation for any non-zero cyclic shift in time, implies that two different time-domain cyclic shifts of the same Zadoff-Chu sequence are orthogonal to each other.

Although Zadoff-Chu sequences of prime length are preferred in order to maximize the number of available sequences, SRS sequences cannot be of prime length as the sequence length has to match the number of resource elements in the set of resource blocks over which the SRS transmission is to take place. In practice this implies that the length of the SRS sequence will be a multiple of twelve. The SRS sequences are therefore *extended* Zadoff-Chu sequences based on the longest prime-length

[f] The inverse obviously holds as well, that is, the inverse DFT of a Zadoff-Chu sequence is also a Zadoff-Chu sequence.

Zadoff-Chu sequence with a length M smaller or equal to the desired SRS-sequence length. The sequence is then cyclically extended in the frequency domain up to the desired SRS sequence length. As the extension is done in the frequency domain, the extended sequence still has constant spectrum, and thus "perfect" cyclic autocorrelation, but the time-domain amplitude properties will be somewhat degraded.

Extended Zadoff-Chu sequences are used as SRS sequences for sequence lengths of 36 or larger, corresponding to an SRS extending over 6 and 12 resource blocks in case of comb-2 and comb-4, respectively. For shorter sequence lengths, special flat-spectrum sequences with good time-domain envelope properties have been found from computer search. The reason is that, for shorter sequences, there would not be sufficient number of Zadoff-Chu sequences available.

The same principle of cyclic extension in the frequency domain will be used also for other cases where Zadoff-Chu sequences are used within the NR specifications, for example, for uplink DM-RS (see Section 9.11.1).

8.3.2 Multi-port SRS resource

In the case of an SRS resources with more than one antenna port, the different ports share the same set of resource elements and the same basic SRS sequence (same root index), at least in case of comb-2 and comb-4. Different phase rotations are then applied to separate the different ports as illustrated in Fig. 8.13. For the special case of comb-8 and four antenna ports, the antenna ports are pair-wise separated by different phase rotations. However, within a pair of antenna ports, the antenna ports use the same phase rotation and are, instead, separated in the frequency domain by using different offsets within the comb.

As described, applying a phase rotation in the frequency domain is equivalent to applying a cyclic shift in the time domain. In the NR specification the above operation is actually referred to as "cyclic shift", although it is mathematically described as a frequency-domain phase shift.

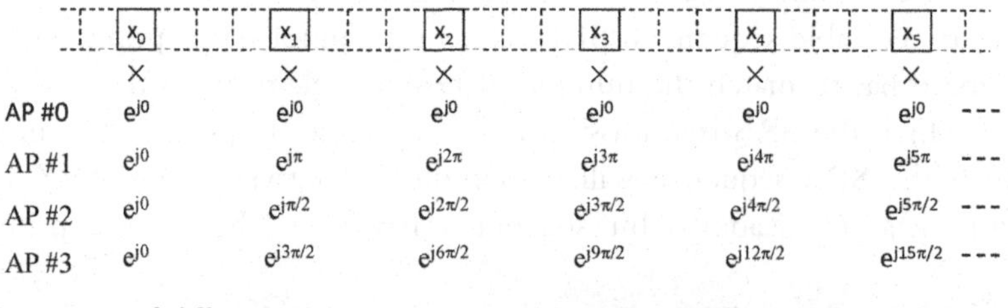

	x_0	x_1	x_2	x_3	x_4	x_5	
	$\times$	$\times$	$\times$	$\times$	$\times$	$\times$	
AP #0	e^{j0}	e^{j0}	e^{j0}	e^{j0}	e^{j0}	e^{j0}	---
AP #1	e^{j0}	$e^{j\pi}$	$e^{j2\pi}$	$e^{j3\pi}$	$e^{j4\pi}$	$e^{j5\pi}$	---
AP #2	e^{j0}	$e^{j\pi/2}$	$e^{j2\pi/2}$	$e^{j3\pi/2}$	$e^{j4\pi/2}$	$e^{j5\pi/2}$	---
AP #3	e^{j0}	$e^{j3\pi/2}$	$e^{j6\pi/2}$	$e^{j9\pi/2}$	$e^{j12\pi/2}$	$e^{j15\pi/2}$	---

Fig. 8.13 Separation of different SRS antenna ports by applying different phase shifts to the basic frequency-domain SRS sequence x_0, x_1, x_2, ... The figure assumes a comb-4 SRS.

8.3.3 SRS frequency hopping

Ideally, a single SRS transmission should cover the entire frequency band of interest. However, due to a limited device transmit power, this may in some cases, for example, for very wide SRS bandwidths and/or a high path loss, lead to a too low received power spectral density and corresponding low-quality channel measurements.

To enable higher instantaneous power spectral density, NR supports *SRS frequency hopping* where the SRS transmission, in each time instant, covers only a limited frequency range and then "hops" in frequency to cover the entire frequency band of interest. As illustrated in Fig. 8.14, the hopping can be done on slot basis (inter-slot hopping) or within a slot (intra-slot hopping).

The SRS frequency hopping introduced in release 15, only supports a minimum per-hop bandwidth of four resource blocks. To allow for even more narrow instantaneous SRS bandwidth, release 17 introduced the concept of *partial SRS transmission*. With partial SRS transmission, the instantaneous bandwidth can be reduced by a frequency scaling factor P_F that can take the values two or four, thereby allowing for an instantaneous SRS bandwidth as small as a single resource block. Although partial SRS transmission is formally seen as separate and a complement to SRS frequency hopping, it can, in practice, be seen as frequency hopping with a minimum per-hop bandwidth of one resource block, rather than four resource blocks.

8.3.4 Time-domain structure of SRS transmissions

Similar to CSI-RS, an SRS resource can be configured for *periodic, semi-persistent*, or *aperiodic* transmission:

- in case of a periodic SRS resource, SRS transmission takes place with a certain configured periodicity and slot offset;
- in case of semi-persistent SRS there is also a configured periodicity/offset. However, actual SRS transmission is activated and deactivated by means of MAC CE signaling;
- in case of aperiodic SRS, each SRS transmission is explicitly triggered by means of DCI signaling.

In case of aperiodic SRS transmission, release 17 introduced some flexibility in the sense that the triggering DCI now includes a time offset ("SRS offset indicator", see

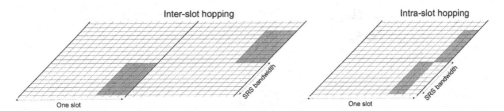

Fig. 8.14 SRS frequency hopping.

Tables 10.2 and 10.3 in Chapter 10) that indicates exactly in what slot the SRS is to be transmitted, given the slot in which the triggering DCI is received. In earlier releases, this timing relation had a fixed specified value.

It should be pointed out that, similar to CSI-RSI, activation/deactivation and triggering for semi-persistent and aperiodic transmission of SRS respectively is actually not done for a specific SRS resource but rather done a *SRS resource set* which, in the general case, includes (pointers to) multiple configured SRS resources (see below).

8.3.5 SRS resource sets

Similar to CSI-RS, a device can be configured with one or several *SRS resource sets* where each resource set includes one or several configured SRS. As described earlier, an SRS can be configured for periodic, semi-persistent, or aperiodic transmission. All SRS included within a configured SRS resource set have to be of the same type. In other words, periodic, semi-persistent, or aperiodic transmission can also be seen as a property of an SRS resource set.

A device can be configured with multiple SRS resource sets that can be used for different purposes, including both downlink and uplink multi-antenna precoding and downlink and uplink beam management

The transmission of aperiodic SRS, or more accurately, transmission of the set of configured SRS included in an aperiodic SRS resource set, is triggered by DCI. More specifically, as will be described in Sections 10.1.4 and 10.1.5, some DCI formats, including both downlink scheduling assignment and uplink scheduling grants, include a 2-bit *SRS-request* that can trigger the transmission of one out of three different aperiodic SRS resource sets configured for the device (the fourth bit combination corresponds to "no triggering").

8.3.6 Mapping to physical antennas

Similar to CSI-RS, SRS ports are often not mapped directly to the device physical antennas but via some spatial filter F that maps M SRS ports to N physical channels, see Fig. 8.15

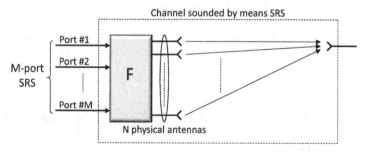

Fig. 8.15 SRS applied to spatial filter (F) before mapping to physical antennas.

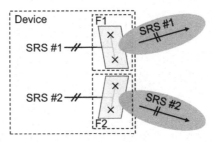

Fig. 8.16 Different spatial filters applied to different SRS.

In order to provide connectivity regardless of the rotational direction of the device, NR devices supporting high-frequency operation will typically include multiple antenna panels pointing in different directions. The mapping of SRS to one such panel is an example of a transformation F from SRS antenna ports to the set of physical antennas. Transmission from different panels will then correspond to different spatial filters F as illustrated in Fig. 8.16.

Similar to the downlink the spatial filtering F has a real impact despite the fact that it is never explicitly visible to the network receiver but just seen as an integrated part of the overall channel. As an example, the network may sound the channel based on a device-transmitted, SRS and then decide on a precoder matrix that the device should use for uplink transmission. The device is then assumed to use that precoder matrix *in combination with the spatial filter F applied to the SRS*. In other cases, a device may be explicitly scheduled for data transmission using the antenna ports defined by a certain SRS. In practice this implies that the device is assumed to transmit using the same spatial filter F that has been used for the SRS transmission. In practice, this may imply that the device should transmit using the same beam or panel that has been used for the SRS transmission.

CHAPTER 9

Transport-channel processing

This chapter will provide a more detailed description of the downlink and uplink physical-layer functionality such as coding, modulation, multi-antenna precoding, resource-block mapping, and reference signal structure.

9.1 Overview

The physical layer provides services to the MAC layer in the form of transport channels as described in Section 6.4.5. In the downlink, there are three different types of transport channels defined for NR: the Downlink Shared Channel (DL-SCH), the Paging Channel (PCH), and the Broadcast Channel (BCH) although the latter two are not used in non-stand-alone operation. In the uplink, there is only one uplink transport-channel type carrying transport blocks in NR,[a] the Uplink Shared Channel (UL-SCH). The overall transport channel processing for NR follows a similar structure as for LTE (see Fig. 9.1). The processing is mostly similar in uplink and downlink and the structure in Fig. 9.1 is applicable for the DL-SCH, BCH, and PCH in the downlink, and the UL-SCH in the uplink. The part of the BCH that is mapped to the PBCH follows a different structure, described in Chapter 16, as does the RACH, described in Chapter 17.

Within each *transmission time interval* (TTI), up to two transport blocks of dynamic size are delivered to the physical layer and transmitted over the radio interface for each component carrier. Two transport blocks are only used in the case of spatial multiplexing with more than four layers, which as a baseline is supported in the downlink direction only and mainly useful in scenarios with very high signal-to-noise ratios.[b] Hence, at most a single transport block per component carrier and TTI is a typical case in practice.

A CRC for error-detecting purposes is added to each transport block, followed by error-correcting coding using LDPC codes. Rate matching, including physical-layer hybrid-ARQ functionality, adapts the number of coded bits to the scheduled resources. The coded bits are scrambled and fed to a modulator, and finally the modulation symbols are mapped to the physical resources, including the spatial domain. For the uplink there is also a possibility of a DFT-precoding. The differences between uplink

---

[a] Strictly speaking, the Random-Access Channel is also defined as a transport-channel type, see Chapter 17. However, RACH only includes a layer-1 preamble and carries no data in form of transport blocks.

[b] In release 18, more than four layers are supported also in the uplink.

5G/5G-Advanced
https://doi.org/10.1016/B978-0-443-13173-8.00010-4

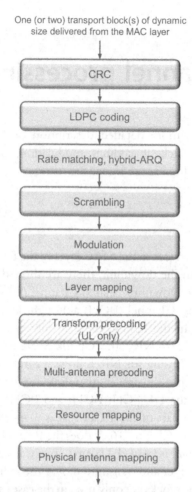

One (or two) transport block(s) of dynamic
size delivered from the MAC layer

CRC

LDPC coding

Rate matching, hybrid-ARQ

Scrambling

Modulation

Layer mapping

Transform precoding
(UL only)

Multi-antenna precoding

Resource mapping

Physical antenna mapping

Fig. 9.1 General transport-channel processing.

and downlink is, apart from DFT-precoding being possible in the uplink only, mainly around multi-antenna mapping and the associated reference signals.

In the following, each of the processing steps will be discussed in more detail. For carrier aggregation, the processing steps are duplicated for each of the carriers and the description herein is applicable to each of the carriers. Since most of the processing steps are identical for uplink and downlink, the processing will be described jointly and any differences between uplink and downlink explicitly mentioned when relevant.

9.2 Channel coding

An overview of the channel coding steps is provided in Fig. 9.2 and described in more detail in the following sections. First, a CRC is attached to the transport block to facilitate

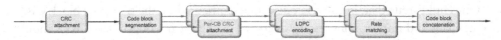

Fig. 9.2 Channel coding.

error detection, followed by code block segmentation. Each code block is LDPC-encoded and rate matched separately, including physical-layer hybrid-ARQ processing, and the resulting bits are concatenated to form the sequence of bits representing the coded transport block.

9.2.1 CRC attachment per transport block

In the first step of the physical-layer processing, a CRC is calculated for and appended to each transport block. The CRC allows for receiver-side detection of errors in the decoded transport block and can, for example, be used by the hybrid-ARQ protocol as a trigger for requesting retransmissions.

The size of the CRC depends on the transport-block size. For transport blocks larger than 3824 bits, a 24-bit CRC is used, otherwise a 16-bit CRC is used to reduce overhead.

9.2.2 Code-block segmentation

The LDPC coder in NR is defined up to a certain code-block size (8424 bits for base graph 1 and 3840 bits for base graph 2). To handle transport block sizes larger than this, code-block segmentation is used where the transport block, including the CRC, is split into multiple equal-sized[c] code blocks as illustrated in Fig. 9.3.

As can be seen in Fig. 9.3, code-block segmentation also implies that an additional CRC (also of length 24 bits but based on a different polynomial compared to the transport-block CRC described above) is calculated for and appended to each code block. In the case of a single code block transmission no additional code-block CRC is applied.

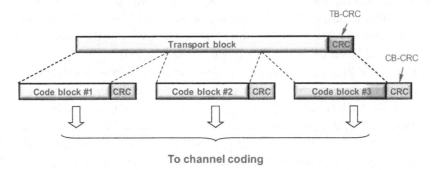

Fig. 9.3 Code block segmentation.

---

[c] The set of possible transport-block sizes are such that it is always possible to split a too large transport block into smaller equally-sized code blocks.

One could argue that, in case of code-block segmentation, the transport-block CRC is redundant and implies unnecessary overhead as the set of code-block CRCs should indirectly provide information about the correctness of the complete transport block. However, to handle *code-block group* (CBG) retransmissions as discussed in Chapter 13, a mechanism to detect errors per code block is necessary. CBG retransmission means that only the erroneous code block groups are retransmitted instead of the complete transport block to improve the spectral efficiency. The per-CB CRC can also be used by the device to limit processing. In case of a retransmission only those CBs whose CRCs did not check after a previous transmissions need to be processed, even if per-CBG retransmission is not configured. This helps reducing the device processing load. The transport-block CRC also adds an extra level of protection in terms of error detection. Note that code-block segmentation is only applied to large transport blocks for which the relative extra overhead due to the additional transport-block CRC is small.

9.2.3 Channel coding

Channel coding is based on LDPC codes, a code design which was originally proposed in the 1960s [32] but forgotten for many years. They were "rediscovered" in the 1990s [56] and found to be an attractive choice from an implementation perspective. From an error-correcting capability point of view, turbo codes, as used in LTE, can achieve similar performance, but LDPC codes can offer lower complexity, especially at higher code rates, and were therefore chosen for NR.

The basis for LDPC codes is a sparse ("low density") parity check matrix H where for each valid code word c the relation $Hc^T=0$ holds. Designing a good LDPC code to a large extent boils down to finding a good parity check matrix H which is sparse (the sparseness implies relatively simple decoding). It is common to represent the parity-check matrix by a graph connecting n variable nodes at the top with $(n-k)$ constraint nodes at the bottom of the graph, a notation that allows a wide range of properties of an (n, k) LDPC code to be analyzed. This explains why the term *base graph* is used in the NR specifications. A detailed description of the theory behind LDPC codes is beyond the scope of this book, but there is a rich literature in the field (for example, see [64]).

Quasi-cyclic LDPC codes with a dual-diagonal structure of the kernel part of the parity check matrix are used in NR, which gives a decoding complexity which is linear in the number of coded bits and enables a simple encoding operation. Two base graphs are defined, BG1 and BG2, representing the two base matrices. The reason for two base graphs instead of one is to handle the wide range of payload sizes and code rates in an efficient way. Supporting a very large payload size at a medium to high code rate, which is the case for very high data rates, using a code designed to support a very low code rate is not efficient. At the same time, the lowest code rates are necessary to provide good performance in challenging situations. In NR, BG1 is designed for code rates from 1/3 to

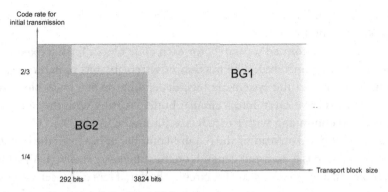

Fig. 9.4 Selection of base graph for the LDPC code.

22/24 (approximately 0.33 to 0.92) and BG 2 from 1/5 to 5/6 (approximately 0.2 to 0.83). Through puncturing, the highest code rate can be increased somewhat, up to 0.95, beyond which the device is not required to decode. The choice between BG1 and BG2 is based on the transport block size and code rate targeted for the first transmission (see Fig. 9.4).

The base graphs, and the corresponding base matrices, define the general structure of the LDPC code. To support a range of payload sizes, 51 different *lifting sizes* and sets of *shift coefficients* are defined and applied to the base matrices. In short, for a given lifting size Z, each '1' in the base matrix is replaced by the $Z \times Z$ identify matrix circularly shifted by the corresponding shift coefficient and each '0' in the base matrix is replaced by the $Z \times Z$ all-zero matrix. Hence, a relatively large number of parity-check matrices can be generated to support multiple payload sizes while maintaining the general structure of the LDPC code. To support payload sizes that are not a native payload size of one of the 51 defined parity check matrices, known filler bits can be appended to the code block before encoding. Since the NR LDPC codes are systematic codes, the filler bits can be removed before transmission.

9.3 Rate matching and physical-layer hybrid-ARQ functionality

The rate-matching and physical-layer hybrid-ARQ functionality serves two purposes, namely to extract a suitable number of coded bits to match the resources assigned for transmission and to generate different redundancy versions needed for the hybrid-ARQ protocol. The number of bits to transmit on the PDSCH or PUSCH depends on a wide range of factors, not only the number of resource blocks and the number of OFDM symbols scheduled, but also on the amount of overlapping resource elements used for other purposes and such as reference signals, control channels, or system information. There is also a possibility to, in the downlink, define *reserved resources* as a tool to provide future compatibility (see Section 9.10), which affects the number of resource elements usable for the PDSCH.

Rate matching is performed separately for each code block. First, a fixed number of the systematic bits are punctured. The fraction of systematic bits punctured can be relatively high, up to 1/3 of the systematic bits, depending on the code block size. The remaining coded bits are written into a circular buffer, starting with the non-punctured systematic bits and continuing with parity bits as illustrated in Fig. 9.5. The selection of the bits to transmit is based on reading the required number of bits from the circular buffer where the exact set of bits to transmit depends on the *redundancy version* (RV) corresponding to different starting positions in the circular buffer. Hence, by selecting different redundancy versions, different sets of coded bits representing the same set of information bits can be generated, which is used when implementing hybrid-ARQ with incremental redundancy. The starting points in the circular buffer are defined such that both RV0 and RV3 are self-decodable, that is, includes the systematic bits under typical scenarios. This is also the reason RV3 is located after "nine o'clock" in Fig. 9.5 as this allows more of the systematic bits to be included in the transmission.

In the receiver, *soft combining* is an important part of the hybrid-ARQ functionality as described in Section 13.1. The soft values representing the received coded bits are buffered and, if a retransmission occurs, decoding is performed using the buffered bits combined with the retransmitted coded bits. In addition to a gain in accumulated received E_b/N_0, with different coded bits in different transmission attempts, additional parity bits are obtained and the resulting code rate after soft combining is lower with a corresponding coding gain obtained.

Soft combining requires a buffer in the receiver. Typically, a fairly high probability of successful transmission on the first attempt is targeted and hence the soft buffer remains

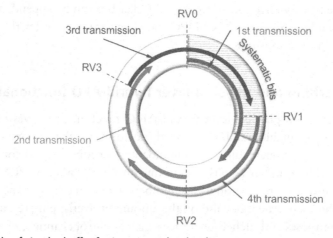

Fig. 9.5 Example of circular buffer for incremental redundancy.

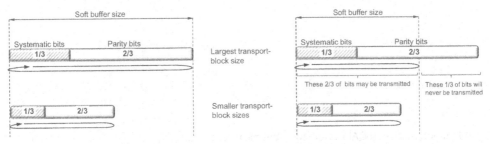

Fig. 9.6 Limited-buffer rate matching.

unused most of the time. Since the soft buffer size is fairly large for the largest transport block sizes, requiring the receiver to buffer all soft bits even for the largest transport block sizes is suboptimal from a cost-performance tradeoff perspective. Hence, limited-buffer rate-matching is supported as illustrated in Fig. 9.6. In principle, only bits the device can buffer are kept in the circular buffer, that is, the size of the circular buffer is determined based on the receiver's soft buffering capability.

For the downlink, the device is not required to buffer more soft bits than corresponding to the largest transport block size coded at rate 2/3. Note that this only limits the soft buffer capacity for the highest transport block sizes, that is, the highest data rates. For smaller transport block sizes, the device is capable of buffering all soft bits down to the mother code rate.

For the uplink, full-buffer rate matching, where all soft bits are buffered irrespective of the transport block size is supported given sufficient gNB memory. Limited-buffer rate matching using the same principles as for the downlink can be configured using RRC signaling.

The final step of the rate-matching functionality is to interleave the bits using a block interleaver and to collect the bits from each code block. The bits from the circular buffer are written row-by-row into a block interleaver and read out column-by-column. The number of rows in the interleaver is given by the modulation order and hence the bits in one column corresponds to one modulation symbol[d] (see Fig. 9.7). This results in the systematic bits spread across the modulation symbols, which improves performance. Bit collection concatenates the bits for each code block.

9.4 Scrambling

Scrambling is applied to the block of coded bits delivered by the hybrid-ARQ functionality by multiplying the sequence of coded bits with a bit-level *scrambling sequence*. Without scrambling, the channel decoder at the receiver could, at least in principle, be equally

[d] This structure improves the performance for higher-order modulation.

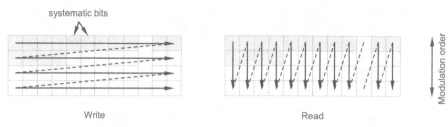

Fig. 9.7 Bit interleaver (16QAM assumed in this example).

matched to an interfering signal as to the target signal, thus being unable to properly suppress the interference. By applying different scrambling sequences for neighboring cells in the downlink or for different devices in the uplink, the interfering signal(s) after descrambling is (are) randomized, ensuring full utilization of the processing gain provided by the channel code.

The scrambling sequence in both downlink (PDSCH) and uplink (PUSCH) depends on the identity of the device, that is, the C–RNTI, and a *data scrambling identity* configured in each device. If no data scrambling identity is configured, the physical layer cell identity is used as a default value to ensure that neighboring devices, both in the same cell and between cells, use different scrambling sequences. Furthermore, in case of two transport blocks being transmitted in the downlink to support more than four layers, different scrambling sequences are used for the two transport blocks.

9.5 Modulation

The modulation step transforms the block of scrambled bits to a corresponding block of complex modulation symbols. The modulation schemes supported include QPSK, 16QAM, 64QAM, and 256QAM in both uplink and downlink. In addition, for the uplink $\pi/2$-BPSK is supported in the case the DFT-precoding is used, motivated by a reduced cubic metric [57] and hence improved power-amplifier efficiency, in particular for coverage limited scenarios. Note that $\pi/2$-BPSK is neither supported nor useful in the absence of DFT-precoding as the cubic metric in this case is dominated by the OFDM waveform. In release 17, the set of downlink modulation schemes supported is extended with 1024-QAM in FR1.

9.6 Layer mapping

The purpose of the layer-mapping step is to distribute the modulation symbols across the different transmission layers. This is done by mapping every n:th symbol to the n:th layer. One coded transport block can be mapped on up to four layers. In the case of five to eight layers, a second transport block is mapped to layers five to eight following the same principle as for the first transport block.

Multi-layer transmission is only supported in combination with OFDM, the baseline waveform in NR. With DFT-precoding in the uplink, only a single transmission layer is supported. This is motivated both by the receiver complexity, which in the case of multi-layer transmission would be significantly higher with a DFT-precoder than without, and the use case originally motivating the additional support of DFT-precoding, namely handling of coverage-limited scenarios. In such a scenario, the received signal-to-noise ratio is too low for efficient usage of spatial multiplexing and there is no need to support spatial multiplexing from a single device.

9.7 Uplink DFT precoding

DFT precoding can be configured in the uplink only.[e] In the downlink, as well as the case of OFDM in the uplink, the step is transparent.

In the case that DFT-precoding is applied in the uplink, blocks of M symbols, are fed through a size-M DFT as illustrated in Fig. 9.8, where M corresponds to the number of subcarriers assigned for the transmission. The reason for the DFT precoding is to reduce the cubic metric for the transmitted signal, thereby enabling higher power-amplifier efficiency. From an implementation complexity point-of-view the DFT size should preferably be constrained to a power of 2. However, such a constraint would limit the scheduler flexibility in terms of the amount of resources that can be assigned for an uplink transmission. Rather, from a flexibility point-of-view all possible DFT sizes should preferably be allowed. For NR, a middle-way has been adopted where the DFT size, and thus also the size of the resource allocation, is limited to products of the integers 2, 3, and 5. Thus, for example, DFT sizes of 60, 72, and 96 are allowed but a DFT size of 84 is not allowed.[f]

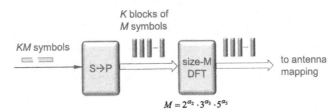

Fig. 9.8 DFT-precoding.

[e] Release 18 introduces the possibility to *dynamically* switch between OFDM and DFT-precoded OFDM as part of the uplink scheduling grant.
[f] As uplink resource assignments are always done in terms of resource blocks of size 12 subcarriers, the DFT size is always a multiple of 12.

In this way, the DFT can be implemented as a combination of relatively low-complex radix-2, radix-3, and radix-5 FFT processing.

9.8 Multi-antenna precoding

The purpose of multi-antenna precoding is to map the different transmission layers to a set of antenna ports using a precoder matrix. In NR, the precoding and multi-antenna operation differs between downlink and uplink. From a specification perspective, the codebook-based precoding step is, except for CSI reporting, only visible in the uplink direction. For a detailed discussion on how the precoding step is used to realize beamforming and different multi-antenna schemes see Chapters 11 and 12.

9.8.1 Downlink precoding

In the downlink, the demodulation reference signal (DM-RS) used for channel estimation is subject to the same precoding as the PDSCH (see Fig. 9.9). Thus, the precoding is not explicitly visible to the receiver but is seen as part of the overall channel. This is similar to the receiver-transparent spatial filtering discussed in the context of CSI-RS and SRS in Chapter 8. In essence, in terms of actual downlink transmission, any multi-antenna precoding can be seen as part of such, to the device transparent, spatial filtering.

However, for the purpose of CSI reporting, the device may assume that a specific precoding matrix W is applied at the network side. The device is then assuming that the precoder maps the signal to the antenna ports of the CSI-RS used for the measurements on which the reporting was done. The network is still free to use whatever precoder it finds advantageous for data transmission.

To handle receiver-side beamforming, or in general multiple reception antennas with different spatial characteristics, QCL relations between a DM-RS port group, which is

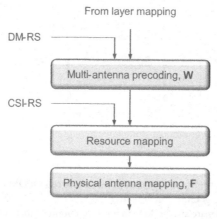

Fig. 9.9 Downlink precoding.

the antenna ports used for PDSCH transmission[g], and the antenna ports used for CSI-RS or SS block transmission can be configured. The *Transmission Configuration Index* (TCI) provided as part of the scheduling assignment indicates the QCL relations to use, or in other words, which reception beam to use. This is described in more detail in Chapter 12.

Demodulation reference signals are, as discussed in Section 9.11, transmitted in the scheduled resource blocks and it is from those reference signals that the device can estimate the channel, including any precoding W and spatial filtering F applied for PDSCH. In principle, knowledge about the correlation between reference signal transmissions, both in terms of correlation introduced by the radio channel itself and correlation in the use of precoder, is useful to know and can be exploited by the device to improve the channel estimation accuracy.

In the time domain, the device is not allowed to make any assumptions on the reference signals being correlated between PDSCH scheduling occasions. This is necessary to allow full flexibility in terms of beamforming and spatial processing as part of the scheduling process.

In the frequency domain, the device can be given some guidance on the correlation. This is expressed in the form of *physical resource-block groups* (PRGs). Over the frequency span of one PRG, the device may assume the downlink precoder remains the same and may exploit this in the channel-estimation process, while the device may not make any assumptions in this respect between PRGs. From this it can be concluded that there is a trade-off between the precoding flexibility and the channel-estimation performance – a large PRG size can improve the channel-estimation accuracy at the cost of precoding flexibility and vice versa. Hence, the gNB may indicate the PRG size to the device where the possible PRG sizes are two resource blocks, four resource blocks, or the scheduled bandwidth as shown in the bottom of Fig. 9.10. A single value may be configured, in which case this value is used for the PDSCH transmissions. It is also possible to dynamically, through the DCI, indicate the PRG size used. In addition, the device can be configured to assume that the PRG size equals the scheduled bandwidth in the case that the scheduled bandwidth is larger than half the bandwidth part.

9.8.2 Uplink precoding

Similar to the downlink, uplink demodulation reference signals used for channel estimation are subject to the same precoding as the uplink PUSCH. Thus, also for the uplink the precoding is not directly visible from a receiver perspective but is seen as part of the overall channel (see Fig. 9.11).

However, from a scheduling point-of-view, the multi-antenna precoding of Fig. 9.1 is visible in the uplink as the network may provide the device with a specific precoder

---

[g] For multi-TRP support in release 16, *two* DM-RS port groups can be used. Some of the PDSCH layers belong to one DM-RS port group and the other layers to the other DM-RS port group.

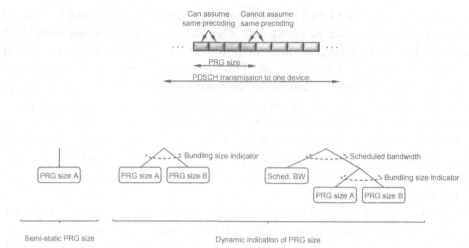

Fig. 9.10 Physical resource-block groups (top) and indication thereof (bottom).

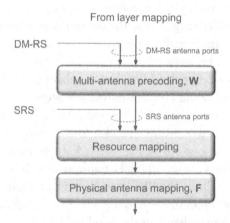

Fig. 9.11 Uplink precoding.

matrix W the receiver should to use for the PUSCH transmission. This is done through the *precoding information* and *antenna port* fields in the DCI. The precoder is then assumed to map the different layers to the antenna ports of a configured SRS indicated by the network. In practice this will be the same SRS as the network used for the measurement on which the precoder selection was made. This is known as *codebook-based* precoding since the precoder W to use is selected from a codebook of possible matrices and explicitly signaled. Note that the spatial filter F selected by the device also can be seen as a precoding operation, although not explicitly controlled by the network. The network can however restrict the freedom in the choice of F through the *SRS resource indicator* (SRI) provided as part of the DCI.

There is also a possibility for the network to operate with *non-codebook-based* precoding. In this case W is equal to the identity matrix and precoding is handled solely by the spatial filter F based on recommendations from the device.

Both codebook-based and non–codebook-based precoding are described in detail in Chapter 11.

9.9 Resource mapping

The resource-block mapping takes the modulation symbols to be transmitted on each antenna port and maps them to the set of available resource elements in the set of resource blocks and OFDM symbols assigned by the MAC scheduler for the transmission. As described in Section 7.3, a resource block is 12 subcarriers wide and typically multiple resource blocks and multiple OFDM symbols are used for the transmission. The set of time-frequency resources used for transmission is determined by the scheduler. However, some or all of the resource elements within the scheduled resource blocks may not be available for the transport-channel transmission as they are used for:

- Demodulation reference signals (potentially including reference signals for *other* co-scheduled devices in case of multi-user MIMO) as described in Section 9.11;
- Other types of reference signals such as CSI-RS and SRS (see Chapter 8);
- Downlink L1/L2 control signaling (see Chapter 10);
- Synchronization signals and system information as described in Chapter 16;
- Downlink reserved resources as a means to provide forward compatibility as described in Section 9.10.

The time-frequency resources to be used for transmission are signaled by the scheduler as set of *virtual resource blocks* and a set of OFDM symbols. The time-domain resources – that is, the OFDM symbols – may span up to one slot. This does not restrict transmissions to slots – a transmission can in principle start at any OFDM symbol – but as a baseline the transmission ends no later than at the slot boundary. In release 17, the possibility for mapping an uplink transport block over multiple slots (sometimes referred to with the abbreviation TBoMS) is introduced. The purpose is to improve uplink coverage by extending the transmission duration and enable channel estimation spanning across multiple slots.

The modulation symbols are mapped to resource elements in the scheduled resource in a frequency first, time second manner. The frequency-first, time-second mapping is chosen to achieve low latency and allows both the transmitter and receiver to process the data "on the fly." For high data rates, there are multiple code blocks in each OFDM symbol and the device can decode those received in one symbol while receiving the next OFDM symbol. Similarly, assembling an OFDM symbol can take place while transmitting the previous symbols, thereby enabling a pipelined implementation. This would not be possible in the case of a time-first mapping as the complete slot needs to be prepared before the transmission can start.

The virtual resource blocks containing the modulation symbols are mapped to *physical resource blocks* in the bandwidth part used for transmission. Depending on the bandwidth part used for transmission, the *common resource blocks* can be determined and the exact frequency location on the carrier determined (see Fig. 9.12 for an illustration). The reason for this, at first sight somewhat complicated, mapping process with both virtual and physical resource blocks is to be able to handle a wide range of scenarios.

There are two methods for mapping virtual resource blocks to physical resource blocks, non-interleaved mapping (Fig. 9.12: top) and interleaved mapping (Fig. 9.12: bottom). The mapping scheme to use can be controlled on a dynamic basis using a bit in the DCI scheduling the transmission.

Non-interleaved mapping means that a virtual resource block in a bandwidth part maps directly to the physical resource block in the same bandwidth part. This is useful in cases when the network tries to allocate transmissions to physical resource with instantaneously favorable channel conditions. For example, the scheduler might have determined that physical resource blocks six to nine in Fig. 9.12 have favorable radio channel properties and are therefore preferred for transmission and a non-interleaved mapping is used.

Interleaved mapping maps virtual resource blocks to physical resource blocks using an interleaver spanning the whole bandwidth part and operating on pairs or quadruplets of resource blocks. A block interleaver with two rows is used, with pairs/quadruplets of

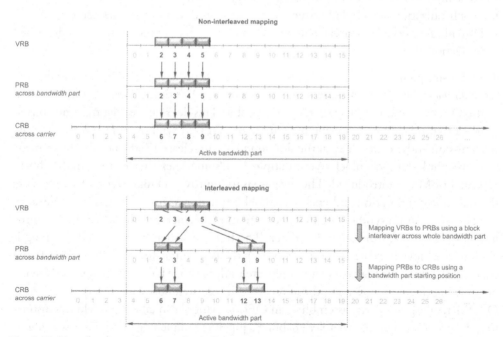

Fig. 9.12 Mapping from virtual to physical to carrier resource blocks.

resource blocks written column-by-column and read out row-by-row. Whether to use pairs or quadruplets of resource blocks in the interleaving operation is configurable by higher-layer signaling.

The reason for interleaved resource-block mapping is to achieve frequency diversity, the benefits of which can be motivated separately for small and large resource allocations.

For small allocations, for example voice services, channel-dependent scheduling may not be motivated from an overhead perspective due to the amount of feedback signaling required, or may not be possible due to channel variations not being possible to track for a rapidly moving device. Frequency diversity by distributing the transmission in the frequency domain is in such cases an alternative way to exploit channel variations. Although frequency diversity could be obtained by using resource *allocation type 0* (see Section 10.1.10), this resource allocation scheme implies a relatively large control signaling overhead compared to the data payload transmitted as well as limited possibilities to signal very small allocations. Instead, by using the more compact *resource allocation type 1*, which is only capable of signaling contiguous resource allocations, combined with an interleaved virtual to physical resource block mapping, frequency diversity can be achieved with a small relative overhead. Since resource allocation type 0 can provide a high degree of flexibility in the resource allocation, interleaved mapping is supported for resource allocation type 1 only.

For larger allocations, possibly spanning the whole bandwidth part, frequency diversity can still be advantageous. In the case of a large transport block, that is, at very high data rates, the coded data are split into multiple code blocks as discussed in Section 9.2.2. Mapping the coded data directly to physical resource blocks in a frequency-first manner (remember, frequency-first mapping is beneficial from an overall latency perspective) would result in each code block occupying only a fairly small number of contiguous physical resource blocks. Hence, if the channel quality varies across the frequency range used for transmission, some code blocks may suffer worse quality than other code blocks, possibly resulting in the overall transport block failing to decode despite almost all code blocks being correctly decoded. The quality variations across the frequency range may occur even if the radio channel is flat due to imperfections in RF components. If an interleaved resource-block mapping is used, one code block occupying a contiguous set of virtual resource blocks would be distributed in the frequency domain across multiple, widely separated physical resource blocks, similar to what is the case for the small allocations discussed in the previous paragraph. The result of the interleaved VRB-to-PRB mapping is a quality-averaging effect across the code blocks, resulting in a higher likelihood of correctly decoding very large transport blocks.

The discussion holds in general and in particular for the downlink. In the uplink, RF requirements are specified for contiguous allocations only and therefore interleaved mapping is only supported for downlink transmissions. To obtain frequency diversity also in the uplink, frequency hopping can be used where the data in the first set of OFDM symbols in the slot are transmitted on the resource block as indicated by the scheduling grant. In the remaining OFDM symbols, data are transmitted on a different set of resource blocks given by a configurable offset from the first set. Uplink frequency hopping can be dynamically controlled using a bit in the DCI scheduling the transmission.

9.10 Downlink reserved resources

One of the key requirements on NR was to ensure forward compatibility, that is, to allow future extensions and technologies to be introduced in a simple way without causing backward-compatibility problems with, at that point in time, already deployed NR networks. Several NR technology components contribute to meeting this requirement, but the possibility to define *reserved resources* in the downlink is one of the more important tools. Reserved resources are semi-statically configured time-frequency resources around which the PDSCH can be rate matched.

Reserved resources can be configured in three different ways:

- By referring to an LTE carrier configuration, thereby allowing for transmissions on an NR carrier deployed on top of an LTE carrier (LTE/NR spectrum co-existence) to avoid the cell-specific reference signals of the LTE carrier (see further details in Chapter 18);
- By referring to a CORESET;
- By configuring resource sets using a set of bitmaps.

There are no reserved resources in the uplink; avoiding transmission on certain resources can be achieved through scheduling.[h]

Configuring reserved resources by referring to a configured CORESET is used to dynamically control whether control signaling resources can be reused for data or not (see Section 10.1.2). In this case the reserved resource is identical to the CORESET configured and the gNB may dynamically indicate whether these resources are usable for PDSCH or not. Thus, reserved resources do not have to be periodically occurring but can be used when needed.

The third way to configure reserved resources is based on bitmaps. The basic building block for a resource-set configuration covers one or two slots in the time domain and can be described by two bitmaps as illustrated in Fig. 9.13:

- A first time-domain bitmap, which in the NR specifications is referred to as "bitmap-2," indicates a set of OFDM symbols within the slot (or within a pair two slots).

---

[h] One reason is that only frequency-contiguous allocations are supported in the uplink, resulting in that 'bitmap-1' cannot be used as this may result in non-contiguous frequency-domain allocations.

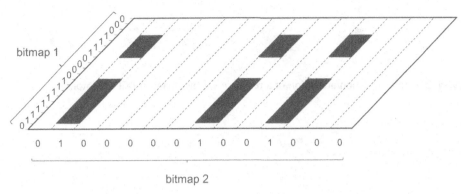

Fig. 9.13 Configuring reserved resources.

- Within the set of OFDM symbols indicated by bitmap-2, an arbitrary set of resource blocks, that is, blocks of 12 resource elements in the frequency domain, may be reserved. The set of resource blocks is indicated by a second bitmap, in the NR specifications referred to as "bitmap-1."

If the resource set is defined on a carrier level, bitmap-1 has a length corresponding to the number of resource blocks within the carrier. If the resource set is bandwidth-part specific, the length of bitmap-1 is given by the bandwidth of the bandwidth part.

The same bitmap-1 is valid for all OFDM symbols indicated by bitmap-2. In other words, the same set of resource elements are reserved in all OFDM symbols indicated by bitmap-2. Furthermore, the frequency-domain granularity of the resource-set configuration provided by bitmap-1 is one resource block. In other words, all resource elements within a (frequency-domain) resource block are either reserved or not reserved.

Whether or not the resources configured as reserved resources are actually reserved or can be used for PDSCH can either be semi-statically or dynamically controlled.

In the case of semi-static control, a third bitmap (bitmap-3) determines whether or not the resource-set defined by the bitmap-1/bitmap-2 pair or the CORSET is valid for a certain slot or not. The bitmap-3 has a granularity equal to the length of bitmap-2 (either one or two slots) and a length of 40 slots. In other words, the overall time-domain periodicity of a semi-static resource set is defined by the triplet {bitmap-1, bitmap-2, bitmap-3} is 40 slots in length.

In the case of dynamic activation of a rate-matching resource set, an indicator in the scheduling assignment indicates if the semi-statically configured pattern is valid or not for a certain dynamically scheduled transmission. Note that, although Fig. 9.14 assumes scheduling on a slot basis, dynamic indication is equally applicable to transmission durations shorter than a slot. The indicator in the DCI should not be seen as corresponding to a certain slot. Rather, it should be seen as corresponding to a certain scheduling

Fig. 9.14 Dynamic activation/activation of a resource set by means of DCI indicator.

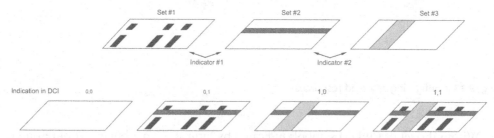

Fig. 9.15 Dynamic activation/activation in case of multiple configured resource sets.

assignment. What the indicator does is simply indicate if, for a given scheduling assignment defined by a given DCI, a configured resource set should be assumed active or not during the time over which the assignment is valid.

In the general case, a device can be configured with up to eight different resource sets. Each resource set is configured either by referring to a CORSEST or by using the bitmap approach described here. By configuring more than one resource-set configuration, more elaborate patterns of reserved resources can be realized as illustrated in Fig. 9.15.

Although a device can be configured with up to eight different resource-set configurations each of which can be configured for dynamic activation, the configurations cannot be independently activated in the scheduling assignment. Rather, to maintain a reasonable overhead, the scheduling assignment includes at most two indicators. Each resource set configured for dynamic activation is assigned to either one or both of these indications and jointly activates or disables all resource set assigned to that indicator. Fig. 9.15 illustrates an example with three configured resources sets where resource set #1 and resource set #3 are assigned to indicator #1 and indicator #2 respectively while resource set # 2 is assigned to both indicators. Note that the patterns in Fig. 9.15 are not necessarily realistic, but rather chosen for illustrative purposes.

9.11 Reference signals

Reference signals are predefined signals occupying specific resource elements within the downlink time–frequency grid. The NR specification includes several types of reference signals transmitted in different ways and intended to be used for different purposes by a receiving device.

Unlike LTE, which relies heavily on always-on, cell-specific reference signals in the downlink for coherent demodulation, channel quality estimation for CSI reporting, and general time-frequency tracking, NR uses different downlink reference signals for different purposes. This allows for optimizing each of the reference signals for their specific purpose. It is also in line with the overall principle of ultra-lean transmission as the different reference signals can be transmitted only when needed. Later release of LTE took some steps in this direction, but NR was designed around different reference signals for different purposes from the start.

The NR reference signals include:

- *Demodulation reference signals* (DM-RS) for PDSCH are intended for channel estimation at the device as part of coherent demodulation. They are present only in the resource blocks used for PDSCH transmission. Similarly, the DM-RS for PUSCH allows the gNB to coherently demodulate the PUSCH. The DM-RS for PDSCH and PUSCH is the focus of this section; DM-RS for PDCCH and PBCH are described in Chapters 10 and 16, respectively.
- *Phase-tracking reference signals* (PT-RS) can be seen as an extension to DM-RS for PDSCH/PUSCH and are intended for phase-noise compensation. The PT-RS is denser in time but sparser in frequency than the DM-RS, and, if configured, occurs only in combination with DM-RS. A discussion of the phase-tracking reference signal is found later in this chapter.
- *CSI reference signals* (CSI-RS) are downlink reference signals intended to be used by devices to acquire downlink channel-state information (CSI). Specific instances of CSI reference signals can be configured for time/frequency tracking and mobility measurements. CSI reference signals are described in Section 8.1.
- *Tracking reference signals* (TRS) are sparse reference signals intended to assist the device in time and frequency tracking. A specific CSI-RS configuration serves the purpose of a TRS (see Section 8.1.7).
- *Sounding reference signals* (SRS) are uplink reference signals transmitted by the devices and used for uplink channel-state estimation at the base stations. Sounding reference signals are described in Section 8.3.
- *Positioning reference signals* (PRS), are downlink reference signals intended for positioning support, a feature introduced in release 16 and described in Chapter 27.

In the following, the demodulation reference signals intended for coherent demodulation of PDSCH and PUSCH are described in more detail, starting with the reference signal structure used for OFDM. The same DM-RS structure is used for both downlink and uplink in the case of OFDM. For DFT-spread OFDM in the uplink, a reference signal based on Zadoff-Chu sequences is used to improve the power-amplifier efficiency but supporting contiguous allocations and single-layer transmission only as discussed in a later section. Finally, a discussion on the phase-tracking reference signal is provided.

9.11.1 Demodulation reference signals for OFDM-based downlink and uplink

The DM-RS in NR provides quite some flexibility to cater for different deployment scenarios and use cases: a front-loaded design to enable low latency, support for up to 12 orthogonal antenna ports for MIMO (increased to 24 in release 18), transmissions durations from 2 to 14 symbols, and up to four reference-signal instances per slot to support very high-speed scenarios.

To achieve low latency, it is beneficial to locate the demodulation reference signals early in the transmission, sometimes known as front-loaded reference signals. This allows the receiver to obtain a channel estimate early and, once the channel estimate is obtained, process the received symbols on the fly without having to buffer a complete slot prior to data processing. This is essentially the same motivation as for the frequency-first mapping of data to the resource elements.

Two main time-domain structures are supported, differencing in the location of the first DM-RS symbol:

- *Mapping type A*, where the first DM-RS is located in symbol 2 or 3 *of the slot* and the DM-RS is mapped relative to the start of the slot boundary, regardless of where in the slot the actual data transmission starts. This mapping type is primarily intended for the case where the data occupy (most of) a slot. The reason for symbol 2 or 3 in the downlink is to locate the first DM-RS occasion after a CORESET located at the beginning of a slot.
- *Mapping type B*, where the first DM-RS is located in the first symbol *of the data allocation*, that is, the DM-RS location is not given relative to the slot boundary but rather relative to where the data are located. This mapping is originally motivated by transmissions over a small fraction of the slot to support very low latency and other transmissions that benefit from not waiting until a slot boundary starts but can be used regardless of the transmission duration.

The mapping type for PDSCH transmission can be dynamically signaled as part of the DCI (see Section 9.11 for details), while for the PUSCH the mapping type is semi-statically configured.

Although front-loaded reference signals are beneficial from a latency perspective, they may not be sufficiently dense in the time domain in the case of rapid channel variations. To support high-speed scenarios, it is possible to configure up to three *additional* DM-RS occasions in a slot. The channel estimator in the receiver can use these additional occasions for more accurate channel estimation, for example to use interpolation between the occasions within a slot. As a baseline, it is not possible to interpolate between slots, or in general different transmission occasions, as different slots may be transmitted to different devices and/or in different beam directions. Interpolating channel estimates across different slots directed to the same user may improve channel estimation at the cost of reducing the multi-antenna and beamforming flexibility. With this in mind, release 17 provides the possibility to perform uplink channel estimation across reference symbols in multiple, subsequent slots to improve coverage in some situations when the multi-antenna flexibility is not needed. From a specification perspective, this boils down to configuring the device to ensure properties such as phase continuity across the slots, while the channel estimation as such is an implementation aspects handled in the base station.

The different time-domain allocations for PUSCH DM-RS are illustrated in Fig. 9.16, including both singe symbol and double symbol DM-RS. The purpose of the double-symbol DM-RS is primarily to provide a larger number of antenna ports than what is possible with a single-symbol structure as discussed later. Note that the time-domain location of the DM-RS depends on the scheduled data duration. Furthermore, the DM-RS patterns are at most one slot long. Interpolation between DM-RS symbols should not extend beyond the scheduled transmission as different transmissions may, for example, use different precoders or beam patterns, and the receive must take this into account. However, release 17 introduces the possibility for uplink DM-RS bundling across longer durations in combination with PUSCH/PUCCH repetition or transport-blocks mapped across multiple slots. In this case, the gNB receiver can, after configuring DM-RS bundling in the device, perform channel estimation across multiple slots, which can be beneficial from a coverage perspective.

For the downlink, the same patterns as for the uplink are used, albeit with some restrictions. For example, mapping type B for PDSCH only supports duration 2, 4, and 7 in release 15, a restriction that has been lifted in release 16 to support any lengths from 2 to 13 in order to better support unlicensed spectra as discussed in Chapter 20, as well as to improve the support for dynamic spectrum sharing between NR and LTE on the same carrier. For some of these lengths, the PDSCH mapping is slightly different than the PUSCH mapping.[i]

Multiple orthogonal reference signals can be created in each DM-RS occasion. The different reference signals are separated in the frequency and code domains, and, in the case of a double-symbol DM-RS, additionally in the time domain. Two different types of demodulation reference signals can be configured, type 1 and type 2, differing in the mapping in the frequency domain and the maximum number of orthogonal reference signals. Type 1 can provide up to four orthogonal signals using a single-symbol DM-RS and up to eight orthogonal reference signals using a double-symbol DM-RS. The corresponding numbers for type 2 are six and twelve. The numbers for both type 1 and type 2 can optionally be doubled in release 18 by trading the ability to handle frequency selectivity for MU-MIMO multiplexing capacity. The reference signal types

---

[i] In case of LTE coexistence using reserved resources based on LTE CRS configurations, the PDSCH DM-RS positions can sometimes be shifted in time not to collide with the LTE CRS positions.

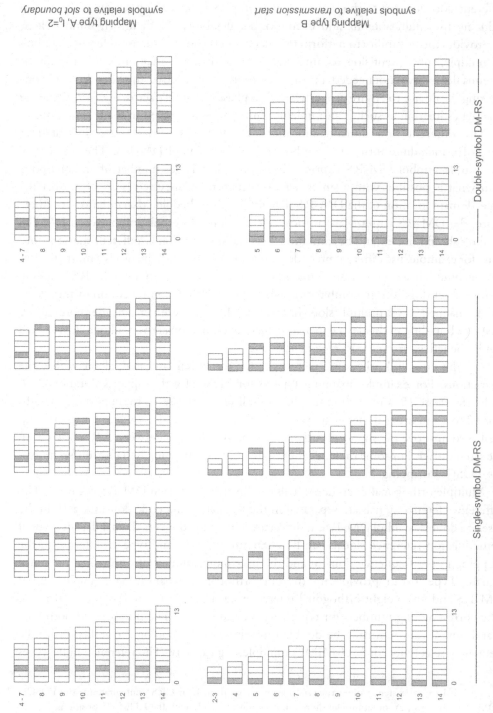

Fig. 9.16 Time-domain location of PUSCH DM-RS.

(1 or 2) should not be confused with the mapping types (A or B); different mapping types can be combined with different reference signal types.

Reference signals should preferably have small power variations in the frequency domain to allow for a similar channel-estimation quality for all frequencies spanned by the reference signal. Note that this is equivalent to a well-focused time–domain auto-correlation of the transmitted reference signal. For OFDM-based modulation, a pseudo-random sequence is used, more specifically a length $2^{31} - 1$ Gold sequence, which fulfills the requirements on a well-focused auto correlation. The sequence is generated across all the common resource blocks (CRBs) in the frequency domain but transmitted only in the resource blocks used for data transmission as there is no reason for estimating the channel outside the frequency region used for transmission. Generating the reference signal sequence across all the resource blocks ensures that the same underlying sequence is used for multiple devices scheduled on overlapping time-frequency resource in the case of multi–user MIMO (see Fig. 9.17, orthogonal sequences are used on top of the pseudo-random sequence to obtain multiple orthogonal reference signals from the same pseudo-random sequence as discussed later). If the underlying pseudo-random sequence would differ between different co-scheduled devices, the resulting reference signals would not be orthogonal. The pseudo-random sequence is generated using a configurable identity, similar to the virtual cell ID in LTE. If no identity has been configured, it defaults to the physical-layer cell identity.[j]

Returning to the type 1 reference signals, the underlying pseudo-random sequence is mapped to every second subcarrier in the frequency domain in the OFDM symbol used for reference signal transmission, see Fig. 9.18 for an illustration assuming only front-loaded reference signals being used. Antenna ports[k] 1000 and 1001 use even–numbered subcarriers in the frequency domain and are separated from each other by multiplying the underlying pseudo-random sequence with different length-2 orthogonal sequences in the frequency domain, resulting in transmission of two orthogonal reference signals for the two antenna ports (in release 18, length-4 sequences can be configured as an

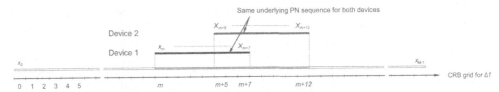

Fig. 9.17 Generating DM-RS sequences based on common resource block 0.

---

[j] In release 16, the CDM group number can also be included in the generation of the sequence in order to lower the cubic metric of the transmitted signal.

[k] The downlink antenna port numbering is assumed in this example. The uplink structure is similar but with different antenna port numbers.

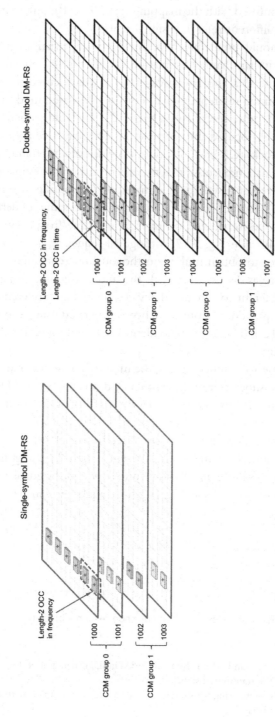

Fig. 9.18 Demodulation reference signals type 1.

alternative to increase the number of reference signals). As long as the radio channel is flat across four consecutive subcarriers, the two reference signals will be orthogonal also at the receiver. Antenna ports 1000 and 1001 are said to belong to *CDM group 0* as they use the same subcarriers but are separated in the code-domain using different orthogonal sequences. Reference signals for antenna ports 1002 and 1003 belong to CDM group 1 and are generated in the same way using odd-numbered subcarriers, that is, separated in the code domain within the CDM group and in the frequency domain between CDM groups. If more than four orthogonal antenna ports are needed, two consecutive OFDM symbols are used instead. The structure above is used in each of the OFDM symbols and a length-2 orthogonal sequence is used to extend the code-domain separation to also include the time domain, resulting in up to eight orthogonal sequences in total.

Demodulation reference signals type 2 (see Fig. 9.19) have a similar structure as type 1, but there are some differences, most notably the number of antenna ports supported. Each CDM group for type 2 consists of two neighboring subcarriers over which a length-2 orthogonal sequences used to separate the two antenna ports sharing the same set of subcarriers. Two such pairs of subcarriers are used in each resource block for one CDM group. Since there are 12 subcarriers in a resource block, up to three CDM groups with two orthogonal reference signals each can be created using one resource block in one OFDM symbol. By using a second OFDM symbol and a time-domain length-2 sequence in the same way as for type 1, a maximum of 12 orthogonal reference signals can be created with type 2. Although the basic structures of type 1 and type 2 have many similarities, there are also differences. Type 1 is denser in the frequency domain, while type 2 trades the frequency-domain density for a larger multiplexing capacity, that is, a larger number of orthogonal reference signals. This is motivated by the support for multi-user MIMO with simultaneous transmission to a relatively large number of devices.

Release 18 provides enhancements doubling the number of possible antenna ports, resulting in up to 16 antenna ports for type 1 and 24 antenna ports for type 2. This is achieved by using longer orthogonal cover codes in the frequency domain, length-4 instead of length-2, resulting in a larger number of antenna ports without increasing the DM-RS overhead. To fully exploit these enhancements, the radio channel needs to be less frequency selective, otherwise the orthogonality between the different OCCs is lost. Semi-static signaling is used to determine whether to use the release 15 reference signals or the enhancements introduced in release 18.

The reference signal structure to use is determined based on a combination of dynamic scheduling and higher-layer configuration. If a double-symbol reference signal is configured, the scheduling decision, conveyed to the device using the downlink control information, indicates to the device whether to use single-symbol or double-symbol reference signals. The scheduling decision also contains information for the device which reference signals (more specifically, which CDM groups) that are intended for *other*

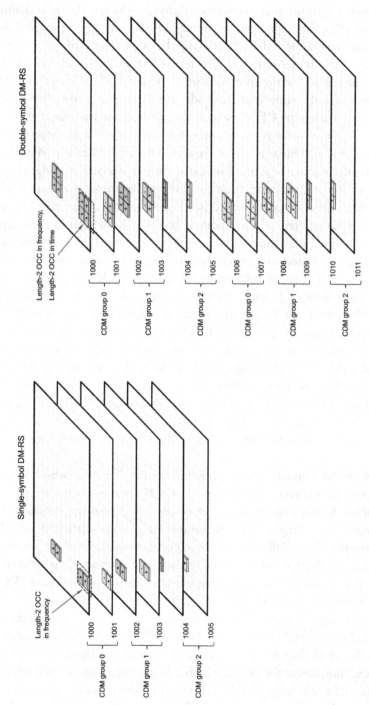

Fig. 9.19 Demodulation reference signals type 2.

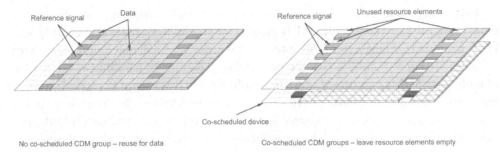

Reference signal   Data

Reference signal   Unused resource elements

Co-scheduled device

No co-scheduled CDM group – reuse for data

Co-scheduled CDM groups – leave resource elements empty

Fig. 9.20 Rate matching data around co-scheduled CDM groups.

devices. The scheduled device maps the data around both its own reference signals as well as the reference signals intended for another device (see Fig. 9.20). This allows for a dynamic change of the number of co-scheduled devices in case of multi-user MIMO. In the case of spatial multiplexing of multiple layers for the same device (also known as single-user MIMO), the same approach is used – each layer leaves resource elements corresponding to another CDM group intended for the same device unused. This is to avoid inter-layer interference for the reference signals.

The reference signal description above is applicable to both uplink and downlink. Note though, that for precoder-based uplink transmissions, the uplink reference signal is applied *before* the precoder (see Fig. 9.11). Hence, the reference signal transmitted is not the structure above, but the precoded version of it.[1]

9.11.2 Demodulation reference signals for DFT-precoded OFDM uplink

DFT-precoded OFDM supports single-layer transmission only and is primarily designed with coverage-challenged situations in mind. Due to the importance of low cubic metric and corresponding high power-amplifier efficiency for uplink DFT-precoded OFDM, the reference signal structure is somewhat different compared to the OFDM case. In essence, transmitting reference signals frequency multiplexed with other uplink transmissions from the same device is not suitable for the uplink as that would negatively impact the device power-amplifier efficiency due to increased cubic metric. Instead, certain OFDM symbols within a slot are used exclusively for DM-RS transmission – that is, the reference signals are *time multiplexed* with the data transmitted on the PUSCH from the same device. The structure of the reference signal itself then ensures a low cubic metric within these symbols as described later.

[1] In general, the reference signal transmitted is in addition subject to any implementation-specific multi-antenna processing, captured by the spatial filter F in Section 9.8, and the word "transmitted" should be understood from a specification perspective, that is, before the spatial filter F.

In the time domain, the reference signals follow the same mapping as configuration type 1. As DFT-precoded OFDM is capable of single-layer transmission only and DFT-precoded OFDM is primarily intended for coverage-challenged situations, there is no need to support configuration type 2 and its capability of handling a high degree of multi-user MIMO. Furthermore, since multi-user MIMO is not a targeted scenario for DFT-precoded OFDM, there is no need to define the reference signal sequence across all common resource blocks as for the corresponding OFDM case, but it is sufficient to define the sequence for the transmitted physical resource blocks only.

Uplink reference signals should preferably have small power variations in the frequency domain to allow for similar channel-estimation quality for all frequencies spanned by the reference signal. As already discussed, for OFDM transmission it is fulfilled by using a pseudo-random sequence with good auto-correlation properties. However, for the case of DFT-precoded OFDM, limited power variations as a function of time are also important to achieve a low cubic metric of the transmitted signal. Furthermore, a sufficient number of reference-signal sequences of a given length, corresponding to a certain reference-signal bandwidth, should be available in order to avoid restrictions when scheduling multiple devices in different cells. A type of sequence fulfilling these two requirements are Zadoff–Chu sequence, discussed in Chapter 8. From a Zadoff–Chu sequence with a given group index and sequence index, additional reference-signal sequences can be generated by applying different linear phase rotations in the frequency domain, as illustrated in Fig. 9.21.

Although the Zaodoff-Chu-based sequence provides a low cubic metric, the cubic metric for the PUSCH data is even lower in case of $\pi/2$-BPSK modulation. Thus, uplink coverage would in this case be limited by the demodulation reference signals, partially offsetting the gains obtained with the low-cubic-metric $\pi/2$-BPSK modulation (for

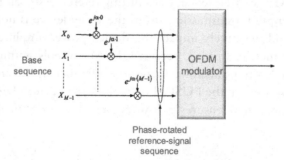

Fig. 9.21 Generation of uplink reference-signal sequence from phase-rotated base sequence.

QPSK and higher modulation schemes, this is not the case). To mitigate this, release 16 allows for an alternative demodulation reference signal sequence. This alternative sequence is based on a $\pi/2$-BPSK-modulated pseudo-random sequence, except for the shortest sequence lengths where carefully selected computer-generated sequences are used.

9.11.3 Phase-tracking reference signals (PT-RS)

Phase-tracking reference signals (PT-RS) can be seen as an extension to DM-RS, intended for tracking phase variations across the transmission duration, for example, one slot. These phase variations can come from phase noise in the oscillators, primarily at higher carrier frequencies where the phase noise tends to be higher. It is an example of a reference signal type existing in NR but with no corresponding signal in LTE. This is partially motivated by the lower carrier frequencies used in LTE, and hence less problematic phase noise situation, and partly it is motivated by the presence of cell-specific reference signals in LTE which can be used for tracking purposes. Since the main purpose is to track phase noise, the PT-RS needs to be dense in time but can be sparse in frequency. The PT-RS only occurs in combination with DM-RS and only if the network has configured the PT-RS to be present. Depending on whether OFDM or DFT-precoded OFDM is used, the structure differs.

For OFDM, the first reference symbol (prior to applying any orthogonal sequence) in the PDSCH/PUSCH allocation is repeated every Lth OFDM symbol, starting with the first OFDM symbol in the allocation. The repetition counter is reset at each DM-RS occasion as there is no need for PT-RS immediately after a DM-RS. The density in the time-domain is linked to the scheduled MCS in a configurable way.

In the frequency domain, phase-tracking reference signals are transmitted in every second or fourth resource block, resulting in a sparse frequency domain structure. The density in the frequency domain is linked to the scheduled transmission bandwidth such that the higher the bandwidth, the lower the PT-RS density in the frequency domain. For the smallest bandwidths, no PT-RS is transmitted.

To reduce the risk of collisions between phase-tracking reference signals associated with different devices scheduled on overlapping frequency-domain resources, the subcarrier number and the resource blocks used for PT-RS transmission are determined by the C-RNTI of the device. The antenna port used for PT-RS transmission is given by the lowest numbered antenna port in the DM-RS antenna port group. Some examples of PT-RS mappings are given in Fig. 9.22.

For DFT-precoded OFDM in the uplink, the samples representing the phase-tracking reference signal are inserted prior to DFT-precoding. The time domain mapping follows the same principles as the pure OFDM case.

Fig. 9.22 Examples of PT-RS mapping in one resource block and one slot.

CHAPTER 10

Physical-layer control signaling

To support the transmission of downlink and uplink transport channels, there is a need for certain *associated control signaling*. This control signaling is often referred to as *L1/L2 control signaling*, indicating that the corresponding information partly originates from the physical layer (layer 1) and partly from MAC (layer 2).

In this chapter, the downlink control signaling, including scheduling grants and scheduling assignments, will be described, followed by the uplink control signaling carrying the necessary feedback from the device.

10.1 Downlink

Downlink L1/L2 control signaling consists of downlink scheduling assignments, including information required for the device to be able to properly receive, demodulate, and decode the DL-SCH on a component carrier, and uplink scheduling grants informing the device about the resources and transport format to use for uplink (UL-SCH) transmission. In addition, the downlink control signaling can also be used for special purposes such as conveying information about the symbols used for uplink and downlink in a set of slots, preemption indication, and power control.

In NR, there is only a single control channel, the *physical downlink control channel* (PDCCH). On a high level, the principles of the PDCCH processing in NR are similar to LTE, namely that the device tries to blindly decode candidate PDCCHs transmitted from the network using one or more search spaces. However, there are some differences compared to LTE based on the different design targets for NR as well as experience from LTE deployments:

- The PDCCH in NR does not necessarily span the full carrier bandwidth. This is a natural consequence of the fact that not all NR devices may be able to receive the full carrier bandwidth as discussed in Chapter 5, and led to the design of a more generic control channel structure in NR.
- The PDCCH in NR is designed to support device-specific beamforming, in line with the general beam-centric design of NR and a necessity when operating at very high carrier frequencies with a corresponding challenging link budget.

These two aspects were to some extent addressed in the LTE EPDCCH design in release 11 although in practice EPDCCH has not been used extensively except as a basis for the control signaling for eMTC.

5G/5G-Advanced
https://doi.org/10.1016/B978-0-443-13173-8.00028-1

Two other control channels present in LTE, the PHICH and the PCFICH, are not needed in NR. The former is used in LTE to handle uplink retransmissions and is tightly coupled to the use of a synchronous hybrid-ARQ protocol, but since the NR hybrid-ARQ protocol is asynchronous in both uplink and downlink the PHICH is not needed in NR. The latter channel, the PCFICH, is not necessary in NR as the size of the *control resource sets* (CORESETs) does not vary dynamically and reuse of control resources for data is handled in a different way than in LTE as discussed further below.

In the following sections, the NR downlink control channel, the PDCCH, will be described, including the notion of CORESETs, the time-frequency resources upon which the PDCCH is transmitted. First, the PDCCH processing including coding and modulation, will be discussed, followed by a discussion on the CORESET structure. There can be multiple CORESETs on a carrier and part of the control resource set is the mapping from resource elements and *resource-element groups* (REGs) to *control channel elements* (CCEs). One or more CCEs from one CORESET are aggregated to form the resources used by one PDCCH. Blind detection, the process where the device attempts to detect if there are any PDCCHs transmitted to the device, is based on search spaces. There can be multiple search spaces using the resources in a single CORESET as illustrated in Fig. 10.1. Finally, the contents of the *downlink control information* (DCI) will be described.

10.1.1 Physical downlink control channel

The payload transmitted on a PDCCH is known as *Downlink Control Information* (DCI) which is processed and transmitted on a PDCCH. The PDCCH processing steps are illustrated in Fig. 10.2. The first step is to calculate and attach a 24-bit CRC to detect transmission errors and to aid the decoder in the receiver. The reason for the somewhat large CRC size of 24 bits (as a comparison LTE uses 16 bits) is to reduce the risk of incorrectly received control information and to assist early termination of the decoding operation in the receiver.

The RNTI (which could be the device identity) modifies the CRC transmitted through a scrambling operation. Upon receipt of the DCI, the device will compute a scrambled CRC on the payload part using the same procedure and compare it against

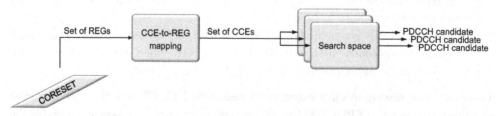

Fig. 10.1 Overview of PDCCH processing in NR from a device perspective.

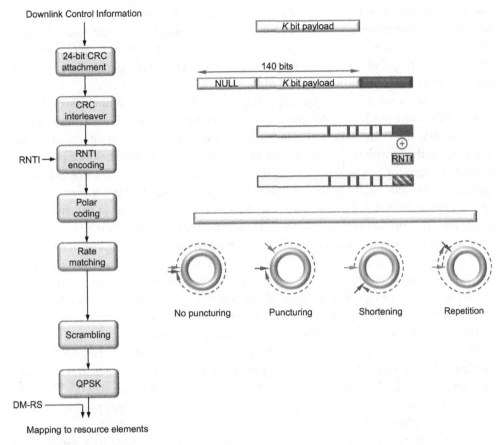

Fig. 10.2 PDCCH processing.

the received CRC. If the CRC checks, the message is declared to be correctly received and intended for the device. Thus, the identity of the device that is supposed to receive the DCI message is implicitly encoded in the CRC and not explicitly transmitted. This reduces the number of bits necessary to transmit on the PDCCH as, from a device point of view, there is no difference between a corrupt message whose CRC will not check, and a message intended for another device. Note that the RNTI does not necessarily have to be the identity of the device, the C-RNTI, but can also be different types of group or common RNTIs, for example to indicate paging or a random-access response.

Channel coding of the PDCCH is based on Polar codes, a relatively new form of channel coding. The basic idea behind Polar codes is to transform several instances of the radio channel into a set of channels that are either noiseless or completely noisy and then transmit the information bits on the noiseless channels. Decoding can be done in several ways, but a typical approach is to use successive cancellation and list decoding.

List decoding uses the CRC as part of the decoding process, which means that the error-detecting capabilities are reduced. For example, list decoding of size eight results in a loss of three bits from an error-detecting perspective, resulting in the 24-bits CRC providing error-detecting capabilities corresponding to a 21-bit CRC. This is part of the reason for using a 24-bit CRC and not a 16-bit CRC.

In contrast to several other coding schemes, for example tail biting convolutional codes, Polar codes need to be designed with a maximum number of information bits in mind. In NR, the Polar code has been designed to support 512 coded bits (prior to rate matching) in the downlink. Up to 140 information bits can be handled, which provides a sufficient margin for future extensions as the DCI payload size in current NR releases is significantly less. For small payloads, below 12 bits, padding up to 12 bits is used. To assist early termination in the decoding process, the CRC is not attached at the end of the information bits, but inserted in a distributed manner, after which the Polar code is applied. Early termination can also be achieved by exploiting the path metric in the decoder.

Rate matching is used to match the number of coded bits to the resources available for PDCCH transmission. This is a somewhat intricate process and is based on a one of shortening, puncturing, or repetition of the coded bits after subblock interleaving of 32 blocks. The set of rules selecting between shortening, puncturing, and repetition, as well as when to use which of the schemes, is designed to maximize performance.

Finally, the coded and rate matched bits are scrambled, modulated using QPSK, and mapped to the resource elements used for the PDCCH, the details of which will be discussed below. Each PDCCH has its own reference signal which means that the PDCCH can make full use of the antenna setup, for example be beamformed in a particular direction. The complete PDCCH processing chain is illustrated in Fig. 10.2.

The mapping of the coded and modulated DCI to resource elements is subject to a certain structure, based on *control-channel elements* (CCEs) and *resource-element groups* (REGs).[a]

A PDCCH is transmitted using 1, 2, 4, 8, or 16 contiguous control-channel elements with the number of control-channel elements used referred to as the *aggregation level*. The control-channel element is the unit upon which the search spaces for blind decoding are defined as will be discussed in Section 10.1.3. A control-channel element consists of six resource-element groups, each of which is equal to one resource block in one OFDM symbol. After accounting for the DM-RS overhead, there are 54 resource elements (108 bits) available for PDCCH transmission in one control-channel element.

The CCE-to-REG mapping can be either interleaved or non-interleaved. The motivation for having two different mapping schemes is to be able to provide frequency diversity by using an interleaved mapping or to facilitate interference coordination and

[a] Although the names are borrowed from LTE, the size of the two differs from their LTE counterparts, as does the CCE-to-REG mapping.

frequency-selective transmission of control channels by using non-interleaved mapping. The details of the CCE-to-REG mapping will be discussed in the next section as part of the overall CORESET structure.

10.1.2 Control resource set

Central to downlink control signaling in NR is the concept of CORESETs. A control resource set is a time-frequency resource in which the device tries to decode candidate control channels using one or more search spaces. The size and location of a CORESET in the time-frequency domain is semi-statically configured by the network and can thus be set to be smaller than the carrier bandwidth. This is especially important in NR as a carrier can be very wide, up to 400 MHz, and it is not reasonable to assume all devices can receive such a wide bandwidth.

In LTE, the concept of a CORESET is not explicitly present. Instead, downlink control signaling in LTE uses the full carrier bandwidth in the first 1–3 OFDM symbols (four for the most narrowband case). This is known as the control region in LTE and in principle this control region would correspond to the "LTE CORESET" if that term would have been used. Having the control channels spanning the full carrier bandwidth was well motivated by the desire for frequency diversity and the fact that all LTE devices support the full 20 MHz carrier bandwidth (at least at the time of specifying release 8). However, in later LTE releases this leads to complications when introducing support for devices not supporting the full carrier bandwidth, for example the eMTC devices introduced in release 12. Another drawback of the LTE approach is the inability to handle frequency-domain interference coordination between cells for the downlink control channels. To some extent, these drawbacks with the LTE control channel design were addressed with the introduction of the EPDCCH in release 11, but the EPDCCH feature has so far not been widely deployed in practice as an LTE network still needs to provide PDCCH support for initial access and to handle non-EPDCCH-capable LTE devices. Therefore, a more flexible structure is used in NR from the start.

Up to three CORESETs can be configured for each of the up to four bandwidth parts configured. The location in the frequency domain can be anywhere within the bandwidth part with a granularity of six resource blocks (see Fig. 10.3). However, a device is not expected to handle CORESETs outside its active bandwidth part.

The first CORESET, CORESET 0, is handled in a special way. Its location in the frequency domain is not restricted to multiples of six resource blocks and is signaled as part of the master information block (MIB) for stand-alone operation. CORESET 0 is used to receive the rest of the system information. For non-standalone operation, the location of CORESET 0 is signaled on the LTE carrier. After connection setup, a device can be configured with multiple additional, potentially overlapping, CORESETs by RRC signaling.

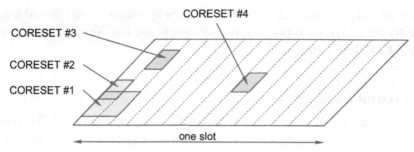

Fig. 10.3 Examples of CORESET configurations.

In the time domain, a CORESET can be up to three OFDM symbols in duration and located anywhere within a slot[b] although a common scenario, suitable for traffic scenarios when a scheduling decision is taken once per slot, is to locate the CORESET at the beginning of the slot. This is similar to the LTE situation with control channels at the beginning of each LTE subframe. However, configuring a CORESET at other time instances can be useful, for example to achieve very low latency for transmissions occupying only a few OFDM symbols without waiting for the start of the next slot. It is important to understand that a CORESET is defined from a *device* perspective and only indicates where a device *may* receive PDCCH transmissions. It does not say anything on whether the gNB actually transmits a PDCCH or not.

Depending on where the front-loaded DM-RS for PDSCH are located, in the third or fourth OFDM symbol of a slot (see Section 9.11.1), the maximum duration for a CORESET is two or three OFDM symbols. This is motivated by the typical case of locating the CORESET before the start of downlink reference signals and the associated data.

A CORESET in NR is of fixed size, that is, it does not vary dynamically over time. This is beneficial from an implementation perspective, both for the device and the network. From a device perspective, a pipelined implementation is simpler if the device can directly start to process the PDCCH without having to first decode another channel to determine the CORESET size.[c] Having a streamlined and implementation-friendly structure of the PDCCH is important in order to realize the very low latency targeted by NR. However, from a spectral efficiency point-of-view, it is beneficial if resources can be shared flexibly between control and data in a dynamic manner. Therefore, NR provides the possibility to start the PDSCH data *before* the end of a CORESET. It is also possible to, for a given device, reuse unused CORESET resources as illustrated in Fig. 10.4. To handle this, the general mechanism of reserved resources is used

---

[b] The time-domain location of a CORESET is obtained from the search space configuration using the CORESET in question.

[c] In LTE, the PCFICH is used to indicate the size of the control region.

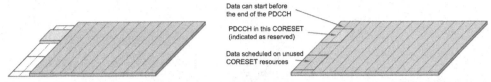

Fig. 10.4 No reuse (left) and reuse (right) of CORESET resources for data transmission (the device is configured with two CORESETs in this example).

(see Section 9.10). Reserved resources that overlap with the CORESET are configured and information in the DCI indicates to the device whether the reserved resources are used by the PDSCH or not. If they are indicated as reserved, the PDSCH is rate matched around the reserved resources overlapping with the CORESET, and if the resources are indicated as available, the PDSCH uses the reserved resources for data except for the resources used by the PDCCH upon which the device received the DCI scheduling the PDSCH.

For each CORESET there is an associated CCE-to-REG mapping, a mapping that is described using the term *REG bundle*. A REG bundle is a set of REGs across which the device can assume the precoding is constant. This property can be exploited to improve the channel-estimation performance in a similar way as resource-block bundling for the PDSCH.

As already mentioned, the CCE-to-REG mapping can be either interleaved or non-interleaved depending on whether frequency-diverse or frequency-selective transmission is desired. There is only one CCE-to-REG mapping for a given CORESET, but since the mapping is a property of the CORESET, multiple CORESETs can be configured with different mappings which can be useful. For example, one or more CORESETs configured with non-interleaved mapping to benefit from frequency-dependent scheduling, and one or more configured with interleaved mapping to act as a fallback in case the channel-state feedback becomes unreliable due to the device moving rapidly.

The non-interleaved mapping is straightforward. The REG bundle size is six for this case, that is, the device may assume the precoding is constant across a whole CCE. Consecutive bundles of six REGs are used to form a CCE.

The interleaved case is a bit more intricate. In this case, the REG bundle size is configurable between two alternatives. One alternative is six, applicable to all CORESET durations, and the other alternative is, depending on the CORESET duration, two or three. For a duration of one or two OFDM symbols, the bundle size can be two or six, and for a duration of three OFDM symbols, the bundle size can be three or six. In the interleaved case, the REG bundles constituting a CCE are obtained using a block interleaver to spread out the different REG bundles in frequency, thereby obtaining frequency diversity. The number of rows in the block interleaver is configurable to handle different deployment scenarios (Fig. 10.5).

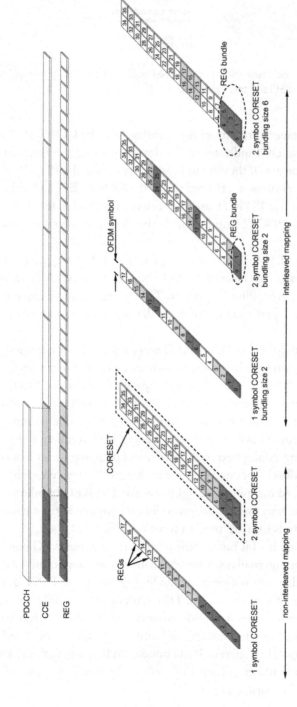

Fig. 10.5 Examples of CCE-to-REG mapping.

As part of the PDCCH reception process, the device needs to form a channel estimate using the reference signals associated with the PDCCH candidate being decoded. A single antenna port is used for the PDCCH, that is, any transmit diversity or multi-user MIMO scheme is handled in a device-transparent manner.

The PDCCH has its own demodulation reference signals, based on the same type of pseudo-random sequence as the PDSCH – the pseudo-random sequence is generated across all the common resource blocks in the frequency domain but transmitted only in the resource blocks used for the PDCCH (with one exception as discussed later). However, during initial access, the location for the common resource blocks is not yet known as it is signaled as part of the system information. Hence, for CORESET 0 configured by the PBCH, the sequence is generated starting from the first resource block in the CORESET instead.

Demodulation reference-signals specific for a given PDCCH candidate are mapped onto every fourth subcarriers in a resource-element group, that is, the reference signal overhead is 1/4. This is a denser reference signal pattern than in LTE which uses a reference signal overhead of 1/6, but in LTE the device can interpolate channel estimates in time and frequency as a consequence of LTE using a cell-specific reference signal common to all devices and present regardless of whether a control-channel transmission takes place or not. The use of a dedicated reference signal per PDCCH candidate is beneficial, despite the slightly higher overhead, as it allows for different types of device-transparent beamforming. By using a beamformed control channel, the coverage and performance can be enhanced compared to the non-beamformed control channels in LTE.[d] This is an essential part of the beam-centric design of NR.

When attempting to decode a certain PDCCH candidate occupying a certain set of CCEs, the device can compute the REG bundles that constitute the PDCCH candidate. Channel estimation must be performed per REG bundle as the network may change precoding across REG bundles. In general, this results in sufficiently accurate channel estimates for good PDCCH performance. However, there is also a possibility to configure the device to assume the same precoding across *contiguous* resource blocks in a CORESET, thereby allowing the device to do frequency-domain interpolation of the channel estimates. This also implies that the device may use reference signals *outside* the PDCCH it is trying to detect, sometimes referred to as wideband reference signals (see Fig. 10.6 for an illustration). In some sense this gives the possibility to partially mimic the LTE cell-specific reference signals in the frequency domain, of course with a corresponding limitation in terms of beamforming possibilities.

Related to channel estimation are, as has been discussed for other channels, the quasi-colocation relations applicable to the reference signals. If the device knows that two reference signals are quasi-colocated, this knowledge can be exploited to improve

---

[d] The LTE EPDCCH introduced device-specific reference signals in order to allow beamforming.

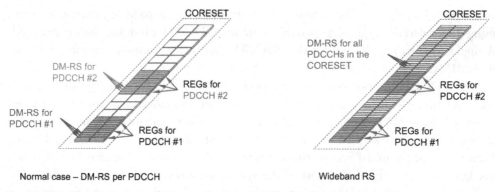

Fig. 10.6 Normal RS structure (left) and wideband RS structure (right).

the channel estimation and, more importantly for the PDCCH, to manage different reception beams at the device (see Chapter 12 for a detailed discussion on beam management and spatial quasi-colocation). To handle this, each CORESET can be configured with a *transmission configuration indication* (TCI) state, that is, providing information of the antenna ports the PDCCH antenna ports are quasi-co-located with. If the device has a certain CORESET spatially co-located with a certain CSI-RS, the device can determine which reception beam is appropriate when attempting to receive PDCCHs in this CORESET, as illustrated in Fig. 10.7. In this example, two CORESETs have been configured in the device, one CORESET with spatial QCL between DM-RS and CSI-RS #1, and one CORESET with spatial QCL between DM-RS and CSI-RS#2. Based on CSI-RS measurements, the device has determined the best reception beam for each of the two CSI-RS:es. When monitoring CORESET #1 for possible PDCCH transmissions, the device knows the spatial QCL relation and uses the appropriate reception beam (similarly for CORESET #2). In this way, the device can handle multiple reception beams as part of the blind decoding framework.

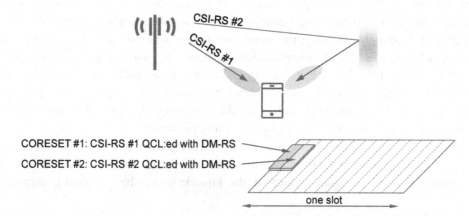

Fig. 10.7 Example of QCL relation for PDCCH beam management.

If no quasi-colocation is configured for a CORESET the device assumes the PDCCH candidates to be quasi-co-located with the SSB with respect to delay spread, Doppler spread, Doppler shift, average delay, and spatial Rx parameters. This is a reasonable assumption as the device has been able to receive and decode the PBCH in order to access the system.

10.1.3 Blind decoding and search spaces

Downlink control signaling is used for multiple purposes and the number of bits required may vary depending on the usage of the control message. A DCI message is therefore characterized by two aspects – the DCI *size* and the DCI *type*. The size and type are a priori unknown to the device, which therefore needs to blindly decode a PDCCH candidate. As part of this process, the code rate needs to be known. By attempting to decode using different hypotheses – combinations of payload sizes and amount of time-frequency resources – and checking the CRC, the device can detect valid control information, if any. The purpose of the control information, that is, the DCI type, also needs to be determined.

One general approach would be to treat DCI size and DCI type separately such that the device blindly decodes a number of different DCI sizes and once a candidate PDCCH was successfully decoded, a header in the first few bits could inform the device how to interpret the remaining payload. However, partly for historical reasons, the NR design inherited the LTE structure where size and type are lumped together into a *DCI format*, although the coupling between the two is somewhat less tight than in LTE. The DCI format thus characterizes not only the type, or purpose, of the DCI but also the size. Different formats could still have the same DCI sizes, but in this case additional bits are required in the payload to indicate the purpose of the DCI.

Blind decoding is a non-negligible processing burden for the device and large parts of the downlink control channel design is related to reducing this complexity. Two important aspects to limit the blind decoding complexity is to limit the number of PDCCH candidates by restricting the location in the time-frequency domain and not allowing arbitrary aggregation levels, and to limit the set of DCI sizes to monitor.

The CCE structure described in the previous section provides a well-defined time-frequency structure useful to limit the number of candidates but is not sufficient. Clearly, from a scheduling point of view, restrictions in the allowed aggregation levels are undesirable as they may reduce the scheduling flexibility and require additional processing at the transmitter side. At the same time, requiring the device to monitor all possible CCE aggregations and locations in all configured CORESETs is not attractive from a device-complexity point of view. To impose as few restrictions as possible on the scheduler while at the same time limit the maximum number of blind decoding attempts in the device, NR defines so-called *search spaces*. A search space is a set of candidate control channels formed by CCEs at a given aggregation level, which the device is supposed to attempt to decode. As there are multiple aggregation levels a device can have multiple search spaces. A *search space set* is a set of search spaces with different aggregation levels linked to the same CORESET. Thus, by configuring a CORESET and a search space set, the device can monitor the presence of control channels with different aggregation levels but using the same time-frequency resource. The purpose of the different aggregation levels is to control the code rate of the PDCCH and therefore be able to perform link adaptation of the PDCCH. The higher the aggregation level, the lower the code rate given a fixed DCI size. Thus, in poor channel conditions, the gNB would select a higher aggregation level than in favorable channel conditions.

Up to 10 search space sets can be configured for each of the four bandwidth parts. For each search space an associated CORESET is configured, together with information on when in time the search space occurs. Thus, time time-domain aspect of a CORESET is not configured as part of the CORESET configuration but obtained from the search space set configuration. The search space set configuration also includes information on the number of PDCCH candidates for each aggregation level and which DCI formats to monitor. A device is not supposed to receive a PDCCH outside its active bandwidth part, which follows from the overall purpose of a bandwidth part. It is also possible to dynamically, through the PDCCH, switch between different groups of search space sets. The purpose of this feature is to reduce the device power consumption during periods of little data activity and is described in more detail in Chapter 14. Another usage of dynamic search space set switching is when operating in unlicensed spectra, see Chapter 20.

At a configured monitoring occasion for a search space set, the devices will attempt to decode the candidate PDCCHs for the search spaces in the search space set. Five different aggregation levels corresponding to 1, 2, 4, 8, and 16 CCEs, respectively, can be configured. The highest aggregation level, 16, was added to NR in case of extreme coverage requirements although the code rate is sufficiently low already at lower aggregation levels and power boosting of, for example, aggregation level 8 or even 4 could be used instead. The number of PDCCH candidates can be configured per search space (and thus also per aggregation level). Hence NR has a more flexible way of spending the blind decoding attempts across aggregation levels than LTE, where the number of blind decodes at each

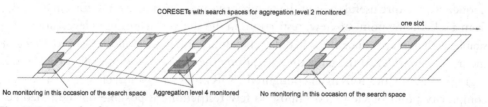

Fig. 10.8 Example of a search space set configuration.

aggregation level was fixed. This is motivated by the wider range of deployments expected for NR. For example, in a small-cell scenario the highest aggregation levels may not be used, and it is better to spend the limited number of blind decoding attempts the device is dimensioned for on the lower aggregation levels than on blind decoding on an aggregation level that is never used.

Search space set 0 is special. It is linked to CORESET 0 and configured based on information in the MIB. The purpose of this search space set is to be able to receive the rest of the system information although it can be used for other purposes as well. The remaining search space sets are configured using RRC signaling, either as part of the system information or as by using dedicated RRC signaling.

Upon attempting to decode a candidate PDCCH, the content of the control channel is declared as valid for this device if the CRC succeeds. The device processes the information (scheduling assignment, scheduling grants, etc.) and acts accordingly. If the CRC does not succeed, the information is either subject to uncorrectable transmission errors or intended for another device and in either case the device ignores that PDCCH transmission.

Having discussed the search spaces, it is clear that the network can only address a device if the control information is transmitted on a PDCCH formed by the CCEs in one of the device's search spaces. For example, device A in Fig. 10.9 cannot be addressed on a PDCCH starting at CCE number 20, whereas device B can. Furthermore, if device A is using CCEs 16–23, device B cannot be addressed on aggregation level 4 as all CCEs in its level-4 search space are blocked by being used for other devices. From this it can be intuitively understood that for efficient utilization of the CCEs in the system, the search spaces should differ between devices (except if all devices are able to monitor all CCEs which is unlikely from a complexity perspective). Each device in the system can therefore have one or more *device-specific* search spaces (also known as UE-specific search spaces, USS) configured. As a device-specific search space for complexity reasons typically cannot contain all the CCEs the network could transmit upon at the corresponding aggregation level, there must be a mechanism determining the set of CCEs in a device-specific search space.

One possibility would be to let the network configure the device-specific search space in each device, similar to the way the CORESETs are configured. However, this would require explicit signaling to each of the devices and possibly reconfiguration at handover. Instead, the location of a device-specific search spaces, expressed as a starting CCE number, is defined without explicit signaling through a function of the C-RNTI, a device identity unique in the cell. Furthermore, the set of CCEs the device should monitor for a certain aggregation level also varies as a function of time to avoid two devices constantly blocking each other. This randomizes the location of a search space over time. If two search spaces collide at one time instant, they are not likely to collide at the next time

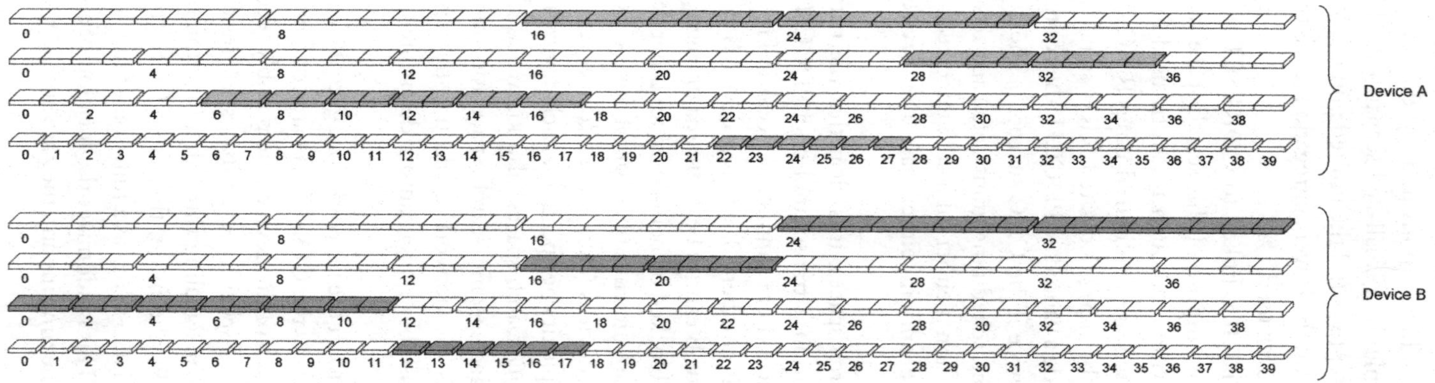

Fig. 10.9 Example of search spaces for two different devices.

instant. In each of these search spaces, the device is attempting to decode the PDCCHs using the device-specific C-RNTI identity.[e] If valid control information is found, for example a scheduling grant, the device acts accordingly.

However, there is also information intended for a group of devices. Furthermore, as part of the random-access procedure, it is necessary to transmit information to a device before it has been assigned a unique identity. These messages are scheduled with different predefined RNTIs, for example the SI-RNTI for scheduling system information, the P-RNTI for transmission of a paging message, the RA-RNTI for transmission of the random-access response, and TPC-RNTI for uplink power control. Other examples are the INT-RNTI used for preemption indication and the SFI-RNTI used for conveying slot-related information. These types of information cannot rely on a device-specific search space as different devices would monitor different CCEs despite the message being intended for all of them. Hence, NR also defines *common search spaces* (CSS).[f] A common search space is similar in structure to a device-specific search space with the difference that the set of CCEs is predefined and hence known to all devices, regardless of their own identity. Whether a search space is common or device specific is provided as part of the search space set configuration.

The different DCI formats, described in more detail in the following section, search spaces, and RNTIs are summarized in Table 10.1. The size of the DCI formats depends heavily on the configuration and which fields that are present, and it is therefore hard to state the DCI size without also stating the corresponding configuration, but the DCI size for uplink and downlink scheduling can be in the order of 70 bits (excluding CRC).

As discussed previously, blind decoding is a non-negligible processing burden for the device. To limit the device complexity, a baseline DCI size budget of "3+1" is therefore used, meaning that a device at most monitors three different DCI sizes using the C-RNTI (and hence being time-critical for scheduling) and one DCI size using other RNTIs (and hence less time critical). Some devices may in some situations be able to perform more blind decodes than what the "3+1" budget might suggest, but not all devices have this capability. Hence, despite there might be a relatively large number of different DCI messages (and DCI formats), the payloads sizes must be aligned to meet the constraints of the device processing capability. The DCI size budget is enforced by an intricate set of size-alignment rules (see [90] for details) where certain DCI formats are padded to match the size of other DCI formats. The result of the size-alignment depends on the configuration and hence the set of DCI formats the device has to monitor, but a

---

[e] There is can also be an additional device-specific identity, the CS-RNTI, used for semi-persistent scheduling as discussed in Chapter 14, and a MCS-CRNTI, used in the same way as the C-RNTI but indicating a more robust transmission as discussed in Section 10.1.17. There are also RNTIs used for semi-persistent scheduling and configured grants.

[f] The NR specifications defines different types of common search spaces depending on the RNTI monitored, but this is not important for understanding the general principle of search spaces.

Table 10.1 Summary of DCI formats, search spaces, and RNTIs (simplified).

DCI format	Search space	Possible RNTIs[a]	Usage
0_0	USS CSS in PCell	C-RNTI	Uplink scheduling (fallback)
	CSS in PCell	TC-RNTI	Random access procedure
0_1	USS	C-RNTI	Uplink scheduling
		SP-CSI-RNTI	Activation of semi-persistent CSI reporting
0_2	USS	C-RNTI	Uplink scheduling
		SP-CSI-RNTI	Activation of semi-persistent CSI reporting
0_3	USS	C-RNTI	Uplink co-scheduling of multiple carriers
1_0	USS CSS in PCell	C-RNTI	Downlink scheduling (fallback)
	CSS in PCell	SI-RNTI	Scheduling of system information
		RA-RNTI, msgB-RNTI	Random-access response
		TC-RNTI	Random access procedure
		P-RNTI	Paging messages
1_1	USS	C-RNTI	Downlink scheduling
1_2	USS	C-RNTI	Downlink scheduling
1_3	USS	C-RNTI	Downlink co-scheduling of multiple carriers
2_0	CSS	SFI-RNTI	Slot format indicator
2_1	CSS	INT-RNTI	Preemption indicator
2_2	CSS	TPC-PUCCH-RNTI	PUCCH power control
		TPC-PUSCH-RNTI	PUSCH power control
2_3	CSS	TPC-SRS-RNTI	SRS power control
2_4	CSS	CI-RNTI	Uplink cancellation indicator
2_5	CSS	AI-RNTI	Soft resource availability for IAB
2_6	CSS	PS-RNTI	Power saving information
2_7	CSS	PEI-RNTI	Paging early indicator, dynamic TRS indication
3_0	USS	SL-RNTI, SL-CS-RNTI	Scheduling of an NR sidelink
3_1	USS	SL-L-CS-RNTI	Scheduling of an LTE sidelink
4_0	CSS	MCCH-RNTI, G-RNTI	Scheduling of broadcast transmissions
4_1	CSS	G-RNTI, G-CS-RNTI	Scheduling of multicast transmissions
4_2	CSS	G-RNTI, G-CS-RNTI	Scheduling of multicast transmissions
5_0	CSS	NCR-RNTI	Beam indication for network-controlled repeaters.

[a]C-RNTI should be read as including all three of C-RNTI, MCS-C-RNTI, and CS-RNTI.

fairly typical setup results in three different DCI sizes for scheduling and one DCI size for various control purposes:

- a size given by DCI format 0_0/1_0, monitored using the C-RNTI in the device specific search spaces, and C-RNTI as well as various other RNTIS in common search spaces;
- a size given by DCI format 0_1, monitored using the C-RNTI in the device specific search spaces;
- a size given by DCI format 1_1, monitored using the C-RNTI in the device specific search spaces; and
- a size given by DCI formats in the 2_x family, monitored in common search spaces and using various RNTIs other than the C-RNTI. Configurations need to ensure that all the monitored formats in the 2_x family are of the same size.

Other configurations are of course also possible, for example using only format 0_0/1_0 for scheduling and thereby allowing a larger freedom when configuring the 2_x formats.

The "3+1" budget is not sufficient to control the device complexity. Even if the number of DCI sizes is limited, there can be multiple search spaces and a limit on the total number of blind decoding attempts is therefore also needed. In NR, the number of blind decoding attempts depends on the subcarrier spacing (and hence the slot duration). For 15/30/60/120 kHz subcarrier spacing, up to 44/36/22/20 blind decoding attempts per slot can be supported across all DCI sizes – a number selected to offer a good tradeoff between device complexity and scheduling flexibility. However, the number of blind decodes is not the only measure of complexity but also channel estimation needs to be accounted for. The number of channel estimates for subcarrier spacings of 15/30/60/120 kHz has been limited to 56/56/48/32 CCEs across all CORESETs in a slot. Depending on the configuration, the number of PDCCH candidates may be limited either by the number of blind decodes, or by the number of channel estimates. CRC checking is of low complexity so monitoring multiple RNTIs all with the same payload size is not costly and almost comes "for free".

In later NR releases, the device can indicate blind decoding capabilities on a finer granularity, moving away from defining it per slot. To obtain very low latency, it may be desirable to configure PDCCH monitoring very frequently, for example every second OFDM symbol. This is possible, but the more frequent the motoring is done, the smaller the number of blind decoding attempts the device can handle. The framework is also used to handle the largest subcarrier spacings, 480 and 960 kHz, where the slot duration is very short and monitoring in every slot may not be possible. Instead, the blind decoding capabilities are defined per group of slots and multi-PDSCH scheduling, described later in the chapter, is used to schedule contiguous data transmission.

Depending on the search space set configuration, the number of blind decoding attempts may vary across slots. In Fig. 10.8 one example is found, where more blind decoding attempts are required in the middle slot compared to the two other slots or, expressed differently, not all blind decoding capability of the device is used in two of the three slots. To avoid wasting blind decoding capabilities, device configuration can be such that the number of blind decoding attempts is larger than the maximum allowed value, known as overbooking. Prioritization rules define how the device should spend the blind decoding attempts in such a case not to violate the blind decoding budget, giving priority to the common search spaces in the slot.

In the case of carrier aggregation, the general blind decoding operation described here is applied per component carrier. The total number of channel estimates and blind decoding attempts is increased compared to the single carrier case, but not in direct proportion to the number of aggregated carriers.

10.1.4 Downlink scheduling assignments – DCI formats 1_0, 1_1, 1_2, and 1_3

Having described the transmission of DCI on PDCCH, the detailed contents of the control information can be discussed, starting with the downlink scheduling assignments. Downlink scheduling assignments use DCI format 1_1, the non-fallback format, or DCI format 1_0, also known as the fallback format. There is also a third format, 1_2, introduced in release 16 as part of the enhanced support for URLLC. The content is basically the same as format 1_1 but with greater configurability of the size of many of the fields. Furthermore, release 18 introduced DCI format 1_3, which is used to schedule downlink transmissions on multiple carriers using a single DCI.

The non-fallback format 1_1 supports all baseline NR features. Depending on the features that are configured in the system, some information fields may or may not be present. For example, if carrier aggregation is not configured, there is no need to include carrier-aggregation-related information in the DCI. Hence the DCI size for format 1_1 depends on the overall configuration but as long as the device knows which features are configured, it also knows the DCI size and blind detection can be performed.

The fallback format 1_0 is typically smaller in size, supports a limited set of NR functionality, and the set of information fields is in general not configurable, resulting in a (more or less) fixed DCI size. One use case of the fallback format is to handle periods of uncertainty in the configuration of a device as the exact time instant when a device applies the configuration information is not known to the network, for example due to transmission errors. Another reason for using the fallback DCI is to reduce control signaling overhead. In many cases the fallback format provides sufficient flexibility for scheduling smaller data packets.

DCI format 1_2 is introduced in release 16 as part of the enhanced support for URLLC traffic, see Chapter 21 for a background. It provides almost the same functionality as formats 1_1, but allows for more flexibility in the configuration of the size of the different fields, including the possibility to configure a size of zero for several of them. Thus, these formats can allow for a smaller DCI size and hence more robust reception in cases where not all DCI information fields are needed and a small number of bits is sufficient. It also omitted some fields less relevant for URLLC-type of traffic.

DCI format 1_3 is introduced in release 18 and enables scheduling of downlink transmissions on up to four carriers with a single DCI. Compared to prior releases, where one DCI per carrier is used, this can reduce the control signaling overhead, especially in scenarios with fragmented spectrum allocations where multiple carriers need to be aggregated. It provides similar information fields as format 1_1, but applicable to multiple carriers. Some of the information fields – for example PUCCH-related information, bandwidth part indication, and channel-access type in unlicensed spectra – are common to all the co-scheduled carriers, while other pieces of information, mainly resource allocation, hybrid-ARQ and some multi-antenna-related information, naturally need to be provided per scheduled carrier.

Parts of the contents are the same for the different DCI formats, as seen in Table 10.2, but there are also differences due to the different capabilities and the release (indicated by a small superscript in the table). The information in the DCI formats used for downlink scheduling can be organized into different groups, with the fields present varying between the DCI formats. The content of DCI formats for downlink scheduling assignments is described here:

- Identifier of DCI format (1 bit). This is a header to indicate whether the DCI is a downlink assignment or an uplink grant, which is important in case the payload sizes of multiple DCI formats are aligned and the size cannot be used to differentiate the DCI formats (one example hereof is the fallback formats 0_0 and 1_0 which are of equal size).
- Resource information, consisting of:
  - Carrier indicator (0 or 3 bit). This field is present if cross-carrier scheduling is configured and is used to indicate the component carrier the DCI relates to. The carrier indicator is not present in the fallback DCI for example used for common signaling to multiple devices as not all devices may be configured with (or capable of) carrier aggregation.
  - Co-scheduled cells, used by DCI format 1_3 only and indicates the set of cells to which this DCI applies
  - Bandwidth-part indicator (0–2 bit), used to activate one of up to four downlink bandwidth parts configured by higher-layer signaling. Not present in in the fallback DCI. In case of co-scheduling of multiple cells the same indicator is applied to all carriers.
  - Frequency-domain resource allocation. This field indicates the resource blocks in the downlink bandwidth part upon which the device should receive the PDSCH. The size of the field depends on the size of the bandwidth and on the resource allocation type, type 0 only, type 1 only, or dynamic switching between the two as discussed in Section 10.1.10. Format 1_0 supports resource allocation type 1 only as the full flexibility in resource allocation is not needed in this case.

Table 10.2 DCI formats 1_0, 1_1, 1_2, and 1_3 for downlink scheduling with C-RNTI. For fields introduced after release 15, the superscript indicates the first release in which the field appeared.

Field		1_0	1_1	1_2	1_3
		DCI format			
Format identifier		•	•	•16	•18
Resource information	CFI		•	•16	
	Co-scheduled cells				•18
	BWP indicator		•	•16	•18
	Frequency domain allocation	•	•	•16	•18
	Time-domain allocation	•	•	•16	•18
	VRB-to-PRB mapping	•	•	•16	•18
	PRB bundling size indicator		•	•16	•18
	Reserved resources		•	•16	•18
	Zero-power CSI-RS trigger		•	•16	•18
	Scheduling offset		•16		•18
	Channel-access type	•16	•16	•17	•18
	Dormancy indication		•16		•18
Transport-block related	MCS	•	•	•16	•18
	NDI	•	•	•16	•18
	RV	•	•	•16	•18
	MCS, 2nd TB		•		•18
	NDI, 2nd TB		•		•18
	RV, 2nd TB		•		•18
	Priority indication		•16	•16	•18
Hybrid-ARQ related	Process number	•	•	•16	•18
	DAI	•	•	•16	•18
	PDSCH-to-HARQ feedback timing	•	•	•16	•18
	CBGTI		•		
	CBGFI		•		
	PDSCH group index		•16		
	One-shot HARQ request		•16	•17	•18
	Number of PDSCH groups		•16		
	New feedback indicator		•16		
	Enhanced type 3 codebook indicator		•17	•17	•18
	HARQ retransmission indicator		•17	•17	•18
Multi-antenna related	Antenna ports		•	•16	•18
	TCI		•	•16	•18
	SRS request		•	•16	•18
	SRS offset indicator		•17	•17	•18
	DM-RS sequence initialization		•	•16	•18
PUCCH-related information	PUCCH power control	•	•	•16	•18
	PUCCH resource indicator		•	•16	•18
	PUCCH cell indicator		•17	•17	•18
PDCCH-related information	PDCCH monitoring control		•17	•17	•18

- Time-domain resource allocation (1–4 bit). This field indicates the resource allocation in the time domain as described in Section 10.1.16. Up to 6 bits in release 17 to support multi–TRP operation.
- VRB-to-PRB mapping (0 or 1 bit) to indicate whether interleaved or non-interleaved VRB-to-PRB mapping should be used as described in Section 9.9. Only present for resource allocation type 1. The same VRB-to-PRB mapping is applied to all co-scheduled cells in case of DCI format 1_3.
- PRB bundling size indicator (0 or 1 bit), used to indicate the PDSCH bundling size as described in Section 9.9. The same bundling size is applied to all co-scheduled cells in case of DCI format 1_3.
- Reserved resources (0–2 bit), used to indicate to the device if the reserved resources can be used for PDSCH or not as described in Section 9.10. Up to 4 bits in case of DCI format 1_3 in which case the field is a pointer into a configured table where each entry contains a list of reserved resource indicators with one indicator for each of the co-scheduled cells.
- Zero-power CSI-RS trigger (0–2 bit), see Section 8.1 for a discussion on CSI reference signals. Up to 3 bits in case of multiple co-scheduled cells, used as a pointer into a configured table providing the CSI-RS trigger for each of the co-scheduled cells.
- Channel-access type and cyclic extension (0–4 bit), used in unlicensed spectra to indicate the channel-access procedure to use as described in Chapter 20.
- Dormancy indicator (0–5 bit), used for cell dormancy and power saving as described in Section 14.5.4.
- Scheduling-offset indicator (0 or 1 bit), used to control cross-slot scheduling for power-saving purpose as described in Section 14.5.3.
- Transport-block related information
  - Modulation-and-coding scheme (5 bit), used to provide the device with information about the modulation scheme, the code rate, and the transport-block size, as described further below.
  - New-data indicator (1 bit), used to clear the soft buffer for initial transmissions as discussed in Section 13.1. Up to 8 bit in release 17 to support multi–TRP operation.
  - Redundancy version (2 bit) (see Section 13.1). Up to 8 bit in release 17 to support multi–TRP operation.
  - If a second transport block is present (only if more than four layers of spatial multiplexing are supported in DCI format 1_1), the three fields above are repeated for the second transport block.
  - Priority indication (0 or 1 bit), introduced in release 16 as part of enhanced URLLC support and used to indicate the priority of the related uplink transmission such as hybrid-ARQ acknowledgments. The same priority is applied to all carriers in case of co-scheduling of multiple cells.

- Hybrid-ARQ related information
  - Hybrid ARQ process number (4 bit), informing the device about the hybrid-ARQ process to use for soft combining. Extended to 5 bit in release 17 to support additional hybrid-ARQ processes needed for non-terrestrial access.
  - Downlink assignment index (DAI, 0, 2, 4, or 6 bit), only present in the case of a dynamic hybrid-ARQ codebook is configured as described in Section 13.1.5. DCI format 1_0 uses 2 bits while formats 1_1 and 1_2 support the full range of bits.
  - HARQ feedback timing (0–3 bit), providing information on *when* the hybrid-ARQ acknowledgment should be transmitted relative to the reception of the PDSCH. The same feedback timing is applied to all carriers in DCI format 1_3.
  - CBG transmission indicator (CBGTI, 0, 2, 4, 6, or 8 bit), indicating the code block groups retransmitted as described in Section 13.1.2. Only present in DCI format 1_1 and only if CBG retransmissions are configured.
  - CBG flush indicator (CBGFI, 0–1 bit), indicating soft buffer flushing as described in Section 13.1.2. Only present in DCI format 1_1 and only if CBG retransmissions are configured.
  - PDSCH group index (0–1 bit), used in unlicensed spectra to indicate the PDSCH group and controlling the hybrid-ARQ codebook, see Chapter 20. Not used in DCI format 1_3.
  - Number of requested PDSCH groups (0–1 bit), used in unlicensed spectra to indicate whether hybrid-ARQ feedback should include only the current PDSCH group or also the other PDSCH group, see Chapter 20. Not used in DCI format 1_3.
  - One-shot hybrid-ARQ request (0–1 bit), used in unlicensed spectra to trigger a hybrid-ARQ report for all hybrid-ARQ processes across all carriers and PDSCH groups, see Chapter 20. Common to all carriers for DCI format 1_3.
  - New feedback indicator (0–2 bit), used in unlicensed spectra to indicate whether the gNB has received the hybrid-ARQ feedback, see Chapter 20.
  - Enhanced type 3 codebook indicator (0–3 bit), introduced in release 17 as part of the industrial IoT enhancements as described in Chapter 21. Common to all carriers for DCI format 1_3.
  - HARQ retransmission indicator (0 or 1 bit), introduced in release 17 as part of the industrial IoT enhancements as described in Chapter 21. Common to all carriers for DCI format 1_3.
- Multi-antenna related information (not present in DCI format 1_0)
  - Antenna ports (4–6 bit), indicating the antenna ports upon which the data are transmitted as well as antenna ports scheduled for other users as discussed in Chapters 9 and 11.

- Transmission configuration indication (TCI, 0 or 3 bit), used to indicate the QCL relations for downlink transmissions as described in Chapter 12. For DCI format 1_3, the field points into a configurable table where the entries contain the TCI state for each of the co-scheduled carriers.
- SRS request (2 bit), used to request transmission of a sounding reference signal as described in Section 8.3. For DCI format 1_3, the size of the field depends on the number of configured carriers possible to co-schedule and the field is used as an index into a configurable table providing the SRS request for each of the co-scheduled carriers.
- SRS offset indicator (0–2 bit), used to control in which slot an aperiodic sounding reference signal is transmitted. For DCI format 1_3, the size of the field depends on the number of configured carriers possible to co-schedule and the field is used as an index into a configurable table providing the SRS offset for each of the co-scheduled carriers.
- DM-RS sequence initialization (0 or 1 bit), used to select between two preconfigured initialization values for the DM-RS sequence.
- PUCCH-related information
  - PUCCH power control (2 bit), used to adjust the PUCCH transmission power. In release 17, two sets of power-control fields, one primary and one secondary, can be configured to support transmissions to different transmission points in case of multi-TRP operation.
  - PUCCH resource indicator (3 bit), used to select the PUCCH resource from a set of configured resources (see Section 10.2.7).
  - PUCCH cell indicator (0–1 bit), to support dynamic selection of PUCCH cell switching.
- PDCCH-related information
  - PDCCH monitoring control (0–2 bit), used to dynamically control the PDCCH monitoring occasions for device power saving purposes as described in Chapter 14.

DCI format 1_0 is also used for paging (together with the P-RNTI), random-access response (together with the RA-RNTI or msgB-RNTI), system-information delivery (together with the SI-RNTI), or for a PDCCH-ordered random-access procedure (together with the C-RNTI). In all these cases the DCI content is (partially) different than what is outlined above although the DCI size is the same.

10.1.5 Uplink scheduling grants – DCI formats 0_0, 0_1, 0_2, and 0_3

Uplink scheduling grants use one of DCI formats 0_1, the non-fallback format, or DCI format 0_0, also known as the fallback format. The reason for having both a fallback and a non-fallback format is the same as for the downlink, namely to handle uncertainties during RRC reconfiguration and to provide a low-overhead format for transmissions not exploiting all uplink features. As for DCI format 1_1, the information fields present in the non-fallback format 0_1 depend on the features that are configured. There is also a third format for uplink scheduling, format 0_2, introduced in release 16 as part of the enhanced support for URLLC. Similarly to the downlink, format 0_2 is based on format 0_1 but with added flexibility in the field sizes, see Chapter 21 for details. Furthermore, format 0_3 is introduced in release 18 as the uplink companion to the downlink format 1_3 and is used for co-scheduling of up to four carriers using a single DCI. Also in this case are some information fields common to all carriers, while other fields provide independent information per carrier. The DCI sizes for the uplink-related DCI format 0_1 and downlink-related DCI format 1_1 are aligned with padding added to the smaller of the two in order to reduce the number of blind decodes.

Parts of the contents are the same for the different DCI formats, as seen in Table 10.3, but there are also differences due to the different capabilities. The information in the DCI formats used for uplink scheduling can be organized into different groups, with the fields present varying between the DCI formats. The content of DCI formats for uplink scheduling are as follows:

- Identifier of DCI format (1 bit), a header to indicate whether the DCI is a downlink assignment or an uplink grant.
- DFI flag (1 bit), present in unlicensed spectra only where it serves as a header to indicate whether the DCI is an activation/release of a configured uplink grant or a request for downlink feedback information (DFI) as described in Chapter 20.
- Resource information, consisting of:
  - Carrier indicator (0 or 3 bit). This field is present if cross-carrier scheduling is configured and is used to indicate the component carrier the DCI relates to. The carrier indicator is not present in DCI format 0_0.
  - Co-scheduled cells, used by DCI format 1_3 only and indicates the set of cells to which this DCI applies
  - UL/SUL indicator (0 or 1 bit), used to indicate whether the grant relates to the supplementary uplink or the ordinary uplink (see Section 7.7). Only present if a supplementary uplink is configured as part of the system information.
  - Bandwidth-part indicator (0–2 bit), used to activate one of up to four uplink bandwidth parts configured by higher-layer signaling. Not present in in DCI format 0_0.
  - Frequency-domain resource allocation. This field indicates the resource blocks in the uplink bandwidth part upon which the device should transmit the PUSCH. The size of the field depends on the size of the bandwidth part and on the resource allocation type, type 0, type 1, or type 2, and whether dynamic switching between type 0 and type 1 is configured as discussed in Section 10.1.10. Format 0_0 supports resource allocation type 1 only.

Table 10.3 DCI formats 0_0, 0_1, 0_2, and 0_3 for uplink scheduling. For fields introduced after release 15, the superscript indicates the first release in which the field appeared.

Field		DCI format			
		0_0	0_1	0_2	0_3
Identifier	DCI format	•	•	•^{16}	•^{18}
	DFI flag		•^{16}		
Resource information	CFI		•	•^{16}	
	Co-scheduled cells				•^{18}
	UL/SUL	•	•	•^{16}	
	BWP indicator		•	•^{16}	•^{18}
	Frequency domain allocation	•	•	•^{16}	•^{18}
	Time-domain allocation	•	•	•^{16}	•^{18}
	Frequency hopping	•	•	•^{16}	•^{18}
	Channel-access type	•^{16}	•^{16}	•^{17}	•^{18}
	Dormancy indication		•^{16}		•^{18}
	Scheduling offset		•^{16}		•^{18}
	Invalid symbol pattern		•^{16}	•^{16}	
	Transform precoding		•^{18}	•^{18}	
Transport-block related	MCS	•	•	•^{16}	•^{18}
	NDI	•	•	•^{16}	•^{18}
	RV	•	•	•^{16}	•^{18}
	UL-SCH indicator		•	•^{16}	•^{18}
	Priority		•^{16}	•^{16}	•^{18}
Hybrid-ARQ related	Process number	•	•	•^{16}	•^{18}
	DAI		•	•^{16}	•^{18}
	SAI		•^{16}		
	CBGTI		•		
Multi-antenna related	DM-RS sequence initialization		•	•^{16}	•^{18}
	Antenna ports		•	•^{16}	•^{18}
	Precoding information		•	•^{16}	•^{18}
	PTRS-DMRS association		•	•^{16}	•^{18}
	SRI		•	•^{16}	•^{18}
	SRS resource set		•^{17}	•^{17}	
	SRS offset		•^{17}	•^{17}	•^{18}
	SRS request		•	•^{16}	•^{18}
	CSI request		•	•^{16}	•^{18}
Power control	PUSCH power control	•	•	•^{16}	•^{18}
	Beta offset		•		•^{18}
	Power control parameter set		•^{16}	•^{16}	•^{18}
PDCCH-related information	PDCCH monitoring control		•^{17}	•^{17}	•^{18}

- Time-domain resource allocation (0–4 bit). This field indicates the resource allocation in the time domain as described in Section 10.1.16. In release 16, up to 6 bits can be used to enhance URLLC support as discussed in Chapter 21.
- Frequency-hopping flag (0 or 1 bit), used to handle frequency hopping for resource allocation type 1.

- Channel-access type and cyclic extension (0–6 bit), used in unlicensed spectra to indicate the channel-access procedure to use as described in Chapter 20.
- Scheduling-offset indicator (1 bit), used to control cross-slot scheduling for power-saving purpose as described in Section 14.5.3.
- Dormancy indicator (0–5 bit), used for cell dormancy and power saving as described in Section 14.5.4.
- Invalid symbol pattern indicator (0 or 1 bit), used for enhanced URLLC support as discussed in Chapter 21.
- Transform precoding (0 or 1 bit), used to dynamically switch between OFDM and DFTS-OFDM for single-layer transmissions. Introduced in release 18.
- Transport-block related information
  - Modulation-and-coding scheme (5 bit), used to provide the device with information about the modulation scheme, the code rate, and the transport-block size, as described further below.
  - New-data indicator (1 bit), used to indicate whether the grant relates to retransmission of a transport block or transmission of a new transport block. In release 16, one DCI message can schedule up to 8 transport blocks to better exploit unlicensed spectra as discussed in Chapter 20 and consequently up to 8 bits are required to provide one new-data indicator per transport block.
  - Redundancy version (2 bit). Similar to the new-data indicator, this field can be enlarged up to 8 bit in release 16 to support multiple transport blocks scheduled by one DCI message.
  - UL-SCH indicator (1 bit), used to indicate whether the PUSCH should contain data from the UL-SCH or not. If no data from the UL-SCH is included, the PUSCH contains the UCI feedback only.
  - Priority indication (0 or 1 bit), introduced in release 16 as part of enhanced URLLC support and used to indicate the priority of the uplink transmission.
- Hybrid-ARQ-related information
  - Hybrid ARQ process number (4 bit, up to 5 bit in release 17), informing the device about the hybrid-ARQ process to (re)transmit.
  - Downlink assignment index (DAI), used for handling of hybrid-ARQ codebooks in case of UCI transmitted on PUSCH. Not present in DCI format 0_0. In release 17 an additional DAI field can be configured to support frequency-domain multiplexing of multicast and unicast.
  - Sidelink assignment index (SAI, 0–2 bit), used when the device reports the sidelink acknowledgement to the gNB using PUSCH.
  - CBG transmission indicator (CBGTI, 0, 2, 4, 6, or 8 bit), indicating the code block groups to retransmit as described in Section 13.1. Only present in DCI format 0_1 and only if CBG retransmissions are configured.
- Multi-antenna-related information (present in DCI format 0_1 only)

- DM-RS sequence initialization (1 bit), used to select between two preconfigured initialization values for the DM-RS sequence when using non-precoded OFDM for PUSCH.
- Antenna ports (2–5 bit), indicating the antenna ports upon which the data are transmitted as well as antenna ports scheduled for other users as discussed in Chapters 9 and 11.
- Precoding information (0–6 bit), used to select the precoding matrix W and the number of layers for codebook-based precoding as described in Section 11.3. The number of bits depends on the number of antenna ports and the maximum rank supported by the device. In release 17, a second precoding field can be configured to support multi-TRP operation.
- PTRS-DMRS association (0 or 2 bit), used to indicate the association between the DM-RS and PT-RS ports. In release 17, a second PTRS-DMRS field can be configured to support multi-TRP operation.
- SRS resource indicator (SRI), used to determine the antenna ports and uplink transmission beam to use for PUSCH transmission as described in Section 11.3. The number of bits depends on the number of SRS groups configured and whether codebook-based or non-codebook-based precoding is used. In release 17, a secondary SRI can be configured to support multi-TRP. For DCI format 0_3, the field in an index into a configured table from which the SRS resources on each of the co-scheduled carriers is obtained.
- SRS resource set indicator (0 or 2 bit), used to select the SRS resource set from which the SRS resource is obtained.
- SRS offset indicator (0–2 bit), used to control when the SRS is transmitted.
- SRS request (2 bit), used to request transmission of a sounding reference signal as described in Section 8.3.
- CSI request (0–6 bit), used to request transmission of a CSI report as described in Section 8.1.
- Power-control related information
  - PUSCH power control (2 bit), used to adjust the PUSCH transmission power. In release 17, two additional bits can be configured for the secondary PUSCH to support multi-TRP operation.
  - Beta offset (0 or 2 bit), used to control the amount of resources used by UCI on PUSCH in case dynamic beta offset signaling is configured for DCI format 0_1 as discussed in Section 10.2.8.
  - Power control parameter set (0–2 bit), used to boost the PUSCH transmission power for uplink preemption as described in Chapter 21.
- PDCCH-related information
  - PDCCH monitoring control (0–2 bit), used for dynamic control of PDCCH monitoring for device power saving as described in Chapter 14.

10.1.6 Slot format indication – DCI format 2_0

DCI format 2_0, if configured, is used to signal the slot format information (SFI) to the device as discussed in Section 7.8.3. It is also used for search space group switching and resource-block set availability relevant for unlicensed spectra (Chapter 20). The SFI is transmitted using the regular PDCCH structure and using the SFI-RNTI, common to multiple devices. To assist the device in the blind decoding process, the device is configured with information about the up to two PDCCH candidates upon which the SFI can be transmitted.

10.1.7 Preemption indication – DCI format 2_1

DCI format 2_1 is used to signal the preemption indicator used for downlink preemption to the device. It is transmitted using the regular PDCCH structure, using the INT-RNTI which can be common to multiple devices. The details and usage of the preemption indicator are discussed in Section 14.1.1.

10.1.8 Uplink power control commands – DCI format 2_2

As a complement to the power-control commands provided as part of the downlink scheduling assignments and the uplink scheduling grants, there is the potential to transmit a power-control command using DCI format 2_2. The main motivation for DCI format 2_2 is to support power control for semi-persistent scheduling and configured grants. In this case there is no dynamic scheduling assignment or scheduling grant which can include the power control information for the PUCCH and PUSCH, respectively. Consequently, another mechanism is needed and DCI format 2_2 fulfills this need. One possibility would be to define a very small power-control message, but this would have resulted in an additional DCI size to monitor. Instead, the size of DCI format 2_2 is aligned with the size of DCI formats 0_0/1_0 to reduce the blind decoding complexity, and can contain power-control bits for multiple devices. Each device is configured with a TPC-related RNTI to use in conjunction with DCI format 2_2 and which of the power control bits in the DCI that are intended for this device.[g]

10.1.9 SRS control commands – DCI format 2_3

DCI format 2_3 is used for power control of uplink sounding reference signals for devices which have not coupled the SRS power control to the PUSCH power control. The structure is similar to DCI format 2_2, but with the possibility to, for each device, configure two bits for SRS request in addition to the two power control bits. DCI format 2_3 is aligned with the size of DCI formats 0_0/1_0 to reduce the blind decoding complexity.

---

[g] The TPC-PUCCH-RNTI is used for PUCCH power control and the TPC-PUSCH-RNTI for PUSCH power control.

10.1.10 Uplink cancellation indicator – DCI format 2_4

The cancellation indicator is used as part of the uplink inter-device preemption in release 16 as described in Chapter 21. It is transmitted using DCI format 2_4 with the CI-RNTI and the regular PDCCH structure.

10.1.11 Soft resource indicator – DCI format 2_5

DCI format 2_5 is introduced in release 16 to support the IAB feature. Its purpose is to indicate whether a certain time resource is available for the access link in the IAB node, see Chapter 24. Messages using DCI-format 2_5 use the AI-RNTI.

10.1.12 DRX activation – DCI format 2_6

To reduce the device power consumption, release 16 introduces a wake-up signal and a mechanism for entering cell dormancy, both which rely on DCI format 2_6 for the necessary control signaling. Wake-up signals and cell dormancy are described in Section 14.5.

10.1.13 Paging early indicator and dynamic TRS control – DCI format 2_7

DCI format 2_7 is introduced in release 17 as part of the device power saving enhancements for idle mode, see Chapter 14. The purpose of the paging early indicator (PEI) is to inform the device whether an upcoming paging occasion is used or not, thereby allowing the device to avoid spending energy on preparing to receive a full paging message in case it is not paged.

10.1.14 Sidelink scheduling – DCI formats 3_0 and 3_1

Sidelink data transmission, where two devices directly exchange data as described in Chapter 26, can either be autonomously handled by the devices or scheduled by the network. In the latter case, DCI format 3_0 is used to convey the sidelink scheduling information. It is also possible for an NR network to schedule LTE sidelink transmissions in which case DCI format 3_1 is used. The two DCI formats are mutually size aligned by padding format 3_1 if necessary until it matches the size of format 3_0.

10.1.15 Multicast/broadcast scheduling – DCI formats 4_0, 4_1, and 4_2

Support for multicast and broadcast transmissions, where the same information is transmitted to multiple devices, is introduced in release 17 and is described in Chapter 23. Scheduling of broadcast transmissions uses DCI format 4_0 with the MCCH-RNTI or the G-RNTI, while scheduling of multicast transmissions uses DCI formats 4_1 or 4_2 with the G-RNTI or G-CS-RNTI.

10.1.16 Beam indication for network-controlled repeaters – DCI format 5_0

DCI format 5_0 is used to control which beam the network-controlled repeater should use when transmitting to/receiving from devices, see Chapter 23. The NCR-RNTI is used on conjunction with this DCI format.

10.1.17 Signaling of frequency-domain resources

To determine the frequency-domain resources to transmit or receive upon, two fields are of interest: the resource-block allocation field and the bandwidth part indicator.

The resources allocation fields determine the resources blocks in the active bandwidth part upon which data are transmitted. There are three possibilities for signaling the resources-block allocation, type 0, type 1, and type 2.

Type 0 is a bitmap-based allocation scheme. The most flexible way of indicating the set of resource blocks the device is supposed to receive the downlink transmission upon is to include a bitmap with size equal to the number of resource blocks in the bandwidth part. This would allow for an arbitrary combination of resource blocks to be scheduled for transmission to the device but would, unfortunately, also result in a very large bitmap for the larger bandwidths. For example, in the case of a bandwidth part of 100 resource blocks, the downlink PDCCH would require 100 bits for the bitmap alone, to which the other pieces of information need to be added. Not only would this result in a large control-signaling overhead, but it could also result in downlink coverage problems as more than 100 bits in one OFDM symbol correspond to a data rate exceeding 1.4 Mbit/s for 15 kHz subcarrier spacing and even higher for the higher subcarrier spacings. Consequently, there is a need to reduce the bitmap size while keeping sufficient allocation flexibility. This can be achieved by pointing not to individual resource blocks in the frequency domain, but to groups of contiguous resource blocks, as shown in the top of Fig. 10.10. The size of such a resource-block group is determined by the size of the bandwidth part. Two different configurations are possible for each size of the bandwidth parts, possibly resulting in different resource-block-group sizes for a given size of the bandwidth part. A third configuration is added in release 18 as part of the multi-carrier enhancements when a single DCI scheduled multiple carriers, see Table 10.4.

Resource allocation type 1 does not rely on a bitmap. Instead, it encodes the resource allocation as a start position and length of the resource-block allocation. Thus, it does not support arbitrary allocations of resource blocks but only frequency-contiguous allocations, thereby reducing the number of bits required for signaling the resource-block allocation.

Resource allocation type 2, was added in release 16 to support interlaced resource allocation in the uplink, see Chapter 20 for a description of interlaces and the associated resource allocation mechanisms.

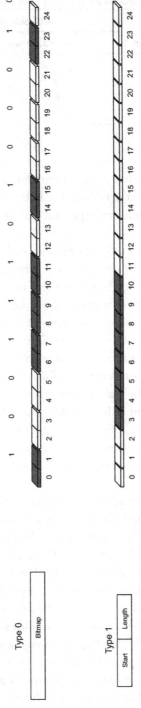

Fig. 10.10 Illustration of resource-block allocation types 0 and 1 (a bandwidth part of 25 resource blocks is used in this example).

Table 10.4 Size of the resource-block groups for resource allocation type 0.

BWP size [RB]	RBG size [RB]		
	Configuration 1	Configuration 2	Configuration 3
1–36	2	4	8
37–72	4	8	16
73–144	8	16	32
145–275	16	16	32

The resource allocation scheme to use is configured: type 0, type 1, type 2, or dynamic selection between types 0 and 1. For the fallback DCIs, only resource block allocation type 1 is supported as a small overhead is more important than the flexibility to configure non-contiguous resources.

All resource-allocation types refer to *virtual* resource blocks (see Section 7.3 for a discussion of resource-block types). For resource-allocation type 0, a non-interleaved mapping from virtual to physical resource blocks is used, meaning that the virtual resource blocks are directly mapped to the corresponding physical resource blocks. Interleaved mapping is not necessary in this case as an interleaved mapping basically corresponds to another bitmap allocation. For resource-allocation type 1, on the other hand, both interleaved and non-interleaved mapping is supported. The VRB-to-PRB mapping bit (if present, downlink only) indicates whether the allocation signaling uses interleaved or non-interleaved mapping. In the uplink, non-interleaved mapping is always used.

Returning to the bandwidth part indicator, this field is used to switch the active bandwidth part. It can either point to the active bandwidth part, or to another bandwidth part to activate. If the field points to the current active bandwidth part, the interpretation of the DCI content is straightforward – the resource allocation applies to the active bandwidth part as described above.

However, if the bandwidth part indicator points to a *different* bandwidth part than the active bandwidth part, the handling becomes more intricate. Many transmission parameters in general are configured per bandwidth part. The DCI payload size may therefore differ between different bandwidth parts. The frequency-domain resource allocation field is an obvious example – the larger the bandwidth part, the larger the number of bits for frequency-domain resource allocation. At the same time, the DCI sizes assumed when performing blind detection were determined by the *currently active* bandwidth part, not the bandwidth part to which the bandwidth part index points. Requiring the device to perform blind detection of multiple DCI sizes matching all possible bandwidth part configurations would be too complex. Hence, the DCI information obtained under the assumption of the DCI format being given by the currently active bandwidth part must be transformed to the new bandwidth part which may have not only a different size in general, but also be configured with a different set of transmission parameters, for example TCI states which are configured per bandwidth part. The transformation is done using padding/truncation for each DCI field to match the requirements of the targeted bandwidth part. Once this is done, the bandwidth part pointed to by the bandwidth part indicator becomes the new active bandwidth part and the scheduling grant is applied to this bandwidth part. Similar transformation is sometimes required for DCI formats 0_0 and 1_0 in situations where the "3+1" DCI size budget otherwise would be violated.

10.1.18 Signaling of time-domain resources

The time-domain allocation for the data to be received or transmitted is dynamically signaled in the DCI, which is useful as the part of a slot available for downlink reception or uplink transmission may vary from slot to slot as a result of the use of dynamic TDD or the amount of resources used for uplink controls signaling. Furthermore, the slot in which the transmission occurs also needs to be signaled as part of the time-domain allocation. Although the downlink data in many cases are transmitted in the same slot as the corresponding assignment, this is frequently not the case for uplink transmissions.

One approach would be to separately signal the slot number, the starting OFDM symbol, and the number of OFDM symbols used for transmission or reception. However, as this would result in an unnecessarily large number of bits, NR has adopted an approach based on configurable tables. The time-domain allocation field in the DCI is used as an index into an RRC-configured table from which the time-domain allocation is obtained as illustrated in Fig. 10.11.

There is one table for uplink scheduling grants and one table for downlink scheduling assignments. Up to 16 rows (a number that can be increased to 64 in later releases) can be configured where each row at least contains

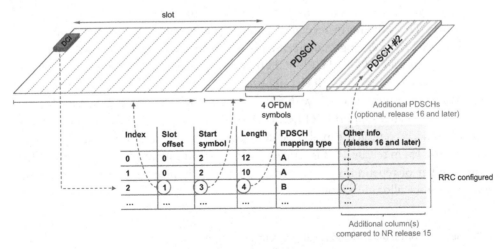

Fig. 10.11 Signaling of time-domain allocation (downlink).

- A slot offset, that is, the slot relative to the one where the DCI was obtained. Typically, the default value 0 is used for the downlink, indicating that data and control are in the same slot, while for the uplink a larger number is used (the default number depends on the subcarrier spacing). The larger value is motivated by the need for scheduling uplink transmissions further into the future for coexistence with (primarily) LTE TDD. The slot offset can range from 0 to 32 in the first release of NR, but for later releases larger numbers are possible in some circumstances in order to handle the longer roundtrip time present for NTN, or to handle the extensions of the subcarrier spacing to 480 and 960 kHz.
- The first OFDM symbol in the slot where the data are transmitted.
- The duration of the transmission in number of OFDM symbols in the slot. Not all combinations of start and length fit within one slot, for example, starting at OFDM symbol 12 and transmit during five OFDM symbols obviously results in crossing the slot boundary and represents an invalid combination. Therefore, the start and length are typically jointly encoded[h] to cover only the valid combinations (although in Fig. 10.11 they are shown as two separate columns for illustrative reasons).
- For the downlink, the PDSCH mapping type, that is, the DM-RS location as described in Section 9.11, is also part of the table. This provides more flexibility compared to separately indicating the mapping type.

It is also possible to configure slot aggregation, that is, a transmission where the same transport block is repeated across up to eight slots. However, in release 15 this is not part of the dynamic signaling using a table but is a separate RRC configuration. Slot aggregation is primarily a tool to handle coverage-challenged deployments and thus there is less need for a fully dynamic scheme.

In release 16, additional columns can be configured in the table to provide additional information. For example, a row in the table may contain multiple sets of time-domain allocation information. The first allocation information in the row is used for the first PDSCH/PUSCH, the second allocation information for the second PDSCH/PUSCH and so forth. This is useful in several scenarios. One example is to support contiguous transmission when operating in unlicensed spectra, see Section 20.6.4.2. Another usage is to enable continuous PDSCH/PUSCH transmission even in cases when the PDCCH is not monitored in every slot, which can be the case in FR2-2. This structure is also used when scheduling multiple TRPs with a single DCI, a feature added in release 17.

Similarly, to better support URLLC, a column indicating the number of times a transmission should be repeated can also be configured as motivated in Chapter 21. Thus, unlike release 15, in release 16 it is actually possible to indicate the slot aggregation – that is, the number of repetitions – in a dynamic manner by properly configure the time-domain resource allocation table.

---

[h] When using TBoMS, the start and length are configured separately.

Using a configurable table as described here results in a very flexible framework; it is possible to support almost any scenario and scheduling strategy by properly configuring the appropriate table. However, it also results in a chicken-and-egg problem – to configure the table downlink data transmission to convey the RRC signaling is required but to transmit data in the downlink a table must have been provided. It is not even possible to receive system information as this is transmitted using the PDSCH and the same allocation principles as for user data. To resolve this problem, the specifications provide default time-domain allocation tables that are used if no table is configured. The entries of the tables are chosen to suit common scenarios used for system information delivery and some typical allocations for user-data transmission, see Chapter 16 for details. These default tables can be used until the necessary table configuration is provided to the device. In many cases the default tables are sufficient, in which case there is no need to configure other values.

Co-scheduling of multiple carriers with a single DCI is possible in release 18 by using DCI formats 0_3/1_3. Consequently, time-domain resources for each of the up to four carriers a group is necessary. This is handled in a two-step manner by an extra configurable table where each row contains one index per carrier in the group scheduled. The time-domain resource allocation field selects a row in this table and the indices in the selected row points into the time-domain resource allocation tables configured for each of the carriers, see Fig. 10.12.

10.1.19 Signaling of transport-block sizes

Proper reception of a downlink transmission requires, in addition to the set of resource blocks, knowledge about the modulation scheme and the transport-block size, information (indirectly) provided by the 5-bit MCS field. In principle, the transport block size as a function of the MCS field and the resource-block allocation could be tabulated in the specifications (this is the approach taken for LTE). However, the significantly larger bandwidths supported in NR, together with a wide range of transmission durations and variations in the overhead depending on other features configured such as CSI-RS, would result in a large number of tables required to handle the large dynamic range in terms of transport block sizes. Such a scheme may also require modifications whenever some of these parameters change. Therefore, NR opted for a formula-based approach combined with a table for the smallest transport-block sizes instead of a purely table-based scheme to achieve the necessary flexibility.

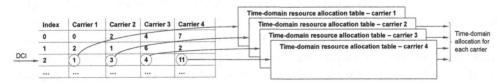

Fig. 10.12 Time-domain resource allocation for co-scheduling of multiple carriers.

The first step is to determine the modulation scheme and code rate from the MCS field. This is done using one of three tables, one table if neither 256QAM nor 1024QAM is configured, one if 256QAM is configured, and one if 1024QAM is configured. For a device not configured with 256QAM or 1024QAM, 29 of the 32 combinations of the 5-bit MCS fields are used to signal the modulation-and-coding scheme whereas three combinations are reserved, the purpose of which is described later. Each of the 29 modulation-and-coding scheme entries represents a particular combination of modulation scheme and channel-coding rate or, equivalently, a certain spectral efficiency measured in the number of information bits per modulation symbol, ranging from approximately 0.2–5.5 bit/s/Hz. For devices configured with support for 256QAM, four of the 32 combinations are reserved and the remaining 28 combinations indicate a spectral efficiency in the range 0.2–7.4 bit/s/Hz. A similar approach is taken for devices with 1024QAM enabled – five combinations are reserved and the remaining table entries cover spectral efficiencies in the range of 0.2–9.3 bit/s/Hz. There is also an alternative table providing lower spectral efficiency values, in the range 0.0586–4.5234 bit/s/Hz. At first glance it might seem strange to lower the spectral efficiency, but this is done in order to improve the robustness and to reduce the error probability to better support reliability-critical information. Whether to use one of the three the regular tables or the more robust table is determined by the RNTI scheduling the device; the C-RNTI implies the use of the regular table and the MCS-C-RNTI (if configured) implies the use of the more robust table.

Given the modulation order, the number of resource blocks scheduled, and the scheduled transmission duration, the number of available resource elements can be computed. From this number the resource elements used for DM-RS are subtracted. A constant, configured by higher layers and modeling the overhead by other signals such as CSI-RS or SRS is also subtracted. The resulting estimate of resource elements available for data is then, together with the number of transmission layers, the modulation order, and the code rate obtained from the MCS, used to calculate an intermediate number of information bits. This intermediate number is then quantized to obtain the final transport block size while at the same time ensuring byte-aligned code blocks, and that no filler bits are needed in the LDPC coding. The quantization also results in the same transport block size being obtained, even if there are modest variations in the amount of resources allocated, a property that is useful when scheduling retransmissions on a different set of resources than the initial transmission.

Returning to the three to five reserved combinations in the modulation-and-coding field mentioned at the beginning of this section, those entries can be used for retransmissions only. In the case of a retransmission, the transport-block size is, by definition, unchanged and fundamentally there is no need to signal this piece of information. Instead, the three to five reserved values represent the modulation scheme – QPSK, 16QAM, 64QAM, 256QAM, or 1024QAM (the last two only if configured by the network) – which allows the scheduler to use an (almost) arbitrary combination of resource blocks

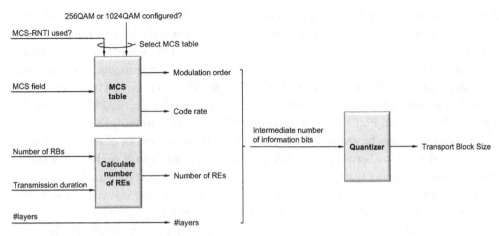

Fig. 10.13 Calculating the transport block size.

for the retransmission. Obviously, using any of the reserved combinations assumes that the device properly received the control signaling for the initial transmission; if this is not the case, the retransmission should explicitly indicate the transport-block size.

The derivation of the transport-block size from the modulation-and-coding scheme and the number of scheduled resource blocks is illustrated in Fig. 10.13.

10.2 Uplink

There is also a need for uplink L1/L2 control signaling to support data transmission on downlink and uplink transport channels. Uplink L1/L2 control signaling consists of:

- Hybrid-ARQ acknowledgments for received DL-SCH transport blocks;
- Channel-state information (CSI) related to the downlink channel conditions, used to assist downlink scheduling, including multi-antenna and beamforming schemes; and
- Scheduling requests, indicating that a device needs uplink resources for UL-SCH transmission.

There is no UL-SCH transport-format information included in the uplink transmission. As mentioned in Section 6.3, the gNB is in complete control of the UL-SCH transmissions and the device always follows the scheduling grants received from the network, including the UL-SCH transport format specified in those grants. Thus, the network knows the transport format used for the UL-SCH transmission in advance and there is no need for any explicit transport-format signaling on the uplink.

The *physical uplink control channel* (PUCCH) is the basis for transmission of uplink control. In principle, the UCI could be transmitted on the PUCCH regardless of whether the device is transmitting data on the PUSCH simultaneously. However, especially if the uplink resources for the PUSCH and the PUCCH are on the same carrier (or, to be more precise, use the same power amplifier) but widely separated in the frequency domain, the device may need a relatively large power back-off to fulfill the spectral emission requirements with a corresponding impact on the uplink coverage. Hence, NR supports *UCI on PUSCH* as the basic way of handling simultaneous transmission of data and control. If the device is transmitting on the PUSCH, the UCI is multiplexed with data on the granted resources instead of being transmitted on the PUCCH. Simultaneous PUSCH and PUCCH is not supported but may be introduced in a later release.

Beamforming can be applied to the PUCCH. This is realized by configuring one or more spatial relations between the PUCCH and downlink signals such as CSI-RS or SSB. In essence, such a spatial relation means that the device can transmit the PUCCH using the same beam as it used for receiving the corresponding downlink signal. For example, if a spatial relation between PUCCH and SSB is configured, the device will transmit PUCCH using the same beam as it used for receiving the SSB. Multiple spatial relations can be configured and MAC control elements are used to indicate which relation to use.

In the case of carrier aggregation, the uplink control information is transmitted on the primary cell as a baseline. This is motivated by the need to support asymmetric carrier aggregation where the number of downlink carriers supported by a device is unrelated to the number of uplink carriers. There are several reasons for devices supporting downlink carrier aggregation but not uplink carrier aggregation being common. The amount of traffic is typically larger in the downlink and implementing uplink carrier aggregation is often more complex than downlink carrier aggregation. Consequently, a large number of downlink component carriers may need to be acknowledged using a single uplink carrier, even for devices supporting uplink carrier aggregation. To avoid overloading a single carrier, it is possible to configure two *PUCCH groups* where feedback relating to the first group of carriers is transmitted in the uplink of the PCell and feedback relating to the second group of carriers are transmitted on another cell known as the PUCCH-SCell, as illustrated in Fig. 10.14.

In release 17, PUCCH transmissions were further enhanced, mainly motivated by further reducing the latency in conjunction with URLLC, see Chapter 21. In short, release 17 allows the PUCCH to be switched to another cell, known as the PUCCH

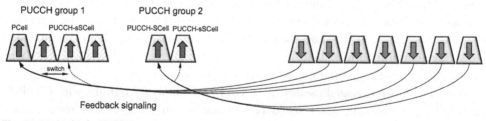

Fig. 10.14 Multiple PUCCH groups.

switching cell (PUCCH-sSCell). There can be one PUCCH-sSCell configured in each of the two PUCCH groups and switching between the PCell and the PUCCH-sSCell (or the PUCCH-SCell and the PUCCH-sSCell) can either be dynamically or semi-statically. In the former case, the PUCCH cell indicator in the DCI selects which of two semi-statically configured cells that should be used for the PUCCH transmission. In the latter case, a semi-static pattern is configured, indicating for each of the PUCCH groups which of the two carriers to use for PUCCH transmission.

In the following section, the basic PUCCH structure and the principles for PUCCH control signaling are described, followed by control signaling on PUSCH.

10.2.1 Basic PUCCH structure

Uplink control information can be transmitted on PUCCH using several different formats.

Two of the formats, 0 and 2, are sometimes referred to as *short PUCCH formats*, as they occupy at most two OFDM symbols. In many cases the last one or two OFDM symbols in a slot are used for PUCCH transmission, for example to transmit a hybrid-ARQ acknowledgment of the downlink data transmission. The short PUCCH formats include:

- PUCCH format 0, capable of transmitting at most two bits and spanning one or two OFDM symbols. This format can, for example, be used to transmit a hybrid-ARQ acknowledgment of a downlink data transmission, or to issue a scheduling request.
- PUCCH format 2, capable of transmitting more than two bits and spanning one or two OFDM symbols. This format can, for example, be used for CSI reports or for multi-bit hybrid-ARQ acknowledgments in the case of carrier aggregation or per-CBG retransmission.

Three of the formats, 1, 3, and 4, are sometimes referred to as *long PUCCH formats* as they occupy from 4 to 14 OFDM symbols. The reason for having a longer time duration than the previous two formats is coverage. If a duration of one or two OFDM symbols does not provide sufficient energy for reliable reception, a longer time duration is necessary and one of the long PUCCH formats can be used. The long PUCCH formats include:

- PUCCH format 1, capable of transmitting at most two bits.
- PUCCH formats 3 and 4, both capable of transmitting more than two bits but differing in the multiplexing capacity, that is, how many devices that can use the same time-frequency resource simultaneously.

Since the PUSCH can be configured to use either OFDM or DFT-spread OFDM, one natural thought would be to adopt a similar approach for the PUCCH. However, to reduce the number of options to specify, this is not the case. Instead, the PUCCH formats are in general designed for low cubic metric, PUCCH format 2 being the exception and using pure OFDM only. Another choice made to simplify the overall design was to only support specification-transparent transmit diversity schemes. In other words, there is only a single antenna port specified for the PUCCH and if the device is equipped with multiple transmit antennas it is up to the device implementation how to exploit these antennas, for example by using some form of delay diversity.

Some of these PUCCH formats – formats 0, 1, 2, and 3 – are also available with interlaced resource mapping where the transmission is spread across a larger number of resource blocks. This is used in unlicensed spectra for regulatory reasons as discussed in Chapter 20.

In the following, the detailed structure of each of the PUCCH formats will be described, assuming non-interlaced mapping. Once the non-interlaced mapping is described, the extension to interlaced mapping as described in Chapter 20 is straight forward.

10.2.2 PUCCH format 0

PUCCH format 0, illustrated in Fig. 10.15, is one of the short PUCCH formats and is capable of transmitting up to two bits. It is used for hybrid-ARQ acknowledgments and scheduling requests.

Sequence selection is the basis for PUCCH format 0. For the small number of information bits supported by PUCCH format 0, the gain from coherent reception is not that large. Furthermore, multiplexing information and reference signals in one OFDM symbol while maintaining a low cubic metric is not possible. Therefore, a different structure where the information bit(s) selects the sequence to transmit is used. The transmitted sequence is generated by different phase rotations of the same underlying length-12 base sequence, where the base sequences are the same base sequences defined for generating the reference signal in the case of DFT-precoded OFDM as described in Section 9.11.2. Thus, the phase rotation applied to the base sequence carries the information. In other words, the information selects one of several phase-rotated sequences.

Twelve different phase rotations are defined for the same base sequence, providing up to twelve different orthogonal sequences from each base sequence. A linear phase

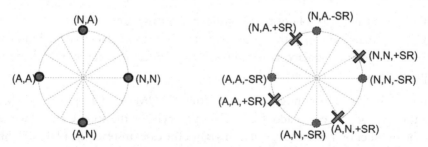

Fig. 10.15 Examples of phase rotations as a function of hybrid-ARQ acknowledgments and scheduling request.

rotation in the frequency domain is equivalent to applying a cyclic shift in the time domain, hence, the term "cyclic shift" is sometimes used with an implicit reference to the time domain.

To maximize the performance, the phase rotations representing the different information bits are separated with $2\pi \cdot 6/12$ and $2\pi \cdot 3/12$ for one and two bits acknowledgments, respectively. In the case of a simultaneous scheduling request, the phase rotation is increased by $3\pi/12$ for one acknowledgment bit and by $2\pi/12$ for two bits as illustrated in Fig. 10.15.

The phase rotation applied to a certain OFDM symbol carrying PUCCH format 0 depends not only on the information to be transmitted as already mentioned, but also on a reference rotation provided as part of the PUCCH resource allocation mechanism as discussed in Section 10.2.7. The intention with the reference rotation is to multiplex several devices on the same time-frequency resource. For example, two devices transmitting a single hybrid-ARQ acknowledgment can be given different reference phase rotations such that one device uses 0 and $2\pi \cdot 6/12$, while the other device uses $2\pi \cdot 3/12$ and $2\pi \cdot 9/12$. Finally, there is also a mechanism for cyclic shift hopping where a phase offset varying between different slots is added. The offset is given by a pseudo-random sequence. The underlying reason is to randomize interference between different devices.

The base sequence to use can be configured per cell using an identity provided as part of the system information. Furthermore, sequence hopping, where the base sequence used varies on a slot-by-slot basis, can be used to randomize the interference between different cells. As seen from this description many quantities are randomized in order to mitigate interference.

PUCCH format 0 is typically transmitted at the end of a slot as illustrated in Fig. 10.16. However, it is possible to transmit PUCCH format 0 also in other positions within a slot. One example is frequently occurring scheduling requests (as frequent as every second OFDM symbol can be configured). Another example when this can be useful is to acknowledge a downlink transmission on a downlink carrier at a high carrier frequency and, consequently, a correspondingly higher subcarrier spacing and shorter downlink slot duration, on an uplink using a much lower carrier frequency. This can be a relevant scenario in the case of carrier aggregation. If low latency is important, the hybrid-ARQ acknowledgment needs to be fed back quickly after the end of the downlink slot, which is not necessarily at the end of the uplink slot if the subcarrier spacing differs between uplink and downlink.

In the case of two OFDM symbols used for PUCCH format 0, the same information is transmitted in both OFDM symbols. However, the reference phase rotation as well as the frequency-domain resources may vary between the symbols, essentially resulting in a frequency-and-sequence-hopping mechanism.

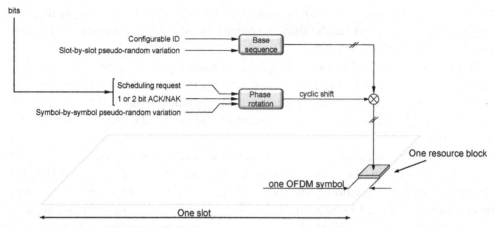

Fig. 10.16 Example of PUCCH format 0.

For operation in FR2-2, PUCCH format 0 can be configured to occupy more than one resource block in the frequency domain. In this case, the sequence lengths above are multiples of 12. Apart from this, the structure is as described above.

10.2.3 PUCCH format 1

PUCCH format 1 is to some extent the long PUCCH counterpart of format 0. It is capable of transmitting up to two bits, using from 4 to 14 OFDM symbols, each one resource block wide in frequency. The OFDM symbols used are split between symbols for control information and symbols for reference signals to enable coherent reception. The number of symbols used for control information vs. the reference signal is a trade-off between channel-estimation accuracy and energy in the information part, respectively. Approximately half the symbols for reference symbols were found to be a good compromise for the payloads supported by PUCCH format 1.

The one or two information bits to be transmitted are BPSK or QPSK modulated, respectively, and multiplied by the same type of length-12 low-PAPR sequence as used for PUCCH format 0. Similar to format 0, sequence and cyclic shift hopping can be used to randomize interference. The resulting modulated length-12 sequence is block-wise spread with an orthogonal code of the same length as the number of symbols used for the control information. The use of the orthogonal code in the time domain increases the multiplexing capacity as multiple devices having the same base sequence and phase rotation still can be separated using different orthogonal codes.

The reference signals are inserted using the same structure, that is, an unmodulated length-12 sequence is block-spread with an orthogonal code and mapped to the OFDM symbols used for PUCCH reference-signal transmission. Thus, the length of the orthogonal code together with the number of cyclic shifts, determines the number of devices

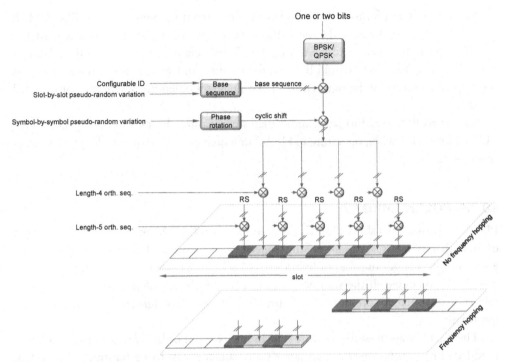

Fig. 10.17 Example of PUCCH format 1 without frequency hopping (top) and with frequency hopping (bottom).

that can transmit PUCCH format 1 on the same resource. An example is shown in Fig. 10.17 where nine OFDM symbols are used for PUCCH transmission, four carrying the information and five used for reference signals. Hence, up to four devices, determined by the shorter of the codes for the information part and the reference-signal part, can share the same cyclic shift of the base sequence, and the same set of time-frequency resources for PUCCH transmission in this particular example. Assuming a cell-specific base sequence and six out of the 12 cyclic shifts being useful from a delay-spread perspective, this results in a multiplexing capacity of at most 24 devices on the same time-frequency resources.

The longer transmission duration of the long PUCCH formats compared to a short single-symbol format opens the possibility for frequency hopping as a mean to achieve frequency diversity. However, LTE-like schemes hopping at the slot boundaries only are not sufficient. Additional flexibility is needed in NR as the PUCCH duration can vary depending on the scheduling decisions and overall system configuration. Furthermore, as the devices are supposed to transmit within their active bandwidth part only, hopping is not necessarily between the edges of the overall carrier bandwidth as in LTE. Therefore, whether to hop or not is configurable and determined as part of the PUCCH resource configuration. The position of the hop is obtained from the length of the PUCCH. If frequency hopping is enabled, one orthogonal block-spreading sequence is used per hop. Using the previous example in Fig. 10.17 two sets of sequences length-2/length-2 and length-2/length-3, would be used for the first and second hops, respectively, as shown at the bottom of the figure instead of a single set of length-4/length-5 orthogonal sequences.

Similar to PUCCH format 0, more than one resource block can be configured for PUCCH format 1 when operating in FR2-2 in which case the sequence lengths are multiples of 12.

10.2.4 PUCCH format 2

PUCCH format 2 is a short PUCCH format based on OFDM and used for transmission of more than two bits, for example simultaneous CSI reports and hybrid-ARQ acknowledgments, or a larger number of hybrid-ARQ acknowledgments. A scheduling request can also be included. If the number of bits to be jointly encoded is too large, the CSI report is dropped to preserve the hybrid-ARQ acknowledgments which are more important.

The overall transmission structure is straightforward. For larger payload sizes, a CRC is added. The control information (after CRC attachment) to be transmitted is coded, using Reed-Muller codes for payloads up to and including 11 bits and Polar[i] coding for larger payloads, followed by scrambling and QPSK modulation. The scrambling sequence is based on the device identity (the C-RNTI) together with the physical-layer cell identity (or a configurable virtual cell identity), ensuring interference randomization across cells and devices using the same set of time-frequency resources. The QPSK symbols are then mapped to subcarriers across multiple resource blocks using one or two OFDM symbols. A pseudo-random QPSK sequence, mapped to every third subcarrier in each OFDM symbol, is used as a demodulation reference signal to facilitate coherent reception at the base station.

The number of resource blocks used by PUCCH format 2 is determined by the payload size and a configurable maximum code rate. The number of resource blocks is thus smaller if the payload size is smaller, keeping the effective code rate roughly constant. The number of resource blocks used is upper bounded by a configurable limit.

PUCCH format 2 is typically transmitted at the end of a slot as illustrated in Fig. 10.18. However, similar to format 0 and for the same reasons, it is possible to transmit PUCCH format 2 also in other positions within a slot.

---

[i] Polar coding is used for the DCI as well, but the details of the Polar coding for UCI are different.

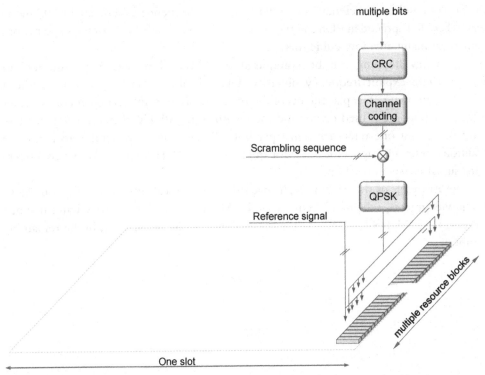

Fig. 10.18 Example of PUCCH format 2 (the CRC is present only for larger payloads).

10.2.5 PUCCH format 3

PUCCH format 3 can be seen as the long PUCCH counterpart to PUCCH format 2. More than two bits can be transmitted using PUCCH format 3 using from 4 to 14 symbols, each of which can be multiple resource blocks wide. Thus, it is the PUCCH format with the largest payload capacity. Similar to PUCCH format 1, the OFDM symbols used are split between symbols for control information and symbols for reference signals.

The control information to be transmitted is coded using Reed–Muller codes for 11 bits or less and Polar codes for large payloads, followed by scrambling and modulation. Using the same principles as PUCCH format 2, a CRC is attached to the control information for the larger payloads. The scrambling sequence is based on the device identity (the C-RNTI) together with the physical-layer cell identity (or a configurable virtual cell identity), ensuring interference randomization across cells and devices using the same set of time-frequency resources. The modulation scheme used is QPSK but it is possible to optionally configure $\pi/2$-BPSK to lower the cubic metric at a loss in link performance.

The resulting modulation symbols are divided into groups corresponding to the OFDM symbols, followed by DFT precoding to reduce the cubic metric and improve the power amplifier efficiency. The reference signal sequence is generated in the same way as for DFT-precoded PUSCH transmissions, see Section 9.11.2, for the same reason, namely to maintain a low cubic metric.

Frequency hopping can be configured for PUCCH format 3 as illustrated in Fig. 10.19 to exploit frequency diversity, but it is also possible to operate without frequency hopping. The placements of the reference signal symbols depend on whether frequency hopping is used or not and the length of the PUCCH transmission as there must be at least one reference signal per hop. There is also a possibility to configure additional reference signal locations for the longer PUCCH durations to get two reference signal instances per hop.

The mapping of the UCI is such that the more critical bits, that is, hybrid-ARQ acknowledgments, scheduling request, and CSI part 1, are jointly coded and mapped close to the DM-RS locations, while the less critical bits are mapped in the remaining positions.

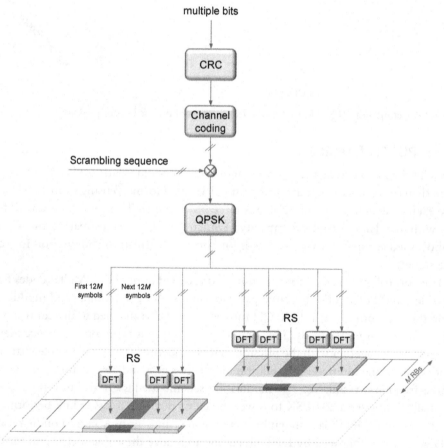

Fig. 10.19 Example of PUCCH format 3 (the CRC is present for large payload sizes only).

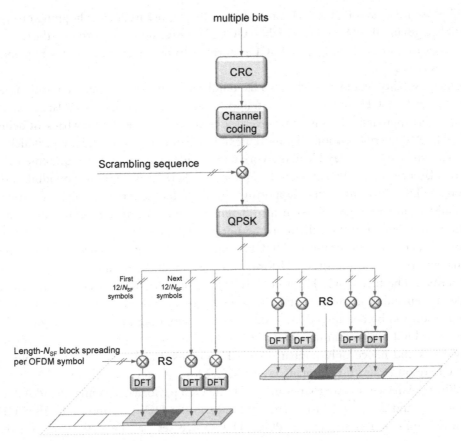

Fig. 10.20 Example of PUCCH format 4.

10.2.6 PUCCH format 4

PUCCH format 4 (see Fig. 10.20) is in essence the same as PUCCH format 3 but with the possibility to code-multiplex multiple devices in the same resource and using at most one resource block in the frequency domain. Each control-information-carrying OFDM symbol carries $12/N_{SF}$ unique modulation symbols. Prior to DFT-precoding, each modulation symbol is block-spread with an orthogonal sequence of length N_{SF}. Spreading factors two and four are supported – that is, N_{SF} equals two or four – implying a multiplexing capacity of two or four devices on the same set of resource blocks.

Similar to PUCCH format 0, more than one resource block can be configured for PUCCH format 4 when operating in FR2-2.

10.2.7 Resources and parameters for PUCCH transmission

In the discussion of the different PUCCH formats, a number of parameters were assumed to be known. For example, the resource blocks to map the transmitted signal to, the initial phase rotation for PUCCH format 0, whether to use frequency hopping or not, and the length in OFDM symbols for the PUCCH transmission. Furthermore, the device also needs to know which of the PUCCH formats to use, and which time-frequency resources to use.

One possibility would be to define a fixed linkage between the uplink control information, the PUCCH format, and the transmission parameters (LTE to a large extent adopted this approach). This is a low-overhead solution but has the drawback of being inflexible. NR therefore adopted a more flexible scheme from the beginning, which is necessary given the very flexible framework with a wide range of service requirements in terms of latency and spectral efficiency, support of no predefined uplink-downlink allocation in TDD, different devices supporting aggregation of different number of carriers, and different antenna schemes requiring different amounts of feedback just to name some motivations. Central in this scheme is the notion of *PUCCH resource sets*. A PUCCH resource set contains one or more PUCCH resource configurations where each resource configuration contains the PUCCH format to use and all the parameters necessary for that format. The first PUCCH resource set can contain up to 32 PUCCH resources while the remaining sets may contain up to eight resources each. Up to four PUCCH resource sets can be configured, each of them corresponding to a certain range of the number of UCI bits to transmit. PUCCH resource set 0 can handle UCI payloads up to two bits and hence only contain PUCCH formats 0 and 1, while the remaining PUCCH resource sets may contain any PUCCH format except format 0 and 1.

When the device is about to transmit UCI, the UCI payload determines the PUCCH resource set and the PUCCH resource indicator in the DCI determines the PUCCH resource configuration within the PUCCH resource set (see Fig. 10.21). Thus, the

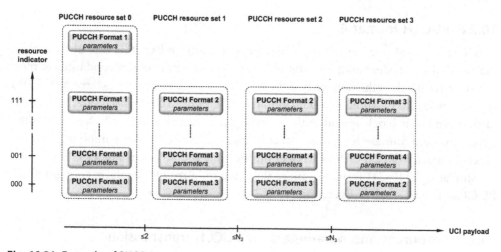

Fig. 10.21 Example of PUCCH resource sets.

scheduler has control of where the uplink control information is transmitted. For the first resource set, which may contain up to 32 resources, there can be more resources than what is possible to indicate with a three-bit PUCCH resource indicator. If this is the case, the index of the first CCE of the PDCCH scheduling the uplink is used together with the PUCCH resource indicator to determine the PUCCH resource within the set.[j] For periodic CSI reports and scheduling request opportunities, which both are semi-statically configured, the PUCCH resources are provided as part of the CSI or SR configuration.

Release 17 further enhances the PUCCH resource allocation by introducing dynamic control of PUCCH repetitions for hybrid-ARQ acknowledgements. This is achieved by including a field in each PUCCH resource to indicate the PUCCH repetition factor. By using different resource indicators, the network can control the amount of PUCCH repetitions dynamically.

10.2.8 Uplink control signaling on PUSCH

If the device is transmitting data on PUSCH – that is, has a valid scheduling grant – simultaneous control signaling could in principle remain on the PUCCH. However, as already discussed, this is not the case as in many cases it is preferable to multiplex data and control on PUSCH and avoid a simultaneous PUCCH. One reason is the increased cubic metric compared to UCI on PUSCH when using DFT-precoded OFDM. Another reason is the more challenging RF implementation if out-of-band emission requirements should be met at higher transmission powers and with PUSCH and PUCCH widely separated in the frequency domain. Hence, UCI on PUSCH is the main mechanism for simultaneous transmission of UCI and uplink data. The same principles are used for both OFDM and DFT-precoded OFDM in the uplink.

Only hybrid-ARQ acknowledgments and CSI reports are rerouted to the PUSCH. There is no need to request a scheduling grant when the device is already scheduled; instead, in-band buffer-status reports can be sent as described in Section 14.2.3.

In principle, the base station knows when to expect a hybrid-ARQ acknowledgment from the device and can therefore perform the appropriate demultiplexing of the acknowledgment and the data part. However, there is a certain probability that the device has missed the scheduling assignment on the downlink control channel. In this case the base station would expect a hybrid-ARQ acknowledgment while the device will not transmit one. If the rate-matching pattern would depend on whether an acknowledgment is transmitted or not, all the coded bits transmitted in the data part could be affected by a missed assignment and are likely to cause the UL-SCH decoding to fail.

One possibility to avoid this error is to puncture hybrid-ARQ acknowledgments onto the coded UL-SCH stream in which case the non-punctured bits are unaffected

---

[j] Thus, by configuring DCI format 1_2 with a zero-bit PUCCH resource indicator, the implicit PUCCH resource allocation scheme of LTE can be mimicked.

by the presence/absence of hybrid-ARQ acknowledgments. This is also the solution adopted in LTE. However, given the potentially large number of acknowledgment bits due to for example carrier aggregation or the use of code block group retransmissions, puncturing is less suitable as a general solution. Instead, NR has adopted a scheme where up to two hybrid-ARQ acknowledgment bits are punctured while for larger number of bits rate matching of the uplink data is used. To avoid the aforementioned error cases, the uplink DAI field in the DCI indicates the amount of resources reserved for uplink hybrid ARQ. Thus, regardless of whether the device missed any previous scheduling assignments or not, the amount of resources to use for the uplink hybrid-ARQ feedback is known.

The mapping of the UCI is such that the more critical bits, that is, hybrid-ARQ acknowledgments, are mapped to the first OFDM symbol after the first demodulation reference signal. Less critical bits, that is CSI reports, are mapped to subsequent symbols.

Unlike the data part, which relies on rate adaptation to handle different radio conditions, this cannot be used for the L1/L2 control-signaling part. Power control could, in principle, be used as an alternative, but this would imply rapid power variations in the time domain, which negatively impact the RF properties. Therefore, the transmission power is kept constant over the PUSCH duration and the amount of resource elements allocated to L1/L2 control signaling – that is, the code rate of the control signaling – is varied. In addition to a semi-static value controlling the amount of PUSCH resources used for UCI, it is also possible to signal this fraction as part of the DCI should a tight control be needed.

CHAPTER 11

Multi-antenna transmission

Multi-antenna transmission is a key component of NR, especially at higher frequencies. This chapter gives a background to multi-antenna transmission in general, followed by a detailed description on NR multi-antenna precoding.

11.1 Introduction

The use of multiple antennas for transmission and/or reception can provide substantial benefits in a mobile-communication system.

Multiple antennas at the transmitter and/or receiver side can be used to provide diversity against fading by utilizing the fact that the channels experienced by different antennas may be at least partly uncorrelated, either due to sufficient inter-antenna distance or due to different polarization between the antennas.

Furthermore, by carefully adjusting the phase, and possibly also the amplitude, of each antenna element, multiple antennas at the transmitter side can be used to provide directivity, that is, to focus the overall transmitted power in a certain direction ("beam forming") or, in the more general case, to specific locations in space. Such directivity can increase the achievable data rates and range due to higher power reaching the target receiver. Directivity will also reduce the interference to other links, thereby improving the overall spectrum efficiency.

Similarly, multiple receive antennas can be used to provide *receiver-side directivity*, focusing the reception in the direction of a target signal while suppressing interference arriving from other directions.

Finally, the presence of multiple antennas at both the transmitter and the receiver sides can be used to enable *spatial multiplexing*, that is, transmission of multiple "layers" in parallel using the same time/frequency resources.

In LTE, multi-antenna transmission/reception for diversity, directivity and spatial multiplexing is a key tool to enable high data rates and high system efficiency. However, multi-antenna transmission/reception is an even more critical component for NR due to the possibility for deployment at much higher frequencies compared to LTE.

There is a well-established and to a large extent correct assumption that radio communication at higher frequencies is associated with higher propagation loss and correspondingly reduced communication range. However, at least part of this is due to an assumption that the dimensions of the receiver antenna scale with the wavelength, that

5G/5G-Advanced
https://doi.org/10.1016/B978-0-443-13173-8.00024-4
269

is, with the inverse of the carrier frequency. As an example, a tenfold increase in the carrier frequency, corresponding to a tenfold reduction in the wave length, is assumed to imply a corresponding tenfold reduction in the physical dimensions of the receiver antenna or a factor of 100 reduction in the physical antenna area. This corresponds to a 20 dB reduction in the energy captured by the antenna.

If the receiver antenna size would instead be kept unchanged as the carrier frequency increases, the reduction in captured energy could be avoided. However, this would imply that the antenna size would increase relative to the wavelength, something that inherently increases the directivity of the antenna.[a] The gain with the larger antenna size can thus only be realized if the receive antenna is well directed toward the target signal.

By also keeping the size of the transmitter-side antenna unchanged, in practice increasing the transmit-antenna directivity, the link budget at higher frequencies can be further improved. Assuming line-of-sight propagation and ignoring other losses, the overall link budget would then actually *improve* for higher frequencies. In practice there are many other factors that negatively impact the overall propagation losses at higher frequencies such as higher atmospheric attenuation and less diffraction leading to degraded non-line-of-sight propagation. Still, the gain from higher antenna directivity at higher frequencies is widely utilized in point-to-point radio links where the use of highly directional antennas at both the transmitter and receiver sides, in combination with line-of-sight links, allows for relatively long-range communication despite operation at very high frequencies.

In a mobile communication system with devices located in many different directions relative to the base station and the devices themselves having an essentially random rotational direction, the use of fixed highly directional antennas is obviously not applicable. However, a similar effect, that is, an extension of the overall receive antenna area enabling higher-directivity transmission, can also be achieved by means of an antenna panel consisting of many small antenna elements. In this case, the dimension of each antenna element, as well as the distance between antenna elements, is proportional to the wave length. As the frequency increases, the size of each antenna element as well as their mutual distances is thus reduced. However, assuming a constant size of the overall antenna configuration, this can be compensated for by increasing the number of antenna elements.

The benefit of such an antenna panel with a large number of small antenna elements, compared to a single large antenna, is that the direction of the transmitter beam can be adjusted by separately adjusting the phase of the signals applied to each antenna element. The same effect can be achieved when a multi-antenna panel is used on the receiver side, that is, the receiver beam direction can be adjusted by separately adjusting the phases of the signals received at each antenna element.

---

[a] The directivity D of an antenna is roughly proportional to the physical antenna area A normalized with the square of the wave length λ.

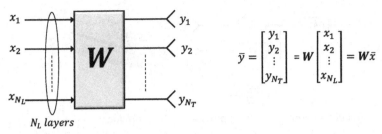

$$\bar{y} = \begin{bmatrix} y_1 \\ y_2 \\ \vdots \\ y_{N_T} \end{bmatrix} = W \begin{bmatrix} x_1 \\ x_2 \\ \vdots \\ x_{N_L} \end{bmatrix} = W\bar{x}$$

Fig. 11.1 General model of multi-antenna transmission mapping N_L layers to N_A antennas.

In general, any linear multi-antenna transmission scheme can be modeled according to Fig. 11.1 with N_L layers, captured by the vector $\bar{x}$, being mapped to N_T transmit antennas (the vector $\bar{y}$) by means of multiplication with a matrix W of size $N_T \times N_L$.

The general model of Fig. 11.1 applies to most cases of multi-antenna transmission. However, depending on implementation there will be various degrees of constraints that will impact the actual capabilities of the multi-antenna transmission.

One such implementation aspect relates to where, within the overall physical transmitter chain, the multi-antenna processing, that is, the matrix W of Fig. 11.1, is applied. On a high level one can distinguish between two cases:

- The multi-antenna processing is applied within the analog part of the transmitter chain, that is, after digital-to-analog conversion (left part of Fig. 11.2)
- The multi-antenna processing is applied within the digital part of the transmitter chain, that is, before digital-to-analog conversion (right part of Fig. 11.2).

The main drawback of digital processing according to the right part of Fig. 11.2 is the implementation complexity, especially the need for one digital-to-analog converter per antenna element. In the case of operation at higher frequencies with a large number of closely-spaced antenna elements, analog multi-antenna processing according to the left part Fig. 11.2 will therefore be the most common case, at least in the short- and medium-term perspectives. In this case, the multi-antenna transmission will typically be limited to per-antenna phase shifts providing beam forming (see Fig. 11.3).

Analog multi-antenna processing

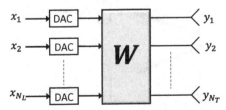

Digital multi-antenna processing

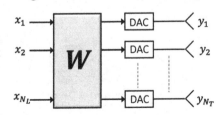

Fig. 11.2 Analog vs. digital multi-antenna processing.

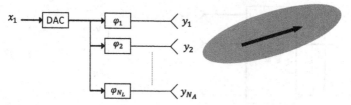

Fig. 11.3 Analog multi-antenna processing providing beam forming.

It should be noted that this may not be a severe limitation as operation at higher frequencies is typically more often power limited than bandwidth limited, making beam-forming more important than, for example, high-order spatial multiplexing. The opposite is often true for lower frequency bands where the spectrum is a more sparse resource with less possibility for wide transmission bandwidths.

Analog processing typically also implies that any beam-forming is carried out on a per-carrier basis. For the downlink transmission direction, this implies that it is not possible to frequency multiplex beamformed transmissions to devices located in different directions relative to the base station. In other words, beam-formed transmissions to different devices located in different directions must be separated in time as illustrated in Fig. 11.4.

In other cases, especially in the case of a smaller number of antenna elements at lower frequencies, multi-antenna processing can be applied in the digital domain according to the right part of Fig. 11.2. This enables much higher flexibility in the multi-antenna processing with a possibility for high-order spatial multiplexing and with the transmission matrix W being a general $N_T \times N_L$ matrix where each element may include both a phase shift and a scale factor. Digital processing also allows for independent multi-antenna processing for different signals within the same carrier, enabling simultaneous beam-formed transmission to multiple devices located in different directions relative to the base station also by means of frequency multiplexing as illustrated Fig. 11.5.

In the case of digital processing, or more generally, in the case where the antenna weights can be flexibly controlled, the transmission matrix W is often refererred to as a *precoder matrix* and the multi-antenna processing is often referred to as *multi-antenna precoding*.

Fig. 11.4 Time-domain (non-simultaneous) beam-forming in multiple directions.

Fig. 11.5 Simultaneous (frequency-multiplexed) beam-forming in multiple directions.

The difference in capabilities between analog and digital multi-antenna processing also applies to the receiver side. In the case of analog processing, the multi-antenna processing is applied in the analog domain before analog-to-digital conversion. In practice, the multi-antenna processing is then limited to receiver-side beam forming where the receiver beam can only be directed in one direction at a time. Reception from two different directions must then take place at different time instances.

Digital implementation, on the other hand, provides full flexibility, supporting reception of multiple layers in parallel and enabling simultaneous beam-formed reception of multiple signals arriving from different directions.

Similar to the transmitter side, the drawback of digital multi-antenna processing on the receiver side is in terms of complexity, especially the need for one analog-to-digital converter per antenna element.

For the remainder of this chapter we will focus on multi-antenna precoding, that is, multi-antenna transmission with full control over the precoder matrix. The limitations of analog processing and how those limitations are impacting the NR design are discussed in Chapter 12.

One important aspect of multi-antenna precoding is whether or not the precoding is also applied to the demodulation reference signals (DM-RS) used to support coherent demodulation of the precoded signal.

If the DM-RS are not precoded, the receiver needs to be informed about what precoder is used at the transmitter side to enable proper coherent demodulation of the precoded data transmission.

On the other hand, if the reference signals are precoded together with the data, the precoding can, from a receiver point-of-view, be seen as part of the overall multidimensional channel (see Fig. 11.6). Simply speaking, instead of the "true" $N_R \times N_T$ channel matrix H, the receiver will see a channel H' of size $N_R \times N_L$ that is the concatenation of the channel H with whatever precoding W is applied at the transmitter side. The precoding is thus transparent to the receiver implying that the transmitter can, at least in principle, select an arbitrary precoder matrix and does not need to inform the receiver about the selected precoder.

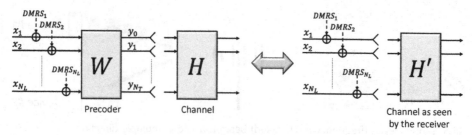

Fig. 11.6 DM-RS precoded jointly with data implying that any multi-antenna precoding is transparent to the receiver.

11.2 NR downlink multi-antenna precoding

All NR downlink physical channels rely on DM-RS to support coherent demodulation. Furthermore, a device can assume that the DM-RS are jointly precoded with the data in line with Fig. 11.6. Consequently, any downlink multi-antenna precoding is transparent to the device and the network can, in principle, apply any transmitter-side precoding with no need to inform the device what precoding is applied. Note though that the device must still know the number of transmission layers, that is, the number of columns in the precoder matrix applied at the transmitter side.

The specification impact of downlink multi-antenna precoding is therefore mainly related to the measurements and reporting done by the device to support network selection of precoder for downlink PDSCH transmission. These precoder-related measurements and reporting are part of the more general CSI reporting framework *based on report configurations* as described in Section 8.2. As described there, a CSI report may consist of one or several of the following quantities:

- A *rank indicator* (RI), indicating what the device believes is a suitable transmission rank, that is, a suitable number of transmission layers N_L for the downlink transmission;
- A *precoder-matrix indicator* (PMI), indicating what the device believes is a suitable precoder matrix, or just *precoder*, given the selected rank;
- A *channel-quality indicator* (CQI), in practice indicating what the device believes is a suitable channel-coding rate and modulation scheme, given the selected precoder matrix.

As mentioned, the PMI reported by a device indicates what the device believes is a suitable precoder to use for downlink transmission to the device. Each possible value of the PMI thus corresponds to one specific precoder. The set of possible PMI values thus corresponds to a set of different precoders, jointly referred to as the *precoder codebook*, that the device can select between when reporting PMI. Note that the device selects PMI based on a certain number of antenna ports N_T, given by the number of antenna ports of the configured CSI-RS associated with the report configuration, and the selected rank N_L. There is thus at least one codebook for each valid combination of N_T and N_L.

It is important to understand that the precoder codebooks for downlink multi-antenna precoding are only used in the context of CSI reporting and do not impose any restrictions on what precoder is eventually used by the network for downlink transmission to the reporting device. The network can use whatever precoder it wants and the precoder selected by the network does not have to be part of any defined codebook.

In many cases it obviously makes sense for the network to use the precoder indicated by the reported PMI. However, in other cases the network may have additional input that speaks in favor of a different precoder. As an example, multi-antenna precoding can be used to enable simultaneous downlink transmission to multiple devices using the same time/frequency resources, so called *multi-user MIMO* (MU-MIMO). The basic principle of MU-MIMO based on multi-antenna precoding is to choose a precoder that does not only focus the energy towards the target device but also takes the interference to other simultaneously scheduled devices into account. Thus, the selection of precoding for transmission to a specific device should not only take into account the PMI reported by that device (which only reflects the channel experienced by that device). Rather, the selection of precoding for transmission to a specific device should, in the general case, take into account the PMI reported by all simultaneously scheduled devices.

To conclude on suitable precoding in the MU-MIMO scenario typically also requires more detailed knowledge of the channel experienced by each device, compared to precoding in the case of transmission to single device. For this reason, NR defines two types of CSI that differ in the structure and size of the precoder codebooks, *Type-I CSI* and *Type-II CSI*.

- Type-I CSI primarily targets scenarios where a single user is scheduled within a given time/frequency resource (no MU-MIMO), potentially with transmission of a relatively large number of layers in parallel (high-order spatial multiplexing)
- Type-II CSI primarily targets MU-MIMO scenarios with multiple devices being scheduled simultaneously within the same time/frequency resource but with a more limited number of spatial layers per scheduled device

The codebooks for Type-I CSI are relatively simple and primarily aim at focusing the transmitted energy at the target receiver. Interference between the potentially large number of parallel layers is assumed to be handled primarily by means receiver processing utilizing multiple receive antennas.

The codebooks for Type-II CSI are significantly more extensive allowing for the PMI to provide channel information with much higher spatial granularity. The more extensive channel information allows the network to select a downlink precoder that not only focuses the transmitted energy at the target device but also limits the interference to other devices scheduled in parallel on the same time/frequency resource. The higher spatial granularity of the PMI feedback comes at the cost of significantly higher signaling overhead. While a PMI report for Type-I CSI will consist of at most a few tens of bits, a PMI report for Type-II CSI may consist of several hundred bits. Type-II CSI is therefore primarily applicable for low-mobility scenarios where the feedback periodicity in time can be reduced.

11.2.1 Type-I CSI

There are two sub-types of Type-I CSI, referred to as *Type-I single-panel CSI* and *Type-I multi-panel CSI*, corresponding to different codebooks. As the names suggest, the codebooks have been designed assuming different antenna configurations on the network/transmitter side.

11.2.1.1 Single-panel CSI

The codebooks for Type-I single-panel CSI are designed assuming a single antenna panel with $N_1 \times N_2$ dual-polarized antenna elements, see, as an example, Fig. 11.7 for the case of $N_1 = 4$ and $N_2 = 2$. There is also an assumption that each of the per-polarization antenna elements corresponds to a CSI-RS antenna port, that is, the number of antenna ports equals $P_{CSI-RS} = 2 \cdot N_1 \cdot N_2$. Thus, for the example of Fig. 11.7 there would be a total of 16 CSI-RS ports.

The NR specifications support different combinations of N_1 and N_2, with N_1 ranging from 1 to 16 and N_2 ranging from 1 to 4, and with a limitation of $N_1 \cdot N_2 \leq 16$, that is, a maximum of 32 CSI-RS ports. A device configured to report Type-I single-panel CSI is configured with a specific combination of N_1 and N_2 with the NR specification defining one or multiple codebooks for each supported (N_1, N_2) combination.

In general, each precoder matrix W in the codebooks for Type-I single-panel CSI can be expressed as the product of two matrices, that is,

$$W = W_1 W_2 \qquad (11.1)$$

with information about W_1 and W_2 reported separately as different parts of the overall PMI.

The matrix W_1 is assumed to capture long-term frequency-independent characteristics of the channel. A single W_1 is therefore selected and reported for the entire reporting bandwidth (wideband reporting).

In contrast, the matrix W_2 is assumed to capture more short-term and frequency-dependent characteristics of the channel. W_2 is therefore selected and reported on a subband basis, where a subband covers a fraction of the overall reporting bandwidth. Alternatively, the device may not report W_2 at all in which case the device, when

Fig. 11.7 Example of assumed antenna structure for Type-I single-panel CSI with $(N_1, N_2) = (4, 2)$.

subsequently selecting CQI, should assume that the network randomly selects W_2 on a per PRG (Physical Resource Block Group, see Section 9.8) basis. Note that this does not impose any restrictions on the actual precoding applied at the network side but is only about the assumptions made by the device when reporting CQI.

On a high-level, the matrix W_1 can be seen as defining a set of beams pointing in different directions. More specifically, the matrix W_1 can be written as

$$W_1 = \begin{bmatrix} V & 0 \\ 0 & V \end{bmatrix} = \begin{bmatrix} (\bar{v}_0 \cdots \bar{v}_{L-1}) & 0 \\ 0 & (\bar{v}_0 \cdots \bar{v}_{L-1}) \end{bmatrix} \qquad (11.2)$$

where each column vector $\bar{v}_k$ of size $N_1 \cdot N_2 = P_{CSI-RS}/2$ defines a specific DFT beam and the 2×2 block structure is due to the two polarizations. Selecting the matrix W_1 or, equivalently, the vectors $\bar{v}_0, \ldots, \bar{v}_{L-1}$ can thus be seen as selecting a limited set of beams from a larger set of candidate beams pointing in different directions. Note that, as the matrix W_1 is assumed to only capture long-term frequency-independent channel characteristics, the same set of beams are selected for the two polarizations.

In the case of rank-1 or rank-2 transmission, either a single beam ($L=1$) or four neighbor beams ($L=4$) are defined/selected by the matrix W_1. In the case of four neighbor beams, the matrix W_2 then selects the exact beam to be used for the transmission. As W_2 can be reported on a subband basis, it is thus possible to fine-tune the beam direction per subband. In addition, W_2 provides co-phasing between the two polarizations. In the case when W_1 only defines a single beam, the matrix W_2 only provides co-phasing between the two polarizations.

For transmission ranks larger than 2, the matrix W_1 defines N neighbor orthogonal beams where $N = \lceil R/2 \rceil$ and R is the number of parallel layers. The N beams together with the two polarizations of each beam are then used for transmission of the R layers, with the matrix W_2 only providing co-phasing between the two polarizations. In case of Type-I CSI, up to eight layers can be transmitted to the same device.

11.2.1.2 Multi-panel CSI

In contrast to single-panel CSI, codebooks for Type-I multi-panel CSI are designed assuming the use of *multiple* antenna panels at the network side and takes into account that it may be difficult to ensure coherence between transmissions from different panels. More specifically, the design of the multi-panel codebooks assumes an antenna configuration consisting of two or four two-dimensional panels, with each panel consisting of $N_1 \times N_2$ cross-polarized antenna elements. Fig. 11.8 illustrates an example of such multi-panel antenna configuration for the case of four antenna panels and $(N_1, N_2) = (4, 1)$, that is, a total of 32 CSI-RS ports.

The basic principle of Type-1 multi-panel CSI is the same as that of Type-I single-panel CSI, that is, the overall precoder can be expressed as the product of two matrices W_1 and W_2 where the structure of W_1 is the same as for Type-I single-panel CSI, that is, it defines a set of beams assumed to be the same for different polarizations and panels.

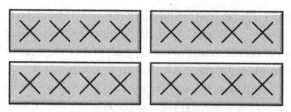

Fig. 11.8 Example of assumed antenna structure for Type-I multi-panel CSI. 32-port antenna with four antenna panels and $(N_1, N_2) = (4, 1)$.

The difference is that, in the multi-panel case, the matrix W_2 provides per-subband co-phasing not only between polarizations but also between panels. The co-phasing between panels based on device reporting is needed due to the assumed lack of coherence between different panels.

Type-I multi-panel CSI supports spatial multiplexing with up to four layers.

11.2.2 Type-II CSI (Release 15)

Type-II CSI was, together with Type-I CSI, introduced as part of the first NR release (3GPP release 15). Some important extensions and enhancements to Type-II CSI were then introduced in release 16. These extensions/enhancements to some extent changed the overall structure of the reported PMI. Thus, we describe the release-15 Type-II CSI in this section and the release-16 enhanced Type-II CSI separately in the next section. There is also a variant of Type-II CSI referred to as *port selection*. This will be separately discussed in Section 11.2.4.

As already mentioned, Type-II CSI provides channel information with significantly higher spatial granularity compared to Type-I CSI. At the same time, as Type-II CSI is targeting the MU-MIMO scenario, it was initially limited to a maximum transmission rank of two. However, as we will see below, this was extended to a maximum rank of four in release 16.

Similar to Type-I CSI, the release-15 Type-II CSI precoder can be expressed as a product of two matrices

$$W = W_1 W_2 \qquad (11.3)$$

where, also similar to Type-I CSI, the matrix W_1 of size $P_{CSI-RS} \times 2L$ is reported wideband and has a structure

$$W_1 = \begin{bmatrix} (\bar{v}_0 \ \cdots \ \bar{v}_{L-1}) & 0 \\ 0 & (\bar{v}_0 \ \cdots \ \bar{v}_{L-1}) \end{bmatrix} \qquad (11.4)$$

where each column vector $\bar{v}_k$ defines a specific DFT beam

In other words, W_1 defines a set of beams that is reported as part of the PMI. In the case of Type-II CSI, $L = 2$ or $L = 4$ beams may be reported. For each of the L reported beams and each of the two polarizations, the matrix W_2 then provides amplitude values (reported partly wideband and partly per subband) and phase values (reported per subband) for the selected beams. Compared to Type-I CSI, this provides much more detailed information about the channel, capturing its main rays and their respective amplitudes and phases.

At the network side, the CSI reported from multiple devices could then be used to identify a set of devices to which transmission could be done simultaneously on a set of time/frequency resources and what precoder to use for each transmission. As an example, if the same beam is reported by multiple devices this would typically speak in favor of not scheduling these devices simultaneously on the same resource (MU-MIMO), alternatively not including the common beam when selecting the final precoders to be used by the network for the downlink transmissions.

11.2.3 Release-16 enhanced Type-II CSI

As already mentioned, release 16 introduced an enhanced Type-II CSI. The basic principle of the release-16 Type-II CSI is the same as that of the release-15 Type-II CSI, that is, the reporting of a set of beams on a wideband basis together with the reporting of a set of combining coefficients on a more narrowband basis. The reported beams are linearly combined by means of the combining coefficients to provide a set of precoder vectors, one for each layer.

For the release-15 Type-II CSI, the combining coefficients, given by the matrix W_2, are reported separately for each subband, despite the fact that the channels of neighbor subbands often have a significant mutual correlation. It is this reporting of a relatively large number of combining coefficients on a per-subband basis that leads to the relatively large reporting overhead for the release-15 Type-II CSI.

An important feature of the release-16 enhanced Type-II CSI is therefore the possibility to utilize correlations in the frequency domain to reduce the reporting overhead. At the same time, the release-16 Type-II CSI allows for a factor of two improvement in the frequency-domain granularity of the PMI reporting. These two properties are jointly enabled by the introduction of the concept of *frequency-domain* (FD) *units*, where each FD unit corresponds to either a subband or half a subband, in combination with a *compression* operation.

In the end, the release-16 Type-II CSI provides the transmitter side with a recommended precoder per FD unit, compared to one precoder *per subband* for the release-15 Type-II CSI, that is, a factor of two improvement in the frequency-domain granularity assuming two FD units per subband. However, in contrast to the release-15 Type-II CSI, actual reporting is not done for each FD unit separately but jointly for all FD units, that is, for the entire frequency band covered by the CSI report.

In more details, for a given layer k the reported precoder vectors $\overline{w}_k^{(0)} \cdots \overline{w}_k^{(N-1)}$ for all FD units can, for the release-16 Type-II CSI, be expressed as

$$\left[\overline{w}_k^{(0)} \quad \cdots \quad \overline{w}_k^{(N-1)} \right] = W_1 \widetilde{W}_{2,k} W_{f,k}^H \qquad (11.5)$$

where N is the number of FD units to be reported.

Note that the matrix $\left[\overline{w}_k^{(0)} \cdots \overline{w}_k^{(N-1)} \right]$ is not a precoder matrix in the sense of mapping layers to antenna ports but just describes the set of precoder vectors for the full set of FD units (one precoder vector for each FD unit) for *a given layer k*. For a total of K layers there are K such sets of precoder vectors, with each set consisting of N precoder vectors. The actual precoder matrix, given by the reported PMI, which maps layers to antenna ports for a given FD unit n, is then given by

$$W^{(n)} = \left[\overline{w}_0^{(n)} \quad \cdots \quad \overline{w}_{K-1}^{(n)} \right] \qquad (11.6)$$

In the above expression for the precoder vectors, W_1 is the same as for the release-15 Type-II CSI, that is,

$$W_1 = \begin{bmatrix} (\overline{v}_0 \quad \cdots \quad \overline{v}_{L-1}) & \mathbf{0} \\ \mathbf{0} & (\overline{v}_0 \quad \cdots \quad \overline{v}_{L-1}) \end{bmatrix} \qquad (11.7)$$

where the column vectors $\overline{v}_0 \cdots \overline{v}_{L-1}$ defines the L selected beams. W_1 is the same for all FD units (wideband reporting) and also the same for all layers.

The main new thing with the release-16 Type-II CSI is the *compression matrix* $W_{f,k}^H$ of size $M \times N$. The compression matrix consists of a set of row vectors from a DFT basis and provides a transformation from the frequency domain of dimension N, corresponding to the N FD units covered by the CSI reporting, into a smaller *delay* domain of dimension M. $W_{f,k}^H$ is frequency independent (one common matrix for all FD units) but reported separately for each layer (thus the index k). The number of rows of $W_{f,k}^H$ equals $M = \left\lceil p \cdot \frac{N}{R} \right\rceil$, where R is the number of FD units per subband ($R=1$ or $R=2$) and p is a configurable parameter that controls the amount of compression.

Finally, the matrix $\widetilde{W}_{2,k}$ (size $2L \times M$) maps from the delay domain to the beam domain. One could have constructed a similar matrix for the release-15 Type-II CSI based on the matrix W_2 in Eq. (11.3) for all the subbands. Such a matrix would map from the frequency (subband) domain to the beam domain. In contrast, $\widetilde{W}_{2,k}$ maps from the smaller delay domain to the beam domain, implying a smaller size of $\widetilde{W}_{2,k}$ and, as a consequence, overall fewer parameters to report. Furthermore, only a fraction β of the total of $2LM$ elements of $\widetilde{W}_{2,k}$ are assumed to be non-zero and thus needs to be reported, where β is a configurable parameter.

Table 11.1 Possible configurations for release-16 Type-II CSI.

Configuration Index	L	p RI = 1 or 2	p RI = 3 or 4	β
1	2	1/4	1/8	1/4
2	2	1/4	1/8	1/2
3	4	1/4	1/8	1/4
4	4	1/4	1/8	1/2
5	4	1/4	1/8	3/4
6	4	1/2	1/8	1/2
7	6	1/4	N/A	1/2
8	6	1/4	N/A	1/2

Thus, the overhead reduction with the release-16 enhanced Type-II CSI is due to two things

- The smaller dimension of the delay space relative to the dimension of the frequency space, given by the parameter p
- The limited fraction of non-zero elements of $\widetilde{W}_{2,k}$, given by the parameter β.

As illustrated in Table 11.1, there can be eight different configurations for the release-16 Type-II CSI. These configurations differ in terms of

- The number of beams to be reported (L)
- The compression factor p, where the exact value for a given configuration depends on the reported rank
- The fraction β of non-zero elements in the matrix $\widetilde{W}_{2,k}$

It can be noted from Table 11.1 that, in addition to the reduced overhead and improved frequency-domain granularity, the release-16 Type-II CSI also extends the maximum reported rank to four and the maximum number of beams to select/report to six. It should be pointed out though that the latter is only supported under limited conditions, namely for 32 antenna ports, reported rank limited to one or two, and no improved frequency-domain granularity, that is, when an FD unit equals a subband.

11.2.4 Port selection

As already mentioned, there is a variant of Type-II CSI referred to as *port selection*. For regular Type-II CSI as described above, as well as for Type-I CSI, there is an assumption that there is no device-specific precoding/beamforming of the CSI-RS. Instead, based on measurements on the configured CSI-RS ports, the device decides on a set of beams described by the matrix W_1 above.

In contrast, with port selection it is assumed that the network has already applied a certain beamforming to the CSI-RS, for example, based on uplink measurements. The task of concluding on a set of suitable beams is then turned into the task of selecting a suitable set of CSI-RS ports, hence the name *port selection*.

Mathematically, port selection implies that the matrix W_1 in, Eqs. (11.3) and (11.5), instead of defining L beams, is a $P_{CSI-RS} \times 2L$ selection matrix consisting of a single "1" per column. Note that, similar to regular Type-II CSI, the same beams are selected for the two polarizations. In release-15 port selection, for each of the up to L selected ports and each of the two polarizations, the matrix W_2 then provides an amplitude value (reported partly wideband and partly per subband) and a phase value (reported per subband) in the same way as for the regular release-15 Type-II CSI.

In release 16, port selection was extended/enhanced along the same lines as the regular enhanced Type-II CSI, including extension to up to rank-4 transmission and overhead reduction by means of frequency-domain units and a *compression matrix* $W_{f,k}^H$ transforming from the frequency domain to the delay domain. Thus, the overall precoding of port selection is identical to that of the release-16 regular Type-II CSI, that is, for a given layer k, the reported precoder vectors can be expressed as

$$W_1 \widetilde{W}_{2,k} W_{f,k}^H \tag{11.8}$$

The only difference to regular Type-II CSI is the structure of the matrix W_1 which is changed from a beam-defining matrix to a port-selection matrix.

Port selection, but not regular Type-II CIS, was then further extended/evolved in release 17. These extensions utilized the fact that even in case of FDD operation in paired spectrum with uplink and downlink in different frequency bands, angle and delay properties of the channel are still to a large extent reciprocal between the downlink and uplink transmission directions. Thus, with precoded CSI-RS, which is the assumption for port selection, the angle and delay properties of the channel can be taken into account in the precoding of the CSI-RS, reducing the amount of information that the device needs to report.

The basic structure of the release-17 port selection is identical to release-16 port selection, that is, for a given layer k, the reported precoder vectors for all FD units can be expressed as

$$W_1 \widetilde{W}_{2,k} W_f^H \tag{11.9}$$

In the above expression, the port selection (W_1), and combination coefficients ($\widetilde{W}_{2,k}$) are identical to release 16. The difference between release 16 and release 17 port selection lies in the frequency-domain compression where, for release 16, the compression matrix is layer specific ($W_{f,k}^H$ depends on k) while, for release 17, it is the same for all layers (W_f^H does not depend on k).

Table 11.2 Evolution of Type-II CSI from release 15 to release 17.

	Codebook structure	FD compression
Rel-15 (regular and port-selection)	$W_1 W_2$	No
Rel-16 (regular and port selection)	$W_1 \widetilde{W}_{2,k} W_{f,k}^H$	Yes, layer specific
Rel 17 (port selection only)	$W_1 \widetilde{W}_{2,k} W_f^H$	Yes, layer common

A main benefit of the release-17 port selection is a reduced signaling overhead. It also allows for a reduced device complexity

Table 11.2 summarizes the evolution of Type-II CSI from release 15 to release 17.

11.3 NR uplink multi-antenna precoding

NR supports uplink (PUSCH) multi-antenna precoding with up to four layers. However, when DFTS-OFDM is used for uplink transmission, only single-layer transmission is supported.

The device can be configured in two different modes for PUSCH multi-antenna precoding, referred to as *codebook-based* transmission and *non-codebook-based* transmission respectively. The selection between these two transmission modes at least partly depends on what can be assumed in terms of uplink/downlink channel reciprocity, that is, to what extent it can be assumed that the detailed uplink channel conditions can be estimated by the device based on downlink measurements.

Like the downlink, any uplink (PUSCH) multi-antenna precoding is also assumed to be applied to the DM-RS used for the PUSCH coherent demodulation. Similar to the downlink transmission direction, uplink precoding is thus transparent to the receiver in the sense that receiver-side demodulation/detection can be carried out without knowledge of the exact precoding applied at the transmitter (device) side. Note though that this does not necessarily imply that the device can freely choose the PUSCH precoder. In the case of codebook-based precoding, the scheduling grant includes information about a precoder, similar to the device providing the network with PMI for downlink multi-antenna precoding. However, in contrast to the downlink, where the network may or may not use the precoder indicated by the PMI, in the uplink direction the device is assumed to use the precoder provided by the network. As we will see in Section 11.3.2, also in the case of non-codebook-based transmission will the network have an influence on the final choice of uplink precoder.

Another aspect that may constrain uplink multi-antenna transmission is to what extent one can assume coherence between different device transmit antennas, that is, to what extent the relative phase between the signals transmitted on two antennas can be well controlled. Coherence is needed in the case of general multi-antenna precoding where antenna-port-specific complex weight factors are applied to the signals transmitted on the different antenna ports. Without coherence between the antenna ports the use of such port-specific weight factors, and especially port-specific phases shift, is obviously meaningless as each antenna port would anyway introduce a more or less random relative phase.

The NR specification allows for different device capabilities with regards to such inter-antenna-port coherence, referred to as *full coherence*, *partial coherence*, and *no coherence*, respectively.

In the case of full coherence, it is assumed that the device can control the relative phase between any of the up to four ports that are to be used for transmission.

In the case of partial coherence, the device is assumed to be capable of *pairwise* coherence, that is, the device can control the relative phase within pairs of ports. However, there is no guarantee of coherence, that is, a controllable phase, between the pairs.

Finally, in the case of no coherence there is no guarantee of coherence between any of the device antenna ports.

11.3.1 Codebook-based transmission

The basic principle of codebook-based transmission is that the network decides on an uplink transmission rank, that is, the number of layers to be transmitted, and a corresponding precoder to use for the transmission. The network informs the device about the selected transmission rank and precoder as part of the uplink scheduling grant. At the device side, the precoder is then applied for the scheduled PUSCH transmission, mapping the indicated number of layers to the antenna ports.

To select a suitable rank and a corresponding precoder, the network needs estimates of the channels between the device antenna ports and the corresponding network receive antennas. To enable this, a device configured for codebook-based PUSCH transmission would typically be configured for transmission of at least one multi-port SRS. Based on measurements on the configured SRS, the network can sound the channel and determine a suitable rank and precoder.

The network cannot select an arbitrary precoder. Rather, for a given combination of number of antenna ports N_T ($N_T = 2$ or $N_T = 4$) and transmission rank N_L ($N_L \leq N_T$), the network selects a precoder from a limited set of available precoders (the "uplink codebook").

As an example, Fig. 11.9 illustrates the available precoders, that is, the code books for the case of two antenna ports.

When selecting the precoder, the network needs to consider the device capability in terms of antenna-port coherence (see above). For devices not supporting coherence, only the first two precoders can therefore be used in the case of single-rank transmission.

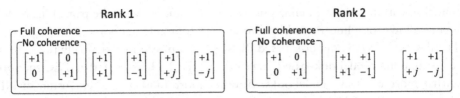

Fig. 11.9 Uplink codebooks for two antenna ports (N_T=2), and N_L=1 and N_L=2 respectively.

It can be noted that restricting the codebook selection to these two precoders is equivalent to selecting either the first or second antenna port for transmission. In the case of such *antenna selection*, a well-controlled phase, that is, coherence between the antenna ports is not required. On the other hand, the remaining precoders of Fig. 11.9 imply linear combination of the signals on the different antenna ports, which requires coherence between the antenna ports.

In the case of rank-2 transmission, (N_L=2). only the first precoder, which does not imply any coupling between the antenna ports, can be selected for devices that do not support coherence. Also note that partial coherence is not relevant for the case of only two antenna ports.

To further illustrate the impact of no, partial, and full coherence, Fig. 11.10 illustrates the full set of rank-1 precoders for the case of four antenna ports. Once again, the precoders corresponding to no coherence are limited to antenna-port selection. The extended set of precoders corresponding to partial coherence allows for linear

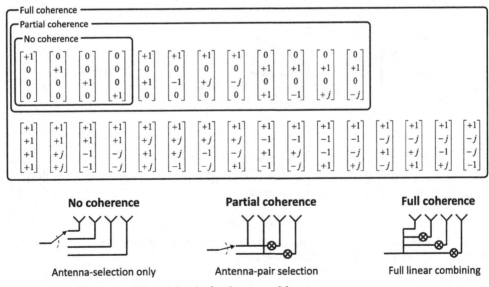

Fig. 11.10 Single-layer uplink codebooks for the case of four antenna ports.

combination within pairs of antenna ports with selection between the pairs. Finally, full coherence allows for a linear combination over all four antenna ports.

The described NR codebook-based PUSCH transmission is essentially the same as the corresponding codebook-based transmission for LTE except that NR supports somewhat more extensive codebooks. Another more fundamental extension of NR codebook-based PUSCH transmission, compared to LTE, is that a device can be configured to transmit *multiple* multi-port SRS, with a corresponding *SRS resource indicator* (SRI) included in the scheduling grant. The device should use the precoder indication provided in the scheduling grant and map the output of the precoding to the antenna ports corresponding to the SRS indicated in the SRI. In terms of the spatial filter F discussed in Chapter 8, the different SRSs would typically be transmitted using different spatial filters, in practice different beam. The device should then transmit the precoded signal using the same spatial filter/beam as used for the SRS indicated by the SRI.

One way to visualize the use of multiple SRS for codebook-based PUSCH transmission is to assume that the device transmits the different multi-port SRS within separate, relatively large beams (see Fig. 11.11). These beams may, for example, correspond to different device antenna panels pointing in different directions, where each panel includes a set of antenna elements, corresponding to the antenna-ports of each multi-port SRS. The SRI received from the network then determines what beam to use for the PUSCH transmission while the precoder information (number of layers and precoder) determines how the transmission is to be done within the selected beam. As an example, in the case

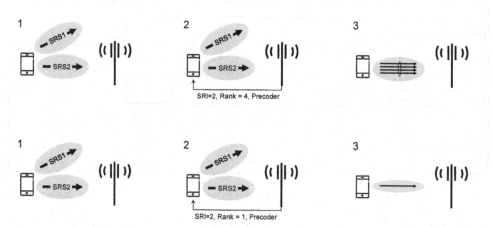

Fig. 11.11 Codebook-based transmission based on multiple SRS. Full-rank (4-layer) transmission (upper part) and single-rank transmission (lower part).

of full-rank transmission the device will do full-rank transmission within the beam corresponding to the SRS selected by the network and signaled by means of SRI (upper part of Fig. 11.11). At the other extreme, in the case of single-rank transmission the precoding will in practice create additional beam-forming within the wider beam indicated by the SRI (lower part of Fig. 11.11).

Codebook-based precoding is typically used when uplink downlink reciprocity does not hold, that is, when uplink measurements are needed in order to determine a suitable uplink precoding.

It should also be mentioned that in case of multi-TRP for PUSCH, see further Section 12.5.3.1, the scheduling grant will include two SIRs and indication of two precoders, one for each TRP.

11.3.2 Non-codebook-based precoding

In contrast to codebook-based precoding, which is based on network measurements and selection of uplink precoder, non-codebook-based precoding is based on device measurements and precoder indications to the network. The basic principle of uplink non-codebook-based precoding is illustrated in Fig. 11.12 with further explanation below.

Based on downlink measurements, in practice measurements on a configured CSI-RS, the device selects what it believes is a suitable uplink multi-layer precoder. Non-codebook-based precoding is thus based on an assumption of channel reciprocity,

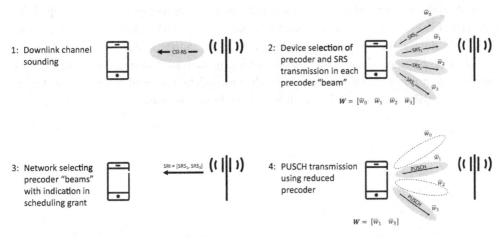

Fig. 11.12 Non-codebook-based precoding.

that is, that the device can acquire detailed knowledge of the uplink channel based on downlink measurements. Note that there are no restrictions on the device selection of precoder, thus the term "non-codebook-based".

Each column of a precoder matrix can be seen as defining a digital "beam" for the corresponding layer. The device selection of precoder for N_L layers can thus be seen as the selection of N_L different beam directions where each beam corresponds to one possible layer.

In principle, PUSCH transmission could be done directly as transmission of N_L layers based on the device-selected precoding. However, device selection of a precoder based on downlink measurements may not necessarily be the best precoder from a network point of view. Thus, the NR non-codebook-based precoding includes an additional step where the network can modify the device-selected precoder, in practice remove some "beams", or equivalently some columns, from the selected precoder.

To enable this, the device applies the selected precoder to a set of configured SRSs, with one SRS transmitted on each layer or "beam" defined by the precoder (step 2 in Fig. 11.12). Based on measurements on the received SRSs, the network can then decide to modify the device-selected precoder for each scheduled PUSCH transmission. This is done by indicating a subset of the configured SRS within the SRS resource indicator (SRI) included in the scheduling grant (step 3).[b] The device then carries out the scheduled PUSCH transmission (step 4) using a reduced precoder matrix where only the columns corresponding to the SRSs indicated within the SRI are included. Note that the SRI then also implicitly defines the number of layers to be transmitted.

It should be noted that the device indication of precoder selection (step 2 in Fig. 11.12) is not done for each scheduled transmission. The uplink SRS transmission indicating device precoder selection can take place periodically (periodic or semi-persistent SRS) or on demand (aperiodic SRS). In contrast, the network indication of precoder, that is in practice the network indication of the subset of beams of the device precoder, is then done for each scheduled PUSCH transmission.

---

[b] For a device configured for non-codebook-based precoding the SRI may thus indicate multiple SRSs, rather than a single SRS which is the case for codebook-based precoding (see Section 11.3.1).

CHAPTER 12

Beam management

Chapter 11 discussed multi-antenna transmission in general and then focused on multi-antenna precoding. A general assumption for the discussion on multi-antenna precoding was the possibility for detailed control, including both phase adjustment and amplitude scaling, of the different antenna elements. In practice this requires that multi-antenna processing at the transmitter side is carried out in the digital domain before digital-to-analog conversion. Likewise, the receiver multi-antenna processing must be carried out *after* analog-to-digital conversion.

However, in the case of operation at higher frequencies with a large number of closely-space antenna elements, the antenna processing will rather be carried out in the analog domain with focus on beam forming. As analog antenna processing will be carried out on a carrier basis, this also implies that beam-formed transmission can only be done in one direction at a time. Downlink transmissions to different devices located in different directions relative to the base station must therefore be separated in time. Likewise, in the case of analog-based receiver-side beam-forming, the receive beam can only focus in one direction at a time.

The ultimate task of beam management is, under these conditions, to establish and retain a suitable *beam pair*, that is, a transmitter-side beam direction and a corresponding receiver-side beam direction that jointly provides good connectivity.

As illustrated in Fig. 12.1, the best beam pair may not necessarily correspond to transmitter and receiver beams that are physically pointing directly toward each other. Due to obstacles in the surrounding environment, such a "direct" path between the transmitter and receiver may be blocked and a reflected path may provide better connectivity, as illustrated in the right-hand part of Fig. 12.1. This is especially true for operation in higher frequency bands with less "around-the-corner" dispersion. The beam-management functionality must be able to handle such a situation and establish and retain a suitable beam pair also in this case.

Fig. 12.1 illustrates the case of beam forming in the downlink direction, with beam-based transmission at the network side and beam-based reception at the device side. However, beam-forming is at least as relevant for the uplink transmission direction with beam-based transmission at the device side and corresponding beam-based reception at the network side.

In many cases, a suitable transmitter/receiver beam pair for the downlink transmission direction will also be a suitable beam pair for the uplink transmission direction and vice

5G/5G-Advanced
https://doi.org/10.1016/B978-0-443-13173-8.00023-2

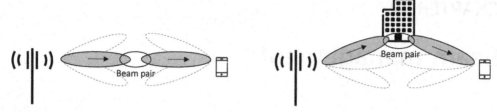

Fig. 12.1 Illustration of beam pairs in the downlink direction. Direct (left hand) and via reflection (right hand).

versa. In that case, it is sufficient to explicitly determine a suitable beam pair in one of the transmission directions. The same pair can then be used also in the opposite transmission direction. Otherwise, beam-pairs have to be established separately for the downlink and uplink transmission directions. Whether or not this is needed is related to so-called *beam-correspondence* between the downlink and uplink direction, see also Chapter 28.

In general, beam management can be divided into different parts:

- Initial *beam establishment*;
- *Beam adjustment*, primarily to compensate for movements and rotations of the mobile device but also for gradual changes in the environment;
- *Beam recovery* to handle the situation when rapid changes in the environment disrupt the current beam pair.

12.1 Initial beam establishment

Initial beam establishment includes the procedures and functions by which a beam pair is initially established in the downlink and uplink transmission directions, for example, when a connection is established. As will be described in more detail in Chapter 16, during initial cell search a device will acquire a so-called *SS block* transmitted from a cell, with the possibility for multiple SS blocks being transmitted in sequence within different downlink beams. By associating each such SS block, in practice the different downlink beams, with a corresponding random-access occasion and preamble (see Section 17.1.4), the subsequent uplink random-access transmission can be used by the network to identify the downlink beam acquired by the device, thereby establishing an initial beam pair

When communication continues after connection set up the device can assume that network transmissions to the device will be done using the same spatial filter, in practice the same transmitter beam, as used for the acquired SS block. Consequently, the device can assume that the receiver beam used to acquire the SS block will be a suitable beam also for the reception of subsequent downlink transmissions. Likewise, subsequent uplink transmissions should be done using the same spatial filter (the same beam) as used for the random-access transmission, implying that the network can assume that the uplink receiver beam established at initial access will remain valid.

12.2 Beam adjustment

Once an initial beam pair has been established, there is a need to regularly re-evaluate the selection of transmitter-side and receiver-side beam directions due to movements and rotations of the mobile device. Furthermore, even for stationary devices, movements of other objects in the environment may block or unblock different beam pairs, implying a possible need to re-evaluate the selected beam directions. This *beam adjustment* may also include refining the beam shape, for example making the beam more narrow compared to a relatively wider beam used for initial beam establishment.

In the general case, beam forming is about beam pairs consisting of transmitter-side beamforming and receiver-side beamforming. Hence, beam adjustment can be divided into two separate procedures:

- Re-evaluation and possible adjustment of the transmitter-side beam direction given the current receiver-side beam direction;
- Re-evaluation and possible adjustment of the receiver-side beam direction given the current transmitter-side beam direction.

As described above, in the general case beam forming, including beam adjustment, needs to be carried out for both the downlink and uplink transmission directions. However, depending on to what extent beam-correspondence can be assumed, explicit beam adjustment may only have to be carried out in one of the directions, for example, in the downlink direction. It can then be assumed that the adjusted downlink beam pair is appropriate also for the opposite transmission direction.

12.2.1 Downlink transmitter-side beam adjustment

Downlink transmitter-side beam adjustment aims at refining the network transmit beam, given the receiver beam currently used at the device side. To enable this, the device can measure on a set of reference signals, corresponding to different downlink beams (see Fig. 12.2). Assuming analog beam forming, transmissions within the different downlink beams must be done in sequence, that is, by means of a beam sweep.

The result of the measurements is then reported to the network which, based on the reporting, may decide to adjust the current beam. Note that this adjustment may not

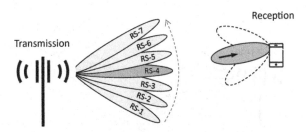

Fig. 12.2 Downlink transmitter-side beam adjustment.

necessarily imply the selection of one of the beams that the device has measured on. The network could, for example, decide to transmit using a beam direction in between two of the reported beams.

Also note that, during measurements done for transmitter-side beam adjustment, the device receiver beam should be kept fixed in order for the measurements to capture the quality of the different transmitter beams *given the current receive beam.*

To enable measurements and reporting on a set of beams as outlined in Fig. 12.2, the reporting framework based on report configurations (see Section 8.2) can be used. More specifically, the measurement/reporting should be described by a report configuration having L1-RSRP as the quantity to be reported.

The set of reference signals to measure on, corresponding to the set of beams, should be included in the NZP-CSI-RS resource set associated with the report configuration. As described in Section 8.1.6, such a resource set may either include a set of configured CSI-RS or a set of SS blocks. Measurements for beam management can thus be carried out on either CSI-RS or SS block. In the case of L1-RSRP measurements based on CSI-RS, the CSI-RS should be limited to single-port or dual-port CSI-RS. In the latter case, the reported L1-RSRP should be a linear average of the L1-RSRP measured on each port.

The device can report measurements corresponding to up to four reference signals (CSI-RS or SS blocks), in practice up to four beams, in a single reporting instance. Each such report would include:

- Indications of the up to four reference signals, in practice beams, that this specific report relates to;
- The measured L1-RSRP for the strongest beam;
- For the remaining up to three beams: The difference between the measured L1-RSRP and the measured L1-RSRP of the best beam.

12.2.2 Downlink receiver-side beam adjustment

Receiver-side beam adjustment aims at finding the best receive beam, given the current transmit beam. To enable this, the device should once again be configured with a set of downlink reference signals that, in this case, are transmitted within *the same* network-side beam (the current serving beam). As outlined in Fig. 12.3, the device can then do a

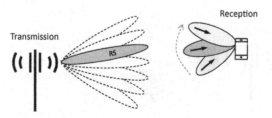

Fig. 12.3 Downlink receiver-side beam adjustment.

receiver-side beam sweep to measure on the configured reference signals in sequence over a set of receiver beams. Based on these measurements the device may adjust its current receiver beam.

Downlink receiver-side beam adjustment can be based on similar report configurations as for transmitter-side beam adjustment. However, as the receiver-side beam adjustment is done internally within the device, there is no report quantity associated with receiver-side beam adjustment. According to Section 8.2, the report quantity should thus be set to "None".

To allow for analog beam-forming at the receiver side, the different reference signals within the resource set should be transmitted in different symbols, allowing for the receiver-side beam to sweep over the set of reference signals. At the same time, the device should be allowed to assume that the different reference signals in the resource set are transmitted using the same spatial filter, in practice the same transmit beam. In general, a configured resource set includes a *"repetition"* flag that indicates whether or not a device can assume that all reference signals within the resource set are transmitted using the same spatial filter. For a resource set to be used for downlink receiver side beam adjustment, the repetition flag should thus be set.

12.2.3 Uplink beam adjustment

Uplink beam adjustment serves the same purpose as downlink beam adjustment, that is, to retain a suitable beam pair which, in the case of uplink beam adjustment implies a suitable transmitter beam at the device side and a corresponding suitable receiver beam at the network side.

As discussed earlier, if sufficient beam correspondence can be assumed and if a suitable downlink beam pair has been established and retained, explicit uplink beam management is not needed. Rather, a suitable beam pair for the downlink transmission direction can be assumed to be suitable also for the uplink direction. Note that the opposite would also be true, that is, if a suitable beam pair is established and retained for the uplink direction, the same beam pair could also be used in the downlink direction without the need for explicit downlink beam management.

If explicit uplink beam adjustment is needed it can be done in essentially the same way as for downlink beam adjustment with the main difference being that measurements are done by the network based on configured SRS, rather than CSI-RS or SS block.

12.3 Beam indication and TCI

As should be obvious from above, key to efficient beam-based operation at higher frequencies is the establishment of a matching beam pair between the transmitter and receiver. This also means that, if the network for some reason changes the beam used for transmission to a device, the device may have to adapt/change the receiver beam to match the change of transmitter beam. To enable this in an efficient manner, NR introduced the concept of TCI (*Transmission Configuration Indicator*) states as a tool to indicate the beam used for downlink transmissions. Based on knowledge of the TCI state of a downlink transmission, a device can select a suitable receiver beam. If the transmitter beam is changed, also the TCI state is changed, and the device can directly adapt/change the receiver beam to match the change of the transmitter beam.

The TCI concept was introduced as part of the first release of NR (release 15) but was then subject to a major update in release 17. Below we will first describe the original release-15/16 TCI concept in Section 12.3.1 and then cover the new release-17 TCI framework in Section 12.3.2.

12.3.1 Release-15/16 TCI states

In the general case, a release-15/16 TCI state corresponds to one or two downlink reference signals, either SSB or configured periodic CSI-RS, where each reference signal is associated with a specific QCL type. For beam indication, the relevant thing is TCI states that include a reference signal associated with QCL type D (spatial QCL). When a downlink transmission (PDSCH or PDCCH) is associated with such a TCI state, a device can assume that the downlink transmission uses the same spatial filtering, in practice uses the same transmitter beam, as is used for the TCI-state reference signal. Thus, the device can assume that it can use the receiver beam established for the reception of the TCI-state reference signal also for the reception of the downlink transmission. If the transmitter beam is changed, the device is informed about the corresponding new TCI state and can adapt/change its receiver beam accordingly. Note that, for this to work properly, the device needs to keep track of suitable receiver beams for a set of candidate TCI states.

To summarize the basic principles of beam indication based on the release-15/16 TCI states:

- The network associates different downlink beams with different TCI states, in practice with different periodic reference signals (SSB or configured periodic CSI-RS)
- The device keeps track of suitable receiver beams for a set of TCI states, in practice beams suitable for the reception of the reference signals of the set of TCI states
- The network provides information to the device about the TCI state of a given downlink transmission (PDSCH or PDCCH)
- The receiver applies the receiver beam established for the provided TCI state to receive the downlink transmission

Note that the release-15/16 TCI states are only applicable to downlink transmissions. For uplink transmissions the concept of *spatial relation* between an uplink transmission and a downlink reference signal is instead used. Configuring such a relation implies that the device can apply the receiver beam established for the reception of the downlink reference signal as transmitter beam for the uplink transmission.

In terms of the network providing the device with information about the current TCI state of a given downlink transmission, this is done somewhat differently for the PDCCH and PDSCH physical channels.

In general, a device can be configured with a relatively large number of TCI states. For the transmission of PDCCH, a subset of the configured TCI states is assigned by RRC signaling to each configured CORESET. By means of MAC signaling (MAC CE), the network can then more dynamically indicate a specific TCI state, within the per-CORESET-configured subset, to be the current TCI state for PDCCH transmissions within that CORESET. The MAC-indicated TCI state will then remain the current TCI state for PDCCH transmissions within the CORESET until a new TCI state is indicated by a MAC CE.

For PDSCH beam indication, there are two alternatives depending on the scheduling offset, that is, the transmission timing of the PDSCH relative to the timing of the corresponding PDCCH that carries the scheduling assignment for the PDSCH.

If this scheduling offset is larger than N symbols, where N can take different values depending on UE capabilities, the DCI of the scheduling assignment may explicitly indicate the TCI state for the scheduled PDSCH transmission. To enable this, a set of up to eight TCI states, from the larger set of configured TCI states, is first activated by means of MAC signaling. A three-bit indicator within the DCI then indicates the exact TCI state for the PDSCH transmission, out of the up to eight activated TCI states.

If the scheduling offset is smaller or equal to N symbols, the device can instead assume that the scheduled PDSCH transmission is spatially QCL with the corresponding PDCCH transmission. In other words, the TCI state for the scheduled PDSCH can be assumed to be the same as the current PDCCH TCI state transmission (indicated by MAC according to above.

The reason for limiting the fully dynamic TCI selection based on DCI signaling to situations when the scheduling offset is larger than a certain value is simply that, for shorter scheduling offsets, there will not be sufficient time for the device to decode the TCI information within the DCI and adapt the receiver beam accordingly before the PDSCH is to be received.

12.3.2 Release-17 unified TCI framework

The release-15/16 TCI concept is very flexible with separate independent beam indication for each downlink physical channel. However, this also leads to a high signaling overhead for the configuration/activation/indication of TCI states. At the same time, in practice the same transmission beam is typically used for different downlink channels targeting the same device. Furthermore, in most cases the same beam pair is used for downlink and uplink transmissions to/from a device.

Due to this, a new/modified TCI concept, referred to as the *unified TCI framework*, was introduced in release 17. The key properties of the release-17 unified TCI framework are

- The merging of the separate indications of PDSCH and PDCCH TCI states of release 15/16, into a *joint downlink TCI state* valid for both PDSCH and PDCCH. Note that this assumes that the same spatial filtering, that is, in practice the same transmitter beam, is used for PDSCH and PCSCH.
- The introduction of a *joint uplink TCI state* valid for both PUSCH and PUCCH, as a replacement to the uplink spatial relation of release 15/16. In contrast to the release-15/16 TCI states, the release-17 TCI framework is thus also used for uplink transmissions.

The release-17 unified TCI framework also allows for combining the downlink and uplink TCI states into a joint downlink/uplink TCI state. More specifically, the unified TCI framework can operate in two different modes:

- *Joint DL/UL TCI* where a single joint TCI state is valid for both downlink and uplink transmissions
- *Separate DL/UL TCI* where there is one TCI state for downlink transmissions and a separate TCI state for uplink transmissions

Operating with joint downlink/uplink TCI states assumes that the same beam pair is used for uplink and downlink transmissions, see left part of Fig. 12.4. In most cases this is a reasonable assumption as the channel properties that impact beam selection are typically the same for the downlink and uplink transmission directions. However, there are cases when it could be beneficial to use different beam pairs for the downlink and uplink transmissions, corresponding to separate downlink and uplink TCI states (right part of Fig. 12.4). One such scenario could be when different antenna panels at the device have different transmit-power capabilities. In such a case, even if one beam pair is better in terms of path gain, a different uplink beam pair may still be preferred as it may allow for higher transmit power and thus overall better link performance.

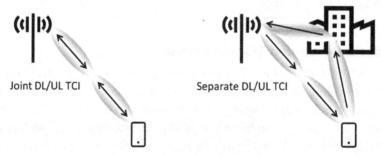

Fig. 12.4 Joint DL/UL TCI vs. separate DL/UL TCI.

Similar to the release-15/16 TCI states, update of a release-17 TCI state can be done either by MAC signaling (MAC CE) or by a combination of MAC and DCI signaling.

In both cases, a device is first configured with a set of TCI states, either a set of joint downlink/uplink TCI states or a set of downlink TCI states and a separate set of uplink TCI states.

- In case of MAC signaling, a MAC CE directly updates the current joint downlink/uplink TCI state, or separately updates the current downlink and uplink TCI states, from the set(s) of configured TCI states.
- In case of combined MAC/DCI signaling, MAC CE first activates up to eight different joint DL/UL TCI states, alternatively up to eight combinations of downlink and uplink TCI states, from the set(s) of configured states. DCI signaling is then used to indicate one of the activated TCI states or combinations of TCI states.

In case of separate downlink and uplink TCI states, each of the up to eight MAC-activated combinations of downlink and uplink TCI states may correspond to

- a combination of a downlink TCI state and an uplink TCI state, in which case the DCI will update both the downlink and uplink TCI states
- only a downlink TCI state, in which case the DCI will update only the downlink TCI state while the uplink TCI state will remain unchanged
- only an uplink TCI state, in which case the DCI will update only the uplink TCI state while the downlink TCI state will remain unchanged

An example of MAC-activated separate downlink and uplink TCI states and their mapping to the TCI field of a DCI is illustrated in Fig. 12.5.

In the example of Fig. 12.5

- if the DCI would signal "3", the device should update the current downlink state to DL TCI 10 and the current uplink TCI state to UL TCI 12,
- if the DCI would signal "0", the device should update the current downlink TCI state to DL TCI 3 but should not update the current uplink TCI state,
- if the DCI would signal "7", the device should update the current uplink TCI state to UL TCI 57 but should not update the downlink TCI state,

From the example above, one can note a difference between the DCI-based TCI-state indication of release-15/16 and release-17.

TCI field	0	1	2	3	4	5	6	7
DL TCI state	DL TCI 3	DL TCI 7	DL TCI 9	DL TCI 10	DL TCI 25	DL TCI 36	-	-
UL DCI state	-	-	UL TCI 1	UL TCI 12	UL TCI 7	UL TCI 20	UL TCI 42	UL TCI 57

Fig. 12.5 Example of mapping of activated DL and UL TCI states to TCI field of DCI.

- For release 15/16, the TCI state indicated in DCI is valid specifically for the PDSCH transmission scheduled by the DCI. For any future PDSCH transmissions, a new TCI state should be indicated in the corresponding scheduling DCI. If no TCI state is indicated, the MAC-CE-provided TCI state for the PDCCH should be assumed to be valid also for the PDSCH transmission.
- For release-17, the TCI state indicated in the DCI *updates* the downlink and/or uplink TCI state and the new TCI state will remain valid until a new TCI state is provided.

In case of DCI-indicated update of TCI states, the network needs to know that the DCI has been properly received by the device. If the DCI carrying the TCI indication is also used to schedule a PDSCH transmission, the Hybrid-ARQ feedback can be used as an indication of correct DCI decoding. There are also means to have Hybrid ARQ feedback also for DCI not scheduling PDSCH transmissions.

12.4 Beam recovery

In some cases, movements in the environment or other events, may lead to a currently established beam pair being rapidly blocked without sufficient time for the regular beam adjustment to adapt. The NR specification includes specific procedures to handle such *beam-failure* events, also referred to as *beam (failure) recovery*.

In many respects, beam failure is similar to the concept of *radio-link failure* (RLF) already defined for earlier radio-access technologies such as LTE and one could in principle utilize already established RLF-recovery procedures to recover also from beam-failure events. However, there are reasons to introduce additional procedures specifically targeting beam failure.

- Especially in the case of narrow beams, beam failure, that is, loss of connectivity due to a rapid degradation of established beam pairs, can be expected to occur more frequently compared to RLF, which typically corresponds to a device moving out of coverage from the currently serving cell;
- RLF typically implies loss of coverage to the currently serving cell in which case connectivity must be reestablished to a new cell, perhaps even on a new carrier. After beam failure, connectivity can often be re-established by means of a new beam pair within the current cell. As a consequence, recovery from beam failure can often be achieved by means of lower-layer functionality, allowing for faster recovery compared to the higher-layer mechanisms used to recover from RLF.

In general, beam-failure/recovery consists of the following steps:

- *Beam-failure detection*, that is, the device detecting that a beam-failure has occurred;
- *Candidate-beam identification*, that is, the device trying to identify a new beam or, more exactly, a new beam pair by means of which connectivity may be restored;

- *Recovery-request transmission*, that is, the device transmitting a beam-recovery request to the network;
- Network response to the beam-recovery request.

12.4.1 Beam-failure detection

Fundamentally, a beam failure is assumed to have happened when the error probability for the downlink control channel (PDCCH) exceeds a certain value. However, similar to radio-link failure, rather than actually measuring the PDCCH error probability the device declares a beam failure based on measurements of the quality of some reference signal. This is often expressed as measuring a *hypothetical error rate*. More specifically, the device should declare beam failure based on measured L1-RSRP of a periodic CSI-RS or an SS block that is spatially QCL with the PDCCH.

By default, the device should declare beam failure based on measurement on the reference signal (CSI-RS or SS block) associated with the PDCCH TCI state. However, there is also a possibility to explicitly configure a different CSI-RS on which to measure for beam-failure detection.

Each time instant the measured L1-RSRP is below a configured value is defined as a *beam-failure instance*. If the number of consecutive beam-failure instances exceeds a configured value, the device declares a beam failure and initiates the *beam-failure-recovery* procedure.

12.4.2 New-candidate-beam identification

As a first step of the beam-recovery procedure, the device tries to find a new beam pair on which connectivity can be restored. To enable this, the device is configured with a resource set consisting of a set of CSI-RS, or alternatively a set of SS blocks. In practice, each of these reference signals is transmitted within a specific downlink beam. The resource set thus corresponds to a set of *candidate beams*.

Similar to normal beam establishment, the device measures the L1-RSRP on the reference signals corresponding to the set of candidate beams. If the L1-RSRP exceeds a certain configured target, the reference signal is assumed to correspond to a beam by means of which connectivity may be restored. It should be noted that, when doing this, the device has to consider different receiver-side beam directions when applicable, that is, what the device determines is, in practice, a candidate beam pair.

12.4.3 Device recovery request and network response

If a beam failure has been declared and a new candidate beam pair has been identified. the device carries out a *beam-recovery request*. The aim of the recovery request is to inform the network that the device has detected a beam failure. The recovery request may also include information about the candidate beam identified by the device.

The beam-recovery request is in essence a contention-free random-access request consisting of preamble transmission and random-access response.[a] Each reference signal corresponding to the different candidate beams is associated with a specific preamble configuration (RACH occasion and preamble sequence, see Chapter 16). Given the identified beam, the preamble transmission should be carried out using the associated preamble configuration. Furthermore, the preamble should be transmitted within the uplink beam that coincides with the identified downlink beam.

It should be noted that each candidate beam may not necessarily be associated with a unique preamble configuration. There are different alternatives:

- Each candidate beam is associated with a unique preamble configuration. In this case, the network can directly identify the identified downlink beam from the received preamble;
- The candidate beams are divided into groups where all beams within the same group correspond to the same preamble configuration while beams of different groups correspond to different preamble configurations. In this case, the received preamble only indicates the group to which the identified downlink beam belongs;
- All candidate beams are associated with the same preamble configuration. In this case, the preamble reception only indicates that beam failure has occurred and that the device requests a beam-failure recovery.

Under the assumption that the candidate beams are originating from the same site it can also be assumed that the random-access transmission is well time-aligned when arriving at the receiver. However, there may be substantial differences in the overall path loss for different candidate beam pairs. The configuration of the beam-recovery-request transmission thus includes parameters for power ramping (see Section 16.2).

Once a device has carried out a beam-recovery-request it monitors downlink for a network response. When doing so, the device may assume that the network, when responding to the request, is transmitting PDCCH QCL with the RS associated with the candidate beam included in the request.

The monitoring for the recovery-request response starts four slots after the transmission of the recovery request. If no response is received within a window of a configurable size, the device retransmits the recovery request according to the configured power-ramping parameters.

12.5 Multi-TRP operation

A TRP, or *Transmission and Reception Point*, is a physical point for network transmission and/or reception. Such physical points could, for example, correspond to different cell sites or to the geographically separated antennas in case of a distributed-antenna system.

[a] See Section 16.2 for more details on the NR random-access procedure including preamble structure.

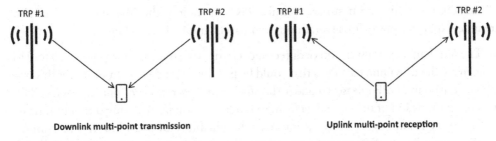

Downlink multi-point transmission   **Uplink multi-point reception**

Fig. 12.6 Downlink multi-point transmission vs. uplink multi-point reception.

In general, multi–TRP operation includes

- *downlink multi-point transmission*, that is, downlink transmission to the same device from multiple TRPs (left part of Fig. 12.6).
- *uplink multi-point reception*, that is, reception of uplink transmissions from a device at multiple TRPs (right part of Fig. 12.6).

There are several potential gains with multi–TRP operation, including

- a power gain in the sense that, by transmitting from multiple points, the total power available for the transmission can be increased;
- a rank gain in the sense that, by transmitting from multiple points, the overall channel may be more "rich", thereby enabling higher-rank transmission and corresponding higher data rates;
- a reliability gain in the sense that, if the link to one TRP is blocked, connectivity could still be maintained via the other TRP.

Note that some forms of multi–TRP operation can be transparent to the device, implying that it can be used without requiring any multi–TRP-specific features in the radio-interface specifications. As an example, if identical transmissions are carried out simultaneously from multiple transmission points, the overall received signal would, at least in principle, be indistinguishable from a single-point transmission subject to multi-path propagation. Likewise, at least in principle, nothing prevents that an uplink transmission is received at multiple reception points without the device even being aware of it. Note though that, to fully utilize these kinds of "transparent" multi–TRP operation, the device would need to simultaneously receive respectively transmit in multiple directions. At higher frequencies, where devices are typically equipped with multiple antenna panels "looking" in different directions, this may not always be possible. Note that the transmission and/or reception in multiple different directions is a fundamental property of all types of multi–TRP operation although in some cases transmissions or receptions in "different directions" does not occur simultaneously.

In addition to device-transparent multi-TRP operation, the NR specifications also include explicit support for several non-transparent multi-TRP schemes:

- The simultaneous transmission of different data from two transmission points, as a tool to increase the downlink data rates that could be provided to a device. Support for this was introduced in 3GPP release 16 under the label *Non-Coherent Joint Transmission* (NCJT).
- The transmission of the same data from two transmission points using different time or frequency resources, as a tool to improve the reliability of downlink transmissions.
- The transmission of the same data towards different reception points using different time resources, as a tool to improve the reliability of uplink transmissions.

As high reliability is especially relevant for so-called URLLC services, see Chapter 1, the two latter cases have, within 3GPP, been jointly referred to as *Multi-TRP for URLLC*. Support for downlink multi-TRP for URLLC was introduced in release 16 but was then limited to the PDSCH physical channel. In release 17, downlink multi-TRP for URLLC was then extended to also cover the PDCCH physical channel. In parallel, release 17 also introduced support for uplink multi-TRP for URLLC, that is, multi-point reception for the PUSCH and PUCCH physical channels.

It should already now be mentioned that the currently supported NR multi-TRP schemes, both non-coherent joint transmission and multi-TRP for URRLC (including schemes introduced in release 17), rely on the release 15/16 TCI (see Section 12.3.1). Extending multi-TRP operation to the release-17 unified TCI framework described in Section 12.3.2 is part of the ongoing work on 3GPP release 18.

12.5.1 Non-coherent joint transmission

As described above, non-coherent joint transmission (NCJT) implies the simultaneous transmission of different downlink data from two transmission points to the same device. Note that, in general, non-coherent joint transmission requires that the device is capable of simultaneous reception from different directions, something which, as already mentioned, may be difficult at higher frequencies such as mm-wave frequencies. At the same time, non-coherent joint transmission, which aims at increasing the achievable data rates for bandwidth-limited scenarios, is probably less relevant for mm-wave frequencies where wide bandwidths are available.

NR supports two variants of non-coherent joint transmission, referred to as *single-DCI-based NCJT* and *multi-DCI-based NCJT* respectively, see also Fig. 12.7. As the names, as well as the figure, suggest, these two variants differ in terms of the scheduling DCI. They also differ in terms of if there is one common PDSCH, or two separate PDSCHs, from the two transmission points.

12.5.1.1 Single-DCI-based NCJT
In case of single-DCI-based NCJT (left part of Fig. 12.7) a single DCI schedules a single multi-layer PDSCH where different PDSCH layers are transmitted from the two transmission points.

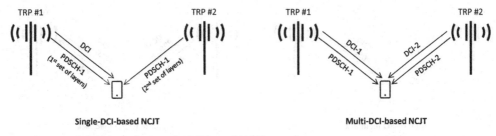

Fig. 12.7 Single-DCI-based vs. multi-DCI-based NCJT.

The main specification impact of single-DCI-based NCJT comes from the fact that PDSCH layers transmitted from different transmission points should have different QCL relations. As already described, the QCL relation for PDSCH can be dynamically indicated in the scheduling DCI by indicating a specific TCI state from a set of up to eight MAC-CE-activated TCI-states. To enable single-DCI-based multi-TRP transmission, release-16 introduced the possibility for the DCI to simultaneously indicate two different TCI states with corresponding QCL relations. These two TCI states then provide the QCL relation for different sets of the PDSCH DMRS ports, that is, in practice for different sets of PDSCH layers. More specifically, the first TCI state provides the QCL relation for the DMRS ports of the lowest CDM group (see Chapter 9) while the second TCI state provides the QCL relation for the remaining DMRS ports.

12.5.1.2 Multi-DCI-based NCJT

In case of multi-DCI-based NCJT (right part of Fig. 12.7), there is one PDSCH with an associated transport block transmitted from each transmission point and with each PDSCH being scheduled by separate DCIs carried by separate PDCCHs. The maximum transmission rank of each PDSCH is limited to four implying that the total number of layers transmitted to a given device is still limited to eight.

As the two PDSCHs can be received independently, one could, in principle, envision completely independent timing of the transmissions from the two transmission points. However, in order to allow for reception of the two transmissions using a single DFT, there is an assumption that also for multi-DCI-based transmission, the transmissions from the two transmission points are tightly aligned in time.

In case of multi-DCI-based NCJT, there are two transport blocks, one from each transmission point. As a consequence, there will also be two separate HARQ feedbacks, one for each transport block. As outlined in Fig. 12.8, there are to two alternatives for this HARQ feedback

- Joint feedback using a single PUCCH
- Separate feedbacks using separate PUCCHs

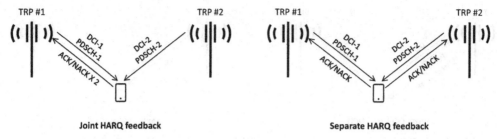

Fig. 12.8 Joint HARQ feedback vs separate HARQ feedback for multi-DCI-based NCJT.

12.5.2 Downlink multi-TRP for URLLC

In case of non-coherent joint transmission as described above, *different data* is transmitted from two different transmission points, either as different layers of the same PDSCH (for single-DSI-based NCJT) or as two different PDSCH (for multi-DCI-based NCJT). In case of non-coherent joint transmission, the multi-point transmission is thus a tool to increase the achievable downlink data rates. However, downlink multi-point transmission can also be used as a tool to improve reliability by transmitting *the same data* from different transmission points. As already mentioned, in 3GPP this has been referred to as (downlink) *multi-TRP for URLLC*.

12.5.2.1 Downlink multi-TRP for URLLC – PDSCH

There are two different approaches to multi-TRP for URLLC for PDSCH, which differ in terms of the time/frequency resources used for the transmissions from the different TRPs.

- Multi-point transmission based on frequency multiplexing
- Multi-point transmission based on time multiplexing

In case of multi-point transmission based on frequency multiplexing, a transport block is transmitted from two TRPs using non-overlapping frequency-domain resources, that is, different non-overlapping sets of resource blocks, see Fig. 12.9. This can be done in two different ways, in the specifications referred to as *scheme A* and *scheme B* respectively.

- In case of scheme A, the transport block is transmitted on a single PDSCH which is split into two parts that are transmitted simultaneously on non-overlapping frequency-domain resources from the two TRPs.
- In case of scheme B, the same transport block is transmitted on two different PDSCHs that are transmitted simultaneously on non-overlapping frequency-domain resources from the two TRPs.

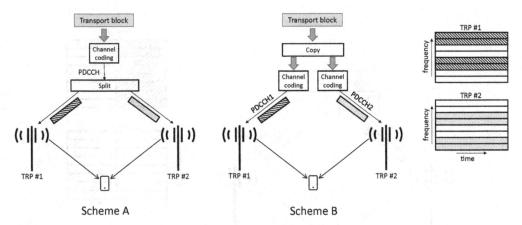

Fig. 12.9 Multi-TRP for URLLC for PDSCH – Frequency multiplexing.

Note that also for Scheme B there is a single DCI scheduling both PDSCH.[b]

To enable device reception of the transmissions from the different TRPs, which typically are located in different directions as seen from the device, the two frequency-domain resources (for scheme A) or the two PDSCH (for scheme B) should be associated with different TCI states. Note that this assumes devices that are capable of simultaneous downlink reception according to two different TCI states, in practice, capable of simultaneous reception from two different directions. As already mentioned, this may not be supported by all devices, especially in case of operation at higher frequencies.

In case of multi-point transmission *based on time multiplexing*, a transport block is instead transmitted from multiple TRPs using non-overlapping *time-domain resources*, see Fig. 12.10. Note that the time multiplexing avoids the problem of simultaneous device reception from multiple directions.

There are two alternatives for the time multiplexing between the TRPs:

- Intra-slot time multiplexing
- Inter-slot time multiplexing

In case of intra-slot time multiplexing, the same transport block is transmitted on two different PDSCHs that are transmitted from the two TRPs using different symbols within a slot. The intra-slot time multiplexing thus follows the same principle as frequency-multiplexing scheme B above, that is, the transport block is transmitted using two different PDSCH from the two TRPs. In the same way as for frequency multiplexing, the two time-domain resources should be associated with different TCI states.

---

[b] The use of a single DCI does not prevent the use of multi-TRP for the PDCCH, that is, the transmission of the same DCI from multiple TRPs as described in Section 12.5.2.2.

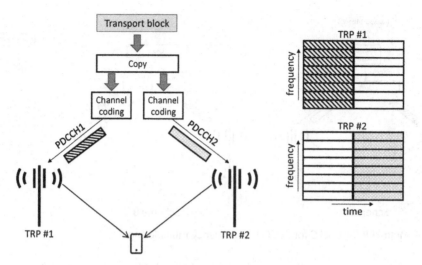

Fig. 12.10 Multi-TRP for URLLC for PDSCH – Time multiplexing.

In case of inter-slot time multiplexing, the PDSCH transmission is instead repeated over multiple slots. There can be up to 16 repetitions, with each repetition associated with one out of two TCI states. Thus, even of there can be up to 16 repetitions, there is still a limitation to two TRPs.

12.5.2.2 Downlink multi-TRP for URLLC – PDCCH

In case of multi-TRP for PDCCH, the same DCI is transmitted on two PDCCHs from different transmission points, where the PDCCHs can either be separated in the frequency domain, that is, transmitted simultaneously on different resource blocks, or separated in the time domain within the same slot.

In more details, multi-TRP for PDCCH is enabled by configuring the device with two search spaces, see Section 10.1.3, that are explicitly configured to be mutually *linked* to each other.[c]

The linked search spaces should be of the same type, periodicity and offset. Furthermore, there should be a one-to-one mapping between monitoring occasions of the two search spaces. This means that, for every PDCCH transmitted within one of the search spaces, there is a corresponding PDCCH, carrying the same DCI, transmitted at a well-defined location within the other search space.

---

[c] The linking is done by assigning a *link identity* to each search space, with two search spaces being linked if they have been assigned the same link identity. Note that a search space can only be linked to one other search space, that is, at most two search spaces can be assigned the same link identity.

To enable multi-TRP transmission, the linked search spaces should be within two different CORESETS that are separated in time and/or frequency and that are assigned different TCI states corresponding to the two TRPs.

To acquire the DCI, it is enough if a device finds a PDCCH within one of the linked search spaces. Furthermore, if the device finds a PDCCH within one of the linked search spaces it knows where, within the second search space, the corresponding PDCCH was transmitted even if the device was not able to correctly decode that PDCCH. This is important as the timing of certain events depends on the timing of the received PDCCH. In case of multi-TRP transmission for PDCCH, these timing relations will, in some cases, be based on the PDCCH that occurs first in time and, in some other cases, on the PDCCH that occurs last in time. Thus, the device needs to know the timing of both PDCCH even if only one of the PDCCH were correctly decoded.

12.5.3 Uplink multi-TRP for URLLC

Release 17 also introduced multi-TRP for URLLC for the uplink direction. More specifically release 17 introduced support for transmission of PUSCH and PUCCH towards two different TRPs using different time-domain resources. Uplink multi-TRP is thus limited to time multiplexing between the TRPs.

12.5.3.1 Uplink multi-TRP for URLLC – PUSCH

Multi-TRP PUSCH transmission extends the release-16 PUSCH repetition which allows a PUSCH transmission to be repeated in up to 16 consecutive slots, see for example Section 10.1.7 or Chapter 21. Release 17 extends this so that the different repetitions may target two different receiving TRPs. In practice, this implies that the different repetitions are transmitted using two different uplink beams matched to the two TRPs, see Fig. 12.11.

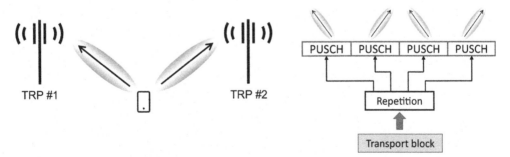

Fig. 12.11 Multi-TRP for URLLC for PUSCH.

There are two different alternatives for the mapping between repetitions and beams:

- Cyclic mapping in which case the beam is changed for each repetition
- Sequential mapping, in which case the beam is change after every second repetition (only relevant for the case of more than two repetitions)

In Section 11.3, uplink multi-antenna transmission was described. This included the configuration of an SRS resource set consisting of one or several SRS resources, as well as the presence of a SRS resource indicator (SRI) and rank/precoder indications (for codebook-based transmission) in the scheduling DCI. To support multi-TRP PUSCH transmission, release 17 extended this so that a device can be configured with two SRS resource sets, that is, one resource set for each receiving TRP. Furthermore, the uplink scheduling grants are extended to include two SRIs and two rank/precoder indicators, corresponding to each of the two SRS resource sets, that is, each of the two receiving TRPs.

12.5.3.2 Uplink multi-TRP for URLLC – PUCCH

Multi-TRP for PUCCH follows the same lines as multi-TRP for PUSCH, that is, the PUCCH transmission is repeated with different repetitions transmitted towards different TRPs. PUCCH repetition is possible for all PUCCH formats of Section 10.2.

The PUCCH repetition can either be inter-slot or intra-slot where, in the former case, there can be two, four or eight repetitions while, for intra-slot repetitions there can only be two repetitions. Note that, in either case, the number of TRPs is limited to two.

CHAPTER 13

Retransmission protocols

Transmissions over wireless channels are subject to errors, for example, due to variations in the received signal quality. To some degree, such variations can be counteracted through link adaptation as will be discussed in Chapter 14. However, receiver noise and unpredictable interference variations cannot be counteracted. Therefore, virtually all wireless communication systems employ some form of Forward Error Correction (FEC), adding redundancy to the transmitted signal allowing the receiver to correct errors and tracing its roots to the pioneering work of Shannon [65]. In NR, LDPC coding is used for error correction as discussed in Section 9.2.

Despite the error-correcting code, there will be data units received in error, for example, due to a too high noise or interference level. Hybrid Automatic Repeat Request (HARQ), first proposed by Wozencraft and Horstein [68] and relying on a combination of error-correcting coding and retransmission of erroneous data units, is therefore commonly used in many modern communication systems. Data units in error despite the error correcting coding are detected by the receiver, which requests a retransmission from the transmitter.

In NR, three different protocol layers offer retransmission functionality – MAC, RLC, and PDCP – as already mentioned in the introductory overview in Chapter 6. The reasons for having a multi-level retransmission structure can be found in the trade-off between fast and reliable feedback of the status reports. The hybrid-ARQ mechanism in the MAC layer targets very fast retransmissions and, consequently, feedback on success or failure of the downlink transmission is provided to the gNB after each received transport block (for uplink transmission no explicit feedback needs to be transmitted as the receiver and scheduler are in the same node). Although it is in principle possible to attain a very low error probability of the hybrid-ARQ feedback, it comes at a cost in transmission resources such as power. In many cases, a feedback error rate of 0.1–1% is reasonable, which results in a hybrid-ARQ residual error rate of a similar order. In many cases this residual error rate is sufficiently low, but there are cases when this is not the case. One obvious case is services requiring ultra-reliable delivery of data combined with low latency. In such cases, either the feedback error rate needs to be decreased and the increased cost in feedback signaling has to be accepted, or additional retransmissions can be performed without relying on feedback signaling, which comes at a decreased spectral efficiency.

5G/5G-Advanced
https://doi.org/10.1016/B978-0-443-13173-8.00022-0

A low error rate is not only of interest for URLLC type-of-services, but is also important from a data-rate perspective. High data rates with TCP may require virtually error-free delivery of packets to the TCP layer. As an example, for sustainable data rates exceeding 100 Mbit/s, a packet-loss probability less than 10^{-5} is required [61]. The reason is that TCP assumes packet errors to be due to congestion in the network. Any packet error therefore triggers the TCP congestion-avoidance mechanism with a corresponding decrease in data rate.

Compared to the hybrid-ARQ acknowledgments, the RLC status reports are transmitted relatively infrequently and thus the cost of obtaining a reliability of 10^{-5} or lower is relatively small. Hence, the combination of hybrid-ARQ and RLC attains a good combination of small round-trip time and a modest feedback overhead where the two components complement each other – fast retransmissions due to the hybrid-ARQ mechanism and reliable packet delivery due to the RLC.

The PDCP protocol is also capable of handling retransmissions, as well as ensuring in-sequence delivery. PDCP-level retransmissions are mainly used in the case of inter-gNB handover as the lower protocols in this case are flushed. Not-yet-acknowledged PDCP PDUs can be forwarded to the new gNB and transmitted to the device. In the case that some of these were already received by the device, the PDCP duplicate detection mechanism will discard the duplicates. The PDCP protocol can also be used to obtain selection diversity by transmitting the same PDUs on multiple carriers. The PDCP in the receiving end will in this case remove any duplicates in case the same information was received successfully on multiple carriers.

In the following sections, the principles behind the hybrid-ARQ, RLC, and PDCP protocols will be discussed in more detail. Note that these protocols are present also in LTE where they to a large extent provide the same functionality. However, the NR versions are enhanced to significantly reduce the delays.

13.1 Hybrid-ARQ with soft combining

The hybrid-ARQ protocol is the primary way of handling retransmissions in NR. In case of an erroneously received packet, a retransmission is requested. However, despite the receiver failing to decode a packet, the received signal still contains information, information which is lost by discarding erroneously received packets. This shortcoming is addressed by *hybrid-ARQ with soft combining*. In hybrid-ARQ with soft combining, the erroneously received packet is stored in a buffer memory and later combined with the retransmission to obtain a single, combined packet that is more reliable than its constituents. Decoding of the error-correction code operates on the combined signal.

Although the protocol itself primarily resides in the MAC layer, there is also physical layer functionality involved in the form of soft combining. Retransmissions of codeblock groups, that is, retransmission of a *part of* the transport block, are handled by the physical layer from a specification perspective, although it could equally well have been described as part of the MAC layer.

The basis for the NR hybrid-ARQ mechanism is a structure with multiple stop-and-wait protocols, each operating on a single transport block. In a stop-and-wait protocol, the transmitter stops and waits for an acknowledgement after each transmitted transport block. This is a simple scheme; the only feedback required is a single bit indicating positive or negative acknowledgement of the transport block. However, since the transmitter stops after each transmission, the throughput is also low. Therefore, *multiple* stop-and-wait processes operating in parallel are used such that, while waiting for acknowledgement from one process, the transmitter can transmit data to another hybrid-ARQ process. This is illustrated in Fig. 13.1; while processing the data received in the first hybrid-ARQ process the receiver can continue to receive using the second process, etc. This structure, multiple hybrid-ARQ processes operating in parallel to form one hybrid-ARQ entity, combines the simplicity of a stop-and-wait protocol with the possibility of continuous data transmission.

The use of multiple parallel hybrid-ARQ processes for a device can result in data being delivered from the hybrid-ARQ mechanism out of sequence. For example, transport block 3 in Fig. 13.1 was successfully decoded before transport block 2 which required retransmissions. For many applications this is acceptable and, if not, in-sequence delivery can be provided through the PDCP protocol. Note that ensuring in-sequence delivery may result in additional latency. In the example in Fig. 13.1, packets numbers 3, 4, and 5 would have to be delayed until packet number 2 is correctly received before delivering them to higher layers, while without in-sequence delivery each packet can be forwarded as soon as it is correctly received.

There is one hybrid-ARQ entity per carrier the receiver is connected to. Spatial multiplexing of more than four layers to a single device in the downlink, where two transport

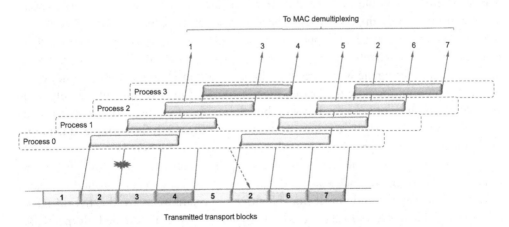

Fig. 13.1 Multiple hybrid-ARQ processes.

blocks can be transmitted in parallel on the same transport channel as described in Section 9.1, is supported by one hybrid-ARQ entity having two sets of hybrid-ARQ processes with independent hybrid-ARQ acknowledgments.

NR uses an *asynchronous* hybrid-ARQ protocol in both downlink and uplink, that is, the hybrid-ARQ process which the downlink or uplink transmission relates to is explicitly signaled as part of the downlink control information (DCI). An alternative would be to use a synchronous hybrid-ARQ protocol where retransmissions occur a fixed time after the initial transmission, but there are several reasons why an asynchronous protocol is used for NR. One reason is that synchronous hybrid-ARQ operation does not allow dynamic TDD. Another reason is that operation in unlicensed spectra, introduced in release 16 and described in Chapter 20, is more efficient with asynchronous operation as it cannot be guaranteed that the radio resources are available at the time for a synchronous retransmission. Thus, NR settled for an asynchronous scheme in both uplink and downlink with up to 16 processes, extended to 32 processes in release 17. Having the possibility for a relatively large number of hybrid-ARQ processes is motivated by the possibility for remote radio heads, which incurs a certain fronthaul delay, and the fact that the slot duration is short at high frequencies. Non-terrestrial access is another scenario with very large delays and thus requiring a large number of hybrid-ARQ processes. It is important though, that the larger number of *maximum* hybrid-ARQ processes does not imply a longer roundtrip time as not all processes need to be used, it is only an upper limit of the number of processes possible to address.

Large transport block sizes are segmented into multiple codeblocks prior to coding, each with its own 24-bit CRC (in addition to the overall transport-block CRC). This was discussed already in Section 9.2 and the reason is primarily complexity; the size of a codeblock is large enough to give good performance while still having a reasonable decoding complexity. Since each codeblock has its own CRC, errors can be detected on individual codeblocks as well as on the overall transport block. A relevant question is if retransmission should be limited to transport blocks or whether there are benefits of retransmitting only the codeblocks that are erroneously received. For the very large transport block sizes used to support data rates of several gigabits per second, there can be hundreds of codeblocks in a transport block. If only one or a few of them are in error, retransmitting the whole transport block results in a low spectral efficiency compared to retransmitting only the erroneous codeblocks. One example where only some codeblocks are in error is a situation with bursty interference where some OFDM symbols are hit more severely than others, as illustrated in Fig. 13.2, for example, due to one downlink transmission preempting another as discussed in Section 14.1.2.

To correctly receive the transport block for the given example, it is sufficient to retransmit the erroneous codeblocks. At the same time, the control signaling overhead would be too large if individual codeblocks can be addressed by the hybrid-ARQ mechanism. Therefore, so-called *codeblock groups* (CBGs) are defined. If per-CBG

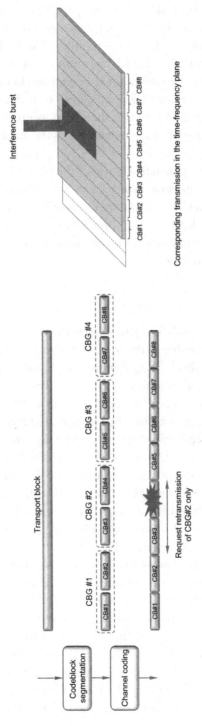

Fig. 13.2 Codeblock-group retransmission.

retransmission is configured, feedback is provided per CBG instead of per transport block and only the erroneously received codeblock groups are retransmitted, which consumes less resources than retransmitting the whole transport block. Two, four, six, or eight codeblock groups can be configured with the number of codeblocks per codeblock group varying as a function of the total number of codeblocks in the initial transmission. Note that the codeblock group a codeblock belongs to is determined from the initial transmission and does not change between the transmission attempts. This is to avoid error cases which could arise if the codeblocks were repartitioned between two retransmissions.

The CBG retransmissions are handled as part of the physical layer from a specification perspective. There is no fundamental technical reason for this but rather a way to reduce the specification impact from CBG-level retransmissions. A consequence of this is that it is not possible, in the same hybrid-ARQ process, to mix transmission of new CBGs belonging to another transport block with retransmissions of CBGs belonging to the incorrectly received transport block.

13.1.1 Soft combining

An important part of the hybrid-ARQ mechanism is the use of *soft combining*, which implies that the receiver combines the received signal from multiple transmission attempts. By definition, a hybrid-ARQ retransmission must represent the same set of information bits as the original transmission. However, the set of coded bits transmitted in each retransmission may be selected differently as long as they represent the same set of information bits. Depending on whether the retransmitted bits are required to be identical to the original transmission or not the soft combining scheme is often referred to as *Chase combining*, first proposed in [20], or *Incremental Redundancy* (IR), which is used in NR. With incremental redundancy, each retransmission does not have to be identical to the original transmission. Instead, *multiple sets* of coded bits are generated, each representing the same set of information bits [63,67]. The rate matching functionality of NR, described in Section 9.3, is used to generate different sets of coded bits as a function of the redundancy version as illustrated in Fig. 13.3.

In addition to a gain in accumulated received E_b/N_0, incremental redundancy also results in a coding gain for each retransmission (until the mother code rate is reached). The gain with incremental redundancy compared to pure energy accumulation (Chase combining) is larger for high initial code rates [22]. Furthermore, as shown in [31], the performance gain of incremental redundancy compared to Chase combining can also depend on the relative power difference between the transmission attempts.

In the discussion so far, it has been assumed that the receiver has received all the previously transmitted redundancy versions. If all redundancy versions provide the same amount of information about the data packet, the order of the redundancy versions is

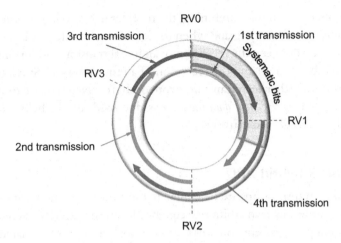

Fig. 13.3 Example of incremental redundancy.

not critical. However, for some code structures, not all redundancy versions are of equal importance. This is the case for the LDPC codes used in NR; the systematic bits are of higher importance than the parity bits. Hence, the initial transmission should at least include all the systematic bits and some parity bits. In the retransmission(s), parity bits not in the initial transmission can be included. This is the background to why systematic bits are inserted first in the circular buffer in Section 9.3. The starting points in the circular buffer are defined such that both RV0 and RV3 are self-decodable, that is, includes the systematic bits under typical scenarios. This is also the reason RV3 is located after nine o'clock in Fig. 13.3 as this allows more of the systematic bits to be included in the transmission. With the default order of the redundancy versions 0, 2, 3, 1, every second retransmission is typically self-decodable.

Hybrid ARQ with soft combining, regardless of whether Chase or incremental redundancy is used, leads to an implicit reduction of the data rate by means of retransmissions and can thus be seen as implicit link adaptation. However, in contrast to link adaptation based on explicit estimates of the instantaneous channel conditions, hybrid ARQ with soft combining implicitly adjusts the coding rate based on the result of the decoding. In terms of overall throughput this kind of implicit link adaptation can be superior to explicit link adaptation, as additional redundancy is only added *when needed* – that is, when previous higher-rate transmissions were not possible to decode correctly. Furthermore, as it does not try to predict any channel variations, it works well regardless of the speed at which the terminal is moving. Since implicit link adaptation can provide a gain in system throughput, a valid question is why explicit link adaptation is necessary at all. One major reason for having explicit link adaptation is the reduced delay. Although relying on implicit link adaptation alone is sufficient from a system throughput perspective, the end-user service quality may not be acceptable from a delay perspective.

For proper operation of soft combining, the receiver needs to know when to perform soft combining prior to decoding and when to clear the soft buffer – that is, the receiver needs to differentiate between the reception of an initial transmission (prior to which the soft buffer should be cleared) and the reception of a retransmission. Similarly, the transmitter must know whether to retransmit erroneously received data or to transmit new data. This is handled by the *new-data indicator* as discussed further below for downlink and uplink hybrid-ARQ, respectively.

13.1.2 Downlink hybrid-ARQ

In the downlink, retransmissions are scheduled in the same way as new data – that is, they may occur at any time and at an arbitrary frequency location within the active bandwidth part. The scheduling assignment contains the necessary hybrid-ARQ-related control signaling – hybrid-ARQ process number, new-data indicator, and CBGTI and CBGFI in case per-CBG retransmission is configured, as well as information to handle the transmission of the acknowledgement in the uplink such as timing and resource indication information.

Upon receiving a scheduling assignment in the DCI, the receiver tries to decode the transport block, possibly after soft combining with previous attempts as described earlier. Since transmissions and retransmissions are scheduled using the same framework in general, the device needs to know whether the transmission is a new transmission, in which case the soft buffer should be flushed, or a retransmission, in which case soft combining should be performed. Therefore, an explicit *new-data indicator* is included for the scheduled transport block as part of the scheduling information transmitted in the downlink. The new-data indicator is toggled for a new transport block – that is, it is essentially a single-bit sequence number. Upon reception of a downlink scheduling assignment, the device checks the new-data indicator to determine whether the current transmission should be soft combined with the received data currently in the soft buffer for the hybrid-ARQ process in question, or if the soft buffer should be cleared.

The new-data indicator operates on the transport-block level. However, if per-CBG retransmissions are configured, the device needs to know which CBGs that are retransmitted and whether the corresponding soft buffer should be flushed or not. This is handled through two additional information fields present in the DCI in case per-CBG retransmission is configured, the *CBG transmission indicator* (CBGTI) and the *CBG flush indicator* (CBGFI). The CBGTI is a bitmap indicating whether a certain CBG is present in the downlink transmission or not (see Fig. 13.4). The CBGFI is a single bit, indicating whether the CBGs indicated by the CBGTI should be flushed or whether soft combining should be performed. The result of the decoding operation – a positive acknowledgement in the case of a successful decoding and a negative acknowledgement in the case of unsuccessful decoding – is fed back to the gNB as part of the uplink control

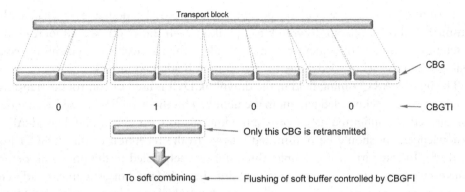

Fig. 13.4 Illustration of per-CBG retransmission.

information. If CBG retransmissions are configured, a bitmap with one bit per CBG is fed back instead of a single bit representing the whole transport block.

13.1.3 Uplink hybrid-ARQ

The uplink uses the same asynchronous hybrid-ARQ protocol as the downlink. The necessary hybrid-ARQ-related information – hybrid-ARQ process number, new-data indicator, and, if per-CBG retransmission is configured, the CBGTI – is included in the scheduling grant.

To differentiate between new transmissions and retransmissions of data, the new-data indicator is used. Toggling the new-data indicator requests transmission of a new transport block, otherwise the previous transport block for this hybrid-ARQ process should be retransmitted (in which case the gNB can perform soft combining). The CBGTI is used in a similar way as in the downlink, namely to indicate the codeblock groups to retransmit in the case of per-CBG retransmission. Note that no CBGFI is needed in the uplink as the soft buffer is located in the gNB which can decide whether to flush the buffer or not based on the scheduling decisions.

13.1.4 Timing of uplink acknowledgments

The device needs to know when to transmit the acknowledgement in the uplink in response to a downlink reception. Having a fixed timing relation between the data reception and the transmission of the acknowledgement is a simple approach and works nicely for FDD. Unfortunately, this type of scheme with predefined timing instants for the acknowledgments does not blend well with dynamic TDD, one of the cornerstones of NR, as an uplink opportunity cannot be guaranteed a fixed time after the downlink transmission due to the uplink-downlink direction being dynamically controlled by the scheduler. Coexistence with other TDD deployments in the same frequency band may also impose restrictions when it is desirable, or possible, to transmit in the uplink.

Furthermore, even if it would be possible, it may not be desirable to change the transmission direction from downlink to uplink in each slot as this would increase the switching overhead. Consequently, a more flexible scheme capable of dynamically controlling when the acknowledgement is transmitted is adopted in NR.

The hybrid-ARQ timing field in the downlink DCI is used to control the transmission timing of the acknowledgement in the uplink. This three-bit field is used as an index into an RRC-configured table providing information on when the hybrid-ARQ acknowledgement should be transmitted relative to the reception of the PDSCH (see Fig. 13.5). In this particular example, three slots are scheduled in the downlink before an acknowledgement is transmitted in the uplink. In each downlink assignment, different acknowledgement timing indices have been used, which in combination with the RRC-configured table result in all three slots being acknowledged at the same time (multiplexing of these acknowledgments in the same slot is discussed below).

NR is designed with very low latency in mind and is therefore capable of transmitting the acknowledgement soon after receiving the downlink data transmission. All devices support the baseline processing times listed in Table 13.1, with even faster processing optionally supported by some devices. The capability is reported per subcarrier spacing. One part of the processing time is constant in symbols across different subcarrier spacings, that is, the time in microseconds scales with the subcarrier spacing, but there is also a part of the processing time fixed in microseconds and independent of the subcarrier spacing.

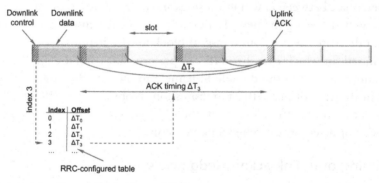

Fig. 13.5 Determining the acknowledgement timing.

Table 13.1 Minimum processing time (PDSCH mapping type A, feedback on PUCCH).

DM-RS configuration	Device capability	Subcarrier spacing				
		15 kHz	30 kHz	60 kHz	120 kHz	LTE rel 8
Front-loaded	Baseline	0.57 ms	0.36 ms	0.30 ms	0.18 ms	2.3 ms
	Aggressive	0.18–0.29 ms	0.08–0.17 ms			
Additional	Baseline	0.92 ms	0.46 ms	0.36 ms	0.21 ms	
	Aggressive	0.85 ms	0.4 ms			

Hence, the processing times listed in the table are not directly proportional to the sub-carrier spacing although there is a dependency. There is also a dependency on the reference signal configuration; if the device is configured with additional reference signal occasions later in the slot, the device cannot start the processing until at least some of these reference signals have been received and the overall processing time is longer. Nevertheless, the processing is much faster than the corresponding LTE case as a result of stressing the importance of low latency in the NR design.

For proper transmission of the acknowledgement, it is not sufficient for the device to know *when* to transmit, which is obtained from the timing field discussed, but also *where* in the resource domain (frequency resources and, for some PUCCH formats, the code domain). Also in this case NR provides a great deal of flexibility. To avoid collisions between multiple devices transmitting their acknowledgements at the same time, it is necessary to provide the devices with separate resources. This is handled through the *PUCCH resource indicator*, which is a three-bit index selecting one of eight RRC-configured resource sets as described in Section 10.2.7.

13.1.5 Multiplexing of hybrid-ARQ acknowledgments

In the previous section, the timing of the hybrid-ARQ acknowledgments in the example was such that multiple transport blocks need to be acknowledged at the same time. Other examples where multiple acknowledgments need to be transmitted in the uplink at the same time are carrier aggregation and per-CBG retransmissions. NR therefore supports multiplexing of acknowledgments for multiple transport blocks (and CBGs) received by a device into one multi-bit acknowledgement message. The multiple bits can be multiplexed using either a semi-static (type 1) codebook or a dynamic (type 2) codebook with RRC configuration selecting between the two. Release 16 in addition introduces a one-shot feedback report where the device is requested to report the status for all the hybrid-ARQ processes. This is also known as a type 3 codebook and used in conjunction with operation in unlicensed spectra as described in more detail in Chapter 20.

The semi-static codebook can be viewed as a matrix consisting of a time-domain dimension and a component-carrier (or CBG or MIMO layer) dimension, both of which are semi-statically configured. The size in the time domain is given by the maximum and minimum hybrid-ARQ acknowledgement timings configured in Table 13.1, and the size in the carrier domain is given by the number of simultaneous transport blocks (or CBGs) across all component carriers. An example is provided in Fig. 13.6, where the acknowledgment timings are one, two, three, and four, respectively, and three carriers, one with two transport blocks, one with one transport block, and one with four CBGs, are configured. Since the codebook size is fixed, the number of bits to transmit in a hybrid-ARQ report is known ($4 \cdot 7 = 28$ bits in the example in Fig. 13.6) and the appropriate format for the uplink control signaling can be selected. Each entry in the matrix represents the decoding outcome, positive or negative acknowledgment, of the

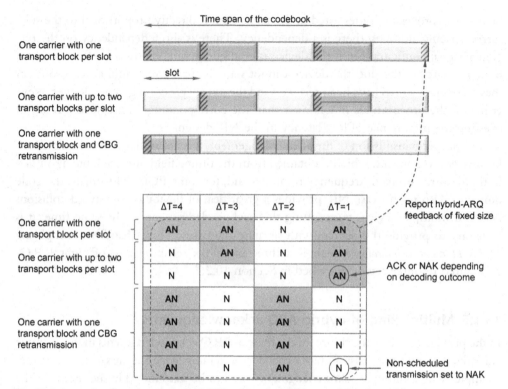

Fig. 13.6 Example of semi-static hybrid-ARQ acknowledgement codebook.

corresponding transmission. Not all transmission opportunities possible with the codebook are used in this example and for entries in the matrix without a corresponding transmission, a negative acknowledgment is transmitted. This provides robustness; in the case of missed downlink assignment a negative acknowledgment is provided to the gNB, which can retransmit the missing transport block (or CBG).

One drawback with the semi-static codebook is the potentially large size of a hybrid-ARQ report. For a small number of component carriers and no CBG retransmissions, this is less of a problem, but if a large number of carriers and codeblock groups are configured out of which only a small number is simultaneously used, this may become more of an issue.

To address the drawback of a potentially large semi-static codebook size in some scenarios, NR also supports a dynamic codebook. In fact, this is the default codebook used unless the system is configured otherwise. With a dynamic codebook, only the acknowledgement information for the *scheduled* carriers[a] is included in the report, instead of all

[a] The description here uses the term "carrier" but the same principle is equally applicable to per-CBG retransmission or multiple transport blocks in case of MIMO and "transmission instant" is a more generic term, albeit the description would be harder to read.

carriers, scheduled or not, as is the case with a semi-static codebook. Hence, the size of the codebook (the matrix in Fig. 13.6) is dynamically varying as a function of the number of scheduled carriers. In essence, only the bold entries in the example in Fig. 13.6 would be included in the hybrid-ARQ report and the non-bold entries with a gray background (which correspond to non-scheduled carriers) would be omitted. This reduces the size of the acknowledgement message.

A dynamic codebook would be straight forward if there were no errors in the downlink control signaling. However, in presence of an error in the downlink control signaling, the device and gNB may have different understanding on the number of scheduled carriers which would lead to an incorrect codebook size and possibly corrupt the feedback report for all carriers, and not only for the ones for which the downlink controls signaling was missed. Assume, as an example, that the device was scheduled for downlink transmission in two subsequent slots but missed the PDCCH and hence scheduling assignment for the first slot. In response the device will transmit an acknowledgement for the second slot only, while the gNB tries to receive acknowledgments for two slots, leading to a mismatch.

To handle these error cases, NR uses the *downlink assignment index* (DAI) included in the DCI containing the downlink assignment. The DAI field is further split into two parts, a counter DAI (cDAI) and, in the case of carrier aggregation, a total DAI (tDAI). The counter DAI included in the DCI indicates the number of scheduled downlink transmissions up to the point the DCI was received in a carrier first, time second manner. The total DAI included in the DCI indicates the total number of downlink transmissions across all carriers up to this point in time, that is, the highest cDAI at the current point in time (see Fig. 13.7 for an example). The counter DAI and total DAI are represented with decimal numbers with no limitation; in practice two bits are used for each and the numbering will wrap around, that is, what is signaled is the numbers in the figure modulo four. As seen in this example, the dynamic codebook needs to account for 17 acknowledgments (numbered 0 to 16). This can be compared with the semi-static codebook which would require 28 entries regardless of the number of transmissions.

Furthermore, in this example, one transmission on component carrier five is lost. Without the DAI mechanism, this would result in misaligned codebooks between the device and the gNB. However, as long as the device receives at least one component carrier, it knows the value of the total DAI and hence the size of the codebook at this point in time. Furthermore, by checking the values received for the counter DAI, it can conclude which component carrier was missed and that a negative acknowledgement should be assumed in the codebook for this position.

In the case that CBG retransmission is configured for some of the carriers, the dynamic codebook is split into two parts, one for the non-CBG carriers and one for the CBG carriers. Each codebook is handled according to the principles outlined. The reason for the split is that for the CBG carriers, the device needs to generate feedback for each of these carriers according to the largest CBG configuration.

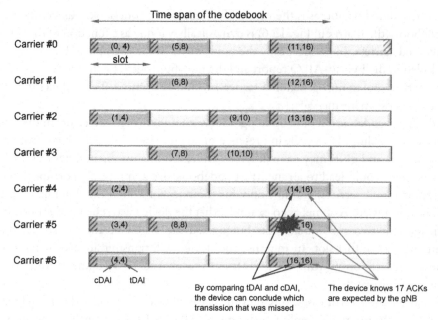

Fig. 13.7 Example of dynamic hybrid-ARQ acknowledgement codebook.

13.2 RLC

The *radio-link control* (RLC) protocol takes data in the form of RLC SDUs from PDCP and delivers them to the corresponding RLC entity in the receiver by using functionality in MAC and physical layers. The relation between RLC and MAC, including multiplexing of multiple logical channels into a single transport channel, is illustrated in Fig. 13.8.

There is one RLC entity per logical channel configured for a device with the RLC entity being responsible for one or more of

- Segmentation of RLC SDUs;
- Duplicate removal; and
- RLC retransmission.

There is no support for concatenation or in-sequence delivery in the RLC protocol. This is a deliberate choice done to reduce the overall latency as discussed further in the following sections. It has also impacted the header design. Also, note that the fact that there is one RLC entity per logical channel and one hybrid-ARQ entity per cell (component carrier) implies that RLC retransmissions can occur on a different cell (component carrier) than the original transmission. This is not the case for the hybrid-ARQ protocol where retransmissions are bound to the same component carrier as the original transmission.

Different services have different requirements; for some services (for example, transfer of a large file), error-free delivery of data is important, whereas for other applications

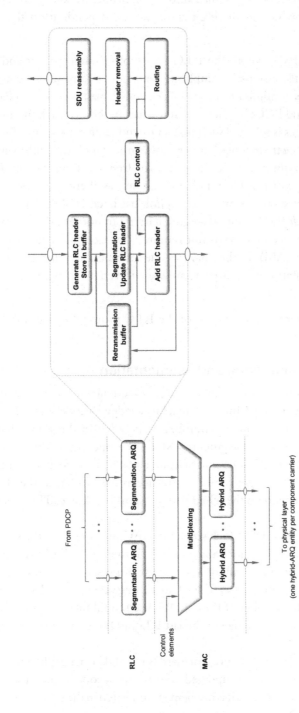

Fig. 13.8 MAC and RLC.

(for example, streaming services), a small amount of missing packets is not a problem. The RLC can therefore operate in three different modes, depending on the requirements from the application:

- *Transparent mode* (TM), where the RLC is completely transparent and is essentially bypassed. No retransmissions, no duplicate detection, and no segmentation/reassembly take place. This configuration is used for control-plane broadcast channels such as BCCH, CCCH, and PCCH, where the information should reach multiple users. The size of these messages is selected such that all intended devices are reached with a high probability and hence there is neither need for segmentation to handle varying channel conditions, nor retransmissions to provide error-free data transmission. Furthermore, retransmissions are not feasible for these channels as there is no possibility for the device to feed back status reports as no uplink has been established.
- *Unacknowledged mode* (UM) supports segmentation but not retransmissions. This mode is used when error-free delivery is not required, for example voice-over IP.
- *Acknowledged mode* (AM) is the main mode of operation for the DL-SCH and UL-SCH. Segmentation, duplicate removal, and retransmissions of erroneous data are all supported.

In the following sections, the operation of the RLC protocol is described, focusing on acknowledged mode.

13.2.1 Sequence numbering and segmentation

In unacknowledged and acknowledged modes, a sequence number is attached to each incoming SDU using 6 or 12 bits for unacknowledged mode and 12 or 18 bits for acknowledged mode. The sequence number is included in the RLC PDU header in Fig. 13.9. In the case of a non-segmented SDU, the operation is straightforward; the RLC PDU is simply the RLC SDU with a header attached. Note that the PDU structure allows the RLC PDUs to be generated in advance as the header, in the absence of segmentation, does not depend on the scheduled transport block size. This is beneficial from a latency perspective.

Depending on the transport-block size after MAC multiplexing, the size of the last of the RLC PDUs in a transport block may not match the RLC SDU size. To handle this, an SDU can be segmented into multiple segments. If no segmentation takes place, padding would need to be used instead, leading to degraded spectral efficiency. Hence, dynamically varying the number of RLC PDUs used to fill the transport block, together with segmentation to adjust the size of the last RLC PDU, ensures the transport block is efficiently utilized.

Segmentation is simple; the last preprocessed RLC SDU can be split into two segments, the header of the first segment is updated, and to the second segment a new header is added (which is not time critical as it is not being transmitted in the current transport block).

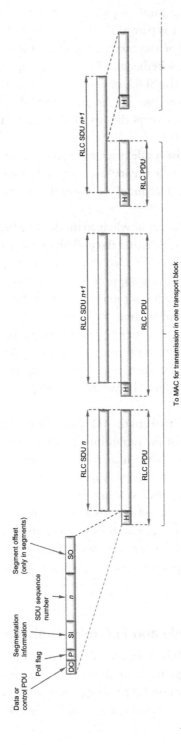

Fig. 13.9 Generation of RLC PDUs from RLC SDUs (acknowledged mode assumed for the header structure).

Each SDU segment carries the same sequence number as the original unsegmented SDU and this sequence number is part of the RLC header. To distinguish whether the PDU contains a complete SDU or a segment, a *segmentation information* (SI) field is also part of the RLC header, indicating whether the PDU is a complete SDU, the first segment of the SDU, the last segment of the SDU, or a segment between the first and last segments of the SDU. Furthermore, in the case of a segmented SDU, a 16-bit *segmentation offset* (SO) is included in all segments except the first one and used to indicate which byte of the SDU the segment represents.

There is also a *poll bit* (P) in the header used to request a status report for acknowledged mode as described further, and a *data/control indicator*, indicating whether the RLC PDU contains data to/from a logical channel or control information required for RLC operation.

The header structure holds for acknowledged mode. The header for unacknowledged mode is similar but does not include either the poll bit or the data/control indicator. Furthermore, the sequence number is included in the case of segmentation only.

As already mentioned, there is no support for concatenation or in-sequence in the RLC in order to reduce latency. If concatenation would be supported, an RLC PDU cannot be assembled until the uplink grant is received as the scheduled transport-block size is not known in advance. Consequently, the uplink grant must be received well in advance to allow sufficient processing time in the device. Without concatenation, the RLC PDUs can be assembled in advance, prior to receiving the uplink grant, and thereby reducing the processing time required between receiving an uplink grant and the actual uplink transmission.

Omitting in-sequence delivery from the RLC also helps reduce the overall latency as later packets do not have to wait for retransmission of an earlier missing packet before being delivered to higher layers, but can be forwarded immediately. This also leads to reduced buffering requirements positively impacting the amount of memory used for RLC buffering. If in-sequence delivery would be supported, an RLC SDU cannot be forwarded to higher layers unless all previous SDUs have been correctly received. A single missing SDU, for example, due to a momentary interference burst, can thus block delivery of subsequent SDUs for quite some time even if those SDUs would be useful to the application, a property which is clearly not desirable in a system targeting very low latency.

13.2.2 Acknowledged mode and RLC retransmissions

Retransmission of missing PDUs is one of the main functionalities of the RLC in acknowledged mode. Although most of the errors can be handled by the hybrid-ARQ protocol, there are, as discussed at the beginning of the chapter, benefits of having a second-level retransmission mechanism as a complement. By inspecting the sequence numbers of the received PDUs, missing PDUs can be detected and a retransmission requested from the transmitting side.

In acknowledged mode, the RLC entity is bidirectional – that is, data may flow in both directions between the two peer entities. This is necessary as the reception of PDUs needs to be acknowledged back to the entity that transmitted those PDUs. Information about missing PDUs is provided by the receiving end to the transmitting end in the form of so-called *status reports*. Status reports can either be transmitted autonomously by the receiver or requested by the transmitter. To keep track of the PDUs in transit, the sequence number in the header is used.

Both RLC entities maintain two windows in acknowledged mode, the transmission and reception windows, respectively. Only PDUs in the transmission window are eligible for transmission; PDUs with sequence number below the start of the window have already been acknowledged by the receiving RLC. Similarly, the receiver only accepts PDUs with sequence numbers within the reception window. The receiver also discards any duplicate PDUs as only one copy of each SDU should be delivered to higher layers.

The operation of the RLC with respect to retransmissions is perhaps best understood by the simple example in Fig. 13.10, where two RLC entities are illustrated, one in the transmitting node and one in the receiving node. When operating in acknowledged mode each RLC entity has both transmitter and receiver functionality, but in this example only one of the directions is discussed as the other direction is identical. In the example, PDUs numbered from n to $n+4$ are awaiting transmission in the transmission buffer. At time t_0, PDUs with sequence number up to and including n have been transmitted and correctly received, but only PDUs up to and including $n-1$ have been acknowledged by the receiver. As seen in the figure, the transmission window starts from n, the first not-yet-acknowledged PDU, while the reception window starts from $n+1$, the next PDU expected to be received. Upon reception of a PDU n, the SDU is reassembled and delivered to higher layers, that is, the PDCP. For a PDU containing a complete SDU, reassembly is simply header removal, but in the case of a segmented SDU, the SDU cannot be delivered until PDUs carrying all the segments have been received.

The transmission of PDUs continues and, at time t_1, PDUs $n+1$ and $n+2$ have been transmitted but, at the receiving end, only PDU $n+2$ has arrived. As soon as a complete SDU is received, it is delivered to higher layers, hence PDU $n+2$ is forwarded to the PDCP layer without waiting for the missing PDU $n+1$. One reason PDU $n+1$ is missing could be that it is under retransmission by the hybrid-ARQ protocol and therefore has not yet been delivered from the hybrid ARQ to the RLC. The transmission window remains unchanged compared to the previous figure, as none of the PDUs n and higher have been acknowledged by the receiver. Hence, any of these PDUs may need to be retransmitted as the transmitter is not aware of whether they have been received correctly or not.

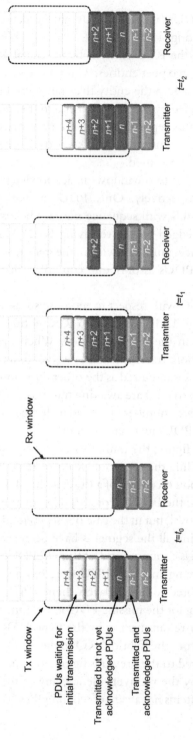

Fig. 13.10 SDU delivery in acknowledged mode.

The reception window is not updated when PDU $n+2$ arrives, the reason being the missing PDU $n+1$. Instead the receiver starts a timer, the *t-Reassembly* timer. If the missing PDU $n+1$ is not received before the timer expires, a retransmission is requested. Fortunately, in this example, the missing PDU arrives from the hybrid-ARQ protocol at time t_2, before the timer expires. The reception window is advanced and the reassembly timer is stopped as the missing PDU has arrived. PDU $n+1$ is delivered for reassembly into SDU $n+1$.

Duplicate detection is also the responsibility of the RLC, using the same sequence number as used for retransmission handling. If PDU $n+2$ arrives again (and is within the reception window), despite it having already been received, it is discarded.

The transmission continues with PDUs $n+3$, $n+4$, and $n+5$ as shown in Fig. 13.11. At time t_3, PDUs up to $n+5$ have been transmitted. Only PDU $n+5$ has arrived and PDUs $n+3$ and $n+4$ are missing. Similar to the earlier case, this causes the reassembly timer to start. However, in this example no PDUs arrive prior to the expiration of the timer. The expiration of the timer at time t_4 triggers the receiver to send a control PDU containing a status report, indicating the missing PDUs, to its peer entity. Control PDUs have higher priority than data PDUs to avoid the status reports being unnecessarily delayed and negatively impacting the retransmission delay. Upon receipt of the status report at time t_5, the transmitter knows that PDUs up to $n+2$ have been received correctly and the transmission window is advanced. The missing PDUs $n+3$ and $n+4$ are retransmitted and, this time, correctly received.

Finally, at time t_6, all PDUs, including the retransmissions, have been delivered by the transmitter and successfully received. As $n+5$ was the last PDU in the transmission buffer, the transmitter requests a status report from the receiver by setting a flag in the header of the last RLC data PDU. Upon reception of the PDU with the flag set, the receiver will respond by transmitting the requested status report, acknowledging all PDUs up to and including $n+5$. Reception of the status report by the transmitter causes all the PDUs to be declared as correctly received and the transmission window is advanced.

Status reports can, as mentioned earlier, be triggered for multiple reasons. However, to control the amount of status reports and to avoid flooding the return link with an excessive number of status reports, it is possible to use a status prohibit timer. With such a timer, status reports cannot be transmitted more often than once per time interval as determined by the timer.

The example basically assumed each PDU carrying a non-segmented SDU. Segmented SDUs are handled the same way, but an SDU cannot be delivered to the PDCP protocol until all the segments have been received. Status reports and retransmissions operate on individual segments; only the missing segment of a PDU needs to be retransmitted.

In the case of a retransmission, all RLC PDUs may not fit into the transport block size scheduled for the RLC retransmission. Resegmentation following the same principle as the original segmentation is used in this case.

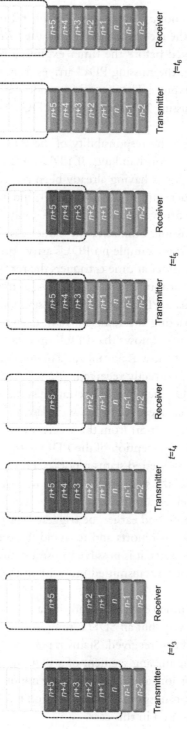

Fig. 13.11 Retransmission of missing PDUs.

13.3 PDCP

The *Packet Data Convergence Protocol* (PDCP) is responsible for

- Header compression;
- Ciphering and integrity protection;
- Routing and duplication for split bearers; and
- Retransmission, reordering, and SDU discard.

Header compression, with the corresponding decompression functionality at the receiver side, can be configured and serves the purpose of reducing the number of bits transmitted over the radio interface. Especially for small payloads, such as voice-over-IP and TCP acknowledgments, the size of an uncompressed IP header is in the same range as the payload itself, 40 bytes for IP v4 and 60 bytes for IP v6, and can account for around 60% of the total number of bits sent. Compressing this header to a couple of bytes can therefore increase the spectral efficiency by a large amount. The header compression scheme in NR is based on Robust Header Compression (ROHC) [36], a standardized header-compression framework also used for several other mobile-communication technologies. Multiple compression algorithms, denoted profiles, are defined, each specific to the particular network layer and transport layer protocol combination such as TCP/IP and RTP/UDP/IP. Header compression is developed to compress IP packets. Hence it is applied to the data part only and not the SDAP header (if present).

Integrity protection ensures that the data originate from the correct source and ciphering protects against eavesdropping. PDCP is responsible for both these functions, if configured. Integrity protection and ciphering are used for both the data plane and the control plane and applied to the payload only and not the PDCP control PDUs or SDAP headers.

For dual connectivity and split bearers (see Chapter 6 for a more in-depth discussion on dual connectivity), PDCP can provide routing and duplication functionality. With dual connectivity, some of the radio bearers are handled by the master cell group while others are handled by the secondary cell group. There is also a possibility to split a bearer across both cell groups. The routing functionality of the PDCP is responsible for routing the data flows for the different bearers to the correct cell groups, as well as handling flow control between the central unit (gNB-CU) and distributed unit (gNB-DU) in the case of a split gNB.

Duplication implies that the same data can be transmitted on two separate logical channels where configuration ensures that the two logical channels are mapped to different carriers. This can be used in combination with carrier aggregation or dual connectivity to provide additional diversity. If multiple carriers are used to transmit the same data, the likelihood that reception of the data on at least one carrier is correct increases. If multiple copies of the same SDU are received, the receiving-side PDCP discards the duplicates. This results in selection diversity which can be essential to providing very high reliability. For the downlink, transmitter-side duplication is up to the implementation, while for the uplink explicit support in the specifications is needed. Duplication of up to two copies can be configured in release 15, a number that is increase to four in release 16 (see Chapter 21).

Retransmission functionality, including the possibility for reordering to ensure in-sequence delivery, is also part of the PDCP. A relevant question is why the PDCP is capable of retransmissions when there are two other retransmission functions in lower layers, the RLC ARQ and the MAC hybrid-ARQ functions. One reason is inter-gNB handover. Upon handover, undelivered downlink data packets will be forwarded by the PDCP from the old gNB to the new gNB. In this case, a new RLC entity (and hybrid-ARQ entity) is established in the new gNB and the RLC status is lost. The PDCP retransmission functionality ensures that no packets are lost as a result of this handover. In the uplink, the PDCP entity in the device will handle retransmission of all uplink packets not yet delivered to the gNB as the hybrid-ARQ buffers are flushed upon handover.

In-sequence delivery is not ensured by the RLC to reduce the overall latency. In many cases, rapid delivery of the packets is more important than guaranteed in-sequence delivery. However, if in-sequence delivery is important, the PDCP can be configured to provide this.

Retransmission and in-sequence delivery, if configured, are jointly handled in the same protocol, which operates similarly to the RLC ARQ protocol except that no segmentation is supported. A so-called count value is associated with each SDU, where the count is a combination of the PDCP sequence number and the hyper-frame number. The count value is used to identify lost SDUs and request retransmission, as well as reorder received SDUs before delivery to upper layers if reordering is configured. Reordering basically buffers a received SDU and does not forward it to higher layers until all lower-numbered SDUs have been delivered. Referring to Fig. 13.10, this would be similar to not delivering SDU $n + 2$ until $n + 1$ has been successfully received and delivered. There is also a possibility to configure a discard timer for each PDCP SDU; when the timer expires the corresponding SDU is discarded and not transmitted.

CHAPTER 14

Scheduling

NR is essentially a scheduled system, implying that the scheduler determines when and to which devices the time, frequency, and spatial resources should be assigned and what transmission parameters, including data rate, to use. Scheduling can be either dynamic or semi-static. Dynamic scheduling is the basic mode-of-operation where the scheduler for each time interval, for example a slot, determines which devices are to transmit and receive. Since scheduling decisions are taken frequently, it is possible to follow rapid variations in the traffic demand and radio-channel quality, thereby efficiently exploiting the available resources. Semi-static scheduling implies that the transmission parameters are provided to the devices in advance and not on a dynamic basis.

In the following, dynamic downlink and uplink scheduling will be discussed, followed by a discussion on non-dynamic scheduling and finally a discussion on power-saving mechanism related to scheduling.

14.1 Dynamic downlink scheduling

Fluctuations in the received signal quality due to small-scale as well as large-scale variations in the environment are an inherent part in any wireless communication system. Historically, such variations were seen as a problem, but the development of *channel-dependent scheduling*, where transmissions to an individual device take place when the radio-channel conditions are favorable, allows these variations to be exploited. Given a sufficient number of devices in the cell having data to transfer, there is a high likelihood of at least some devices having favorable channel conditions at each point in time and able to use a correspondingly high data rate. The gain obtained by transmitting to users with favorable radio-link conditions is commonly known as multiuser diversity. The larger the channel variations and the larger the number of users in a cell, the larger the multiuser diversity gain. Channel-dependent scheduling was introduced in the later versions of the 3G standard known as HSPA [19] and is also used in LTE [26] as well as NR.

There is a rich literature in the field of scheduling and how to exploit variations in the time and frequency domains (see for example [26] and the references therein). Lately, there has also been a large interest in various massive multiuser MIMO schemes [53] where a large number of antenna elements are used to create very narrow "beams", or, expressed differently, isolate the different users in the spatial domain. It can be shown that, under certain conditions, the use of a large number of antennas results in an effect

5G/5G-Advanced
https://doi.org/10.1016/B978-0-443-13173-8.00004-9

333

known as "channel hardening". In essence, the rapid fluctuations of the radio-channel quality disappear, simplifying the time-frequency part of the scheduling problem at the cost of a more complicated handling of the spatial domain.

In NR, the *downlink scheduler* is responsible for dynamically controlling the device(s) to transmit to. Each of the scheduled devices is provided with a *scheduling assignment* including information on the set of time-frequency resources upon which the device's DL-SCH[a] is transmitted, the modulation-and-coding scheme, hybrid-ARQ-related information and multiantenna parameters as outlined in Chapter 10. In most cases the scheduling assignment is transmitted just before the data on the PDSCH, but the timing information in the scheduling assignment can also schedule in OFDM symbols later in the slot or in later slots. One use for this is bandwidth adaptation as discussed below. Changing the bandwidth part may take some time and hence data transmission may not occur in the same slot as the control signaling was received in.

It is important to understand that NR *does not* standardize the scheduling behavior. Only a set of supporting mechanisms are standardized on top of which a vendor-specific scheduling strategy is implemented. The information needed by the scheduler depends on the specific scheduling strategy implemented, but most schedulers need information about at least:

- Channel conditions at the device, including spatial-domain properties;
- Buffer status of the different data flows; and
- Priorities of the different data flows, including the amount of data pending retransmission.

Additionally, the interference situation in neighboring cells can be useful if some form of interference coordination is implemented.

Information about the channel conditions at the device can be obtained in several ways. In principle, the gNB can use any information available, but typically the CSI reports from the device are used as discussed in Section 8.1. There is a wide range of CSI reports that can be configured where the device reports the channel quality in the time, frequency, and spatial domains. The amount of correlation between the spatial channels to different devices is also of interest to be able to estimate the degree of spatial isolation between two devices in the case they are candidates for being scheduled on the same time-frequency resources using multiuser MIMO. Uplink sounding using SRS transmission can, together with assumptions on channel reciprocity, also be used to assess the downlink channel quality. Various other quantities can be used as well, for example signal-strength measurements for different beam candidates.

The buffer status and traffic priorities are easily obtained in the downlink case as the scheduler and the transmission buffers reside in the same node. Prioritization of different

[a] In case of carrier aggregation there is one DL-SCH (or UL-SCH) per component carrier.

traffic flows is purely implementation-specific, but retransmissions are typically priori-tized over transmission of new data, at least for data flows of the same priority. Given that NR is designed to handle a much wider range of traffic types and applications than previous technologies such as LTE, priority handling in the scheduler can in many cases be even more emphasized than in the past. In addition to selecting data from different data flows, the scheduler also has the possibility to select the transmission duration. For exam-ple, for a latency-critical service with its data mapped to a certain logical channel, it may be advantageous to select a transmission duration corresponding to a fraction of a slot, while for another service on another logical channel, a more traditional approach of using the full slot duration for transmission might be a better choice. It may also be the case that, for latency reasons and shortage of resources, an urgent transmission using a small number of OFDM symbols needs to preempt an already ongoing transmission using the full slot. In this case, the preempted transmission is likely to be corrupted and require a retrans-mission, but this may be acceptable given the very high priority of the low-latency trans-mission. There are also some mechanisms in NR which can be used to mitigate the impact on the preempted transmission as discussed in Section 14.1.2.

Different downlink schedulers may coordinate their decisions to increase the overall performance, for example by avoiding transmission on a certain frequency range in one cell to reduce the interference toward another cell. In the case of (dynamic) TDD, the different cells can also coordinate the transmission direction, uplink or downlink, between the cells to avoid detrimental interference situations. Such coordination can take place on different time scales. Typically, the coordination is done at a slower rate than the scheduling decisions in each cell as the requirements on the backhaul connecting differ-ent gNBs otherwise would be too high.

14.1.1 Carrier aggregation

In the case of carrier aggregation, the scheduling decisions are taken per carrier and the scheduling assignments are transmitted separately for each carrier, that is, a device sched-uled to receive data from multiple carriers simultaneously receives multiple PDCCHs. A PDCCH received can either point to the same carrier, known as self-scheduling, or to another carrier, commonly referred to as cross-carrier scheduling (see Fig. 14.1).

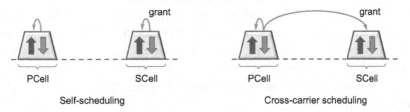

Fig. 14.1 Self scheduling and cross-carrier scheduling.

In the case of cross-carrier scheduling of a carrier with a different numerology than the one upon which the PDCCH was transmitted, timing offsets in the scheduling assignment, for example, which slot the assignment relates to, are interpreted in the PDSCH numerology (and not the PDCCH numerology). In release 17, it is possible to schedule multiple subsequent PDSCH transmissions on the same carrier using a single DCI. This is useful to reduce overhead as well as to enable continuous PDSCH transmission even in cases when the PDCCH is not monitored in every slot, which can be the case in FR2-2. The same release also saw refinements to the carrier aggregation framework with the possibility to use cross-carrier scheduling from an SCell to schedule PCell transmissions, something that was previously not possible. The background to this enhancement is the relatively common scenario with aggregation of a PCell on a lower frequency band with good coverage but modest transmission bandwidth with an SCell on a higher frequency band and hence less good coverage but a relatively large carrier bandwidth. If cross-carrier scheduling is possible only from the PCell, there is a risk of running out of PDCCH resources given the relatively small bandwidth available on the low-frequency PCell. By allowing cross-carrier scheduling also from the SCell this risk is reduced. Self-scheduling on the PCell is still possible in case of coverage problems on the high frequency band.

Further enhancement to scheduling of multiple carriers are introduced in release 18. Groups of up to four carriers can be configured and combinations of carriers in this group can be scheduled using a single DCI message. Compared to prior releases, where one DCI per carrier is used, this can reduce the control signaling overhead, especially in scenarios with fragmented spectrum allocations where multiple carriers need to be aggregated. Resource allocation, multi-antenna information and hybrid-ARQ parameters are provided per carrier, while some of the other pieces of information such as PUCCH-related information, bandwidth part indication, and channel-access type in unlicensed spectra are common to all the co-scheduled carriers.

The scheduling decisions for the different carriers are not taken in isolation. Rather, the scheduling of the different carriers for a given device needs to be coordinated. For example, if a certain piece of data is scheduled for transmission on one carrier, the same piece of data should normally not be scheduled on another carrier as well. However, it is in principle possible to schedule the same data on multiple carriers. This can be used to increase reliability; with multiple carriers transmitting the same data the likelihood of successful reception is increased. At the receiver the RLC (or PDCP) layer can be configured to remove duplicates in case the same data are successfully received on multiple carriers. In the downlink, duplication at the transmitter side is an implementation choice while specification support is required in the uplink. Chapter 21 discusses the release 16 enhancements in this area.

14.1.2 Downlink pre-emption handling

Dynamic scheduling implies, as discussed, that a scheduling decision is taken for each time interval. In many cases the time interval is equal to a slot, that is, the scheduling decisions are taken once per slot. The duration of a slot depends on the subcarrier spacing; a higher subcarrier spacing leads to a shorter slot duration. In principle this could be used to support lower-latency transmission, but as the cyclic prefix also shrinks when increasing the subcarrier spacing, it is not a feasible approach in all deployments. Therefore, as discussed in Section 7.2, NR supports a more efficient approach to low latency by allowing for transmission over a fraction of a slot, starting at any OFDM symbol. This allows for very low latency without sacrificing robustness to time dispersion.

In Fig. 14.2, an example of this is illustrated. Device A has been scheduled with a downlink transmission spanning one slot. During the transmission to device A, latency-critical data for device B arrives to the gNB, which immediately scheduled a transmission to device B. Typically, if there are frequency resources available, the transmission to device B is scheduled using resources not overlapping with the ongoing transmission to device A. However, in the case of a high load in the network, this may not be possible and there is no choice but to use (some of) the resources originally intended for device A for the latency-critical transmission to device B. This is sometimes referred to as the transmission to device B preempting the transmission to device A, which obviously will suffer an impact as a consequence of some of the resources device A assumes contains data for it suddenly containing data for device B.

There are several possibilities to handle this in NR. One approach is to rely on hybrid-ARQ retransmissions. Device A will not be able to decode the data due to the resources being preempted and will consequently report a negative acknowledgment to the gNB, which can retransmit the data at a later time instant. Either the complete

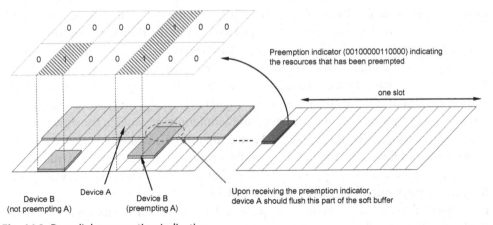

Fig. 14.2 Downlink preemption indication.

transport block is retransmitted, or CBG-based retransmission is used to retransmit only the impacted codeblock groups as discussed in Section 13.1.

There is also a possibility to indicate to device A that some of its resources have been preempted and used for other purposes. This is done by transmitting a *preemption indicator* to device A in a slot after the slot containing the data transmission. The preemption indicator uses DCI format 2_1 (see Chapter 10 for details on different DCI formats) and contains a bitmap of 14 bits for each of the configured cells. Interpretation of the bitmap is configurable such that each bit represents one OFDM symbol in the time domain and the full bandwidth part, or two OFDM symbols in the time domain and one half of the bandwidth part. Furthermore, the monitoring periodicity of the preemption indicator is configured in the device, for example, every n:th slot.

The behavior of the device when receiving the preemption indicator is not specified, but a reasonable behavior could be to flush the part of the soft buffer which corresponds to the preempted time-frequency region to avoid soft-buffer corruption for future retransmissions. From a soft-buffer handling perspective in the device, the more frequent the monitoring of the preemption indicator, the better (ideally, it should come immediately after the preemption occurred).

Uplink preemption, where one device needs to use uplink resources originally intended for another device, is discussed in Chapter 21, including the relevant enhancements part of release 16.

14.2 Dynamic uplink scheduling

The basic function of the *uplink scheduler* in the case of dynamic scheduling is similar to its downlink counterpart, namely to dynamically control which devices are to transmit, on which uplink resources, and with what transmission parameters.

The general downlink scheduling discussion is applicable to the uplink as well. However, there are some fundamental differences between the two. For example, the uplink power resource is *distributed* among the devices, while in the downlink the power resource is *centralized* within the base station. Furthermore, the maximum uplink transmission power of a single device is often significantly lower than the output power of a base station. This has a significant impact on the scheduling strategy. Even in the case of a large amount of uplink data to transmit there might not be sufficient power available – the uplink is basically power limited and not bandwidth limited, while in the downlink the situation can typically be the opposite. Hence, uplink scheduling typically results in a larger degree of frequency multiplexing of different devices than in the downlink.

Each scheduled device is provided with a *scheduling grant* indicating the set of time/frequency/spatial resources to use for the UL-SCH as well as the associated transport format. Uplink data transmissions only take place in the case that the device has a valid grant. Without a grant, no data can be transmitted.

The uplink scheduler is in complete control of the transport format the device shall use, that is, the device has to follow the scheduling grant. The only exception is that the device will not transmit anything, regardless of the grant, if there is no data in the transmission buffer. This reduces the overall interference by avoiding unnecessary transmissions in the case that the network scheduled a device with no data pending transmission.

Logical channel multiplexing is controlled by the device according to a set of rules (see Section 14.2.1) configured by the network. Thus, the scheduling grant does not explicitly schedule a certain logical channel but rather the device as such – uplink scheduling is primarily *per device* and not per radio bearer (although the priority handling mechanism discussed below in principle can be configured to obtain scheduling per radio bearer). Uplink scheduling is illustrated in the right part of Fig. 14.3, where the scheduler controls the transport format and the device controls the logical channel multiplexing. This allows the scheduler to tightly control the uplink activity to maximize the resource usage compared to schemes where the device autonomously selects the data rate, as autonomous schemes typically require some margin in the scheduling decisions. A consequence of the scheduler being responsible for selection of the transport format is that accurate and detailed knowledge about the device situation with respect to buffer status and power availability is accentuated compared to schemes where the device autonomously controls the transmission parameters.

The time during which the device should transmit in the uplink is indicated as part of the DCI as described in Section 10.1.11. Unlike in the downlink case, where the scheduling assignment typically is transmitted close in time to the data, this is not necessarily the case in the uplink. Since the grant is transmitted using downlink control

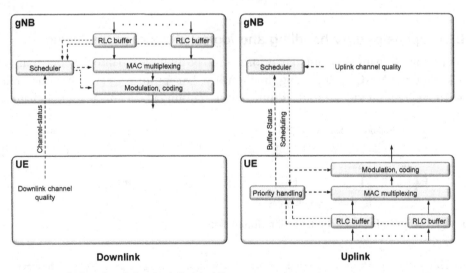

Fig. 14.3 Downlink and uplink scheduling in NR.

signaling, a half-duplex device needs to change the transmission direction before transmitting in the uplink. Furthermore, depending on the uplink-downlink allocation, multiple uplink slots may need to be scheduled using multiple grants transmitted at the same downlink occasion.[b] Hence, the timing field in the uplink grant is important.

The device also needs a certain amount of time to prepare for the transmission as outlined in Fig. 14.4. From an overall performance perspective, the shorter the time the better. However, from a device complexity perspective the processing time cannot be made arbitrarily short. In LTE, more than 3 ms was provided for the device to prepare the uplink transmission. For NR, a more latency-focused design, for example the updated MAC and RLC header structure, as well as technology development in general has considerably reduced this time. The delay from the reception of a grant to the transmission of uplink data is summarized in Table 14.1. As seen from these numbers, the processing time depends on the subcarrier spacing although it is not purely scaled in proportion to the subcarrier spacing. It is also seen that two device capabilities are specified. All devices need to fulfill the baseline requirements, but a device may also declare whether it is capable of a more aggressive processing time line which can be useful in latency-critical applications.

Similar to the downlink case, the uplink scheduler can benefit from information on channel conditions, buffer status, and power availability. However, the transmission buffers reside in the device, as does the power amplifier. This calls for the reporting mechanisms described later to provide the information to the scheduler, unlike the downlink case where the scheduler, power amplifier, and transmission buffers all are in the same node. Uplink priority handling is, as already touched upon, another area where uplink and downlink scheduling differ.

14.2.1 Uplink priority handling and logical-channel multiplexing

Multiple logical channels of different priorities can be multiplexed into the same transport block using the MAC multiplexing functionality. Except for the case when the uplink scheduling grant provides resources sufficient to transmit all data on all logical channels,

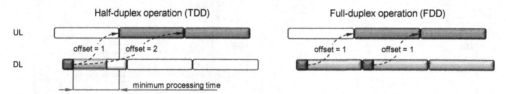

Fig. 14.4 Example of uplink scheduling into future slots.

[b] In release 16 it is possible to schedule multiple uplink transmissions using *one* grant as part of the extensions to unlicensed spectra, see Chapter 20.

Table 14.1 Minimum processing time in OFDM symbols from grant reception to data transmission.

Device capability	Subcarrier spacing				
	15 kHz	30 kHz	60 kHz	120 kHz	LTE rel 8
Baseline	0.71 ms	0.43 ms	0.41 ms	0.32 ms	3 ms
Aggressive	0.18–0.39 ms	0.08–0.2 ms			

the multiplexing needs to prioritize between the logical channels. However, unlike the downlink case, where the prioritization is up to the scheduler implementation, the uplink multiplexing is done according to a set of well-defined rules in the device with parameters set by the network. The reason for this is that a scheduling grant applies to a specific uplink carrier of a device, not explicitly to a specific logical channel within the carrier.

A simple approach would be to serve the logical channels in strict priority order. However, this could result in starvation of lower-priority channels – all resources would go to the high-priority channel until the buffer is empty. Typically, an operator would instead like to provide at least some throughput for low-priority services as well. Furthermore, as NR is designed to handle a mix of a wide range of traffic types, a more elaborate scheme is needed. For example, traffic due to a file upload should not necessarily exploit a grant intended for a latency-critical service.

The starvation problem could in principle be mitigated by assigning a "guaranteed" data rate to each channel. The logical channels are then served in decreasing priority order up to their guaranteed data rate, which avoids starvation as long as the scheduled data rate is at least as large as the sum of the guaranteed data rates.[c] Beyond the guaranteed data rates, channels are served in strict priority order until the grant is fully exploited, or the buffer is empty.

Given the large flexibility of NR in terms of different transmission durations and a wide range of traffic types supported, NR uses a refined version of this principle. One possibility would be to define different profiles, each outlining an allowed combination of logical channels, and explicitly signal the profile to use in the grant. However, in NR the profile to use is implicitly derived from other information available in the grant rather than explicitly signaled.

Upon reception of an uplink grant, two steps are performed. First, the device determines which logical channels are eligible for multiplexing using this grant. Second, the device determines the fraction of the resources that should be given to each of the logical channels.

The first step determines the logical channels from which data can be transmitted with the given grant. This can be seen as an implicitly derived profile. For each logical channel, the device can be configured with:

[c] This is the approach taken in LTE.

- The set of allowed subcarrier spacings this logical channel is allowed to use;
- The maximum PUSCH duration which is possible to schedule for this logical channel; and
- The set of serving cells, that is, the set of uplink component carriers the logical channel is allowed to be transmitted upon.

Additionally, in release 16 it is also possible to dynamically signal the priority of an uplink transmission – 'normal' or 'high' – as discussed in Chapter 21.

Only the logical channels for which the scheduling grant meets the restrictions configured may use the grant, that is, are eligible for multiplexing at this particular time instant. In addition, the logical channel multiplexing can also be restricted for uplink transmissions using a configured grant such that not all logical channels are allowed to use a configured grant.

Coupling the multiplexing rule to the PUSCH duration is in 3GPP motivated by the possibility to control whether latency-critical data should be allowed to exploit a grant intended for less time-critical data.

As an example, assume there are two data flows, each on a different logical channel. One logical channel carries latency-critical data and is given a high priority, while the other logical channel carries non-latency-critical data and is given a low priority. The gNB takes scheduling decisions based on, among other aspects, information about the buffer status in the device provided by the device. Assume that the gNB scheduled a relatively long PUSCH duration based on information that there is only non-time-critical information in the buffers. During the reception of the scheduling grant, time-critical information arrives to the device. Without the restriction on the maximum PUSCH duration, the device would transmit the latency-critical data, possibly multiplexed with other data, over a relatively long transmission duration and potentially not meeting the latency requirements set up for the particular service. Instead, a better approach would be to separately request a transmission using a short PUSCH duration for the latency critical data, something which is possible by configuring the maximum PUSCH duration appropriately. Since the logical channel carrying the latency-critical traffic has been configured with a higher priority than the channel carrying the non-latency-critical service, the non-critical service will not block transmission of the latency-critical data during the short PUSCH duration.

The reason to also include the subcarrier spacing is similar to the duration. In the case of multiple subcarrier spacings configured for a single device, a lower subcarrier spacing implies a longer slot duration and the reasoning above can also be applied in this case.

Restricting the uplink carriers allowed for a certain logical channel is motivated by the possibly different propagation conditions for different carriers and by dual connectivity. Two uplink carriers at vastly different carrier frequencies can have different reliability. Data which are critical to receive might be better to transmit on a lower carrier frequency to ensure good coverage, while less sensitive data can be transmitted on a carrier with a higher carrier frequency and possibly spottier coverage. Another motivation is duplication, that is, the same data transmitted on multiple logical channels, to obtain diversity as mentioned in Section 6.4.2. If both logical channels would be transmitted on the same uplink carrier, the original motivation for duplication – to obtain a diversity effect – would be gone. Uplink duplication has been further extended in release 16, see Chapter 21 for details.

At this point in the process, the set of logical channels from which data are allowed to be transmitted given the current grant is established, based on the mapping-related parameters configured. Multiplexing of the different logical channels also needs to answer the question of how to distribute resources between the logical channels having data to transmit and eligible for transmission. This is done based on a set of priority-related parameters configured for each local channel:

- *Priority*;
- *Prioritized bit rate* (PBR); and
- *Bucket size duration* (BSD).

The prioritized bit rate and the bucket size duration together defines a guaranteed bit rate as discussed above but can account for the different transmission durations possible in NR. The product of the prioritized bit rate and the bucket size duration is in essence a bucket of bits that as a minimum should be transmitted for the given logical channel during a certain time. At each transmission instant, the logical channels are served in decreasing priority order, while trying to fulfill the requirement on the minimum number of bits to transmit. Excess capacity when all the logical channels are served up to the bucket size is distributed in strict priority order.

Priority handling and logical channel multiplexing are illustrated in Fig. 14.5.

14.2.2 Scheduling request

The uplink scheduler needs knowledge of devices with data to transmit and that therefore need to be scheduled. There is no need to provide uplink resources to a device with no data to transmit. Hence, as a minimum, the scheduler needs to know whether the device has data to transmit and should be given a grant. This is known as a *scheduling request*. Scheduling requests are used for devices not having a valid scheduling grant; devices that have a valid grant provide more detailed scheduling information to the gNB as discussed in the next section.

A scheduling request is a flag, raised by the device to request uplink resources from the uplink scheduler. Since the device requesting resources by definition has no PUSCH resource, the scheduling request is transmitted on the PUCCH using preconfigured and periodically reoccurring PUCCH resources dedicated to the device. With a dedicated scheduling-request mechanism, there is no need to provide the identity of the device requesting to be scheduled as the identity is implicitly known from the resources

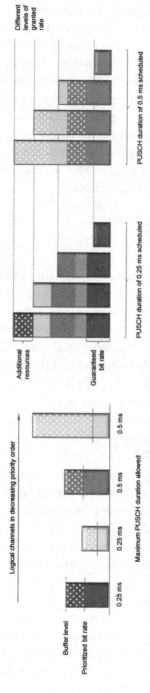

Fig. 14.5 Example of logical channel prioritization for four different scheduled data rates and two different PUSCH durations.

upon which the request is transmitted. When data with higher priority than already existing in the transmit buffers arrive at the device and the device has no grant and hence cannot transmit the data, the device transmits a scheduling request at the next possible instant and the gNB can assign a grant to the device upon reception of the request (see Fig. 14.6).

In NR, *multiple* scheduling requests from a single device can be configured. A logical channel can be mapped to zero or more scheduling request configurations. This provides the gNB not only with information that there are data awaiting transmission in the device, but also *what type of* data are awaiting transmission. This is useful information for the gNB given the wide range of traffic types the NR is designed to handle. For example, the gNB may want to schedule a device for transmission of latency-critical information but not for non-latency-critical information.

Each device can be assigned dedicated PUCCH scheduling request resources with a periodicity ranging from every second OFDM symbol to support very latency-critical services up to every 80 ms for low overhead. Only one scheduling request can be transmitted at a given time, that is, in the case of multiple logical channels having data to transmit a reasonable behavior is to trigger the scheduling request corresponding to the highest-priority logical channel. A scheduling request is repeated in subsequent resources, up to a configurable limit, until a grant is received from the gNB. It is also possible to configure a prohibit timer, controlling how often a scheduling request can be transmitted. In the case of multiple scheduling-request resources in a device, both of these configurations are done per scheduling request resource.

A device which has not been configured with scheduling request resources relies on the random-access mechanism to request resources. This can be used to create a contention-based mechanism for requesting resources. One could even consider transmitting small amounts of data as part of the random-access procedure, an approach that is taken in the SDT enhancement in release 17 as described in Chapter 22. Basically, contention-based designs are suitable for situations where there are a large number of devices in the cell and the traffic intensity, and hence the scheduling intensity, is low. In the case of higher traffic intensities, it is beneficial to set up at least one scheduling request resource for the device.

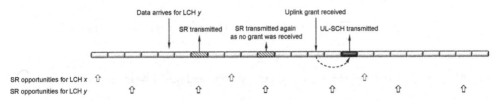

Fig. 14.6 Example of scheduling request operation.

14.2.3 Buffer status reports

Devices that already have a valid grant do not need to request uplink resources. However, to allow the scheduler to determine the amount of resources to grant to each device in the future, information about the buffer situation, discussed in this section, and the power availability, discussed in the next section, is useful. This information is provided to the scheduler as part of the uplink transmission through MAC control elements (see Section 6.4.4.1 for a discussion on MAC control elements and the general structure of a MAC header). The LCID field in one of the MAC subheaders is set to a reserved value indicating the presence of a buffer status report, as illustrated in Fig. 14.7.

From a scheduling perspective, buffer information for each logical channel is beneficial, although this could result in a significant overhead. Logical channels are therefore grouped into up to eight logical-channel groups and the reporting is done per group. The buffer-size field in a buffer-status report indicates the amount of data awaiting transmission across all logical channels in a logical-channel group. Four different formats for buffer status reports are defined, differing in how many logical-channel groups are included in one report and the resolution of the buffer status report. A buffer-status report can be triggered for the following reasons:

- Arrival of data with higher priority than currently in the transmission buffer – that is, data in a logical-channel group with higher priority than the one currently being transmitted – as this may impact the scheduling decision.
- Periodically as controlled by a timer.
- Instead of padding. If the amount of padding required to match the scheduled transport block size is larger than a buffer-status report, a buffer-status report is inserted as it is better to exploit the available payload for useful scheduling information instead of padding if possible.

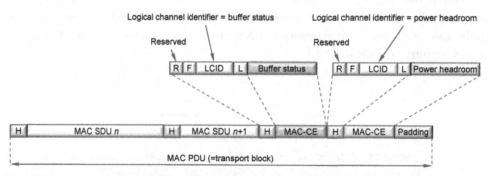

Fig. 14.7 MAC control elements for buffer status reporting and power headroom reports.

14.2.4 Power headroom reports

In addition to buffer status, the amount of transmission power available in each device is also relevant for the uplink scheduler. There is little reason to schedule a higher data rate than the available transmission power can support. In the downlink, the available power is immediately known to the scheduler as the power amplifier is in the same node as the scheduler. For the uplink, the power availability, or *power headroom*, needs to be provided to the gNB. Power headroom reports are therefore transmitted from the device to the gNB in a similar way as the buffer-status reports – that is, only when the device is scheduled to transmit on the UL-SCH. A power headroom report can be triggered for the following reasons:

- Periodically as controlled by a timer;
- Change in path loss (the pathloss difference relative to the time of the last power headroom report is larger than a configurable threshold);
- Instead of padding (for the same reason as buffer-status reports).

It is also possible to configure a prohibit timer to control the minimum time between two power-headroom reports and thereby the signaling load on the uplink.

There are three different types of power-headroom reports defined in NR, *Type 1*, *Type 2*, and *Type 3*. In the case of carrier aggregation or dual connectivity, multiple power headroom reports can be contained in a single message (MAC control element).

Type-1 power headroom reporting reflects the power headroom assuming PUSCH-only transmission on the carrier. It is valid for a certain component carrier, assuming that the device was scheduled for PUSCH transmission during a certain duration, and includes the power headroom and the corresponding value of the *maximum per-carrier transmit power* for component carrier *c*, denoted $P_{CMAX,c}$. The value of $P_{CMAX,c}$ is explicitly configured and should hence be known to the gNB, but since it can be separately configured for a normal uplink carrier and a supplementary uplink carrier, both belonging to the same cell (that is, having the same associated downlink component carrier), the gNB needs to know which value the device used and hence which carrier the report belongs to.

It can be noted that the power headroom is not a measure of the difference between the maximum per-carrier transmit power and the actual carrier transmit power. Rather, the power headroom is a measure of the difference between $P_{CMAX,c}$ and the transmit power that would have been used *assuming that there would have been no upper limit on the transmit power* (see Fig. 14.8). Thus, the power headroom can very well be negative, indicating that the per-carrier transmit power was limited by $P_{CMAX,c}$ at the time of the power headroom reporting – that is, the network has scheduled a higher data rate than the device can support given the available transmission power. As the network knows what modulation-and-coding scheme and resource size the device used for transmission

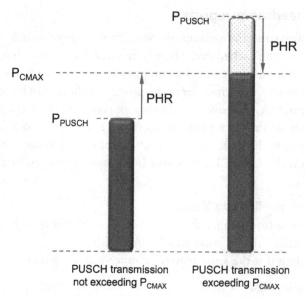

Fig. 14.8 Illustration of power headroom reports.

in the time duration to which the power-headroom report corresponds, it can determine the valid combinations of modulation-and-coding scheme and resource size allocation, assuming that the downlink path loss is constant.

Type-1 power headroom can also be reported when there is no actual PUSCH transmission. This can be seen as the power headroom assuming a default transmission configuration corresponding to the minimum possible resource assignment.

Type-2 power headroom reporting is similar to type 1, but assumes simultaneous PUSCH and PUCCH reporting, a feature that is not fully supported in the NR specifications but planned for finalization in later releases.

Type-3 power headroom reporting is used to handle SRS switching, that is, SRS transmissions on an uplink carrier where the device is not configured to transmit PUSCH. The intention with this report is to be able to evaluate the uplink quality of alternative uplink carries and, if deemed advantageous, (re)configure the device to use this carrier for uplink transmission instead.

Compared to power control, which can operate different power-control processes for different beam-pair links (see Chapter 15), the power-headroom report is per carrier and does not explicitly take beam-based operation into account. One reason is that the network is in control of the beams used for transmission and hence can determine the beam arrangement corresponding to a certain power-headroom report.

14.3 Scheduling and dynamic TDD

One of the key features of NR is the support for *dynamic TDD* where the scheduler dynamically determines the transmission direction. Although the description uses the term dynamic TDD, the framework can in principle be applied to half-duplex operation in general, including half-duplex FDD. Since a half-duplex device cannot transmit and receive simultaneously, there is a need to split the resources between the two directions. As mentioned in Chapter 7, three different signaling mechanisms can provide information to the device on whether the resources are used for uplink or downlink transmission:

- Dynamic signaling for the scheduled device;
- Semi-static signaling using RRC; and
- Dynamic slot-format indication shared by a group of devices, primarily intended for non-scheduled devices.

The scheduler is responsible for the dynamic signaling for the scheduled device, that is, the first of the three bullets above.

In the case of a device capable of full-duplex operation, the scheduler can schedule uplink and downlink independent of each other and there is limited, if any, need for the uplink and downlink scheduler to coordinate their decisions.

In the case of a half-duplex device, on the other hand, it is up to the scheduler to ensure that a half-duplex device is not requested to simultaneously receive and transmit. If a semi-static uplink-downlink pattern has been configured, the schedulers obviously need to obey this pattern as well as it cannot, for example, schedule an uplink transmission in a slot configured for downlink usage only.

14.4 Transmissions without a dynamic grant – semi-persistent scheduling and configured grants

Dynamic scheduling, as described above, is the main mode of operation in NR. For each transmission interval, for example a slot, the scheduler uses control signaling to instruct the device to transmit or receive. It is flexible and can adopt to rapid variations in the traffic behavior, but obviously requires associated control signaling; control signaling that in some situations it is desirable to avoid. NR therefore also supports transmission schemes not relying on dynamic grants.

In the downlink, *semi-persistent scheduling* is supported where the device is configured with a periodicity of the data transmissions using RRC signaling. Activation of semi-persistent scheduling is done using the PDCCH as for dynamic scheduling but with the CS-RNTI instead of the normal C-RNTI.[d] The PDCCH also carries the necessary

---

[d] Each device has two identities, the "normal" C-RNTI for dynamic scheduling and the CS-RNTI for activation/deactivation of semi-persistent scheduling.

information in terms of time-frequency resources and other parameters needed in a similar way as dynamic scheduling. The hybrid-ARQ process number is derived from the time when the downlink data transmission starts according to a formula. Upon activation of semi-persistent scheduling, the device receives downlink data transmission periodically according to the RRC-configured periodicity using the transmission parameters indicated on the PDCCH activating the transmission.[e] Hence, control signaling is only used once and the overhead is reduced. After enabling semi-persistent scheduling, the device continues to monitor the set of candidate PDCCHs for uplink and downlink scheduling commands. This is useful in the case that there are occasional transmissions of large amounts of data for which the semi-persistent allocation is not sufficient. It is also used to handle hybrid-ARQ retransmissions which are dynamically scheduled.

In the uplink, *configured grants* are used to handle transmissions without a dynamic grant. Two types of configured grants are supported, differing in the ways they are activated (see Fig. 14.9):

• *Configured grant type 1,* where an uplink grant is provided by RRC, including activation of the grant; and
• *Configured grant type 2*, where the transmission periodicity is provided by RRC and L1/L2 control signaling is used to activate/deactivate the transmission in a similar way as in the downlink case.

The benefits for the two schemes are similar, namely to reduce control signaling overhead and, to some extent to reduce the latency before uplink data transmission as no scheduling request–grant cycle is needed prior to data transmission.

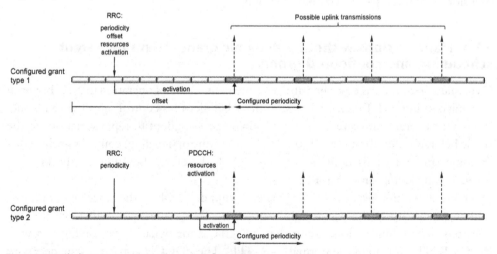

Fig. 14.9 Uplink transmissions using a configured grant.

---

[e] Periodicities of 10 ms and up can be configured in release 15, a number that is reduced in later releases as discussed in Chapter 21.

Type 1 sets all the transmission parameters, including periodicity, time offset, and frequency resources as well as modulation-and-coding scheme of possible uplink transmissions, using RRC signaling. Upon receiving the RRC configuration, the device can start to use the configured grant for transmission in the time instant given by the periodicity and offset. The reason for the offset is to control at what time instants the device is allowed to transmit. There is no notion of activation time in the RRC signaling in general; RRC configurations take effect as soon as they are received correctly. This point in time may vary as it depends on whether RLC retransmissions were needed to deliver the RRC command or not. To avoid this ambiguity, a time offset relative to the SFN is included in the configuration.

Type 2 is similar to downlink semi-persistent scheduling. RRC signaling is used to configure the periodicity, while the transmission parameters are provided as part of the activation using the PDCCH. Upon receiving the activation command, the device transmits according to the preconfigured periodicity if there are data in the buffer. If there are no data to transmit, the device will, similar to type 1, not transmit anything. Note that no time offset is needed in this case as the activation time is well defined by the PDCCH transmission instant.

The device acknowledges the activation/deactivation of the configured grant type 2 by sending a MAC control element in the uplink. If there are no data awaiting transmission when the activation is received, the network would not know if the absence of transmission is due to the activation command not being received by the device or if it is due to an empty transmission buffer. The acknowledgment helps in resolving this ambiguity.

In both these schemes it is possible to configure multiple devices with overlapping time-frequency resources in the uplink. In this case it is up to the network to differentiate between transmissions from the different devices.

When transmissions are dynamically scheduled in either uplink or downlink, the hybrid-ARQ process number is part of the dynamically signaled DCI. Since there is no dynamic signaling of the hybrid-ARQ process number for semi-persistent scheduling and configured grants, the process number to use must be derived in a different manner. This is done by linking the process number to the absolute slot number (downlink semi-persistent scheduling) or symbol number (uplink configured grants) within the configured periodicity. Thus, the device and the gNB have the same understanding of the hybrid-ARQ process number and there is no ambiguity.

14.5 Power-saving mechanisms

Packet-data traffic is often highly bursty, with occasional periods of transmission activity followed by longer periods of silence. From a delay perspective, it is beneficial to monitor the downlink control signaling in each slot (or even more frequently) to receive uplink grants or downlink data transmissions and instantaneously react on changes in the traffic behavior. At the same time this comes at a cost in terms of power consumption at the device; the receiver circuitry in a typical device represents a non-negligible amount of power consumption and battery lifetime is one of the most important end-user metrics.

Modelling the device power consumption is a complex task and depends on a multitude of factors. The lowest power consumption occurs in RRC_IDLE where the device only occasionally checks for paging. In many cases the device is therefore moved to the idle state whenever possible, initiated by either the network or the device itself. However, to transfer data the device needs to be in the connected state, RRC_CONNECTED. Transferring the device between the states takes some time as many parameters need to be signaled and the context re-established when the device moves from idle to connected mode. Therefore, the device typically remains in the connected state for several seconds after the last packet being transmitted before moving to the idle state in case in case there are additional data packets to transmit. The intermediate state, RRC_INACTIVE, can also be useful as the device context is preserved in the network which reduces the amount of signaling when transitioning to connected mode.

Since data reception can occur in connected state only, it is important to consider the device power consumption also in this state – a device constantly being in idle state is not that useful for obvious reasons. Despite its name, most of the time in the connected state the device is typically not active with receiving (or transmitting) data. Rather, it is monitoring the PDCCHs for *potential* scheduling information. The net result of this is that time-wise a device typically spends most of its time in idle mode, but energy-wise, a large fraction of the total energy is spent on monitoring PDCCHs in active mode without any associated data reception (or transmission). A relatively small fraction of the total energy consumption is due to actual reception and transmission of data.

To tackle these partially contradicting requirements – a long battery lifetime and a low delay – NR includes several power-saving mechanisms. *Discontinuous reception* (DRX) is a basic mechanism included already in the first NR release, as are bandwidth adaptation and carrier (de)activation. Additional tools such as wake-up signals, dynamic control of cross-slot scheduling delays, cell dormancy, PDCCH monitoring adaptation, and paging early indicator are introduced in later releases (see [107] for a discussion of the potential savings of the different features).

In addition to these standardized mechanisms, there are also a lot of implementation-specific techniques that can be used. For example, if a device is configured for PDCCH monitoring once per slot but does not receive a valid scheduling command at the beginning of a slot, it can sleep for the remainder of the slot. This is sometimes referred to as *micro sleep*.

14.5.1 Discontinuous reception

The basis for DRX is a configurable DRX cycle in the device. With a DRX cycle configured, the device monitors the downlink control signaling only when active, sleeping with the receiver circuitry switched off the remaining time. This allows for a significant reduction in power consumption – the longer the cycle, the lower the power consumption. Naturally, this implies restrictions to the scheduler as the device can be addressed only when active according to the DRX cycle.

In many situations, if the device has been scheduled and is actively receiving or transmitting data, it is highly likely it will be scheduled again in the near future. One reason could be that it was not possible to transmit all the data in the transmission buffer in one scheduling occasion and hence additional occasions are needed. Waiting until the next activity period according to the DRX cycle, although possible, would result in additional delays. Hence, to reduce the delays, the device remains in the active state for a certain configurable time after being scheduled. This is implemented by the device (re)starting an inactivity timer every time it is scheduled and remaining awake until the time expires, as illustrated in Fig. 14.10. Due to the fact that NR can handle multiple numerologies, the DRX timers are specified in milliseconds in order not to tie the DRX periodicity to a certain numerology.

Hybrid-ARQ retransmissions are asynchronous in both uplink and downlink. If the device has been scheduled a transmission in the downlink it could not decode, the typical situation is that the gNB retransmits the data shortly after the initial transmission. Therefore, the DRX functionality has a configurable timer which is started after an erroneously received transport block and used to wake up the device receiver when it is likely for the gNB to schedule a retransmission. The value of the timer is preferably set to match the roundtrip time in the hybrid-ARQ protocol; a roundtrip time that depends on the implementation.

The mechanism – a (long) DRX cycle in combination with the device remaining awake for some period after being scheduled – is sufficient for most scenarios. However, some services, most notably voice-over IP, are characterized by periods of regular transmission, followed by periods of no or very little activity. To handle these services, a second short DRX cycle can optionally be used in addition to the long cycle described above. Normally, the device follows the long DRX cycle, but if it has recently been scheduled, it follows a shorter DRX cycle for some time. Handling voice-over IP in this

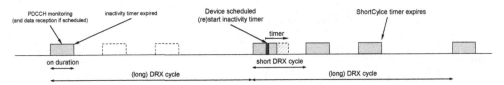

Fig. 14.10 DRX operation.

scenario can be done by setting the short DRX cycle to 20 ms, as the voice codec typically delivers a voice-over-IP packet per 20 ms. The long DRX cycle is then used to handle longer periods of silence between talk spurts.

In addition to the RRC configuration of the DRX parameters, the gNB can terminate an "on duration" and instruct the device to follow the long DRX cycle. This can be used to reduce the device power consumption if the gNB knows that no additional data are awaiting transmission in the downlink and hence there is no need for the device to be active.

14.5.2 Wake-up signals

The DRX mechanism as described here gives significant improvements in device power consumption compared to being continuously active. Nevertheless, further improvements are possible if the network could inform the device to sleep for another long DRX cycle if no downlink data is expected instead of waking up regularly and monitor PDCCHs for a certain time before going back to sleep. Therefore, release 16 introduces the possibility for a *wake-up signal*. If the wake-up signal is configured, the device wakes up a configurable time before the start of the long DRX cycle, checks for the wake-up signal and, if told not to wake up, returns to sleep for the next long DRX cycle; see Fig. 14.11 for an example. The wake-up signal uses DCI format 2_6, introduced to support power saving. Multiple wake-up signals are transmitted together using DCI format 2_6 and the device is configured which of the bits represents the wake-up signal for that particular device. Checking for the wake-up signal typically requires less power than a complete search for many different DCI formats and PDCCH candidates. Together with a significantly shorter duration for checking for the wake-up signal than what is dictated by the on duration in the (long) DRX cycle, there is a gain in power consumption.

14.5.3 Cross-slot scheduling for power saving

NR allows the data to start immediately after the PDCCH, or, with the proper configuration, already at the same time as the PDCCH as described in Chapter 10. From a latency perspective this is clearly beneficial, but it also requires the device to keep the receiver open and buffer the received signal at least until the PDCCH decoding is ready. In many cases, the device is not scheduled and the buffering the received signal is done in vain. From this perspective, cross-slot scheduling, where the PDSCH is transmitted in a

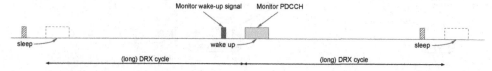

Fig. 14.11 Wake-up signal in release 16.

later slot than the PDCCH, is beneficial as no buffering of the received signal is required. Cross-slot scheduling is supported in NR by configuring the time-domain resource allocation table properly, see Chapter 10. If the table is configured such that all time-domain allocations are in the next slot, the device implementation could in principle exploit this and skip buffering the signal. However, this would increase latency as cross-slot scheduling would be used for all transmissions, also when there is a large amount of data to transmit. Therefore, in release 16 it is possible to *dynamically* signal the minimum scheduling offset, selecting between two preconfigured values using one bit in the DCI. If the minimum slot offset indicated to the device is zero, all entries in the time-domain allocation table are valid and the device need to be prepared to receive a PDSCH starting immediately after (or simultaneously with) the PDCCH and hence require buffering of the received signal. On the other hand, if the minimum slot offset indicated is, as an example, one slot, all entries in the time-domain resource allocation table with a slot offset of zero are invalid. Hence, the device does not need to buffer the received signal and can in principle sleep until the next slot where the PDSCH transmission is located as illustrated in Fig. 14.12, thereby saving power. Dynamic indication of the minimum slot offset is applicable to both uplink and downlink scheduling of unicast data using DCI formats 0_1/0_3 and 1_1/1_3, respectively. It is not applicable to transmissions such as system information and random-access response using the fallback DCI format. Obviously, the indicated minimum slot offset cannot be applied in the same slot as it was signaled but is valid starting at a future slot.

14.5.4 Cell dormancy

To improve the power consumption in scenarios with carrier aggregation, *SCell dormancy* is introduced in release 16. For a dormant cell, the device stops PDCCH monitoring but continues to perform CSI measurements and beam management. Although a dormant cell is still considered as active and is not deactivated, there is considerably less activity from a device perspective which saves power. Deactivating a cell is another possibility to save power but in this case no CSI reports are provided and the activation of an SCell takes longer time than returning from dormancy.

The dormancy mechanism is based on the bandwidth part framework. One dormant bandwidth part without any PDCCH monitoring is configured in addition to the one or more regular bandwidth parts. A dormant cell is thus a cell with the dormant bandwidth part as the active bandwidth part. By switching to any other bandwidth part the cell is taken out of dormancy.

The switching between the dormant bandwidth part and the regular bandwidth parts is done via L1/L2 control signaling. In addition to non-fallback DCI formats scheduling uplink or downlink transmissions, DCI format 2_6 used for the wake-up signal can also be used.

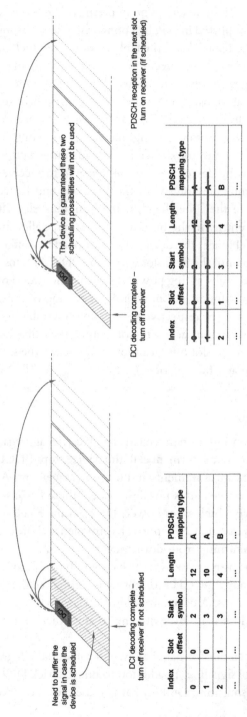

Fig. 14.12 Illustration of dynamic signaling of minimum slot offset to save power.

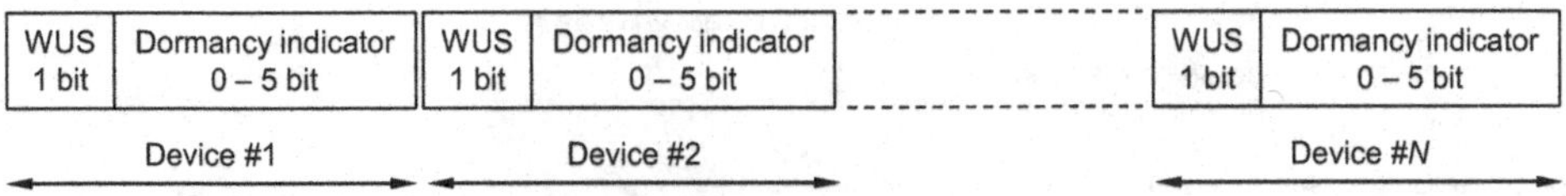

WUS 1 bit	Dormancy indicator 0 – 5 bit	WUS 1 bit	Dormancy indicator 0 – 5 bit			WUS 1 bit	Dormancy indicator 0 – 5 bit

Device #1   Device #2   Device #N

Fig. 14.13 Wake-up signal (WUS) and dormancy indicator in DCI format 2_6.

DCI format 2_6 is used when the device is DRX and monitoring for the wake-up signal. In this case a dormancy indicator of up to five bits can be transmitted in addition to the wake-up signal, see Fig. 14.13. Each of the dormancy indicator bits corresponds to an RRC-configured group of SCells, indicating whether the corresponding group of SCells should enter dormancy or not.

To address devices not being in DRX, DCI formats 0_1, 0_3, 1_1, and 1_3 can be used. The DCI size is increased to include the up to five dormancy indicator bits, using to indicate dormancy for a group of SCells in the same way as for format 2_6. If the increased DCI size is problematic, it is also possible to configure a standalone dormancy indication in DCI format 1_1 for all the up to 15 configured SCells by setting the resource allocation fields to a reserved value and reinterpreting some of the other bits as a bitmap with one bit for each configured SCell. Obviously, in this case it is not possible to simultaneously schedule data.

14.5.5 Bandwidth adaptation

NR support a very wide transmission bandwidth, up to several 100 MHz on a single carrier. This is useful for rapid delivery of large payloads but is not needed for smaller payload sizes or for monitoring the downlink control channels when not scheduled. Hence, as mentioned already in Chapter 5, NR supports *receiver-bandwidth adaptation* such that the device can use a narrow bandwidth for monitoring control channels and only open the full bandwidth when a large amount of data is scheduled, thereby reducing the device power consumption. This can be seen as discontinuous reception in the frequency domain.

Opening the wideband receiver can be done by using the bandwidth-part indicator in the DCI. If the bandwidth part indicator points to a different bandwidth part than the currently active one, the active bandwidth part is changed (see Fig. 14.14). The time it takes to change the active bandwidth part depends on several factors, for example, if the center frequency changes and the receiver needs to retune or not, but can be in the order of a slot. Once activated, the device uses the new, and wider, bandwidth part for its operation.

Upon completion of the data transfer requiring the wider bandwidth, the same mechanism can be used to revert back to the original bandwidth part. There is also a possibility to configure a timer to handle the bandwidth-part switching instead of explicit signaling. In this case, one of the bandwidth parts is configured as the default bandwidth part. If no

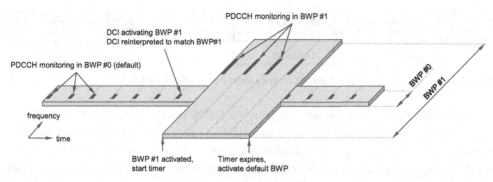

Fig. 14.14 Illustration of bandwidth adaptation principle.

default bandwidth part is explicitly configured, the initial bandwidth part obtained from the random-access procedure is used as the default bandwidth part. Upon receiving a DCI indicating a bandwidth part other than the default one, the timer is started. When the timer expires, the device switches back to the default bandwidth part. Typically, the default bandwidth part is narrower and can hence help reducing the device power consumption.

The introduction of bandwidth adaptation in NR raised several design questions not present in LTE, in particular related to the handling of controls signaling as many transmission parameters are configured per bandwidth part and the DCI payload size therefore may differ between different bandwidth parts. The frequency-domain resource allocation field is an obvious example; the larger the bandwidth part, the larger the number of bits for frequency-domain resource allocation. This is not an issue as long as the downlink data transmission uses the same bandwidth part as the DCI control signaling.[f] However, in the case of bandwidth adaptation this is not true as the bandwidth part indicator in the DCI received in one bandwidth part can point to *another* differently sized bandwidth part for data reception. This raises the issue on how to interpret the DCI if the bandwidth part index points to another bandwidth part than the current one, as the DCI fields in the detected DCI may not match what is needed in the bandwidth part pointed to by the index field.

One possibility to address this would be to blindly monitor for multiple DCI payload sizes, one for each configured bandwidth part, but unfortunately this would imply a large burden on the device. Instead, an approach where the DCI fields detected are reinterpreted to be useful in the bandwidth part pointed to by the index is used. A simple approach has been selected where the bitfields are padded or truncated to match what is assumed by the bandwidth part scheduled. Naturally, this imposes some limitation

---

[f] Strictly speaking, it is sufficient if the size and configuration of the bandwidth part used for PDCCH and PDSCH are the same.

on the possible scheduling decisions, but as soon as the new bandwidth part is activated the device monitors downlink control signaling using the new DCI size and data can be scheduled with full flexibility again.

Although the handling of different bandwidth parts has been described from a downlink perspective, the same approach of reinterpreting the DCI is applied to the uplink.

14.5.6 PDCCH monitoring control

Blind decoding of PDCCHs is, as already described, carried out at regular time instants. The periodicities of the blind decodings are configurable, but a typical configuration is to monitor PDCCHs at the beginning of each slot. However, in most cases the device is not scheduled and the blind decodings are done in vain, resulting in a waste of the device energy. Actually, a relatively large portion of the device energy consumption in connected mode is spent on blind decoding of the PDCCH candidates. To mitigate this and reduce the device energy consumption in connected mode, release 17 introduces PDCCH monitoring adaptation, based on two components: search-space-set-group (SSSG) switching and PDCCH skipping.

Search space sets control when a device should perform blind decodings and the aggregation levels for the different candidates as discussed in Chapter 10. As part of adding support for unlicensed spectrum in release 16, two SSSGs can be configured to allow the device to switch between two different monitoring configurations. The concept of search space set groups is reused in release 17 with some extensions to provide better control of time instants at which the device performs blind decodings. Up to three SSSGs can be configured in release 17 with one of the groups being active. The reason for defining multiple SSSGs is to reduce device power consumption by adapting the monitoring occasions depending on the traffic situation. For example, PDCCHs can be monitored at the beginning of each slot to allow for quick scheduling of transmissions during on ongoing transmission burst. Less frequent monitoring, for example every tenth slot, may be sufficient between data burst or when expecting less latency critical traffic. Search space set groups can be used to achieve this flexibility with group 0, the default group, used for frequent monitoring and group 1 is used for less frequent monitoring. A third group can also be configured for additional flexibility if deemed beneficial. In principle, a similar effect could be achieved by configuring multiple bandwidth parts with different search space configurations and switch between the bandwidth parts as the traffic changes. However, not only would this occupy more BWP resources, it would also be slower as a BWP switch takes a certain amount of time.

Switching between the SSSGs is controlled dynamically, see Fig. 14.15. One or two bits in the DCI determines which SSSG to use. There is also a timer mechanism defined, which is used as a complement to dynamic signaling of the search space set groups. If the

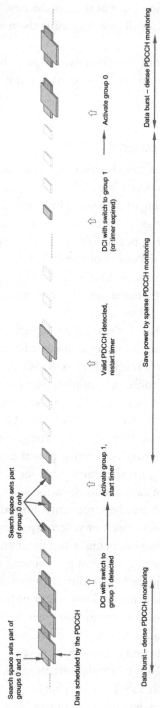

Fig. 14.15 Example of SSSG switching.

Search space sets part of groups 0 and 1

Data scheduled by the PDCCH

DCI with switch to group 1 detected

⇧

Activate group 1, start timer

Data burst – dense PDCCH monitoring

Search space sets part of group 0 only

Valid PDCCH detected, restart timer

⇧

Save power by sparse PDCCH monitoring

DCI with switch to group 1 (or timer expired)

⇧ → Activate group 0

Data burst – dense PDCCH monitoring

Table 14.2 PDCCH monitoring control.

Bits		Skipping only	Switching only	Skipping and switching
One bit	0	No skipping	SSSG#0	–
	1	Skipping duration 1	SSSG#1	–
Two bits	00	No skipping	SSSG#0	SSSG#0
	01	Skipping duration 1	SSSG#1	SSSG#1
	10	Skipping duration 2	SSSG#2	Skipping duration 1
	11	Skipping duration 3	reserved	Skipping duration 2

device is instructed to switch to SSSG 1 or 2, a timer is started. When the timer expires, the device switches back to SSSG 0. Transmission of a scheduling request also forces the device to switch to SSSG 0.

Another mechanism to control the amount of blind decodings is PDCCH skipping. The basic principle is straight forward – when there is no data in the buffer the gNB can instruct the device to skip monitoring for PDCCHs during a preconfigured period of time. Up to three different skipping durations can be configured and the gNB can dynamically select between them using DCI. Once the the skipping duration has passed, the device returns to monitoring PDCCHs at the same time instants as before the skipping.

Search space group switching and PDCCH skipping can be used in combination. If configured, one or two bits in a non-fallback DCI format are used to jointly control switching and skipping as illustrated in Table 14.2. As seen in the table, both switching and skipping can be configured. However, only one mechanism can be triggered at a time. SSSG switching can also be controlled through DCI format 2_0, see Chapter 20, which is useful when operating in unlicensed spectra.

14.5.7 Early indication of paging

In the idle state, the device sleeps most of the time and wakes up at the paging occasions to see whether it is paged or not. Each of these paging occasions is associated with, relative to the sleeping, significant processing in the device and a corresponding energy consumption. Prior to each paging occasion, the device needs to wake up to obtain time/frequencysynchronization and to stabilize oscillators, followed by PDCCH and PDSCH reception. Depending on the signal quality, multiple SSBs may be needed before the time/frequency synchronization is sufficiently accurate for PDSCH reception. This is a large part of the energy consumption in the device, compared to the PDCCH reception, energy that is wasted in the cases when the device is not paged.

To address this, release 17 introduces a mechanism to indicate to the device in advance whether it is likely to be paged in an upcoming paging occasion. The indicator, known as the *paging early indicator* (PEI) is conveyed on a PDCCH using DCI format 2_7 with PEI-RNTI. If no PEI is received prior to a paging occasion, there is no need for the device to spend energy on receiving potential paging messages in the paging occasion, see Fig. 14.16.

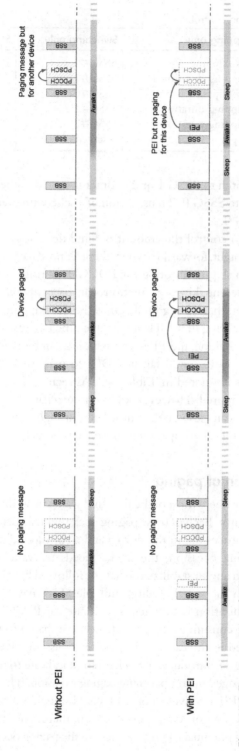

Fig. 14.16 Energy saving with PEI.

Paging subgroups can also be defined and only if the device detects a PEI including the subgroup to which the device belongs is monitors the associated paging occasion.

The time-frequency synchronization was one of the more power consuming procedures in the paging process. One way to reduce this is to provide the device with additional reference signals, thereby speeding up the channel estimation process. Periodic TRSs is one such reference signal, typically broadcasted in a cell if there is at least one device in connected mode. At the same time, to maintain the ultra-lean principle, reference signals should not be transmitted unless they serve a purpose. Therefore, devices can be provided with a TRS configuration in one of the SIBs and the PEI can indicate whether a TRS is present or not. If present, the TRS can be used by the devices to speed up the time-frequency synchronization. TRS presence can also be indicated in normal paging PDCCH in case the PEI is not used.

CHAPTER 15

Uplink power and timing control

Uplink power control and uplink timing control are the topics of this chapter. Power control serves the purpose of controlling the interference, mainly toward other cells as transmissions within the same cell typically are orthogonal. Timing control ensures that different devices are received with the same timing, a prerequisite to maintain orthogonality between different transmissions.

15.1 Uplink power control

NR uplink power control is the set of algorithms and tools by which the transmit power for different uplink physical channels and signals is controlled to ensure that they, to the extent possible, are received by the network at an appropriate power level. In the case of an uplink physical channel, the appropriate power is simply the received power needed for proper decoding of the information carried by the physical channel. At the same time, the transmit power should not be unnecessarily high as that would cause unnecessarily high interference to other uplink transmissions.

The appropriate transmit power will depend on the channel properties, including the channel attenuation and the noise and interference level at the receiver side. It should also be noted that the required received power is directly dependent on the data rate. If the received power is too low one can thus either increase the transmit power or reduce the data rate. In other words, at least in the case of PUSCH transmission, there is an intimate relationship between power control and link adaptation (rate control).

NR uplink power control is based on a combination of:

- *Open-loop* power control, including support for *fractional path-loss compensation*, where the device estimates the uplink path loss based on downlink measurements and sets the transmit power accordingly.
- *Closed-loop* power control based on explicit power-control commands provided by the network. In practice, these power-control commands are determined based on prior network measurements of the received uplink power, thus the term "*closed loop*".

A main new feature of NR uplink power control, not present in, for example, power control in LTE, is the possibility for beam-based power control (see Section 15.1.2).

15.1.1 Baseline power control

Power-control for PUSCH transmissions can, somewhat simplified, be described by the following expression:

$$P_{\text{PUSCH}} = \min\{P_{\text{CMAX}}, P_0(j) + \alpha(j) \cdot PL(q) + 10 \cdot \log_{10}(2^\mu \cdot M_{\text{RB}}) + \Delta_{\text{TF}} + \delta(l)\}$$

$$(15.1)$$

where

- P_{PUSCH} is the PUSCH transmit power;
- P_{CMAX} is the maximum allowed transmit power per carrier;
- $P_0(.)$ is a network-configurable parameter that can, somewhat simplified, be described as a target received power;
- $PL(\cdot)$ is an estimate of the uplink path loss;
- $\alpha(\cdot)$ is a network-configurable parameter (≤ 1) for fractional path-loss compensation;
- μ relates to the sub-carrier spacing Δf used for the PUSCH transmission. More specifically, $\Delta f = 2^\mu \cdot 15$ kHz;
- M_{RB} is the number of resource blocks assigned for the PUSCH transmission;
- Δ_{TF} relates to the modulation scheme and channel-coding rate used for the PUSCH transmission;[a]
- $\delta(\cdot)$ is the power adjustment due to the closed-loop power control.

The expression describes uplink power control *per carrier*. If a device is configured with multiple uplink carriers (carrier aggregation and/or supplementary uplink), power control according to Eq. (15.1) is carried out separately for each carrier. The min $\{P_{\text{CMAX}}, \ldots\}$ part of the power-control expression then ensures that the power per carrier does not exceed the maximum allowed transmit power per carrier. However, there will also be a limit on the total device transmit power over all configured uplink carriers. In order to stay below this limit there will, in the end, be a need to coordinate the power setting between the different uplink carriers (see further Section 15.1.4). Such coordination is needed also in the case of LTE/NR dual-connectivity.

We will now consider the different parts of the above power control expression in somewhat more detail. When doing this we will initially ignore the parameters j, q, and l. The impact of these parameters will be discussed in Section 15.1.2.

The expression $P_0 + \alpha \cdot PL$ represents basic open-loop power control supporting fractional path-loss compensation. In the case of full path-loss compensation, corresponding to $\alpha = 1$, and under the assumption that the path-loss estimate PL is an accurate estimate of the uplink path loss, the open-loop power control adjusts the PUSCH transmit power so that the received power aligns with the "target received power" P_0. The quantity P_0 is

---

[a] The abbreviation TF = Transport Format, a term used in earlier 3GPP technologies but not used explicitly for NR.

provided as part of the power-control configuration and would typically depend on the target data rate but also on the noise and interference level experienced at the receiver.

The device is assumed to estimate the uplink path loss based on measurements on some downlink signal. The accuracy of the path-loss estimate thus partly depends on what extent downlink/uplink reciprocity holds to. Especially, in the case of FDD operation in paired spectra, the path-loss estimate will not be able to capture any frequency-dependent characteristics of the path loss.

In the case of fractional path-loss compensation, corresponding to $\alpha < 1$, the path loss will not be fully compensated for and the received power will even on average vary depending on the location of the device within the cell, with lower received power for devices with higher path loss, in practice for devices at larger distance from the cell site. This must then be compensated for by adjusting the uplink data rate accordingly.

The benefit of fractional path-loss compensation is reduced interference to neighbor cells. This comes at the price of larger variations in the service quality with reduced data-rate availability for devices closer to the cell border.

The term $10 \cdot \log(2^{\mu} \cdot M_{RB})$ reflects the fact that, everything else unchanged, the received power, and thus also the transmit power, should be proportional to the bandwidth assigned for the transmission. Thus, assuming full path-loss compensation ($\alpha = 1$), P_0 can more accurately be described as a *normalized* target received power. Especially, assuming full path-loss compensation, P_0 is the target received power assuming transmission over a single resource block with 15 kHz numerology.

The term Δ_{TF} tries to model how the required received power varies when the number of information bits per resource element varies due to different modulation schemes and channel-coding rates. More precisely

$$\Delta_{TF} = 10 \cdot \log\big((2^{1.25 \cdot \gamma} - 1) \cdot \beta\big) \tag{15.2}$$

where γ is the number of information bits in the PUSCH transmission, normalized by the number of resource elements used for the transmission not including resource elements used for demodulation reference symbols.

The factor β equals 1 in the case of data transmission on PUSCH but can be set to a different value in the case that the PUSCH carries layer-1 control signaling (UCI).[b]

It can be noted that, ignoring the factor β, the expression for Δ_{TF} is essentially a rewrite of the Shannon channel capacity $C = W \cdot \log_2(1 + SNR)$ with an additional factor 1.25. In other words, Δ_{TF} can be seen as modeling link capacity as 80% of Shannon capacity.

The term Δ_{TF} is not always included when determining the PUSCH transmit power.

- The term Δ_{TF} is only used for single-layer transmission, that is, $\Delta_{TF} = 0$ in case of uplink multi-layer transmission

---

[b] Note that one could equally well have described this as a separate term $10 \cdot log(\beta)$ applied when PUSCH carries UCI.

- The term Δ_{TF} can, in general, be disabled. Δ_{TF} should, for example, not be used in combination with fractional power control. Adjusting the transmit power to compensate for different data rates would counteract any adjustment of the data rate to compensate for the variations in received power due to fractional power control as described.

Finally, the term $\delta(\cdot)$ is the power adjustment related to closed-loop power control. The network can adjust $\delta(\cdot)$ by a certain step given by a *power-control command* provided by the network, thereby adjusting the transmit power based on network measurements of the received power. The power control commands are carried in the TPC field within uplink scheduling grants (DCI formats 0_0 through 0_3). Power control commands can also be carried jointly to multiple devices by means of DCI format 2_2. Each power control command consists of 2 bits corresponding to four different update steps (-1 dB, 0 dB, $+1$dB, $+3$ dB). The reason for including 0 dB as an update step is that a power-control command is included in every scheduling grant and it is desirable not to have to adjust the PUSCH transmit power for each grant.

15.1.2 Beam-based power control

In the discussion we ignored the parameter j for the open-loop parameters $P_0(.)$ and $\alpha(.)$, the parameter q in the path loss estimate $PL(.)$, and the parameter l in the closed-loop power adjustment $\delta(.)$. The primary aim of these parameters is to take beam forming into account for the uplink power control.

15.1.2.1 Multiple path-loss-estimation processes

In the case of uplink beam forming, the uplink-path-loss estimate $PL(q)$ used to determine the transmit power according to Eq. (15.1) should reflect the path loss, including the beam-forming gains, of the uplink beam pair to be used for the PUSCH transmission. Assuming beam correspondence, this can be achieved by estimating the path loss based on measurements on a downlink reference signal transmitted over the corresponding downlink beam pair. As the uplink beam used for the transmission pair may change between PUSCH transmissions, the device may thus have to retain multiple path-loss estimates, corresponding to different candidate beam pairs, in practice, path loss estimates based on measurements on different downlink reference signals. When actual PUSCH transmission is to take place over a specific beam pair, the path-loss estimate corresponding to that beam pair is then used when determining the PUSCH transmit power according to the power-control Eq. (15.1).

This is enabled by the parameter q in the path-loss estimate $PL(q)$ of Eq. (15.1). The network configures the device with a set of downlink reference signals (CSI-RS or SS block) on which path loss is to be estimated, with each reference signal being associated with a specific value of q. In order not to put too high requirements on the device, there can be at most four parallel path-loss-estimation processes, each corresponding to a specific value of q. The network also configures a mapping from the possible SRI values provided in the scheduling grant to the up to four different values of q. In the end there is thus a mapping from each of the possible SRI values provided in the scheduling grant to one of up to four configured downlink reference signals and thus, indirectly, a mapping from each of the possible SRI values to one of up to four path-loss estimates reflecting the path loss of a specific beam pair. When a PUSCH transmission is scheduled by a scheduling grant including SRI, the path-loss estimate associated with that SRI is used when determining the transmit power for the scheduled PUSCH transmission.

The procedure is illustrated in Fig. 15.1 for the case of two beam pairs. The device is configured with two downlink reference signals (CSI-RS or SS block) that in practice will be transmitted on the downlink over a first and second beam pair respectively. The device is running two path-loss-estimation processes in parallel, estimating the path loss $PL(1)$ for the first beam pair based on measurements on reference signal RS-1 and the path loss $PL(2)$ for the second beam pair based on measurements on reference-signal RS-2. The parameter q associates SRI = 1 with RS-1 and thus indirectly with $PL(1)$. Likewise, SRI = 2 is associated with RS-2 and thus indirectly with $PL(2)$. When the device is scheduled for PUSCH transmission with the SRI of the scheduling grant set to 1, the transmit power of the scheduled PUSCH transmission is determined based on the path-loss estimate $PL(1)$ that is, the path-loss estimate based on measurements on RS-1. Thus, assuming beam correspondence the path-loss estimate reflects the path loss of the beam pair over which the PUSCH is transmitted. If the device is instead scheduled for PUSCH transmission with SRI = 2, the path-loss estimate $PL(2)$, reflecting the path loss of the beam pair corresponding to SRI=2, is used to determine the transmit power for the scheduled PUSCH transmission.

15.1.2.2 Multiple open-loop-parameter sets

In the PUSCH power-control Eq. (15.1), the open-loop parameters P_0 and α are associated with a parameter j. This simply reflects that there may be multiple open-loop-parameter pairs $\{P_0, \alpha\}$. Partly, different open-loop parameters will be used for different types of PUSCH transmission (random-access "message 3" transmission, see Chapter 17, grant-free PUSCH transmissions, and scheduled PUSCH transmissions). However, there is also a possibility to have multiple pairs of open-loop parameter for scheduled

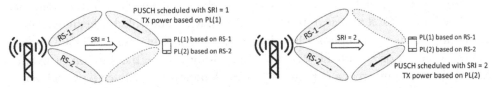

Fig. 15.1 Use of multiple power-estimation processes to enable uplink power control in case of dynamic beam management.

PUSCH transmission, where the pair to use for a certain PUSCH transmission can be selected based on the SRI similar to the selection of path-loss estimates as described above. In practice this implies that the open-loop parameters P_0 and α will depend on the uplink beam.

For the power setting of random-message 3, which in the NR specification corresponds to $j=0$, α always equals 1. In other words, fractional power control is not used for message-3 transmission. Furthermore, the parameter P_0 can, for message 3, be calculated based on information in the random-access configuration.

For other PUSCH transmissions the device can be configured with different open-loop-parameter pairs $\{P_0(j), \alpha(j)\}$, corresponding to different values for the parameter j. Parameter pair $\{P_0(1), \alpha(1)\}$ should be used in the case of grant-free PUSCH transmission while the remaining parameter pairs are associated with scheduled PUSCH transmission. Each possible value of the SRI that can be provided as part of the uplink scheduling grant is associated with one of the configured open-loop-parameter pairs. When a PUSCH transmission is scheduled with a certain SRI included in the scheduling grant, the open-loop parameters associated with that SRI are used when determining the transmit power for the scheduled PUSCH transmission.

Multiple open-loop-parameter sets can also be used for uplink preemption. The release-16 enhancements supporting this are discussed in Chapter 20.

15.1.2.3 Multiple closed-loop processes

The final parameter is the parameter l for the closed-loop process. PUSCH power control allows for the configuration of two independent closed-loop processes, associated with $l=1$ and $l=2$, respectively. Similar to the possibility for multiple path-loss estimates and multiple open-loop-parameter sets, the selection of l, that is, the selection of closed-loop process can be tied to the SRI included in the scheduling grant by associating each possible value of the SRI to one of the closed-loop processes.

15.1.3 Power control for PUCCH

Power control for PUCCH follows essentially the same principles as power control for PUSCH with some minor differences

First, for PUCCH power control, there is no fractional path-loss compensation, that is, the parameter α always equals one

Furthermore, for PUCCH power control, the closed-loop power control commands are carried within DCI formats 1_0 and 1_1, that is, within downlink scheduling assignments rather than within uplink scheduling grants which is the case for PUSCH power control. One reason for uplink PUCCH transmissions is the transmission of hybrid-ARQ acknowledgments as a response to downlink transmissions. Such downlink transmissions are typically associated with downlink scheduling assignments on PDCCH and the corresponding power-control commands could thus be used to adjust the PUCCH transmit power prior to the transmission of the hybrid-ARQ acknowledgments. Similar to PUSCH, power-control commands can also be carried jointly to multiple devices by means of DCI format 2_2.

15.1.4 Power control in case of multiple uplink carriers

The above procedures describe how to set the transmit power for a given physical channel in the case of a single uplink carrier. For each such carrier there is a maximum allowed transmit power P_{CMAX} and the $\min\{P_{CMAX}, \ldots\}$ part of the power-control expression ensures that the per-carrier transmit power of a carrier does not exceed power P_{CMAX}.[c]

In many cases, a device is configured with multiple uplink carriers

- Multiple uplink carriers in a carrier aggregation scenario
- An additional supplementary uplink carrier in case of SUL

In addition to the maximum per-carrier transmit power P_{CMAX}, there is a limit P_{TMAX} on the total transmitted power over all carriers. For a device configured for NR transmission on multiple uplink carriers, P_{CMAX} should obviously not exceed P_{TMAX}. However, the sum of P_{CMAX} over all configured uplink carriers may very well, and often will, exceed P_{TMAX}. The reason is that a device will often not transmit simultaneously on all its configured uplink carriers and the device should then preferably still be able to transmit with the maximum allowed power P_{TMAX}. Thus, there may be situations when the sum of the transmit power of each carrier given by the power-control Eq. (15.1) exceeds P_{TMAX}. In that case, the power of each carrier needs to be scaled down to ensure that the eventual transmit power of the device does not exceed the maximum allowed value.

Another situation that needs to be taken care of is the simultaneous uplink transmission of LTE and NR in the case of a device operating in dual-connectivity between LTE and NR. Note that, at least in an initial phase of NR deployment this will be the normal mode-of-operation as the first release of the NR specifications only support non-standalone NR deployments. In this case, the transmission on LTE may limit the power available for NR transmission and vice versa. The basic principle is that the LTE transmission has priority, that is the LTE carrier is transmitted with the power given by the LTE uplink power control [26]. The NR transmission can then use whatever power is left up to the power given by the power-control Eq. (15.1).

The reason for prioritizing LTE over NR is multifold:

- In the specification of NR, including the support for NR/LTE dual connectivity, there has been an aim to as much as possible avoid any impact on the LTE specifications. Imposing restrictions on the LTE power control, due to the simultaneous transmission on NR would have implied such an impact.

[c] Note that, in contrast to LTE, at least for NR release 15 there is not simultaneous PUCCH and PUSCH transmission on a carrier and thus there is at most one physical channel transmitted on an uplink carrier at a given time instant.

- At least initially, LTE/NR dual-connectivity will have LTE providing the control-plane signaling, that is, LTE is used for the master cell group (MCG). The LTE link is thus more critical in terms of retaining connectivity and it makes sense to prioritize that link over the "secondary" NR link

15.2 Uplink timing control

The NR uplink allows for uplink intra-cell orthogonality, implying that uplink transmissions received from different devices within a cell do not cause interference to each other. A requirement for this *uplink orthogonality* to hold is that the uplink slot boundaries for a given numerology are (approximately) time aligned at the base station. More specifically, any timing misalignment between received signals should fall within the cyclic prefix. To ensure such receiver-side time alignment, NR includes a mechanism for *transmit-timing advance*.

In essence, timing advance is a negative offset, at the device, between the start of a downlink slot n as observed by the device and the start of the corresponding uplink slot n. By controlling the offset appropriately for each device, the network can control the timing of the signals received at the base station from the devices. Devices far from the base station encounter a larger propagation delay and therefore need to start their uplink transmissions somewhat in advance, compared to devices closer to the base station, as illustrated in Fig. 15.2. In this specific example, the first device is located close to the base station and experiences a small propagation delay, $T_{P,1}$. Thus, for this device, a small value of the timing advance offset $T_{A,1}$ is sufficient to compensate for the propagation delay and to ensure the correct timing at the base station. However, a larger value of the timing advance is required for the second device, which is located at a larger distance from the base station and thus experiences a larger propagation delay.

The timing-advance value for each device is determined by the network based on measurements on the respective uplink transmissions. Hence, as long as a device carries out uplink data transmission, this can be used by the receiving base station to estimate the uplink receive timing and thus be a source for the timing-advance commands. Sounding reference signals can be used as a regular signal to measure upon, but in principle the base station can use any signal transmitted from the devices.

Based on the uplink measurements, the network determines the required timing correction for each device. If the timing of a specific device needs correction, the network issues a timing-advance command for this specific device, instructing it to retard or advance its timing relative to the current uplink timing. The user-specific timing-advance command is transmitted as a MAC control element on the DL-SCH. Typically, timing-advance commands to a device are transmitted relatively infrequently – for example, one or a few times per second – but obviously this depends on how fast the device is moving.

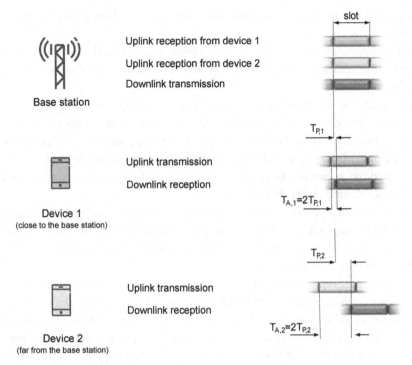

Fig. 15.2 Uplink timing advance.

In more details, the timing of T_{TA} between the start of an uplink slot n and the start of the corresponding downlink slot n is (see also Fig. 15.3)

$$T_{TA} = (N_{TA} + N_{TA,\text{offset}}) \cdot T_c \qquad (15.3)$$

$N_{TA,\text{offset}}$ is a cell-specific parameter that can only take a limited set of values, in practice depending on the frequency band and if the NR carrier needs to co-exist with an LTE carrier (only relevant for FR1). The default values for $N_{TA,\text{offset}}$ are

- $N_{TA,\text{offset}} = 25600$ for FR1 in case of no spectrum-coexistence
- $N_{TA,\text{offset}} = 0$ (FDD) and 39936 (TDD) for FR1 in case of spectrum-coexistence
- $N_{TA,\text{offset}} = 13792$ for FR2

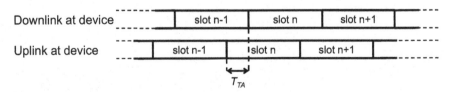

Fig. 15.3 Uplink/downlink timing relation at the device.

N_{TA} is the device-specific timing advance that is updated based on network signaling as described above.

It should be pointed out that, in case of operation in a so-called *non-terrestrial network* (see Chapter 25), additional parameters are included when calculating the uplink-downlink timing offset T_{TA},

As discussed above, the target of timing advance is to keep the timing misalignment within the size of the cyclic prefix and the step size of the timing advance is therefore chosen as a fraction of the cyclic prefix. However, as NR supports multiple numerologies with the cyclic prefix being shorter the higher the subcarrier spacing, the timing advance step size is scaled in proportion to the cyclic prefix length and given by the subcarrier spacing of the active uplink bandwidth part.

If the device has not received a timing-advance command during a (configurable) period, the device assumes it has lost the uplink synchronization. In this case, the device must reestablish uplink timing using the random-access procedure prior to any PUSCH or PUCCH transmission in the uplink.

For carrier aggregation, there may be multiple component carriers transmitted from a single device. A straightforward way of handling this would be to apply the same timing-advance value for all uplink component carriers. However, if different uplink carriers are received at different geographical locations, for example by using remote radio heads for some carriers but not others, different carriers would need different timing advance values. Dual connectivity with different uplink carriers terminated at different sites is an example when this is relevant. To handle such scenarios, uplink carriers are grouped into so-called *timing advanced groups* (TAGs). All component carriers within a TAG are subject to the same timing-advance command. However, different TAGs may be subject to different timing-advance commands, that is, the transmission timing of component carriers of different TAGs may be set differently. If a TAG includes component carriers with different numerology, the timing-advance step size is determined by the highest subcarriers spacing among the carriers.

CHAPTER 16

Cell search and system information

Cell search covers the functions and procedures by which a device finds new cells. Cell search is carried out when a device is initially entering the coverage area of a system. To enable mobility, cell search is also continuously carried out by devices moving within the system, both when the device is connected to the network and when in idle/inactive state. In NR, cell search is primarily based on so-called *synchronization signal blocks* (SSBs).

Once a device has found and connected to a cell, it needs to acquire the *system information* valid for the cell.

In this chapter we will describe the NR cell search, including the detailed structure for SSB transmissions. We will also discuss the structure for the NR system information and how it is provided to devices.

16.1 The SSB

To enable devices to find a cell when entering a system, as well as to find new cells when moving within the system, synchronization signals consisting of two parts, the *Primary Synchronization Signal* (PSS) and the *Secondary Synchronization Signal* (SSS), are periodically transmitted within NR cells. The PSS/SSS together with the *Physical Broadcast Channel* (PBCH), see Section 16.3, are jointly referred to as a *Synchronization Signal Block* or SSB.

The SSB serves a similar purpose and, in many respects, has a similar structure as the PSS/SSS/PBCH of LTE [26]. However, there are some important differences between the LTE PSS/SSS/PBCH and the NR SSB. At least partly, the origin of these differences can be traced back to some NR-specific requirements and characteristics including the aim to reduce the amount of "always-on" signals, as discussed in Section 5.1.2, and the possibility for beamforming during initial access.

16.1.1 Basic structure

As all NR downlink transmissions, SSB transmission is based on OFDM. In other words, the SSB is transmitted on a set of time/frequency resources (resource elements) within the basic OFDM grid discussed in Section 7.3. Fig. 16.1 illustrates the time/frequency structure of a single SSB transmission. As can be seen, the SSB spans four OFDM symbols in the time domain and 240 subcarriers in the frequency domain.

5G/5G-Advanced
https://doi.org/10.1016/B978-0-443-13173-8.00018-9

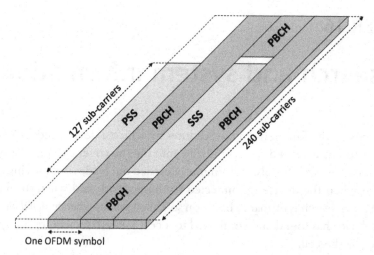

Fig. 16.1 Time/frequency structure of a single SSB consisting of PSS, SSS, and PBCH.

- PSS is transmitted in the first OFDM symbol of the SSB and occupies 127 subcarriers in the frequency domain. The remaining 113 subcarriers within the SSB bandwidth are empty.
- SSS is transmitted in the third OFDM symbol of the SSB and occupies the same set of subcarriers as the PSS. There are eight and nine empty subcarriers on each side of the SSS respectively.
- The PBCH is transmitted within the second and fourth OFDM symbols of the SSB. In addition, PBCH also uses 48 subcarriers on each side of the SSS.

Different numerologies can be used for SSB transmission. However, to limit the need for a device to simultaneously search for SSBs of different numerologies on a given frequency, there is at most two SSB numerologies defined for a given frequency band.

Table 16.1 lists the different numerologies applicable for SSB transmission together with the corresponding SSB bandwidth and time duration, and the frequency range for which each specific numerology applies.[a] Note that 60 kHz numerology cannot be used for SSB transmission regardless of frequency range. In contrast, 240 kHz numerology can be used for SSB transmission although it is currently not supported for other downlink transmissions. The reason to support 240 kHz SSB numerology is to enable a shorter time duration for each SSB. This is relevant in the case of beam-sweeping over many beams with a corresponding large number of time multiplexed SSBs (see further details in Section 16.2). The 480 kHz and 960 kHz SSB numerologies were introduced together with the introduction of NR support for operation above 52.6 GHz (FR2-2).

---

[a] Note that, although the frequency range for 30 kHz SS numerology fully overlaps with the frequency range for 15 kHz numerology, for a given frequency band within the lower frequency range there is in most cases only a single numerology supported.

Table 16.1 SSB numerologies and corresponding frequency ranges.

Numerology (kHz)	SSB bandwidth[a] (MHz)	SSB duration (μs)	Frequency range
15	3.6	≈285	FR1 (<3 GHz)
30	7.2	≈143	FR1
120	28.8	≈36	FR2-1
240	57.6	≈18	FR2-1
480	115.2	≈9	FR2-2
960	230.4	≈4.5	FR2-2

[a]The SSB bandwidth is simply the number of sub-carriers used for SSB (240) multiplied by the SSB sub-carrier spacings.

As mentioned in Section 5.2.15, release 18 introduces the possibility to operate NR in less than 5 MHz of spectrum. In such cases, the full set of SSB subcarriers (240 subcarriers) does not fit within the available spectrum. To support this, no changes are made to the basic SSB structure but the resource elements corresponding to the subcarriers outside the available bandwidth will simply not be transmitted. This will obviously impact the SSB link performance which can be compensated for by operating with a somewhat higher SSB SINR.

16.1.2 Frequency-domain position

In LTE, the PSS and SSS were always located at the center of the carrier. Thus, once an LTE device had found a PSS/SSS, that is, found a carrier, it inherently knew the center frequency of the found carrier. The drawback with this approach, that is, always locating the PSS/SSS at the center of the carrier, is that a device with no a priori knowledge of the frequency-domain carrier position must search for PSS/SSS at all possible carrier positions (the "*carrier raster*").

To allow for faster cell search, a different approach was adopted for NR. Rather than always being located at the center of the carrier, implying that the possible SSB locations coincide with the carrier raster, there are, within each frequency band, a more limited set of possible locations of SSB, referred to as the "*synchronization raster*". Instead of searching for an SSB at each position of the carrier raster, a device thus only needs to search for an SSB on the sparser synchronization raster.

As carriers can still be located at an arbitrary position on the more dense carrier raster, the SSB may not end up at the center of a carrier. The SSB may not even end up aligned with the resource-block grid. Hence, once the SSB has been found, the device must be explicitly informed about the exact SSB frequency-domain position within the carrier. This is done by means of information partly within the SSB itself, more specifically information carried by the PBCH (Section 16.3), and partly within the remaining broadcast system information (see further Section 16.5).

16.1.3 SSB periodicity

SSB is transmitted periodically with a period that may vary from 5 ms up to 160 ms. However, devices doing initial cell search, as well as devices in inactive/idle state doing cell search for mobility, can assume that the SSB is occurs at least once every 20 ms. This allows for a device that searches for SSB in the frequency domain to know how long it must stay on a given frequency before concluding that there is no SSB present on the frequency and thus that it should move on to the next candidate frequency within the synchronization raster.

The 20 ms NR SSB periodicity is four times longer than the corresponding 5 ms periodicity of LTE PSS/SSS transmission. The longer SSB period was selected to allow for enhanced NR network energy performance and in general to follow the ultra-lean design paradigm described in Section 5.1.2. The drawback with a longer SSB period is that, compared to LTE, a device must stay on each frequency for a longer time before it can conclude that there is no SSB on the frequency. However, this is compensated for by the more sparse synchronization raster which reduces the number of frequency-domain locations on which a device must search for SSB.

Even though devices doing initial cell search can assume that SSB is repeated at least once every 20 ms, there are situations when there may be reasons to use either a shorter or longer SSB periodicity:

- A shorter SSB periodicity may be used to enable faster cell search for devices in connected mode as such devices can be explicitly informed about the shorter SSB period by the network.
- A longer SSB periodicity may be used to further enhance network energy performance. A carrier with an SSB periodicity larger than 20 ms may not be found by devices doing initial access. However, such a carrier could still be used by devices in connected mode, for example, as a secondary carrier in a carrier-aggregation scenario.

It should be pointed out that the discussion above, with a constant SSB period, is not fully accurate for the case of operation in unlicensed spectrum. In case of operation is unlicensed spectrum there will typically be some more or less random variations in the SSB transmission timing due to the channel-access procedure needed for operation in unlicensed spectrum, see further Section 20.8.2. Thus one cannot, in this case, really talk about a specific SSB periodicity.

16.1.4 Detailed structure of PSS and SSS

Above we have described the overall structure of an SSB and how it consists of three parts: PSS, SSS, and PBCH. In this section we will describe the detailed structure of the PSS and SSS. The PBCH is further discussed in Section 16.3.

16.1.4.1 The primary synchronization sequence (PSS)

The PSS is the first signal that a device entering the system will search for. At that stage, the device has no knowledge of the system timing. Furthermore, even though the device searches for a cell at a given carrier frequency, there may, due to inaccuracy of the device internal frequency reference, be a relatively large deviation between the device and network carrier frequency. The PSS has been designed to be detectable despite these uncertainties.

Once the device has found the PSS, it has found synchronization up to the periodicity of the PSS. It can then also use transmissions from the network as a reference for its internal frequency generation, thereby to a large extent eliminating any frequency deviation between the device and the network.

As already mentioned, the PSS extends over 127 resource elements, onto which a *PSS sequence* $\{x_n\} = x_n(0), x_n(1), \ldots, x_n(126)$ is mapped (see Fig. 16.2).

There are three different PSS sequences $\{x_0\}$, $\{x_1\}$, and $\{x_2\}$, derived as different cyclic shifts of a basic length–127 M-sequence [66] $\{x\} = x(0), x(1), \ldots, x(126)$ generated according to the recursive formula (see Fig. 16.3).

$$x(n) = x(n - 7) \oplus x(n - 3)$$

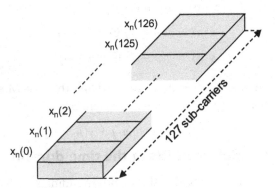

Fig. 16.2 PSS structure.

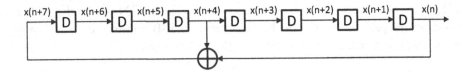

Initial value: [x(6) x(5) x(4) x(3) x(2) x(1) x(0)] = [1 1 1 0 1 1 0]

Fig. 16.3 Generation of basic *M*-sequence from which three different PSS sequences are derived.

By applying different cyclic shifts to the basic M-sequence $x(n)$, three different PSS sequences $x_0(n)$, $x_1(n)$ and $x_2(n)$ can be generated according to

$$x_0(n) = x(n); \quad x_1(n) = x(n + 43 \ mod \ 127); \quad x_2(n) = x(n + 86 \ mod \ 127)$$

Which of the three PSS sequences to use in a certain cell is determined by the *physical cell identity* (PCI) of the cell. When searching for new cells, a device thus must search for all three PSSs.

16.1.4.2 *The secondary synchronization sequence (SSS)*

Once a device has detected a PSS it knows the transmission timing of the SSS. By detecting the SSS, the device can determine the PCI of the detected cell. There are 1008 different PCIs. However, already from the PSS detection the device has reduced the set of candidate PCIs by a factor 3. There are thus 336 different SSSs which, together with the already-detected PSS, provide the full PCI. Note that, since the timing of the SSS is known to the device, the per-sequence search complexity is reduced compared to the PSS, enabling the larger number of SSS sequences.

The basic structure of the SSS is the same as that of the PSS, that is, the SSS consists of 127 subcarriers to which an SSS sequence is applied.

On an even more detailed level, each SSS is derived from two basic M-sequences generated according to the recursive formulas

$$x(n) = x(n - 7) \oplus x(n - 3)$$
$$y(n) = y(n - 7) \oplus y(n - 6)$$

The actual SSS sequence is then derived by adding the two M sequences together, with different shifts being applied to the two sequences.

$$x_{m_1,m_2}(n) = x(n + m_1) + y(n + m_2)$$

16.2 SS burst set – Multiple SSBs in the time domain

One key difference between the SSB and the corresponding signals for LTE is the possibility to apply beam-sweeping for SSB transmission, that is, the possibility to transmit SSBs in different, more narrow, beams in a time-multiplexed fashion (see Fig. 16.4). The set of SSBs within a beam-sweep is referred to as an *SS burst set*.[b] Note that the SSB period discussed in the previous section is the time between SSB transmissions *within a specific beam*, that is, it is actually the periodicity of the SS burst set.

---

[b] The term SS *burst set* originates from early 3GPP discussions on NR when SSBs were assumed to be grouped into *SS bursts* and the SS bursts then grouped into *SS burst sets*. The intermediate SS-burst grouping was eventually not used but the term *SS burst set* for the full set of SSBs was retained.

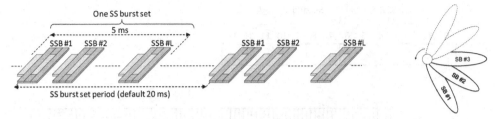

Fig. 16.4 Multiple time-multiplexed SSBs within an SS-burst-set period.

By applying beam-sweeping for the SSB, the coverage of a single SSB transmission can be increased. Beam-sweeping for SSB transmission also enables receiver-side beam-sweeping for the reception of uplink random-access transmissions as well as downlink beam-forming for the *random-access response*. This will be further discussed as part of the description of the NR random-access procedure in Chapter 17.

Although the periodicity of the SS burst set is flexible with a minimum period of 5 ms and a maximum period of 160 ms, each SS burst set is always confined to a 5 ms time interval, either in the first or second half of a 10 ms frame.

The maximum number of SSBs within an SS burst set is different for different frequency bands.

- For frequency bands below 3 GHz, there can be up to four SSBs within an SS burst set, enabling SSB beam sweeping over up to four beams;
- For frequency bands between 3 GHz and 6 GHz, there can be up to eight SSBs within an SS burst set, enabling beam sweeping over up to eight beams;
- For higher-frequency bands (FR2-1/FR2-2) there can be up to 64 SSBs within an SS burst set, enabling beam-sweeping over up to 64 beams.

There are two reasons why the maximum number of SSBs within an SS burst set, and thus also the maximum number of beams over which the SSB can be swept, is larger for higher frequency bands.

- The use of a large number of beams with a corresponding more narrow beam-width is typically more relevant for higher frequencies.
- As the absolute duration of an SSB depends on the SSB numerology, see Table 16.1, a large number of SSBs within an SS burst set would imply a very large SSB overhead for lower frequencies for which lower SSB numerology (15 or 30 kHz) must be used.

The set of possible SSB locations in the time domain differ somewhat between different SSB numerologies. As an example, Fig. 16.5 illustrates the possible SSB locations within an SS-burst-set period for the case of 15 kHz numerology. As can be seen, there may be SSB transmission in any of the first four slots.[c] Furthermore, there can be up to two SSB

---

[c] For operation below 3 GHz, the SSB can only be located within the first two slots.

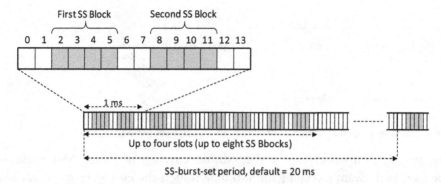

Fig. 16.5 Possible time-domain locations of SSB within an SS burst set for 15 kHz numerology.

transmissions in each of these slots, with the first possible SSB location corresponding to symbol two to symbol five and the second possible SSB location corresponding to symbol eight to symbol eleven. Finally, note that the first and last two OFDM symbols of a slot are unoccupied by SSB transmission. This allows for these OFDM symbols to be used for downlink and uplink control signaling, respectively, for devices already connected to the network. The same is true for all SSB numerologies.

It should be noted that the SSB locations outlined in Fig. 16.5 are *possible* SSB locations, that is, an SSB is not necessarily transmitted in all the locations outlined in Fig. 16.5. There may be anything from one single SSB transmission up to the maximum number of SSBs within an SS burst set depending on the number of beams over which the SSB is to be beam-swept.

Furthermore, if less than the maximum number of SSBs is transmitted, the transmitted SSBs do not have to be transmitted in consecutive SSB locations. Rather, any subset of the possible set of SSB locations outlined in Fig. 16.5 can be used for actual SSB transmission. In the case of four SSBs within an SS burst set these may, for example, be located as two SSBs within each of the two first slots or as one SSB in each of the four slots of Fig. 16.5.

The PSS and SSS of an SSB only depend on the physical cell identity (see later). Thus, the PSS and SSS of all SSBs within a cell are identical and cannot be used by the device to determine the relative location of an acquired SSB within the set of possible SSB locations. For this reason, each SSB, more specifically, the PBCH, includes a "time index" that explicitly provides the relative location of the SSB within the sequence of possible SSB locations (see further details in Section 16.3). Knowing the relative location of the SSB is important for several reasons:

- It makes it possible for the device to determine frame timing, see Section 16.3;
- It makes it possible to associate different SSBs, in practice different beams, with different so called *RACH occasions*. This, in turn, is a prerequisite for the use of network-side beam forming for random-access reception (see further details in Chapter 17).

Note that, similar to the discussion on the basic SSB periodicity in Section 16.1.3, the above discussion on SS burst set and its time-domain structure is strictly speaking only valid for the case of operation in licensed spectrum. In case of operation is unlicensed spectrum the structure will differ somewhat, once again due to the need for some flexibility in the exact SSB transmission timing taking into account the channel-access procedure for unlicensed spectrum, see further Section 20.8.2.

16.3 PBCH and MIB

While the PSS and SSS are physical signals with specific structures, the PBCH is a more conventional physical channel on which explicit channel-coded information is transmitted. The PBCH carries the so-called *master information block* (MIB), which contains a small amount of information that the device needs for the acquisition of the remaining system information broadcast by the network.[d]

Table 16.2 lists the different information carried within the PBCH together with the corresponding size in number of bits for each piece of information. Note that the information differs slightly depending on if the carrier is operating in lower-frequency bands (FR1) or higher-frequency bands (FR2). Also note that the table assumes operation in licensed spectrum. For operation in unlicensed spectrum there are some modifications to the PBCH information.

As already mentioned, the *SSB time index* identifies the SSB location within an SS burst set. As described in Section 16.2, each SSB has a well-defined position within an SS burst set which, in turn, is contained within the first or second half of a 5 ms frame.

Table 16.2 Information carried within the PBCH.

Information	Size (bits)
SSB time index	0 (FR1) / 3 (FR2)
CellBarred flag	1
Intra-frequency-reselection flag	1
DMRS Type A position	1
SIB1 numerology	1
SIB1 PDCCH configuration	8
CRB grid offset	5 (FR1)/4 (FR2)
Half-frame bit	1
System Frame Number (SFN)	10
Cyclic Redundancy Check (CRC)	24

[d] Some of the information on the PBCH is strictly speaking not part of the MIB, see below.

From the SSB time index, in combination with the *half-frame bit* (see also below), the device can thus determine the frame boundary.

The SSB time index is provided to the device as two parts:

- An implicit part encoded in the scrambling applied to the PBCH;
- An explicit part included in the PBCH payload.

Eight different scrambling patterns can be used for the PBCH allowing for the implicit indication of up to eight different SSB time indices. This is sufficient for operation in lower frequency bands (FR1) where there can be at most eight SSBs within an SS burst set.[e]

For operation in the higher NR frequency range (FR2) there can be up to 64 SSBs within an SS burst set, implying the need for three additional bits to indicate the SSB time index. These three bits, which are thus only needed for operation above 10 GHz, are included as explicit information within the PBCH payload.

The *CellBarred and Intra-frequency-reselection flags* are related to whether or not devices are allowed to access cells

- The CellBarred flag indicates whether or not devices are allowed to access the specific cell;
- Assuming devices are not allowed to access the cell, that is, the CellBarred flag set to TRUE, the Intra-frequency-reselection flag indicates whether or not access is allowed to other cells on the same frequency.

If detecting that a cell is barred and that access to other cells on the same frequency is not allowed, a device can and should immediately reinitiate cell search on a different carrier frequency.

It may seem strange to deploy a cell and then prevent devices from accessing it. Historically this kind of functionality has been used to temporarily prevent access to a certain cell during maintenance. However, the functionality has additional usage within NR due to the possibility for non-standalone NR deployments for which devices should access the network via an associated LTE carrier. By setting the CellBarred flag to TRUE for the NR carrier in an NSA deployment, the network prevents NR devices from trying to access the system via the NR carrier.

The *DMRS Type A position* indicates the time-domain position of the first DMRS symbol assuming DMRS Mapping Type A (see Section 9.11).

The *SIB1 numerology* provides information about the subcarrier spacing used for the transmission of SIB1 (see Section 16.5). The same numerology is then also used for *Message 2/4* and *Message B* that are part of the NR random-access procedures (see Chapter 17). For each of FR-1 and FR-2-1 there are two possible SIB1 numerologies

---

[e] Only up to four SSBs for operation below 3 GHz.

(15/30 kHz for FR1 and 60/120 kHz for FR2-1 while, for FR2-2, the SIB1 numerology is always the same as the SSB numerology.[f] Thus, one bit is sufficient to signal the SIB1 numerology.

The *SIB1 PDCCH configuration* provides information about the search space, corresponding CORESET, and other PDCCH-related parameters that a device needs in order to monitor for scheduling of SIB1, see Section 16.5.

The *CRB grid offset* provides information about the frequency offset between the SSB and the common resource block grid. As discussed in Section 16.1.2, the frequency-domain position of the SSB relative to the carrier is flexible and does not even have to be aligned with the carrier CRB grid. However, for SIB1 reception, the device needs to know the CRB grid. Thus, information about the frequency offset between the SSB and the CRB grid must be provided within the PBCH in order to be available to devices prior to SIB1 reception.

Note that the CRB grid offset only provides the offset between the SSB and the CRB grid. Information about the absolute position of the SSB within the overall carrier is then provided within SIB1.

The *half-frame bit* indicates if the SSB is located in the first or second 5 ms part of a 10 ms frame. As mentioned earlier, the half-frame bit, together with the SSB time index, allows for a device to determine the cell frame boundary.

Although all the information above is carried within the PBCH and is jointly channel coded and CRC-protected, some of the information is strictly speaking not part of the MIB. The MIB is assumed to be the same over an 80 ms time interval (eight frames) as well as for all SSBs within an SS burst set. Thus, the SSB time index, which is inherently different for different SSBs within an SS burst set, the half-frame bit and the four least significant bits of the SFN are PBCH information that is formally carried outside of the MIB.

16.4 Cell-defining and non-cell-defining SSBs

Fundamentally, the SSB is, as described above, used to obtain time/frequency synchronization and acquisition of (parts of) system information. This is essential to access a system and for the terminal to transfer from idle to connected mode. However, the SSB is also used once the terminal is in connected mode with neighboring cell mobility measurements as one example hereof. It is also useful as a QCL root for several other signals. Typically, antenna ports assume QCL with the SSB unless configured otherwise. In the latter cases, when the device already is connected to the network, there is no (or very limited) need for system information and an SSB without an accompanying SIB1 is sufficient. Thus, it is possible to indicate that no SIB1 is present by setting the CRB grid

---

[f] The same is true for operation in unlicensed spectrum, in which case the "SIB1 numerology" bit is used for a completely different purpose, namely, to derive the QCL relation between SSBs, see further Chapter 20.

offset in the MIB to a "too large" value.[g] Clearly, an SSB without SIB1 cannot be used for initial access and is sometimes referred to as a *non-cell-defining SSB* (NCD-SSB) to differentiate it from a *cell-defining SSB* (CD-SSB) that can be used also for initial access. Idle mode devices finding a NCD-SSB will simply continue the cell search until an CD-SSB is found, while a device in connected (or inactive) mode through RRC signaling can be told to use a NCD-SSB. Non-cell-defining SSBs can be useful, for example, when a carrier is used for SCells only. In this case a device connects to a regular carrier, uses that as a PCell and is told to acquire an SCell without SIB1. Another situation where non-cell-defining SSBs are useful is when not all devices on a carrier support the full carrier bandwidth. RedCap devices, described in Chapter 22, is one example hereof. In such cases, the network may not want all device to camp on the cell-defining SSB but rather spread the out across the full carrier bandwidth. This can be achieved by configuring additional non-cell-defining SSBs on the carrier and configuring the devices to camp on one of these SSBs. In case the device need to reacquire the system information in connected or inactive mode, they first need to move to the cell-defining SSB in order to obtain the MIB and SIB1.[h]

16.5 Providing remaining system information

System information is a joint name for all the common (non-device-specific) information that a device needs in order to properly operate within the network. In general, the system information is carried within different *System Information Blocks* (SIBs), each consisting of different types of system information. Delivering the SIBs is done in different ways depending on whether the device already is connected to the network or not:

- If the device is already connected to the network, dedicated RRC signaling is used. An obvious example is operation in non-standalone mode, where LTE is used for initial access and mobility. In this case the device is obviously already connected via LTE and system information is delivered through that connection when setting up the NR carrier. Another example is adding a carrier in a carrier aggregation scenario in which case the existing NR connection is used for the dedicated RRC signaling.
- If the device has no connection to the network, broadcast signaling is used. This is the case for standalone operation when the device is in idle mode and has no valid system information.

Broadcasting of system information is used also in LTE but NR takes this approach one step further. In LTE, all system information is periodically broadcast over the entire cell area making it always available but also implying that it is transmitted even if there is no device within the cell.

---

[g] In this case, the SIB1 configuration field indicates where in frequency the device may find a CD-SSB.
[h] It is possible to use paging to indicate to connected-mode devices that system information has been updated and needs to be reacquired.

For NR, a different approach has been adopted where the system information, beyond the very limited information carried within the MIB, is divided into two parts.

SIB1 consists of the system information that a device needs to know before it can access the system. SIB1 is always periodically broadcast over the entire cell area. One important task of SIB1 is to provide the information a device needs to be able to carry out an initial random access (see Chapter 17).

SIB1 is provided by means of ordinary scheduled PDSCH transmissions with a periodicity of 160 ms. As described earlier, the PBCH/MIB provides information about the numerology used for SIB1 transmission as well as the search space and corresponding CORESET used for scheduling of SIB1. The *SIB1 configuration* field in the MIB is used as an index into predefined tables. There are multiple tables provided in the specifications depending the frequency band. From the appropriate table, information on the CORESET is obtained. Within that CORESET, referred to as CORESET#0, the device then monitors for scheduling of SIB1 which is indicated by a special *System Information RNTI* (SI-RNTI).

The location of CORESET#0 and the search space is given relative to the detected SSB by the tables. Three different possibilities for multiplexing SSB and CORESET#0 are possible, see Fig. 16.6, although not all multiplexing patterns are available in all frequency bands.

Pattern 1 is used in FR1. The size of CORESET#0 is signaled such that it fits within the carrier bandwidth; the smallest CORESET#0 size if around 5 MHz and the largest 20 MHz which are reasonable values for the lower frequency bands. It also means that around 5 MHz is the smallest possible carrier bandwidth as the CORET#0 would not fit otherwise.

Patterns 2 and 3 are intended for FR2, although pattern 1 can also be used. The reason for patterns 2 and 3 is to allow for efficient beam sweeping for SSB and system information delivery. By frequency multiplexing the SSB and CORESET#0, the duration in time can be reduced, and hence more rapid beam sweeping can be supported, at the price of requiring a wider minimum carrier bandwidth.

Scheduling of SIB1 on the PDSCH is done using a PDCCH with the SI-RNTI in one of the search spaces using CORESET#0. However, as discussed in Chapter 7, data

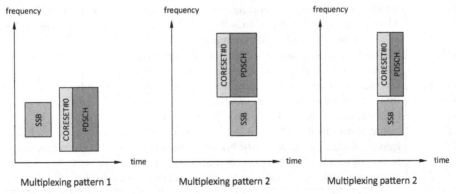

Fig. 16.6 Multiplexing of SSB, CORESET#0, and PDSCH for SIB1.

transmission in NR uses the concept of bandwidth parts. Multiple bandwidth parts can be configured, but at least one is needed in order to receive data. The initial downlink bandwidth part therefore equals the resource blocks covered by CORESET#0. This allows an initial downlink bandwidth part of up to 96 resource blocks, depending on the frequency band and subcarrier spacing. Although additional bandwidth parts can be configured once the device is connected, a single bandwidth part is in many cases sufficient. In such cases, it should preferably cover the full carrier bandwidth. It is therefore possible to signal the initial downlink bandwidth part in SIB1, thereby avoiding being limited by the relatively limited.

The remaining SIBs, not including SIB1, consist of the system information that a device does not need to know before accessing the system. These SIBs can also be periodically broadcast similar to SIB1, in which case SIB1 includes information about exactly when and how the remaining SIBS are being transmitted. Alternatively, these SIBs can be transmitted *on demand*, that is, only transmitted when explicitly requested by a connected device. This implies that the network can avoid periodic broadcast of these SIBs in cells where no device is currently camping, thereby allowing for enhanced network energy performance.

Table 16.3 lists all the currently (in release 17) defined SIBs together with an indication of what kind of information is contained within each SIB.

Table 16.3 System Information Blocks (SIBs) and their content.

SIB	Information
SIB1	General information about the cell including identities of the networks supported by the cell (in case of network sharing) and information about how to access the cell, for example, information related to random access. Also provides information about the scheduling of remaining broadcast SIBs.
SIB2	Information related to cell re-selection in general
SIB3	Information specifically related to intra-frequency cell re-selection
SIB4	Information specifically related to inter-frequency cell re-selection
SIB5	Information specifically related to inter-RAT cell re-selection
SIB6/7	Information related ETWS (Earthquake and Tsunami Warning System)
SIB8	Information related CMAS (Commercial Mobile Alert Service)
SIB9	Information related to GPS time and Coordinated Universal Time (UTC)
SIB10	The names, formally referred to as the HRNNs or *Human Readable Network Names*, of non-public networks (NPNs) announced in SIB 1
SIB11	Information related to idle/inactive mode measurements
SIB12	Information related for NR sidelink communication (see Chapter 26)
SIB13/14	Information related to LTE sidelink communication under NR coverage
SIB15	Information related to disaster roaming
SIB16	Information related to slice-based cell re-selection
SIB17	Information about TRS resources (see Section 8.1.7) for idle/in-active UEs
SIB18	Information related to access to NPNs by means of downloaded credentials
SIB19	Information related to Non-Terrestrial Networks (see Chapter 25)
SIB20–21	Information related to multicast/broadcast services (see Chapter 23)

CHAPTER 17

Random access

In most cases, an NR uplink transmission takes place using a dedicated resource that is assigned by the network/cell for that specific transmission. As a consequence, there is no risk for collision with transmissions from other devices within the cell. This is true for scheduled data transmission on PUSCH as well as for control signaling on PUCCH. It is also true for the transmission of sounding-reference signals SRS.

Note that when we say that there is a dedicated resource assigned for an uplink transmission, this does not necessarily mean that uplink transmissions from different devices take place on non-overlapping time/frequency resources. As an example, PUCCH transmissions from different devices may, in some cases, share the same time/frequency resource with the transmissions instead being separated by means of different sequences. The same is true for SRS transmissions, see Chapter 8. Different uplink transmissions may also be separated in the spatial domain by means of multiple receive antennas. In such case, the spatial separation is typically enabled by using different demodulation reference signals (DM-RS) for the different transmissions, allowing the network to estimate the channel from each device separately without being interfered by DM-RS transmissions by other devices. Based on these channel estimates, the receiver can utilize the multiple receive antennas to separate the different overlapping transmissions. In this case, one is sometimes referring to the different DM-RS as *different DM-RS resources*.

In most cases, the transmission timing of an uplink transmission is also controlled by the network, in a closed-loop manor, to ensure that different uplink transmissions are received within a narrow time window, see Section 15.2.

However, in case of a device making an initial access to the network from the idle/inactive state, there is not yet any connection by which a dedicated resource can be assigned for the initial transmission from the device. Rather, the device must make the initial transmission on an uplink resource that is shared with other devices, that is, a resource on which there may be a collision if multiple devices, by chance, happens to use the resource at the same time.

Furthermore, at initial access there is no way for the network to accurately control the transmission timing of the device based on previously received transmissions. Rather, the device transmission timing can only be determined by the transmitting device itself based on the timing of broadcast signals received from the network, in the NR case the received timing of the SSB. The receive timing of the uplink transmissions will then have an uncertainty of at least two times the propagation time. This has two impacts

5G/5G-Advanced
https://doi.org/10.1016/B978-0-443-13173-8.00012-8

- Except for the case of very small cells, the maximum misalignment between signals received from different devices may exceed the cyclic prefix. The frequency-domain orthogonality between neighbor OFDM subcarriers will then no longer be retained, leading to inter-user interference or the need for extra guardbands.
- Additional guard times may be needed to avoid overlap, and corresponding interference, between received signals transmitted within different time-domain resources.

The *random-access procedure* is specifically designed to handle this situation of collision risk and lack of accurate timing control. The basic NR random-access procedure consists of four steps, see also Fig. 17.1.

- Step 1: Device transmission of a *preamble*, also referred to as the *physical random-access channel* (PRACH). The preamble is specifically designed for relatively low-complexity reception despite a lack of accurate timing control. As described further below, the preamble transmission may be carried out repeatedly with stepwise increased transmit power until a random-access response is received (Step 2).
- Step 2: Network transmission of a *random-access response* (RAR) indicating reception of the preamble and providing a time-alignment command adjusting the transmission timing of the device based on the timing of the received preamble.
- Steps 3/4: Device and network exchange of messages (uplink "*message 3*" and subsequent downlink "*message 4*") with the aim of resolving potential collisions, also referred to as contention resolution, due to simultaneous transmission of the same preamble from multiple devices.

In the more detailed description of the NR random-access procedure given here we will assume the initial-access scenario. However, as will be discussed further in Section 17.5, the NR random-access procedure can also be used in other situations such as

- During handover, when synchronization needs to be established to a new cell.
- To reestablish uplink synchronization to the current cell if synchronization has been lost due to, for example, a too long period without any uplink transmission from the device.

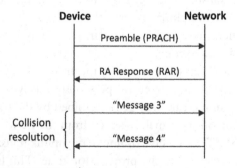

Fig. 17.1 Four-step random-access procedure.

- To request uplink scheduling if no dedicated scheduling-request resource has been assigned to the device.
- To request the transmission of non-broadcast system information as was briefly discussed already in Chapter 16.

Parts of the basic random-access procedure are also used within the *beam-recovery* procedure (see Section 12.3).

Note that, in some of these situations, a device can actually by configured with a dedicated resource, in practice with a dedicated preamble, for the random-access transmission. In this case one is talking about *contention-free random access* (CFRA), in contrast to the *contention-based random access* (CBRA) carried out using a common resource, in practice a preamble, shared with other users devices.

17.1 Step 1 – Preamble transmission

As mentioned, the random-access preamble is also referred to as the *physical random access channel,* indicating that, in contrast to steps 2–4 of the random-access procedure, the preamble transmission (step 1) corresponds to a special physical channel.

17.1.1 RACH configuration and RACH resources

The details of the preamble transmission are given by the *random-access configuration* provided as part of SIB1. The random-access configuration, for example, provides information about the time/frequency resources in which preamble transmission can take place within a cell. It also provides information about what preambles are available within a cell as well as parameters related to the preamble transmit power. The RACH configuration also provides the mapping from SSB indices to RACH occasions, something which is a critical for the initial beam establishment in case of operation in mm-wave spectrum, see also Chapter 12.

Due to the lack of detailed transmission-timing control for the preamble transmission there is an uncertainty in terms of when the preamble is received at the target cell. The range of this uncertainty depends on the maximum propagation delay within the cell. For cell sizes in the order of a few hundred meters, this uncertainty will be in the order of few microseconds. However, for large cells the uncertainty could be in the order of 100 μs or even more.

In general, it is up to the network scheduler to ensure that there are no other transmissions in the uplink resources in which preamble may be received. When doing this, the network needs to take the uncertainty in the preamble reception timing into account. In practice the scheduler needs to provide an extra guard time that captures this uncertainty (see Fig. 17.2). Note that the presence of the guard time is not part of the NR specifications but just a result of scheduling restrictions. Consequently, different guard

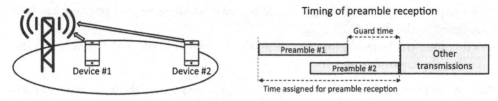

Fig. 17.2 Guard-time needs for preamble transmission.

times can easily be provided to match different uncertainties in the preamble reception timing, for example, due to different cell sizes.

Fig. 17.3 illustrates the structure of the overall RACH resource, that is, the time/frequency resource in which preamble transmission can take place. Within a cell, preamble transmission can take place within a configurable subset of slots (the *RACH slots*) within a specific frame. The set of RACH slots is then repeated every Nth frame where N can range from $N=1$, that is, there are RACH slots in every frame, to $N=16$ (RACH slots in every 16th frame). The number of RACH slots within a frame can range from one to eight depending on the cell RACH configuration.

Furthermore, within the RACH slots there may be multiple frequency-domain *RACH occasions* jointly covering $K \cdot M$ consecutive resource blocks, where M is the size of a frequency-domain RACH occasion, that is, the size of the frequency resource assigned for each preamble measured in number of resource blocks and K is the number of frequency-domain RACH occasions. Thus, up to K preamble transmissions from different devices can be frequency multiplexed within one RACH slot.

The size of a frequency-domain RACH occasion, given by M, depends on the preamble type (*long* vs. *short* preambles, see below). Furthermore, as will also be seen below, the actual number of subcarriers for a given preamble does not fully match an integer number of resource blocks, implying that the number of subcarriers for a frequency-domain

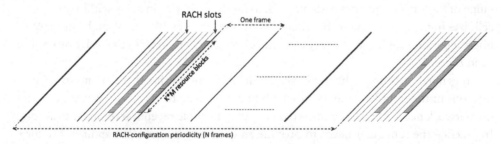

Fig. 17.3 Overall RACH resource consisting of a set of consecutive resource blocks within a set of RACH slots and where the slot pattern repeats every *RACH-configuration period*.

RACH occasion is somewhat larger than the actual number of subcarriers of the preamble transmitted within the RACH occasion.

For a given preamble type, corresponding to a certain preamble bandwidth, the overall available time/frequency RACH resource within a cell can thus be described by:

- A configurable *RACH periodicity* that can range from 10 ms up to 160 ms.
- A configurable set of RACH slots within the RACH period (all within the same frame).
- A configurable frequency-domain RACH resource given by the index of the first resource block in the resource and the number of frequency-multiplexed RACH occasions.

As we will see, depending on the exact set of preambles used in a cell, there may also be multiple RACH occasions in the time domain within a RACH slot.

17.1.2 Basic preamble structure

Fig. 17.4 illustrates the basic structure for generating NR random-access preambles. A preamble is generated based on a length-L *preamble sequence* $p_0, p_1, ..., p_{L-1}$ which is DFT precoded before being applied to a conventional OFDM modulator. The preamble can thus be seen as a DFTS-OFDM signal. It should be noted though that one could equally well see the preamble as a conventional OFDM signal based on a frequency-domain sequence $P_0, P_1, ..., P_{L-1}$ being the discrete Fourier transform of the sequence $p_0, p_1, ..., p_{L-1}$.

The output of the OFDM modulator is then repeated N times after which a cyclic prefix is inserted. For the preamble, the cyclic prefix is thus not inserted per OFDM symbol but only once for the block of N repeated symbols.

Different preamble sequences can be used for the NR preambles. Similar to, for example, uplink SRS, the preamble sequences are based on Zadoff-Chu sequences [23]. As described in Section 8.3.1, for prime-length ZC sequences, which is the case for the sequences used as a basis for the NR preamble sequences, there are $L-1$ different sequences, with each sequence corresponding to a unique *root-sequence index*.

Different preamble sequences can be generated from different Zadoff-Chu sequences corresponding to different root-sequence indices. However, different preamble

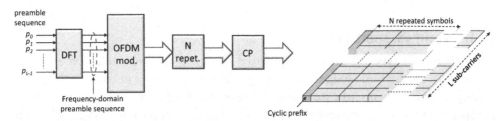

Fig. 17.4 Basic structure for generation of NR random-access preamble.

sequences can also be generated from different cyclic shifts of the same root sequence. As described in Section 8.3.1, such sequences are inherently orthogonal to each other. However, this orthogonality is retained at the receiver side only if the relative cyclic shift between two sequences is larger than any difference in their respective receive timing. Thus, in practice only a subset of the cyclic shifts can be used to generate different preambles, where the number of available shifts depends on the maximum uncertainty in receive timing which, in turn, depends on the cell size. For small cell sizes a relatively large number of cyclic shifts can be used while, larger cells, only a small number of cyclic shifts may be possible to use.

The set of cyclic shifts that can be used within a cell is given by the so-called *zero-correlation-zone* parameter which is part of the cell random-access configuration provided within SIB1. In practice, the zero-correlation-zone parameter points to a specified table where each row corresponds to the set of cyclic shifts available for a given zero-correlation-zone parameter. The name "zero-correlation zone" comes from the fact that the different zero-correlation-zone parameter are associated with cyclic-shift sets with different distances between the cyclic shifts, thus providing larger or smaller "zones" in terms of timing misalignment for which orthogonality (=zero correlation) is retained.

Within a cell there can be up to 64 different preambles available, with each preamble identified by a *preamble index* in the range 0 to 63. The specific preambles available in a cell are given by a root-sequence index provided as part of the cell RACH configuration. The up to 64 different preambles are generated by first using all the possible cyclic shifts, constrained by the zero-correlation zone, of the root sequence defined by the provided root-sequence index. If a sufficient number of preambles cannot be generated, that is, if sufficiently many cyclic shifts are not available with the given zero-correlation zone, additional preambles are generated from cyclic shifts of the next root sequence. This may then proceed with yet another root sequence until the required up to 64 preambles are generated.

As we will see, not all of the up to 64 preambles available in a cell may be used for normal contention-based random access. If so, the remaining preambles can be used for contention-free random access, for example for mobility/handover, see Section 17.5.

17.1.3 Long vs. short preambles

NR defines two types of preambles, referred to as *long preambles* and *short preambles*, respectively. As the name suggests, the two preamble types differ in terms of the length of the preamble sequence (the parameter L). They also differ in the numerology (subcarrier spacing) used for the preamble transmission. The type of preamble is part of the cell random-access configuration, that is, within a cell only one type of preamble can be used for initial access.

17.1.3.1 Long preambles

Long preambles are based on a sequence length $L=839$ and a subcarrier spacing of either 1.25 kHz or 5 kHz. The long preambles thus use a numerology different from any other NR transmissions. The long preambles originate from the preambles used for LTE random-access [26]. Long preambles can only be used for frequency bands below 6 GHz (FR1).

As illustrated in Table 17.1 there are four different formats for the long preamble where each format corresponds to a specific numerology (1.25 kHz or 5 kHz), a specific number of repetitions (the parameter N in Fig. 17.4), and a specific length of the cyclic prefix. The preamble format is part of the cell random-access configuration, that is, each cell is limited to a single preamble format. It could be noted that the two first formats of Table 17.1 are identical to the LTE preamble formats 0 and 3 [14].

In the previous section it was described how the overall RACH resource consists of a set of slots and resource blocks in the time-domain and frequency-domain, respectively. For long preambles, which use a numerology that is different from other NR transmissions, the slot and resource block should be seen from a 15 kHz numerology point-of-view. In the context of long preambles, a slot thus has a length of 1 ms, while a resource-block has a bandwidth of 180 kHz. A long preamble with 1.25 kHz numerology thus occupies six resource blocks in the frequency domain, while a preamble with 5 kHz numerology occupies 24 resource blocks.

It can be observed that preamble format 1 and preamble format 2 in Table 17.1 correspond to a preamble length that exceeds a slot. This may appear to contradict the assumption of preamble transmissions taking place in RACH slots of length 1 ms as discussed in Section 17.1.1. However, the RACH slots only indicate the possible starting positions for preamble transmission. If a preamble transmission extends into a subsequent slot, this only implies that the scheduler needs to ensure that no other transmissions take place within the corresponding frequency-domain resources within that slot.

17.1.3.2 Short preambles

Short preambles are based on a sequence length $L=139$ and use a subcarrier spacing aligned with the normal NR subcarrier spacing. More specifically, short preambles use a subcarrier spacing of:

- 15 kHz or 30 kHz in the case of operation below 6 GHz (FR1).
- 60 kHz or 120 kHz in the case of operation in the higher NR frequency bands (FR2).

Table 17.1 Preamble formats for long preambles.

Format	Numerology (kHz)	Number of repetitions	CP length (µs)	Preamble length (not incl. CP) (µs)
0	1.25	1	≈ 100	800
1	1.25	2	≈ 680	1600
2	1.25	4	≈ 15	3200
3	5	1	≈ 100	800

Table 17.2 Preamble formats for short preambles.

Format	Number of repetitions	CP length (μs)	Preamble length (not incl. CP) (μs)
A1	2	9.4	133
A2	4	18.7	267
A3	6	28.1	400
B1	2	7.0	133
B2	4	11.7	267
B3	6	16.4	400
B4	12	30.5	800
C0	1	40.4	66.7
C2	4	66.7	267

In the case of short preambles, the RACH resource described in Section 17.1.1 is based on the same numerology as the preamble. A short preamble thus always occupies 12 resource blocks in the frequency domain regardless of the preamble numerology.

Table 17.2 lists the preamble formats available for short preambles. The labels for the different preamble formats originate from the 3GPP standardization discussions during which an even larger set of preamble formats were discussed. The table assumes a preamble subcarrier spacing of 15 kHz. For other numerologies, the length of the preamble as well as the length of the cyclic prefix scale correspondingly, that is, with the inverse of the subcarrier spacing.

The short preambles are, in general, shorter than the long preambles and often span only a few OFDM symbols. In most cases it is therefore possible to have multiple preamble transmissions multiplexed in time within a single RACH slot. In other words, for short preambles there may not only be multiple RACH occasions in the frequency domain but also in the time domain within a single RACH slot (see Table 17.3).

It can be noted that Table 17.3 includes columns labeled A1/B1, A2/B2, and A3/B3. These columns correspond to the use of a mix of the "A" and "B" formats of Table 17.2 where the A format is used for all except the last RACH occasion within a RACH slot. Note that the A and B preamble formats are identical except for a somewhat shorter cyclic prefix for the B formats.

Table 17.3 Number of RACH time-domain occasions within a RACH slot for short preambles.

	A1	A2	A3	B1	B4	C0	C2	A1/B1	A2/B2	A3/B3
Number of RACH occasions	6	3	2	7	1	7	2	7	3	2

For the same reason there are no explicit formats B2 and B3 in Table 17.3 as these formats are always used in combination with the corresponding A formats (A2 and A3) according to the above.

17.1.3.3 "Short" preambles for unlicensed spectrum

As will be described in Chapter 20, release 16 extended NR with support for operation in unlicensed spectrum. As part of this, additional preamble types where introduced. These preambles have the same structure as the short preambles discussed above but with a different sequence length L, more specifically sequence lengths $L=571$ and $L=1151$. There are some restrictions on the combinations of preamble lengths and subcarrier spacings that can be used as not all combinations are needed in practice.[a]

The larger sequence lengths imply a corresponding wider preamble bandwidth. The reason for introducing these more wideband preambles is to be able to provide sufficient preamble transmit power despite the limitations in allowed transmit power density (W/Hz) in case of operation in unlicensed spectrum,

17.1.4 Mapping from SSB indices to RACH occasions and preambles

As described in the previous chapter, there are multiple SSB transmissions within an SSB burst set, where each SSB transmission associated with an SSB index signaled within the MIB/PBCH. In practice, the different SSB transmissions, or SSB indices, correspond to different downlink beams within which SSB is transmitted.

A key feature of the NR initial access is the possibility to establish a suitable beam pair already during the initial-access phase and to apply receiver-side analog beam-sweeping for the preamble reception. This is enabled by the mapping from SSB index to RACH occasions and/or preamble. As different SSB time indices in practice correspond to SSB transmissions in different downlink beams, this means that the network, based on the received preamble, will be able to determine the downlink beam in which the corresponding device is located. This beam can then be used as an initial beam for subsequent downlink transmissions to the device.

Furthermore, if the association between SSB time index and RACH occasion is such that a given time-domain RACH occasion corresponds to one specific SSB time index, the network will know when, in time, preamble transmission from devices within a specific downlink beam will take place. Assuming beam correspondence, the network can then focus the uplink receiver beam in the corresponding direction for beam-formed preamble reception. In practice this implies that the receiver beam will be swept over the coverage area synchronized with the corresponding downlink beam sweep for the SSB transmission.

[a] $L=139$ can be used with subcarrier spacings of 15, 30, 60, 120, 480, and 960 kHz; $L=571$ can be used with 30, 120, and 480 kHz; and $L=1151$ can be used with 15 and 120 kHz.

Note that beam-sweeping for preamble transmission is only relevant when analog beam forming is applied at the receiver side. If digital beam-forming is applied, beam-formed preamble reception can be done from multiple directions simultaneously.

To associate a certain SSB time index with a specific random-access occasion and a specific set of preambles, the random-access configuration of the cell specifies the number of SSB time indices per RACH time/frequency occasion. It also specifies the number of preambles per SSB time index.

The number of SSB time indices per RACH time/frequency can be larger than one, indicating that multiple SSB time indices correspond to a single RACH time/frequency occasion. However, it can also be smaller than one, indicating that one single SSB time index corresponds to multiple RACH time/frequency occasions.

In the latter case, each RACH occasion corresponds to a single SSB index, that is the RACH occasion in itself indicates the SSB index. In this case, each SSB index can be mapped to the same set of preambles.

In contrast, in the former case, each RACH occasion corresponds to multiple SSB indices and these different SSB indices are mapped to different sets of preambles.

The mapping from SSB time indices to RACH occasions in the following order:

- First in the frequency domain.
- Then in the time domain within a slot, assuming the preamble format configured for the cell allows for multiple time-domain RACH occasions within a slot (only relevant for short preambles).
- Finally in the time domain between RACH slots.

Fig. 17.5 exemplifies the association between SSB time indices and RACH occasions under the following assumptions:

- Two RACH frequency occasions.
- Three RACH time occasions per RACH slot.
- Each SSB time index associated with four RACH occasions.

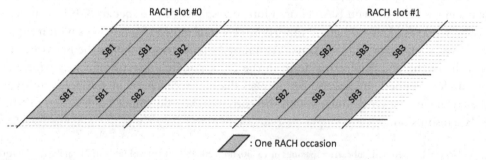

Fig. 17.5 Association between SSB time indices and RACH occasions assuming (example).

17.1.5 Preamble power control and power ramping

As discussed above, preamble transmission will take place with a relatively large uncertainty in the required preamble transmit power. Preamble transmission therefore includes a *power-ramping* mechanism where the preamble may be repeatedly transmitted with a transmit power that is increased between each transmission.

The device selects the initial preamble transmit power based on estimates of the downlink path loss in combination with a target received preamble power configured by the network. The path loss should be estimated based on the received power of the SS block that the device has acquired and from which it has determined the RACH resource to use for the preamble transmission. This is aligned with an assumption that if the preamble transmission is received by means of beam-forming the corresponding SS block is transmitted with a corresponding beam-shape. If no random-access response (see below) is received within a predetermined window, the device can assume that the preamble was not correctly received by the network, most likely due to the fact that the preamble was transmitted with too low power. If this happens, the device repeats the preamble transmission with the preamble transmit power increased by a certain configurable offset. This power ramping continues until a random-access response has been received or until a configurable maximum number of retransmissions has been carried out, alternatively a configurable maximum preamble transmit power has been reached. In the two latter cases, the random-access attempt is declared as a failure.

17.1.6 Preamble transmission in case of NTN

For so-called *non-terrestrial networks* or NTN (see Chapter 25), the propagation time is much too long for it to be handled by a guard time as discussed in Section 17.1.6. Rather, in the case of NTN, the device is assumed to be able to estimate the propagation time and compensate for it at the transmitter side, see Chapter 25 for more details.

17.2 Step 2 – Random-access response

Once a device has transmitted a random-access preamble, it waits for a random-access response, that is, a response from the network that it has properly received the preamble. The random-access response is transmitted as a conventional downlink PDCCH/ PDSCH transmission with the corresponding PDCCH transmitted within the common search space.

The random-access response includes the following:

- The index of the random-access preamble the network detected and for which the response is valid.
- A timing correction calculated by the network based on the preamble receive timing. The device should update the uplink transmission timing according to the correction before further uplink transmissions.

- A scheduling grant, indicating what resource the device should use for the transmission of the subsequent message 3 (see below).
- A temporary identity, the TC-RNTI, used for further communication between the device and the network.

If the network detects multiple random-access attempts (from different devices), the individual response messages can be combined in a single transmission. Therefore, the response message is scheduled on the DL-SCH and indicated on a PDCCH using an identity reserved for random-access response, the RA-RNTI. The use of the RA-RNTI is also necessary as a device may not have a unique identity in the form of a C-RNTI allocated. All devices that have transmitted a preamble monitor the L1/L2 control channels for random-access response within a configurable time window. The timing of the response message is not fixed in the specification in order to be able to respond to many simultaneous accesses. It also provides some flexibility in the base-station implementation. If the device does not detect a random-access response within the time window, the preamble will be retransmitted with higher power according to the preamble power ramping described above.

As long as the devices that performed random access in the same resource used different preambles, no collision will occur and from the downlink signaling it is clear to which device(s) the information is related. However, there is a certain probability of contention – that is, multiple devices using the same random-access preamble at the same time. In this case, multiple devices will react upon the same downlink response message and a collision occurs. Resolving these collisions is part of the subsequent steps, as discussed below.

Upon reception of the random-access response, the device will adjust its uplink transmission timing and continue to the third step. If contention-free random access using a dedicated preamble is used, then this is the last step of the random-access procedure as there is no need to handle contention in this case. Furthermore, the device already has a unique identity allocated in the form of a C-RNTI.

In the case of downlink beam-forming, the random-access response should follow the beam-forming used for the SS block which was acquired during the initial cell search. This is important as the device may use receive-side beam-forming and it needs to know how to direct the receiver beam. By transmitting the random-access response using the same beam as the SS block, the device knows that it can use the same receiver beam as identified during the cell search.

17.3 Step 3/4 – Contention resolution

17.3.1 Message 3

After the second step, the uplink of the device is time synchronized. However, before user data can be transmitted to/from the device, a unique identity within the cell, the

C-RNTI, must be assigned to the device (unless the device already has a C-RNTI assigned). Depending on the device state, there may also be a need for additional message exchange for setting up the connection.

In the third step, the device transmits the necessary messages to the gNB using the UL-SCH resources assigned in the random-access response in the second step.

An important part of the uplink message is the inclusion of a device identity, as this identity is used as part of the contention-resolution mechanism in the fourth step. If the device is already known by the radio-access network, that is, in RRC_CONNECTED or RRC_INACTIVE state, the already assigned C-RNTI is used as the device identity.[b] Otherwise, a core-network device identifier is used and the gNB needs to involve the core network prior to responding to the uplink message in step 4 (see following content).

17.3.2 Message 4

The last step in the random-access procedure consists of a downlink message for contention resolution. Note that, from the second step, multiple devices performing simultaneous random-access attempts using the same preamble sequence in the first step listen to the same response message in the second step and therefore have the same temporary identifier. Hence, the fourth step in the random-access procedure is a contention-resolution step to ensure that a device does not incorrectly use another device's identity. The contention resolution mechanism differs somewhat depending on whether the device already has a valid identity in the form of a C-RNTI or not. Note that the network knows from the uplink message received in step 3 whether the device has a valid C-RNTI or not.

If the device already had a C-RNTI assigned, contention resolution is handled by addressing the device on the PDCCH using the C-RNTI. Upon detection of its C-RNTI on the PDCCH the device will declare the random-access attempt successful and there is no need for contention-resolution-related information on the DL-SCH. Since the C-RNTI is unique to one device, unintended devices will ignore this PDCCH transmission.

If the device does not have a valid C-RNTI, the contention resolution message is addressed using the TC-RNTI and the associated DL-SCH contains the contention-resolution message. The device will compare the identity in the message with the identity transmitted in the third step. Only a device which observes a match between the identity received in the fourth step and the identity transmitted as part of the third step will declare the random-access procedure successful and promote the TC-RNTI from the second step to the C-RNTI. Since uplink synchronization has already been established, hybrid ARQ is applied to the downlink signaling in this step and devices with

---

[b] The device identity is included as a MAC control element on the UL-SCH.

a match between the identity they transmitted in the third step and the message received in the fourth step will transmit a hybrid-ARQ acknowledgment in the uplink.

Devices that do not detect PDCCH transmission with their C-RNTI or do not find a match between the identity received in the fourth step and the respective identity transmitted as part of the third step are considered to have failed the random-access procedure and need to restart the procedure from the first step. No hybrid-ARQ feedback is transmitted from these devices. Furthermore, a device that has not received the downlink message in step 4 within a certain time from the transmission of the uplink message in step 3 will declare the random-access procedure as failed and need to restart from the first step.

17.4 Random access for supplementary uplink

Section 7.7 discussed the concept of supplementary uplink (SUL), that is, that a downlink carrier may be associated with two uplink carriers (the non-SUL carrier and the SUL carrier) where the SUL carrier is typically located in lower-frequency bands thereby providing enhanced uplink coverage.

That a cell is an SUL cell, that is, includes a complementary SUL carrier, is indicated as part of SIB1. Before initially accessing a cell, a device will thus know if the cell to be accessed is an SUL cell or not. If the cell is an SUL cell and the device supports SUL operation for the given band combination, initial random-access may be carried out using either the SUL carrier or the non-SUL uplink carrier. The cell system information provides separate RACH configurations for the SUL carrier and the non-SUL carrier and a device capable of SUL determines what carrier to use for the random-access by comparing the measured RSRP of the selected SS block with a *carrier-selection threshold* also provided as part of the cell system information.

- If the RSRP is above the threshold, random-access is carried out on the non-SUL carrier.
- If the RSRP is below the threshold, random access is carried out on the SUL carrier.

In practice the SUL carrier is thus selected by devices with a (downlink) pathloss to the cell that is larger than a certain value.

The device carrying out a random-access transmission will transmit the random-access message 3 on the same carrier as used for the preamble transmission.

For other scenarios when a device may do a random access, that is, for devices in connected mode, the device can be explicitly configured to use either the SUL carrier or the non-SUL carrier for the uplink random-access transmissions.

17.5 Random access beyond initial access

As already mentioned in the introduction to this chapter, the NR random-access procedure is not only used when a device is initially accessing the network from the idle/inactive state. This section will briefly discuss some other situations when the random-access procedure can be applied.

17.5.1 Random access at handover

When a device in connected state is to make a handover to a new cell it may not yet be sufficiently well synchronized to that cell. This is especially the case in an asynchronous network deployment where the cells are not tightly synchronized to each other. In such a case, the device accesses the new cell by first carrying out a random access to establish synchronization and an RRC connection to the cell. In this case of an already connected device doing a random access, the device may be assigned a dedicated preamble index, corresponding to a specific preamble sequence and/or a specific sequence shift, to use for the random access to the new cell. The random access to the new cell will then be contention-free, avoiding the risk for collision when accessing the new cell.

Note that if the new cell consists of multiple beams with different SSBs, the actual preamble to use, as well as the exact PRACH (time and frequency) occasion depends on the preamble index in combination with the selected SSB index in a similar way as for initial random access as described in Section 17.1.4.

17.5.2 Random-access for SI request

In the previous chapter it was described how system information was provided to a device in form of system information (SI) messages. It was also described how an SI message could be broadcast and thus always be available also for devices in idle/inactive state. Alternatively, the SI message is not broadcast and a device in idle/inactive state has to explicitly request its transmission by means of an *SI request*.

One way for a device to request the transmission of an SI message is to first enter connected state by means of a conventional random access and then explicitly request the SI message by conventional RRC signaling.[c] However, NR also allows for devices in idle/inactive state to use the random-access procedure to directly request the transmission of SI messages without the device having to enter connected state.

As described in the previous chapter, SIB1 includes information about the mapping of the remaining System Information Blocks to SI messages, information about the transmission periodicity of each SI message and whether or not the SI message is broadcast. If a certain SI message is not broadcast, SIB1 also includes, in form of a *request configuration*, a random-access configuration and a preamble index. By carrying out a random-access with the given random-access configuration and the preamble index, the device is directly indicating a request for SI transmission.

[c] Note that SIB1, which is the only SIB needed before accessing the system, is always broadcast.

SIB1 may either provide a single request configuration valid for all non-broadcast SI messages or separate request configurations, with different preamble indices, for each non-broadcast SI message. In the former case, the network will, upon detection of the SI request, transmit all non-broadcast system information while, in the latter case, only the requested system-information message will be transmitted. The choice between these two alternatives depends on the tradeoff between, on one hand, the overhead in terms of RACH resources for different SI requests and, on the other hand, the overhead of having to transmit all SI messages although, perhaps, only one specific SI message was actually needed by the requesting device.

Once again, in case of a cell with multiple beam-formed SSBs, the preamble index (or indices) associated with system-information request will be combined with the detected SSB to determine the exact preamble and exact PRACH occasions, in a similar way as for initial random access as described in Section 17.1.4.

17.5.3 Re-establishing synchronization by means of PDCCH order

If a device in connected state has been inactive, that is, not carried out any uplink transmission for a certain time, the synchronization to the network may be lost. If the network detects such a loss of uplink synchronization it may trigger a random access from the device by means of a so-called *PDCCH order*.

The PDCCH order is provided using DCI format 1_0 with the frequency-domain assignment set to all-ones, indicating that the DCI is not providing a downlink scheduling assignment but rather a PDCCH order for random access. The DCI includes a dedicated preamble index which the device should use for a contention-free random-access. It also includes an SSB index indicating the SSB that should be used to determine the RACH occasion for the random-access transmission.

17.6 Two-step RACH

The NR random-access procedure discussed until now consists of four steps:

- Step 1: Uplink preamble/PRACH transmission.
- Step 2: Downlink random-access-response (on PDSCH).
- Step 3/4: Uplink message-3 transmission (on PUSCH) with a corresponding network response (message-4) to resolve collisions (on PDSCH).

This four-step random-access procedure, also referred to as *four-step RACH*, was introduced as part of the first NR release, that is, 3GPP release 15.

A complementary *two-step* random-access procedure, or *two-step RACH*, was introduced in release 16.[d] Somewhat simplified, the two-step random-access procedure

---

[d] In the specifications, the four-step RACH and two-step RACH are referred to as *Type-1 random access* and *Type-2 random access* respectively.

combines step 1 and step 3 into a single *Step A*. It also combines step 2 and 4 into a single *Step B*. More specifically, the two-step random-access procedure consists of

- Step A: Uplink preamble/PRACH transmission together with a PUSCH data transmission, jointly referred to as *message A*.
- Step B: A single downlink transmission (referred to as *message B*) indicating reception of message A, providing time alignment and resolving any collision that may have occurred in Step A. Alternatively, as we will see in Section 17.6.2, the message B may include an *fallback indication* for fallback to four-stage RACH.

Similar to Step 1 of the four-step random-access procedure, Step A, that is, transmission of preamble together with a message-A PUSCH, may be carried out repeatedly with stepwise increased transmit powers until a response is received (Step B).

The main benefit of two-step RACH, compared to four-step RACH, is a shorter random-access procedure enabling faster access. It should be noted though that this benefit is quickly diminishing if message-A has to be repeated several times before being detected by the network.

There are also additional benefits with two-step RACH in case of operation in unlicensed spectrum (see Chapter 20) as the simultaneous transmission of preamble and PUSCH, as well as the combination of steps 2 and 4 into a single step B, may imply a reduced number of LBT operations with a corresponding reduction in overhead and delay. The two-step RACH procedure is also used in conjunction with small-data transmission, see Chapter 22, as a way to quickly transfer small amounts of data while the device still remains in the inactive state.

The main drawback of two-step RACH is the extra overhead of transmitting message-A PUSCH (corresponding to message 3 of four-step RACH) for each preamble transmission, that is potentially multiple times, until a random-access response is received in Step B. Also, the message-A PUSCH transmission is, in itself, somewhat less efficient compared to message-3 PUSCH due to the lack of tight time alignment for the message-A PUSCH transmission.

17.6.1 Two-step RACH – Step A

As described, Step A of the two-step random-access procedure consists of a preamble transmission in combination with a PUSCH transmission.

17.6.1.1 Preamble transmission

The preamble transmission of two-step RACH is essentially identical to the preamble transmission of four-step RACH.

- The same type of preambles is used as for four-step RACH.
- The principle of associating SSB indices with RACH occasions and preamble indices is the same as for four-step RACH.

The RACH occasions for two-step RACH may be configured to be the same or different from the RACH occasions for four-step RACH. If the same random-access configuration is used for four-step RACH and two-step RACH, implying the same set of RACH occasions, the preambles for two-step RACH are taken from the set of contention-free preambles associated with each SSB index (preambles which then becomes contention-based preambles for two-step RACH). In this way, the network will be able to distinguish between contention-based preamble transmissions associated with two-step RACH and four-step RACH respectively. This is important as the network response to a received preamble transmission differs between two-step RACH and four-step RACH. If the RACH occasions for two-step RACH are different from those for four-step RACH, the same set of preambles are used for four-step and two-step RACH.

17.6.1.2 PUSCH transmission

In many respects, the message-A PUSCH transmission is like any other PUSCH transmission. However, there are some differences.

- A message-A PUSCH transmission is not scheduled in the sense that the device is not granted a dedicated resource for the transmission. Rather, as we will see below, the resource to use for a message-A PUSCH transmission is given by the combination of RACH occasion and preamble index selected for the corresponding preamble transmission.
- When doing a message-A PUSCH transmission, there is not yet any closed-loop uplink timing control. Consequently, the message-A PUSCH transmissions may arrive with a relatively large timing misalignment relative to other uplink transmissions. Extra guard times and guard bands may therefore be needed to handle intra-cell interference to/from message-A PUSCH transmissions.

Similar to PRACH transmissions taking place in RACH occasions, message-A PUSCH transmissions take place in *PUSCH occasions*, see Fig. 17.6.

In the time domain, the size of each PUSCH occasion may range from a minimum of one symbol to a maximum of 14 symbols (three symbols assumed in Fig. 17.6). In the

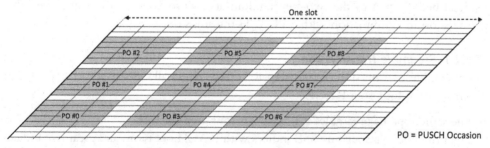

Fig. 17.6 PUSCH occasions (POs) within a slot.

frequency domain, the size of a PUSCH occasion may range from a minimum of one resource block to a maximum of 32 resource blocks (five resource blocks assumed in Fig. 17.6).

Within a slot, up to six PUSCH occasions can be multiplexed in the time domain, with the possibility for a guard space between consecutive PUSCH occasions (three time-multiplexed PUSCH occasions assumed in Fig. 17.6).[e] The length of the guard space may range from zero symbols, that is, no guard space, up to a maximum of three symbols (one-symbol guard space assumed in Fig. 17.6). The purpose of the guard space is to avoid overlap, at the receiver side, between message-A PUSCH transmissions in consecutive PUSCH occasions, taking into account the lack of tight control of the message-A PUSCH transmission timing.

PUSCH occasions can also be multiplexed in the frequency domain, with the possibility to configure a one-resource-block guard band between frequency-multiplexed PUSCH occasions (guard band assumed to be configured in Fig. 17.6). Similar to the time-domain guard symbols, the guard band may be needed to handle interference due to the possible misalignment in terms of reception timing between message-A PUSCH transmissions from different devices. If this misalignment exceeds the cyclic prefix, frequency-domain orthogonality will not be retained, and a guard band may be needed to avoid or at least reduce the inter-PUSCH interference.

In addition to separation in the time/frequency domain by transmission in different PUSCH occasions, message-A PUSCH transmissions can also be separated in the spatial domain. This is enabled by the possibility to use different DM-RS ports/sequences for the different message-A PUSCH transmissions. In addition to PUSCH occasions, the concept of a *PUSCH resource unit* (PRU), defined as a combination of a specific PUSCH occasion and a specific DM-RS port/sequence, was used during the 3GPP work on two-step RACH. The term PRU was eventually not used in the final specifications but replaced by the more complex term "a PUSCH occasion with a DM-RS resource". To simplify the description, we will here use the term PRU though.

17.6.1.3 Mapping from PRACH slots to PUSCH resources

As described in Section 17.1.1, PRACH/preamble transmissions take place in RACH slots where, within each RACH slot, there are typically multiple random-access occasions in the time and frequency domain. Although earlier described in the context of four-step RACH, this is equally valid for two-step RACH.

For two-step RACH, each RACH slot is associated with a set of message-A PUSCH occasions with associated DM-RS ports/sequences, herein referred to as a *PRU set*. The

---

[e] The maximum number of time-multiplexed PUSCH occasions within a slot will obviously also be limited by the duration, in number of symbols, of the message-A PUSCH, as well as the size of the configured guard space.

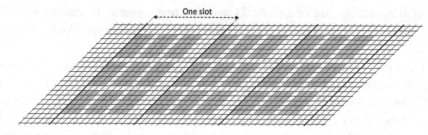

Fig. 17.7 A PRU set spanning three slots with nine PUSCH occasions within each slot.

structure of such a PRU set is illustrated in Fig. 17.7. The PRU set can stretch over up to four consecutive slots (three slots assumed in Fig. 17.7) where, as described above, each slot may contain multiple PUSCH occasions in both the time and frequency domain. Furthermore, as also described above, each PUSCH occasion corresponds to N_{DMRS} PRUs where N_{DMRS} is the number of available DM-RS ports/sequences.

A PRU set and its associated PRUs and PUSCH occasions is thus characterized by

- The number of slots within the PRU set (up to 4).
- The number of PUSCH occasions in the time domain within one slot.
- The length (in number of symbols) of each PUSCH occasion.
- The number of guard symbols between each PUSCH occasion within a slot.
- The number of PUSCH occasions in the frequency domain.
- The bandwidth (in number of resource blocks) of each PUSCH occasion.
- The frequency-domain location of the first PUSCH occasion.
- Whether or not frequency multiplexed PUSCH occasions are separated by a one-resource-block guard band.
- The number of different DM-RS ports/sequences, that is, the number of PRUs per PUSCH occasion.

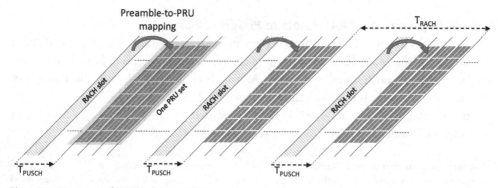

Fig. 17.8 Mapping of RACH slots to PRU sets.

Each RACH slot, that is, each slot in which there are RACH occasions for two-step RACH, corresponds to one specific PRU set located a configurable time offset T_{PUSCH} from the RACH slot as illustrated in Fig. 17.8. Within that RACH slot, each RACH-occasion/preamble combination maps to one specific PRU within the corresponding PRU set.

In other words, once a device has selected the RACH occasion and exact preamble to use for the preamble transmission, it knows what PRU to use for the corresponding message-A PUSCH transmission. Likewise, once the network has detected a preamble within a RACH occasion, it knows in what PRU the corresponding message-A PUSCH transmission is to be received.

If the number of RACH-occasion/preamble combinations within a RACH slot exceeds the number of available PRUs, multiple RACH-occasion/preamble combinations may map to the same PRU. In that case, two devices could, in principle, carry out two-step RACH using different RACH-occasion/preamble combinations but "collide" in the corresponding message-A PUSCH transmission. At least in principle, the PUSCH transmissions could still be detectable but this would require spatial separation and that channel estimation for the PUSCH demodulation is not based on the DM-RS (which would be the same for the two transmissions as they are using the same PRU) but from the received preambles.

17.6.2 Two-step RACH – Step B

Step B of the two-step random-access procedure is, that is, the message-B transmission is a conventional PDCCH/PDSCH transmission with the PDCCH encoded with a new MsgB-RNTI different from the RA-RNTI used for the random-access response for four-step RACH (Section 17.2). Note that, similar to the four-step-RACH random-access response, multiple 2-step-RACH random-access responses to different devices can be provided within the same message-B PDSCH transmission.

There are two alternatives for the two-step-RACH random-access response, depending on whether or not the network is able to detect and decode the message-A PUSCH transmission.

If the network is able to decode the message-A PUSCH transmission, it provides a *Success RAR* including the following information:

- A timing-adjustment (TA) command [12 bits].
- A C-RNTI [16 bits].
- A contention resolution identity [48 bits].

There is also a possibility to include an RRC signaling message, for example for connection set up, within the message-B PDSCH. However, within a message-B PDSCH, there may only be one such RRC signaling message, that is, it is not possible to multiplex RRC signaling messages to multiple devices within the same message-B PDSCH.

If the network detects the preamble transmission but is not able to correctly decode the message-A PUSCH, it may instead provide a *Fallback RAR*. The Fallback RAR contains the same information as the random-access response of four-step RACH and indicates that the device should continue the random-access procedure as a four-step RACH, that is, with the uplink transmission of a message 3 using the scheduling grant included with the Fallback RAR.

17.6.3 Selection between two-step and four-step RACH

In order to support legacy (pre-release-16) devices, there must always be a possibility for four-step RACH, at least for initial access, within a cell. If there is also a possibility for two-step RACH, there must be a way for a device capable of two-step RACH to select between using four-step RACH or two-step RACH.

The selection between two-step RACH and four-step RACH could be based on the device proximity to the cell site in the sense that the device should use two-step RACH if the received signals strength (RSRP) of the cell exceeds a certain configurable value. Otherwise, four-step RACH should be used.

The network could also configure a maximum number of message A transmissions for two-step RACH. If the maximum number of transmissions is exceeded without any random-access response being received, the device should switch to four-step RACH.

CHAPTER 18

LTE/NR interworking and coexistence

The initial deployment of a new generation of mobile-communication technology typically takes place in areas with high traffic density and with high demands for new service capabilities. This is then followed by a gradual further build-out that can be more or less rapid depending on the operator strategy. During this subsequent gradual deployment, ubiquitous coverage to the operator network will be provided by a mix of new and legacy technology, with devices continuously moving in and out of areas covered by the new technology. Seamless handover between new and legacy technology has therefore been a key requirement at least since the introduction of the first 3G networks.

Furthermore, even in areas where a new technology has been deployed, earlier generations must typically be retained and operated in parallel for a relatively long time in order to ensure continued service for legacy devices not supporting the new technology. The majority of users will migrate to new devices supporting the latest technology within a few years. However, a limited amount of legacy devices may remain for a long time. This becomes even more the case with an increasing number of mobile devices not being directly used by persons but rather being an integrated part of other equipment, such as parking meters, card readers, surveillance cameras, etc. Such equipment may have a life time of more than 10 years and will be expected to remain connectable during this life time. This is actually one important reason why many second-generation GSM networks are still in operation even though both 3G and 4G, and even 5G, networks have subsequently been deployed.

However, the interworking between NR and LTE goes further than just enabling smooth handover between the two technologies and allowing for their parallel deployment.

- NR allows for *dual-connectivity* with LTE, implying that devices may have simultaneous connectivity to both LTE and NR. As already mentioned in Chapter 5, the first release of NR actually relies on such dual-connectivity, with LTE providing the control plane and NR only providing additional user-plane capacity;
- NR can be deployed in the same spectrum as LTE in such a way that the overall spectrum capacity can be dynamically shared between the two technologies. Such *spectrum coexistence* allows for a more smooth introduction of NR in spectrum already occupied by LTE.

18.1 LTE/NR dual-connectivity

The basic principle of LTE/NR dual-connectivity is the same as LTE dual-connectivity [26], see also Fig. 18.1,

- A device has simultaneous connectivity to multiple nodes within the radio-access network (eNB in the case of LTE, gNB in the case of NR);
- There is one *master node* (in the general case either an eNB or a gNB) responsible for the radio-access control plane. In other words, on the network side the signaling radio bearer terminates at the master node which then also handles all RRC-based configuration of the device;
- There is one, or in the general case multiple, *secondary node(s)* (eNB or gNB) that provides additional user-plane links for the device.

18.1.1 Deployment scenarios

In the case of LTE dual-connectivity, the multiple nodes to which a device has simultaneous connectivity are typically geographically separated. The device may, for example, have simultaneous connectivity to a small-cell layer and an overlaid macro layer.

The same scenario, that is, simultaneous connectivity to a small-cell layer and an overlaid macro layer, is a highly relevant scenario also for LTE/NR dual-connectivity. Especially, NR in higher frequency bands may be deployed as a small-cell layer under an existing macro layer based on LTE (see Fig. 18.2). The LTE macro layer would then provide the master nodes, ensuring that the control plane is retained even if the connectivity to the high-frequency small-cell layer is temporarily lost. In this case, the NR layer provides very high capacity and very high data rates, while dual-connectivity to the lower-frequency LTE-based macro layer provides additional robustness to the inherently less robust high-frequency small-cell layer. Note that this is essentially the same scenario as the LTE dual-connectivity scenario described above, except for the use of NR instead of LTE in the small-cell layer.

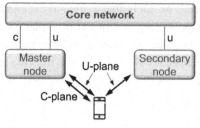

Fig. 18.1 Basic principle of dual connectivity.

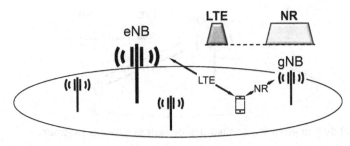

Fig. 18.2 LTE/NR dual-connectivity in a multi-layer scenario.

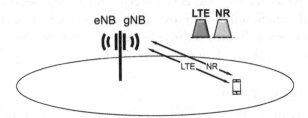

Fig. 18.3 LTE/NR dual-connectivity, co-sited deployment.

However, LTE/NR dual-connectivity is also relevant in the case of co-sited LTE and NR network nodes (Fig. 18.3).[a] As an example, for initial NR deployment an operator may want to reuse an already deployed LTE site grid also for NR to avoid the cost of deploying additional sites. In this scenario, dual-connectivity enables higher end-user data rates by allowing for aggregation of the throughput of the NR and LTE carriers. In the case of a single radio-access technology, such aggregation between carriers transmitted from the same node would be more efficiently realized by means of *carrier aggregation* (see Section 7.6). However, NR does not support carrier aggregation with LTE and thus dual-connectivity is needed to support aggregation of the LTE and NR throughput.

Co-sited deployments are especially relevant when NR is operating in lower-frequency spectrum, that is, in the same or similar spectrum as LTE. However, co-sited deployments can also be used when the two technologies are operating in very different spectra, including the case when NR is operating in mm-wave bands (Fig. 18.4). In this case, NR may not be able to provide coverage over the entire cell area. However, the NR part of the network could still capture a large part of the overall traffic, thereby allowing for the LTE part to focus on providing service to devices in poor-coverage locations.

---

[a] Note that there would in this case still be two different logical nodes (an eNB and a gNB) although these could very well be implemented in the same physical hardware.

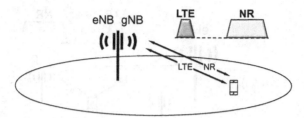

Fig. 18.4 LTE/NR dual-connectivity, co-sited deployment in different spectrum.

In the scenario of Fig. 18.4, the NR carrier would typically have much wider bandwidth compared to LTE. As long as there is coverage, the NR carrier would therefore, in most cases, provide significantly higher data rates compared to LTE, making throughout aggregation less important. Rather, the main benefit of dual-connectivity in this scenario would, once again, be enhanced robustness for the higher-frequency deployment.

18.1.2 Architecture options

Due to the presence of two different radio-access technologies (LTE and NR) as well as the availability of a new 5G core network as an alternative to the legacy 4G core network (EPC), there are several different alternatives, or *options*, for the architecture of LTE/NR dual-connectivity (see Fig. 18.5). The labeling of the different options in Fig. 18.5 originates from early 3GPP discussions on possible NR architecture options where a number of different alternatives were on the table, a subset of which was eventually agreed to be supported (see Chapter 6 for some additional, non-dual-connectivity, options).

18.1.3 Single-TX operation

In the case of dual-connectivity between LTE and NR there will be multiple uplink carriers (at least one LTE uplink carrier and one NR uplink carrier) transmitted from the same device. Due to non-linearities in the RF circuitry, simultaneous transmission on two carriers will create intermodulation products at the transmitter output. Depending on the specific carrier frequencies of the transmitted signals, some of these intermodulation products may end up within the device receiver band causing "self-interference", also referred to as *intermodulation distortion* (IMD). The IMD will add to the receiver noise and lead to a degradation of the receiver sensitivity. The impact from IMD can be

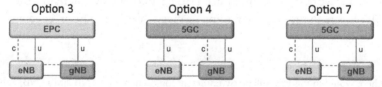

Fig. 18.5 Different architecture options for LTE/NR dual-connectivity.

reduced by imposing tighter linearity requirements on the device. However, this will have a corresponding negative impact on device cost and energy consumption.

To reduce the impact of IMD without imposing very tight RF requirements on all devices, NR allows for *single-TX* dual-connectivity for "difficult band combinations". In this context, difficult band combinations correspond to specifically identified combinations of LTE and NR frequency bands for which lower-order intermodulation products between simultaneously transmitted LTE and NR uplink carriers may fall into a corresponding downlink band. Single-TX operation implies that there will not be simultaneous transmission on the LTE and NR uplink carriers within a device even though the device is operating in LTE/NR dual-connectivity.

It is the task of the LTE and NR schedulers to jointly prevent simultaneous transmission on the LTE and NR uplink carriers in the case of single-TX operation. This requires coordination between the schedulers, that is between an eNB and a gNB. The 3GPP specifications include explicit support for the interchange of standardized inter-node messages for this purpose.

Single TX operation inherently leads to time multiplexing between the LTE and NR uplink transmissions within a device, with none of the uplinks being continuously available. However, it is still desirable to be able to retain full utilization of the corresponding downlink carriers.

For NR, with its high degree of scheduling and hybrid-ARQ flexibility, this can easily be achieved with no additional impact on the NR specifications. For the LTE part of the connection the situation is somewhat more elaborate though. LTE FDD is based on synchronous HARQ, where uplink HARQ feedback is to be transmitted a specified number of subframes after the reception of the corresponding downlink transmission. With a single-TX constraint, not all uplink subframes will be available for transmission of HARQ feedback, potentially restricting the subframes in which downlink transmission can take place.

However, the same situation may already occur within LTE itself, more specifically in the case of FDD/TDD carrier aggregation with the TDD carrier being the primary cell [28]. In this case, the TDD carrier, which is inherently not continuously available for uplink transmission, carries uplink HARQ feedback corresponding to downlink transmissions on the FDD carrier. To handle this situation, LTE release 13 introduced so-called DL/UL reference configurations [26] allowing for a TDD-like timing relation, for example for uplink feedback, for an FDD carrier. The same functionality can be used to support continuous LTE downlink transmission in the case of LTE/NR dual-connectivity constrained by single-TX operation.

In the LTE FDD/TDD carrier-aggregations scenario, the uplink constraints are due to cell-level downlink/uplink configurations. On the other hand, in the case of single-TX dual-connectivity the constraints are due to the need to avoid simultaneous transmission on the LTE and NR uplink carriers, but without any tight interdependency between different devices. The set of unavailable uplink subframes may thus not need to be the same for different devices. To enable a more even load on the LTE uplink, the DL/UL reference configurations in the case of single-TX operation can therefore be shifted in time on a per-device basis.

18.2 LTE/NR coexistence

The introduction of earlier generations of mobile communication has always been associated with the introduction of a new spectrum in which the new technology can be deployed. This is the case also for NR, for which the support for operation in mm-wave bands opens up for the use of a spectrum range never before applied to mobile communication.

Even taking into account the use of antenna configurations with a large number of antenna elements enabling extensive beam-forming, operation in such high-frequency spectrum is inherently disadvantageous in terms of coverage though. Rather, to provide truly wide-area NR coverage, lower-frequency spectrum must be used.

However, a large fraction of lower-frequency spectrum is already occupied by current technologies, primarily LTE. In many cases NR deployments in lower-frequency spectrum will therefore need to take place in spectrum already used by LTE.

The most straightforward way to deploy NR in a spectrum already used by LTE is static frequency-domain sharing, where part of the LTE spectrum is migrated to NR (see Fig. 18.6).

There are two drawbacks with this approach though.

At least at an initial stage, the main part of the traffic will still be via LTE. At the same time, the static frequency-domain sharing reduces the spectrum available for LTE, making it more difficult to satisfy the traffic demands.

Furthermore, static frequency-domain sharing will lead to less bandwidth being available for each technology, leading to a reduced peak data rate per carrier. The possible use of LTE/NR dual-connectivity may compensate for this for new devices capable of such operation. However, at least for legacy LTE devices there will be a direct impact on the achievable data rates.

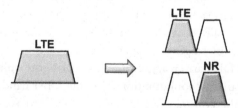

Fig. 18.6 Migration of LTE spectrum to NR.

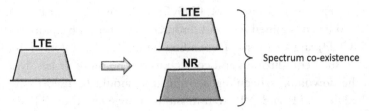

Fig. 18.7 LTE/NR spectrum coexistence.

A more attractive solution is to have NR and LTE dynamically share the same spectrum as illustrated in Fig. 18.7. Such spectrum coexistence will retain the full bandwidth and corresponding peak data rates for each technology. Furthermore, the overall spectrum capacity could be dynamically assigned to match the traffic conditions on each technology.

The fundamental tool to enable such LTE/NR spectrum coexistence is the dynamic scheduling of both LTE and NR. However, there are several other NR features that play a role in the overall support for LTE/NR spectrum coexistence:

- The availability of the LTE–compatible 15 kHz NR numerology that allows for LTE and NR to operate on a common time/frequency grid;
- The general NR forward-compatibility design principles listed in Section 5.1.3. This also includes the possibility to define reserved resources based on bitmaps or LTE carrier configurations as described in Section 9.10;
- A possibility for NR PDSCH mapping to avoid resource elements corresponding to LTE cell-specific reference signals (see further details below).

As already mentioned in Section 5.1.11 there are two main scenarios for LTE/NR coexistence (see also Fig. 18.8):

- Coexistence in both downlink and uplink;
- Uplink–only coexistence.

A typical use case for uplink-only coexistence is the deployment of a supplementary uplink carrier (see Section 7.7).

In general, coexistence in the uplink direction is more straightforward compared to the downlink direction and can, to a large extent, be supported by means of scheduling coordination/constraints. NR and LTE uplink scheduling should be coordinated to

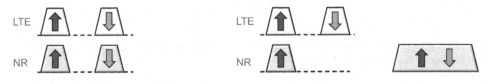

Downlink and uplink co-existence   Uplink-only co-existence

Fig. 18.8 Downlink/uplink coexistence vs. uplink-only coexistence.

avoid collision between LTE and NR PUSCH transmissions. Furthermore, the NR scheduler should be constrained to avoid resources used for LTE uplink layer 1 control signaling (PUCCH) and vice versa. Depending on the level of interaction between the eNB and gNB, such coordination and constraints can be more or less dynamic.

Also for the downlink, scheduling coordination should be used to avoid collision between scheduled LTE and NR transmissions. However, the LTE downlink also includes several non-scheduled "always-on" signals that cannot be readily scheduled around. This includes (see [26] for details):

- The LTE PSS and SSS, which are transmitted over two OFDM symbols and six resource blocks in the frequency domain once every fifth subframe;
- The LTE PBCH which is transmitted over four OFDM symbols and six resource blocks in the frequency domain once every frame (10 subframes);
- The LTE CRS which is transmitted regularly in the frequency domain and in four or six symbols in every subframe depending on the number of CRS antenna ports.[b]

Rather than being avoided by means of scheduling, the concept or reserved resources (see Section 9.10) can be used to rate match the NR PDSCH around these signals.

Rate matching around the LTE PSS/SSS can be done by defining reserved resources according to bitmaps as described in Section 9.10. More specifically a single reserved resource given by a {bitmap-1, bitmap-2, bitmap-3} triplet could be defined as follows (see also Fig. 18.9):

- A bitmap-1 of a length equal to the number of NR resource blocks in the frequency domain, indicating the six resource blocks within which LTE PSS and SSS are transmitted;
- A bitmap-2 of length 14 (one slot), indicating the two OFDM symbols within which the PSS and SSS are transmitted within an LTE subframe;

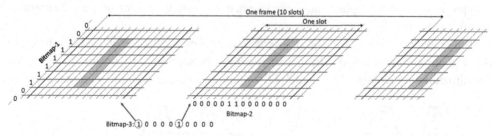

Fig. 18.9 Configuration of reserved resource to enable PDSCH rate matching around LTE PSS/SS. Note that the figure assumes 15 kHz NR numerology.

[b] Only one or two symbols in case of so-called MBSFN subframes.

- A bitmap-3 of length 10 indicating the two subframes within which the PSS and SSS are transmitted within a 10 ms frame.

This assumes a 15 kHz NR numerology. Note though that the use of reserved resources based on bitmaps is not limited to 15 kHz numerology and, in principle, a similar approach to rate match around LTE PSS and SSS could be used also with, for example, a 30 kHz NR numerology.

The same approach can be used to rate match around the LTE PBCH with the only difference that bitmap-2 would, in this case, indicate the four symbols within which PBCH is transmitted, while bitmap-3 would indicate a single subframe.

Regarding the LTE CRS, the NR specification includes explicit support for PDSCH rate matching around resource elements corresponding to CRS of an overlaid LTE carrier. In order to be able to properly receive such a rate-matched PDSCH, the device is configured with the following information:

- The LTE carrier bandwidth and frequency domain location, to allow for LTE/NR coexistence even though the LTE carrier may have a different bandwidth and a different center-carrier location, compared to the NR carrier;
- The LTE MBSFN subframe configuration, as this will influence the set of OFDM symbols in which CRS transmission takes place within a given LTE subframe;
- The number of LTE CRS antenna ports as this will impact the set of OFDM symbols on which CRS transmission takes place as well as the number of CRS resource elements per resource block in the frequency domain;
- The LTE CRS shift, that is, the exact frequency-domain position of the LTE CRS.

In release 16, NR rate matching around LTE CRS transmissions was extended to support multiple LTE CRS patterns which is useful in case of carrier aggregation. Note though that rate matching around LTE CRS is only possible for the 15 kHz NR numerology.

CHAPTER 19

Interference handling in TDD networks

NR already from the start provides flexibility in terms of the duplexing schemes supported – FDD for paired spectrum and TDD for unpaired spectrum. The two duplexing schemes have different properties in terms of the interference scenarios and the handling thereof. In contrast to an FDD network, where uplink and downlink uses separate frequencies and therefore are fairly isolated, a TDD network uses the same frequency for uplink and downlink transmission and separates the two in the time domain. This can result in interference scenarios not present in an FDD network. Downlink-to-uplink interference, illustrated to the left in Fig. 19.1, refers to a situation where the downlink transmission in one cell impact the uplink reception in another cell, and uplink-to-downlink interference, illustrated to the right in Fig. 19.1, refers to uplink transmissions from one terminal interfering with downlink reception in a neighboring terminal. These two interference scenarios need to be handled and the solutions may be different in wide-area and small-cell deployments.

In a wide-area macro-type deployment, the base station antennas are often located above rooftop for coverage reasons – that is, relatively far above the ground compared to the devices. This can result in (close to) line-of-site propagation between the cell sites. Coupled with the relatively large difference in transmission power between uplink and downlink in these types of networks, high-power downlink transmissions from one cell site would significantly impact the ability to receive a weak uplink signal in a neighboring cell. The classical way of handling this is to (semi-)statically split the resources between uplink and downlink in the same way across all the cells in the network. In particular, uplink reception in one cell never overlaps in time with downlink transmission in a neighboring cell. This can be achieved by semi-statically configuring uplink and downlink resources in all terminals as described in Chapter 7. The set of slots (or, in general, time-domain resources) allocated for a certain transmission direction, uplink or downlink, is identical across the whole networks and can be seen as a simple form of inter-cell coordination, albeit on a semi-static basis.

In a small-cell network, where uplink and downlink transmissions use similar power levels and the antennas are located indoors or below rooftop, the downlink-to-uplink interference may be less of an issue given the lower transmission power and the larger isolation between the sites. Uplink-to-downlink interference between two closely located devices, illustrated in the right part of Fig. 19.1, can sometimes be an issue but also in this case a semi-static split between uplink and downlink across all cells helps.

Fig. 19.1 Interference scenarios in TDD networks.

In some scenarios it might even be possible to use dynamic TDD where downlink transmission in one cell simultaneously with uplink reception in another cell is possible as neighboring cells, depending on the deployment details, can be fairly isolated. In release18, 3GPP is studying the feasibility of various duplex enhancements, including dynamic TDD and subband full duplex [108].

The interference scenarios outlined here are not unique to NR and can be handled by proper network implementation and deployment. Nevertheless, release 16 introduces enhancements to more efficiently handle TDD-specific interference scenarios,

- *remote interference management* (RIM), addressing downlink-to-uplink interference in wide-area large-cell networks, and
- *cross-link interference* (CLI) mitigation, addressing uplink-to-downlink interference handling in small-cell deployments using dynamic TDD.

These two enhancements will be described in more detail in the following sections.

19.1 Remote interference management

Remote interference management refers to a set of tools to handle BS-to-BS interference from very distant base stations in a wide-area TDD network. As described in the introduction, the classical way of handling this in macro networks is to (semi-)statically split the resources between uplink and downlink in the same way across all the cells in the network to ensure that downlink and uplink never overlaps in time between neighboring cells. A sufficiently large guard period, covering the propagation delays from neighboring cells, is configured. This is sufficient most of the time. However, in certain weather conditions, atmospheric ducts may be formed [85], efficiently serving as a waveguide between base stations located very far apart. The ducts can be a couple of hundred meters up in the atmosphere. Downlink transmissions from one base station can in these conditions travel very large distances, up to 150 km is not uncommon and distances up to up to 400 km sometimes occur, with very little attenuation. At the receiving end of the duct, the delayed but strong copy of the downlink signal will interfere with the uplink reception at another base station, see Fig. 19.2. Note that a guard period designed to handle interference from neighboring base stations typically is in the order of a few OFDM symbols – that is, a few hundred microseconds at most – is far from sufficient in these rare scenarios. This can be compared with the guard time required, 0.5–1.3 ms, corresponding to distances in the range of 150 km to 400 km. Ducting may be a rare event,

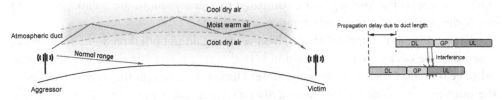

Fig. 19.2 Remote interference due to atmospheric ducts.

but when it occurs thousands of base stations may be affected and the impact on the network performance is significant.

Unlike many other types of uplink interference with fairly constant level over time, remote interference due to atmospheric ducts has a decaying profile with stronger interference at the beginning of the uplink period than at the end, see Fig. 19.3. This is intuitively understandable as the interference originates from the end of the preceding downlink transmission. At the beginning of the uplink period, there are many downlink transmissions "still in the air", potentially also stronger as they are located closer to the victim – the base station subject to the interference – while further into the uplink period the interference from downlink transmissions has vanished. Thus, by comparing the interference level at the beginning of an uplink period with level at a later point in time, it is possible to detect the presence or absence of remote interference. The higher the

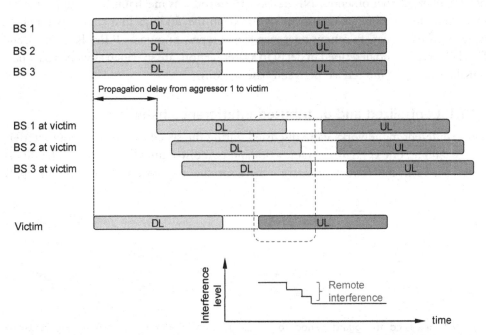

Fig. 19.3 Schematic illustration of remote interference as a function of time in the uplink period.

interference level is at the beginning compared to a later point in the uplink period, the larger the amount of remote interference.

The duct itself is reciprocal, that is, has the same gain in both directions. This means, when ducting occurs, that a base station is both an aggressor – causing interference to other cells – and a victim at the same time. However, although the duct is reciprocal, the interference situation is not necessarily reciprocal and depends on the transmission activity in the different cells involved.

Handling the remote interference in these (rare) events can be done in multiple ways. Beamforming and using interference-cancelling receivers can in some scenarios be helpful, but the most common way is to adjust the guard period such that virtually no overlap between downlink transmission from remote cells and the uplink occur at the victim. The additional guard period can be obtained either by the transmitter (aggressor) ending the downlink transmission earlier or by the receiver (victim) starting the uplink reception later, see Fig. 19.4. Doing this in a static manner and constantly having a sufficiently large guard period would solve the remote interference problem but is very costly in terms of overhead most of the time and is therefore not a viable solution. Instead, in 3G and 4G macro TDD networks, adaptive schemes are used where the guard period in the affected cells is increased but only during the periods when the ducting phenomenon occurs.

To simplify and automate the handling of the remote interference occurring as a result of such ducting phenomenon, NR release 16 introduces mechanisms to support automatic remote interference management based on the findings in a study documented in [85]. In particular, two new reference signal types, known as RIM-RS type 1 and RIM-RS type 2 and described in detail in a later section, are specified as well as backhaul signaling on the Xn interface between base stations.

19.1.1 Centralized and distributed interference handling

Three different frameworks were discussed during the specification of remote interference handling, one centralized framework and two distributed frameworks. The differences between the frameworks are where in the network the decision to apply

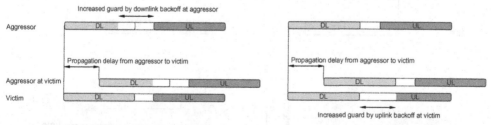

Fig. 19.4 Increasing the guard period to handle remote interference by shortening downlink transmissions at the aggressor (left) or by shortening the uplink reception at the victim (right).

interference mitigation is taken and how the decisions are signaled between the nodes. It is important to understand though that the exact operation on how to handle remote interference is not specified by 3GPP but left for implementation to allow for a range of algorithms suitable for different scenarios. Variants of the frameworks discussed later can easily be thought of and artificial intelligence and machine learning can be used as well.

In the centralized framework, all decisions related to mitigation of remote interference are taken by a centralized node, typically the operation and management (OAM) system which is responsible for configuring and operating the network. Upon detection of remote interference, for example by detecting a decaying interference profile typical for remote interference as described earlier, the victim node starts transmitting RIM-RS type 1. This reference signal, which will be described in more detail later, serves multiple purposes. Not only does it indicate that the cell is experiencing remote interference, it also contains the identity of the node (or a group of nodes) transmitting the reference signal and information on how many OFDM symbols in the uplink period that are affected. This information is implicitly encoded in the reference signal as described in a later section. Since the atmospheric ducts are reciprocal, the aggressor cells contributing to the remote interference at the victim will receive the RIM-RS.

Upon reception of such a reference signal, an aggressor node will report the detected RIM-RS, including the information encoded in the reference signal, to the OAM system for further decision on how to resolve the interference problem. The central OAM system will take a decision on a suitable mitigation scheme, for example to request the aggressor cells to stop downlink transmissions earlier in order to increase the guard period. When the ducting phenomenon disappears, the aggressor cells will no longer detect the RIM-RS and report this to the OAM system, which in turn can request the victim to stop RIM-RS transmission and the aggressors to restore the original configuration of the guard period (Fig. 19.5).

A well-designed centralized framework is likely to have superior performance compared to distributed approaches, but in many cases a distributed implementation is

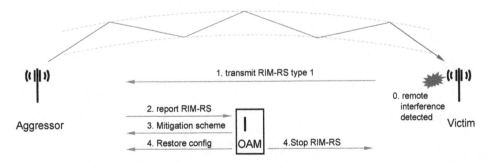

Fig. 19.5 Centralized remote interference management.

preferable for simplicity reasons and to reduce OAM signaling. In a distributed scheme, the victim transmits RIM-RS type 1 upon detecting remote interference, similar to the centralized scheme. However, instead of the aggressor cell informing a centralized node, the aggressor autonomously decides to apply a suitable mitigation scheme as long as RIM-RS type 1 is present. For example, it could end the downlink transmissions earlier in the downlink period in order to increase the guard period.

Two flavors of the distributed schemes are supported, differing in whether over-the-air signaling or backhaul signaling is used to inform the victim that the aggressor has received the RIM-RS type 1 and applied a suitable mitigation scheme.

For the case of over-the-air signaling, the aggressor starts transmitting RIM-RS type 2 once it has detected RIM-RS type 1. The purpose of this is to allow the victim to detect whether the ducting phenomenon is still present; when the victim does not detect RIM-RS type 2 it concludes that the duct has disappeared and stops transmitting RIM-RS type 1 and the aggressor returns to normal operation whenever RIM-RS type 1 disappears (Fig. 19.6).

Alternatively, backhaul signaling over the Xn interface can be used to inform the victim cell(s) about the detection of RIM-RS type 1 at the aggressor, as well as when the RIM-RS type 1 is no longer detected by the aggressor. The latter piece of information can be used by the victim cell(s) to determine that the ducting phenomenon is no longer present, and that RIM-RS type 1 transmission could stop. In response to RIM-RS type 1

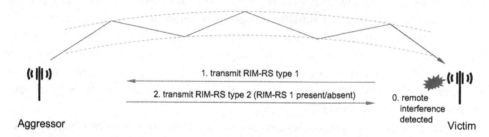

Fig. 19.6 Distributed remote interference management with over-the-air signaling.

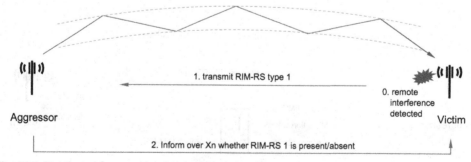

Fig. 19.7 Distributed framework using backhaul signaling.

no longer being received by the aggressor, the guard period configuration will be restored to its normal value (Fig. 19.7).

19.1.2 RIM reference signals

The primary enhancement in the physical layer to support RIM is the introduction of two new reference signals, RIM-RS type 1 and type 2. The usage of these two reference signals was briefly described in the previous section with this section devoted to the detailed structure of these reference signals.

The two reference signal types share the same basic design although they serve different purposes:

- RIM-RS type 1 is transmitted by a victim cell, intended to be received by aggressor cells, and is used to signal that remote interference is present at the victim cell – that is, a ducting phenomenon exists. In addition to an identity of the (group of) cells that causes the interference, it can convey information about the number of OFDM symbols at the beginning of the uplink period that is affected by remote interference, information that is useful to determine by how much the guard period should be increased. It can also convey information about whether the amount of mitigation applied by an aggressor cell is sufficient or not from the perspective of the victim cell.
- RIM-RS type 2 is transmitted by the aggressor cell and is used to indicate that a ducting phenomenon exists. Unlike type 1, it does not carry any additional information.

The RIM-RS of either type is designed to fulfill a number of requirements. First, it should be different from any other uplink reference signal. This is important as other reference signals in a neighboring cell otherwise might trigger the RIM mechanism even if there is no ducting phenomenon from a distant cell. Differentiating between the different reference siganls is easily achieved by using a pseudo-random sequence that differs from any uplink reference signal as will be described further later.

Second, it should be possible to detect the RIM-RS without having to obtain OFDM symbol synchronization with the aggressor at the receiver which would increase

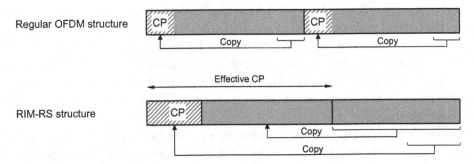

Fig. 19.8 Structure of RIM-RS.

complexity. This is achieved by using two consecutive OFDM symbols for the RIM-RS where useful part of the first and second symbols is identical. Furthermore, unlike other downlink transmissions, all the samples used for cyclic prefix are located in the first symbol, see Fig. 19.8. The net effect of this structure is a very long cyclic prefix for the last RIM-RS OFDM symbol which allows detection without having to estimate the OFDM symbol timing of the aggressor.

The RIM-RS is specified for 15 and 30 kHz subcarrier spacing. The reason for this is that RIM is a feature intended for wide-area deployments with relatively large cells. The higher subcarrier spacings, 60 kHz and above, are, on the other hand, mainly intended for high carrier frequencies and small cells, and not for wide-area large-cell deployments.

19.1.3 Resources for RIM-RS

A RIM reference signal is transmitted using a RIM-RS resource, defined by a triplet of indices in time, frequency, and sequence domains. From this index triplet, the actual location in time and frequency is computed, as well as the part of the QPSK-modulated length $2^{31} - 1$ Gold sequence to use to the RIM-RS.

- In the time domain, RIM-RSs are transmitted periodically with the periodicity of a specific RIM-RS being in the order of seconds or even minutes.
- In the frequency domain, up to four RIM-RS resources can be configured (depending on the carrier bandwidth). A RIM-RS with 15 kHz subcarrier spacing occupies the full carrier bandwidth or 96 resource blocks, whichever is smallest. For 30 kHz subcarrier spacing, the RIM-RS can be limited to 48 or 96 resource blocks.
- In the sequence domain up to eight different sequences can be configured. The set of sequences used changes over time for resilience against jammer repetition attacks.

For RIM-RS type 1, information on the number of OFDM symbols affected by remote interference, as well as an indication whether the mitigation applied by the aggressor cell is sufficient, also affects the computation of time resource from the index triplet. In other words, this information can, if desirable, be implicitly encoded in a type 1 RIM-RS. Each index triplet is also linked to a configured identity of a cell (or set of cells). Thus, the identity of the cell or group of cells that experience remote interference is also implicitly included in the resources used for type 1 RIM-RS. Note that the configured identity is not necessarily the same identity as the physical layer cell identity. This is needed as the set of physical-layer cell identities is relatively small, 1008, and the remote interference management mechanisms must be able to operate over very large areas with thousands of cells.

In Fig. 19.9, the mapping from an index triplet to the time, frequency and sequence resources to use is illustrated. Note that the RIM-RS is transmitted at the computed time location, regardless of whether the node has applied interference mitigation or not. In

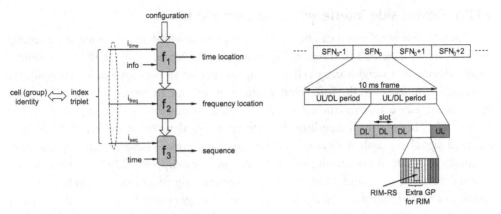

Fig. 19.9 Illustration of RIM-RS resources.

other words, the RIM-RS may sometimes be transmitted in the extended guard interval. This is necessary as the receiving side otherwise would not be able to detect whether the absence of a RIM-RS is a result of the duct disappearing or an extended guard period at the transmitter side.

19.2 Cross-link interference

Cross-link interference mitigation refers to ways to control the downlink-to-uplink interference and uplink-to-downlink interference, particularly in small-cell networks with small inter-site distances. The classical way of handling these interference problems is, as discussed at the beginning of the chapter, to use a semi-static split between uplink and downlink across all cells. However, this would contradict the basic intention with dynamic TDD, part of the basic NR framework, where the transmission direction in each cell is dynamically selected based on the traffic scenario in that cell. In many scenarios, especially if the cells are relatively isolated, dynamic TDD works fine without further enhancements. However, to expand the set of scenarios where dynamic TDD can be applied, release 16 introduces enhancements to better handle cross-link interference. These enhancements consist of interference measurements at the device side and inter-cell coordination using the Xn interface, both of which will be discussed in the following. Note that the scheduling behavior and how the measurements and coordination mechanism should be used are not specified but left for implementation.

19.2.1 Device-side interference measurements

Various scheduling solutions can be used to mitigate cross-link interference, for example to avoid scheduling uplink transmission from one device in a cell at the same time a nearby device in a neighboring cell is trying to receive in the downlink. To assist the scheduler to understand the interference situation, release 16 introduces enhancements such that a device can be instructed to measure on transmissions originating from another device. This is done by extending the reference signal received power (RSRP) and received signal strength indicator[a] (RSSI) measurements. These measurements were originally introduced for, among other things, mobility support. RSRP is the received power (excluding noise and interference) of a reference signal, either the synchronization signals or a CSI-RS, and is typically averaged over a longer period of time, in the order of hundreds of milliseconds. RSSI is the total received power, including noise and interference, over a given number of resource blocks. In release 16, these measurements are extended such that it is possible to measure not only on downlink signals – synchronization signals and CSI-RS – but also on the uplink SRS. Thus, by requesting a device to measure SRS-RSRP the network will obtain knowledge about how well that device can hear transmissions from another device or, in other words, the amount of device-to-device interference from activity in neighboring cells (see Fig. 19.10).

Note that these measurements provide information about the average, long-term basis interference, in the range of a few hundred milliseconds. They do not reflect the instantaneous situation. In a small-cell scenario, the devices are typically relatively stationary and this is less of an issue. The measurements may also provide some information about out-of-band interference, for example from a neighboring operator running in a different frequency band, which might be useful.

Base-station-to-base-station interference measurements are not standardized but left for implementation. In principle one base station can measure on any downlink signal

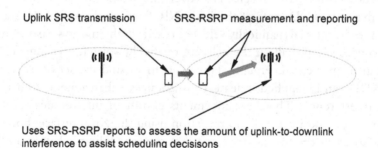

Uses SRS-RSRP reports to assess the amount of uplink-to-downlink interference to assist scheduling decisions

Fig. 19.10 SRS-RSRP to estimate cross-link interference.

---

[a] The RSSI measurement is not explicitly defined in NR release 15 but occurs as part of other measurements.

from a neighboring base station; CSI-RS, synchronization signals, or data transmission just to mention a few examples.

19.2.2 Inter-cell coordination

The other area where CLI impact the specification is the inter-gNB signaling over the Xn interface (or inter-CU signaling over the F1 interface in case of a split architecture). In essence, the resources are split into fixed resources, where the gNB promises to use the resource in a certain direction only, and flexible resources, where the gNB indicates it may use the resources in either transmission direction (uplink or downlink). With this knowledge about the scheduling behavior in neighboring cells, the scheduler may schedule more sensitive transmission in a fixed resource, where the transmission direction and hence the interference characteristics are known. Flexible resources can instead be used for less critical data where occasional retransmissions due to strong interference is less of an issue.

CHAPTER 20

NR in unlicensed spectrum

The first release of NR was primarily designed with licensed spectra in focus although extensions to unlicensed spectra in later releases was considered already from the start. Licensed spectrum implies the operator has an exclusive license for a certain frequency range which offers many benefits since the operator can plan the network and control the interference. It is thus instrumental for providing quality-of-service guarantees and wide-area coverage. However, the amount of licensed spectra an operator has access to may not be sufficient and there is typically a cost associated with obtaining a spectrum license.

Unlicensed spectra, on the other hand, is open for anyone to use at no cost, subject to a set of rules, for example on maximum transmission power. Since anyone can use the spectra, the interference situation is typically much more unpredictable than for licensed spectra. Consequently, quality-of-service and availability cannot be guaranteed as the interference cannot be controlled. Furthermore, the maximum transmission power is modest, making it less suitable for wide-area coverage. Wi-Fi and Bluetooth are two examples of communication systems exploiting unlicensed spectra.

In release 16, NR is extended to support unlicensed spectra in addition to licensed spectra, primarily targeting the 5 GHz and (later) 6 GHz bands. The frequency range is extended to include also the 60 GHz band in release 17. Both license-assisted access (LAA) and stand-alone operation is supported, see Fig. 20.1. In the LAA framework, a carrier operating in a licensed frequency band is used for initial access and mobility, combined with one or more carriers in unlicensed bands used to boost the capacity and data rates. This is similar to LTE-LAA, see Chapter 4. For NR, the dual-connectivity framework is used when the licensed carrier is using LTE, which is the same approach as for non-stand-alone NR. If the licensed carrier is using NR, either the dual connectivity or the carrier aggregation framework can be used.

Stand-alone operation, on the other hand, implies that NR operates in unlicensed spectra without support of a licensed carrier. Initial access and mobility are handled entirely using unlicensed spectra. This allows for deployments of NR without having access to licensed spectra, which can be valuable for, for example, local deployments in factories.

In the remainder of the chapter, the frequency bands targeted and the regulatory requirements on operation in these bands are discussed, followed by a description of the enhancements added to NR in order to support operation in unlicensed spectra. The NR specifications uses the term "shared spectrum access" when referring to operation in unlicensed spectra, but herein the more common term "unlicensed spectra" will be used.

https://doi.org/10.1016/B978-0-443-13173-8.00027-X

Copyright © 2024 Elsevier Ltd.
All rights reserved.   **433**

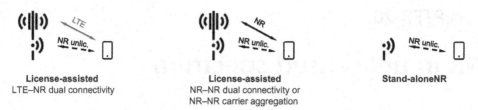

License-assisted
LTE–NR dual connectivity

License-assisted
NR–NR dual connectivity or
NR–NR carrier aggregation

Stand-aloneNR

Fig. 20.1 License-assisted (left and middle) and stand-alone (right) operation of NR in unlicensed spectra.

20.1 Unlicensed spectrum for NR

Unlicensed spectra exist in multiple frequency bands. In principle, any unlicensed band could be exploited by NR but the release 16 work focused on the 5 and 6 GHz bands. One reason is the availability of fairly large amounts of bandwidth in the 5 GHz band and a reasonable load compared to the congested 2.4 GHz band. Release 17 extended the frequency range to include also the 60 GHz band.

20.1.1 The 5 GHz band

The 5 GHz band is available in most parts of the world, see Fig. 20.2, although there are some differences between the different regions, see [86] and the references therein for details. The regulatory situation is fairly mature and the 5 GHz band has been in use for many years by, among other technologies, Wi-Fi and LTE/LAA. The band is typically divided into 20 MHz wide pieces known as channels, a term that will be used heavily when describing the channel-access mechanisms.

The lower part of the band, 5150–5350 MHz, is typically intended for indoor usage with a maximum transmission power of 23 dBm in most regions. In total 200 MHz is available, divided in two parts of 100 MHz each. In the part of the band above 5470 MHz, transmission powers up to 30 dBm and outdoor usage is allowed in many regions. The amount of spectra differs across regions but up to 255 MHz can be available.

In addition to the limitations of the maximum output power, typically given as an EIRP value, there are additional requirements in some of the band and in some of the regions. These requirements have, as will be discussed in later sections, an impact on the technical design on the radio interface. Some of these regulatory requirements are limitations on the power-spectral density, the maximum channel-occupancy time, the minimum occupied bandwidth, and requirements on dynamic frequency selection, transmit-power control, and listen-before-talk.

Power spectral density (PSD) limitations may be applicable, implying that the device cannot transmit at its full power when using smaller bandwidths. For example, in the 5150–5350 MHz range the European regulations limit the power spectral density to 10 dBm/MHz. Hence, a device cannot transmit at it maximum allowed transmit power

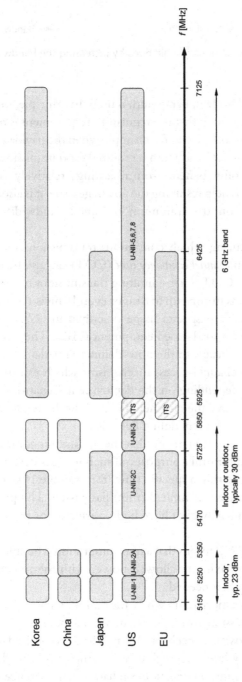

Fig. 20.2 Overview of unlicensed frequency bands in FR1 in different regions.

Fig. 20.3 Fulfilling power-spectral-density limitations by increasing the bandwidth.

of 23 dBm unless at least 20 MHz of bandwidth is used. In principle, one approach would be to limit the output power such that the regulatory requirement is met. However, this would limit the coverage in some cases, for example when the payload to transmit is small and only a fraction of the carrier bandwidth is required for transmission. Therefore, operation in unlicensed spectra often benefits from allocating a relatively large bandwidth, see Fig. 20.3. Not only does this help resolving the coverage issue, it is also beneficial in terms of fulfilling requirements on the minimum occupied bandwidth defined in some frequency bands.

The maximum time during which continuous transmission is allowed, sometimes referred to as the maximum *channel occupancy time* (COT) can also be limited. For example, in Japan the maximum COT is 4 ms, limiting transmissions to at most this duration, while in other regions such as Europe up to 8 ms or even 10 ms is allowed. Europe also has two sets of rules for unlicensed spectra usage described in [87], one for frame-based equipment (FBE) and one for load-based equipment (LBE). The two sets of rules were originally specified to be applicable to the now–defunct Hiperlan/2 standard and Wi-Fi, respectively. NR includes channel-access mechanisms which can support either of the FBE or LBE frameworks, depending on the deployment scenario. Finally, the fraction of time a transmitter must leave the channel idle may also be regulated. These requirements all have an impact on the scheduling behavior.

Dynamic frequency selection (DFS) means that the transmitter continuously must assess whether the spectra is used for other purposes. If such usage is detected, the transmitter must vacate the frequency within a specific time (for example 10 s) and not use it again until at least a certain time (for example 30 min) has passed. The purpose is to protect other systems, primarily radars, which have higher priority for the usage of unlicensed spectra.

Transmit power control (TPC) means that a transmitter should be able to reduce its transmission power below the maximum allowed power with the intention to reduce the overall interference level when needed.

Listen before talk (LBT) is a mechanism where the transmitter listens to any activity on the channel prior to each transmission in order not to transmit if the channel is occupied. It is thus a much more dynamic coexistence mechanism than DFS. It is required in some regions, for example Europe and Japan, while other regions, for example the United States, do not have any LBT requirements but instead regulations on harmful interference.

20.1.2 The 6 GHz band

The 6 GHz band provides a very large amounts of spectra, see Fig. 20.2, with 500 MHz being available in Europe and 1200 MHz in the United States. The regulatory details remain to be settled in some regions, but one major difference between the 5 GHz and 6 GHz bands is the presence of legacy mobile technologies in the former but not in the latter. The 5 GHz band has been around for several years and is already used by, among other technologies, Wi-Fi and LTE/LAA. On the technical side it may affect, for example, the need to reuse existing energy thresholds when declaring the channel available to an NR transmitter or not. NR uses the same energy thresholds as LTE-LAA and specified by ETSI BRAN [87]. The 6 GHz band, on the other hand, has recently become available and the regulatory framework is done in a technology-neutral manner.

20.1.3 The 60 GHz band

The 60 GHz band is available in most parts of the world although the size differs between regions, see Fig. 20.4. Regardless of the region, the frequencies are all within FR2-2. Power levels varies between regions but are typically in the range of 40–50 dBm EIRP and with channel bandwidths of 2 GHz. Given the propagation conditions, the 60 GHz band is primarily useful in scenarios such as short range communication within a room, and outdoor fixed links with highly directive antennas, but not for wide-area coverage in general. The 60 GHz band has been around for several years with IEEE 802.11ad being one standard designed for this frequency range, although with modest commercial uptake.

20.2 Technology components for unlicensed spectrum

Accessing unlicensed spectra is, as seen here, different than using licensed spectra and these differences impact the technical solutions. Unlike LTE, where support for unlicensed spectra was added at a relatively late stage, the requirements for accessing

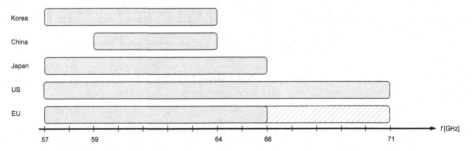

Fig. 20.4 The 60 GHz unlicensed band.

unlicensed spectra were taken into account already from the start of the NR work as part of the general emphasis on forward compatibility. Consequently, there are a number of NR technology components in release 15 that fit very well with unlicensed spectra. For example, ultra-lean transmission, flexible frame structure, and dynamic TDD are all technology components valuable for operation in unlicensed spectra.

Ultra-lean design is an important aspect. Since channel access in unlicensed spectra typically requires listen-before-talk, "always on" signals can be difficult to accommodate, especially if they are frequent and has to be transmitted at a specific time. The cell-specific reference signals in LTE is an example of such a signal. However, in NR the amount of always on signals is very small due to the ultra-lean design. The SSB is the only fundamental always-on signal and in a stand-alone deployment the device only expects this signal once per 20 ms.[a] In non-standalone deployments this time can be even longer.

The flexible frame structure of NR is another example. The subcarrier spacing is chosen to 30 kHz, which is a suitable value for the unlicensed 5 GHz and 6 GHz bands. For SCells, 15 kHz subcarrier spacing can be configured as an alternative to 30 kHz. However, more important is the possibility already in release 15 for a transmission to cover only a part of a slot, sometimes referred to as a "mini-slot", as described in Chapter 7 and illustrated in Fig. 20.5. This is highly beneficial in conjunction with the channel-access procedures. Once access to the channel has been obtained, the transmission should start immediately to avoid another device occupying the channel. In addition to transmissions starting at any OFDM symbols, cyclic prefix extension can be used to obtain sub-OFDM-symbol granularity of the starting time. The use of front-loaded reference signals for demodulation and in general short processing times in NR are also highly beneficial for efficient exploitation of unlicensed spectra.

Dynamic TDD, which is the baseline in NR when handling unpaired spectra, is beneficial for unlicensed operation. With a semi-static uplink-downlink allocation, the system would be severely hampered in when channel access can be attempted. Furthermore,

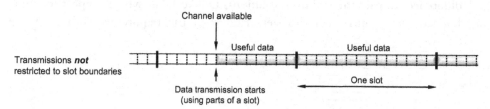

Fig. 20.5 Decoupling transmissions from slot boundaries to achieve better support for unlicensed spectra.

---

[a] In the connected state the device may also expect presence of the tracking reference signal.

once the gNB successfully accesses the channel, dynamic TDD allows the channel to be shared in a flexible way between downlink and uplink without being restricted by a semi-static uplink-downlink allocation.

Nevertheless, despite NR being well-prepared for exploiting unlicensed spectra, there are some enhancements needed to make the support complete. Dynamic frequency-selection transmit power control, channel access procedures, resource-block mapping, configured grant transmission, and hybrid-ARQ feedback are some of the areas impacted when operating in unlicensed spectra. There are also some smaller enhancements added in release 16 in order to support operation in unlicensed spectra, for example removing the restriction of PDSCH mapping type B supporting 2, 4, and 7 OFDM symbols only to support any duration from 2 to 13 symbols. In the following, a quick overview of some of these components is given with a more detailed description being provided in the following sections.

Dynamic frequency selection is used to vacate the channel upon detecting interference from radar systems, which is a requirement in some frequency bands. DFS is also used when activating the node, for example at power up, in order to find an unused or lightly used portion of the spectra for future transmissions. No specification enhancements are needed to support DFS; implementation-specific algorithms in the gNB are sufficient.

Transmit power control is required in some bands and regions, requiring the transmitter to be able to lower the power by 3 or 6 dB relative to the maximum output power. This is purely an implementation aspect and is not visible in the specifications.

Channel-access procedures, including listen-before-talk, ensures that the carrier is free to use prior to transmission. It is a vital feature that allows fair sharing of the spectra between NR and other technologies such as Wi-Fi. In some regions, in particular Europe and Japan, it is a mandatory feature. The channel-access procedures added to NR are very similar to the one in LTE/LAA as well as the one used in Wi-Fi.

20.3 Channel access in unlicensed spectra

Accessing the radio channel in unlicensed spectra is different from licensed spectra in some respects. In licensed spectra, the scheduler is in control of all the transmission activity in the cell and can coordinate the spectra usage across multiple devices. Periodic transmission such as the SSB can also occur at regular intervals. In unlicensed spectra, on the other hand, the situation is different. Accommodating multiple, uncoordinated users which may use completely different radio-access technologies require additional mechanisms with two approaches defined for NR:

- *Dynamic channel-access* (also known as LBE) relies on listen-before-talk, where the transmitter listens to potential transmission activity on the channel, applies a random back-off before, and in general follows the same underlying principles as Wi-Fi. In some regulatory regions, for example Japan and Europe, listen-before-talk is mandated in unlicensed bands with some other regions being more relaxed. Dynamic channel-access is supported by NR in FR1 and FR2-2 and described in more detail in Section 20.3.1.

• *Semi-static channel-access* (also known as FBE) does not use a random back-off but instead allows transmissions to start at specific points in time only, subject to the channel being available as discussed in Section 20.3.2. It can be used if absence of any other technology sharing a channel can be guaranteed on a long-term basis, for example through regulation or operation in a limited, controlled area such a specific building. Hence, the semi-static channel-access is particularly useful in many industrial scenarios. NR supports semi-static channel access in FR1.[b]

Following a successful procedure according to either the dynamic or semi-static channel-access procedures, the channel can be used during a period referred to as the *channel occupancy time* (COT). During a COT, one or more *transmission bursts* can be exchanged between the communicating nodes where a transmission burst is a downlink or uplink transmission.

The purpose of the channel-access procedure is to detect whether the channel is in use or not, typically employing some form of sensing such as LBT. An underlying assumption is that a node trying to access the channel can hear any other node which might already use the channel, see the left part of Fig. 20.6. However, in some situations this is not the case as illustrated in the middle of Fig. 20.6. This is often referred to as the *hidden node problem* and the channel-access procedures includes mechanism to mitigate this problem. Note that, in FR2-2 where beamforming with narrow beams is extensively used, the hidden node problem is less pronounced as nodes are better isolated in the spatial domain (see right part of Fig. 20.6). Compared to FR1, the gain with LBT is therefore small in FR2-2 and in principle LBT can be omitted without degrading the overall system performance.

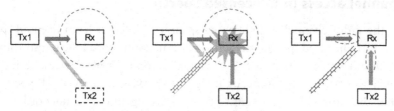

Fig. 20.6 Illustration of hidden node problem mitigation by beamforming.

[b] There is no fundamental technical reason prohibiting FR2-2 from being supported but semi-static channel access for FR2-2 is not included in the release 17 specifications.

In the following, both dynamic and semi-static channel access procedures will be described, starting with dynamic channel access.

20.3.1 Dynamic channel-access procedures (LBE)

Dynamic channel-access procedures are based on listen-before-talk, where the transmitter listens to potential transmission activity on the channel and uses a random back-off time prior to transmitting. This is the same principle as used for Wi-Fi and LTE-LAA and results in fair sharing of the unlicensed spectra with these technologies. Note that LBT is a much more dynamic operation than DFS as it is performed prior to each transmission burst. It can therefore follow variations in the channel usage on a very fast time scale, fractions of milliseconds. Following a successful channel-access procedure, the channel can be used during the COT.

Three main types of dynamic channel-access procedures are defined in NR:

- Type 1 (also known as "LBT cat4"), used for starting uplink or downlink data transmission at the beginning of a COT[c];
- Type 2, used for COT sharing as described in Section 20.3.1.2 and transmission of the discovery burst. In FR2-2, there is a single version of type 2, while in FR1 there are three flavors of channel access type 2 – 2A, 2B, and 2C – depending on the duration of the gap in the COT
- Type 3, used for transmission without sensing the channel (FR2-2 only)

The different channel-access procedures are described in more detail in the following.

20.3.1.1 Channel-access procedure type 1 and listen-before-talk

Channel-access procedure type 1 ("LBT cat4") is the procedure used to initiate one or more transmission within the same COT. The initiator, which can be either the gNB or the device, assesses whether the channel is available or not by performing the LBT procedure with a random back-off as illustrated in Fig. 20.7. In the following, the access procedure for FR1 is described. The procedure for FR2-2 is similar but with slightly different values of some parameters as discussed at the end of the section.

First, the initiator listens and waits until the frequency channel is available during at least a period referred to as the defer period. The defer period consist of 16 μs and a number of 9 μs slots where the defer period depends on the priority class as shown in Table 20.1. A channel is declared to be available if the received energy during at least 4 μs of each 9 μs slot is below a threshold. The defer period, which always is at least 25 μs long,[d] serves the purpose of avoiding collisions with, for example,

[c] During the standardization discussions, "LBT cat3" was also discussed but eventually not included in the specifications. Cat3 is similar to cat4 but uses a fixed-size contention window.
[d] The time 25 μs is known as AIFS (arbitration inter-frame space) in Wi-Fi and equals the sum of the 16 μs SIFS (short inter-frame space) and the 9 μs slot duration.

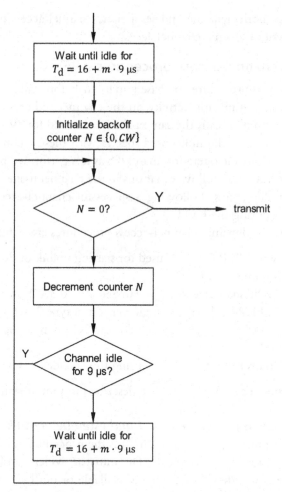

Fig. 20.7 Channel-access procedure type 1.

acknowledgments from other nodes in response to data reception. If a receiving node transmits the acknowledgment at most 16 μs after data reception, it will not run the risk of having some other node grabbing the channel before the acknowledgment is transmitted.

Once the channel has been declared available during (at least) the defer period, the transmitter starts the back-off during which it waits a random period of time. The back-off procedure starts by initializing the back-off counter with a random number within the *contention window* (CW). The number is drawn from a uniform distribution [0, CW] and represents the duration in multiples of 9 μs the channel must be available for before transmission can take place. The larger the contention window, the larger the average back-off value, and the lower the likelihood of collisions. The back-off timer is decreased by one for each 9 μs slot the channel is sensed idle, whereas whenever the channel is sensed busy the back-off timer is put on hold until the channel has been idle for a defer period. The idle sensing in each 9 μs time slot is subject to the same rules as described earlier, that is, the received energy should be below a threshold. Once the timer has expired, the random back-off procedure is completed, and the transmitter has acquired the channel and can use it for transmission up to the maximum channel occupancy time for that priority class.

The reason for the random back-off procedure is to avoid collisions between multiple transmitters. Without the random back-off, two nodes waiting for the channel to become available would start transmitting at the same time, resulting in a collision and most likely both transmissions being corrupted. With the random back-off, the likelihood of multiple transmitters simultaneously trying to access the channel is greatly reduced.

There are four different priority classes defined, each with individual contention windows and with different maximum and minimum values of the contention window as shown in Table 20.1. The intention with different priority classes is to use a smaller contention window to get faster access to the channel for high-priority traffic while low-priority traffic uses a larger contention window, increasing the likelihood of high-priority data being transmitted before low-priority data. The priority class is configured per logical channel. Likewise, different defer periods are used for the different priority classes, resulting in high-priority traffic sensing the channel for a shorter period of time and grabbing the channel quicker than low-priority traffic. Furthermore, as seen in the table, downlink transmissions have a shorter defer time than uplink transmissions. Thus, at high load, downlink transmissions have a higher probability to happen compared to uplink transmissions. In other words, a gNB potentially serving multiple users has a higher likelihood of winning the channel contention compared to a device UE transmitting in the uplink and serving that user only.

Table 20.1 Contention-windows sizes for different priority classes in FR1.

Priority class		Defer period $T_d = 16+m \cdot 9$ μs	Possible *CW* values {CW_{min}, ..., CW_{max}]	Max COT[a,b]
1	DL	25 μs	{3, 7}	2 ms
	UL	34 μs		2 ms
2	DL	25 μs	{7, 15}	3 ms
	UL	34 μs		4 ms
3	DL	43 μs	{15, 31, 63}	8 ms or 10 ms
	UL		{15, 31, 63, 127, 255, 511, 1023}	6 ms or 10 ms
4	DL	79 μs	{15, 31, 63, 127, 255, 511, 1023}	8 ms or 10 ms
	UL			6 ms or 10 ms

[a]Regulatory requirements may limit the burst length to smaller values than in the table. If no other technology is sharing the channel, 10 ms is used, otherwise 6 ms.
[b]The 6 ms may be increase to 8 ms by inserting one or more gaps. The minimum duration of a gap shall be 100 μs. The maximum duration before including any such gap shall be 6 ms.

The size of the contention window is, except for FR2-2, adjusted based on hybrid ARQ acknowledgments received from the transmitter during a reference interval which (somewhat simplified) covers the beginning of the COT. For each received hybrid-ARQ report, the contention window CW is (approximately) doubled up to the limit CW_{max} if a negative hybrid-ARQ is received.[e] For a positive hybrid-ARQ acknowledgment, the contention window is reset to it minimum value, $CW = CW_{min}$. The motivation for this procedure is to be less aggressive in using the channel when transmissions does not succeed, which most likely is due to collisions with other transmissions and thus an indication of a highly loaded system. The intention of only considering acknowledgments at the beginning of the COT is that a negative acknowledgment for the first transmission in a COT may be triggered by a collision, in which case the contention-window size should be updated, while any negative acknowledgments later in the COT are not due to collisions and hence should not affect the window size.

For downlink transmissions and uplink transmissions with configured grants, the contention-window adjustment described above can directly use the acknowledgments transmitted in the uplink and on the downlink feedback channel (see Section 20.5.3), respectively. For dynamically scheduled uplink transmissions, where no explicit acknowledgment is transmitted in the downlink, the new-data indicator, toggled for each new transmission and not toggled for retransmissions, is used instead.

As discussed above, the channel is declared to be available if the energy measured in at least 4 μs of each 9 μs slot is below a threshold. The threshold depends on several parameters such as the channel bandwidth,[f] but most important is whether the carrier frequency on a long-term basis is shared with other radio-access technologies, for example Wi-Fi, or if the deployment is such that it can be guaranteed to be used by NR only. It can also depend on the frequency band upon which the system operates [88].

In the former case, coexistence with other technologies on the same carrier in the 5 GHz band, the maximum threshold is set somewhat conservative to −72 dBm for a 20 MHz carrier. This can be compared to Wi-Fi which uses two thresholds, −62 dBm if no Wi-Fi preamble is detected and −82 dBm in case a Wi-Fi preamble is detected. The choice of −72 dBm for NR (and LTE-LAA) can thus be seen as a compromise in between the two Wi-Fi levels. It also means that NR tend to yield to Wi-Fi.

In the latter case, when NR is the only technology using the carrier, the threshold is set to −62 dBm for a 20 MHz channel, unless regulations require a lower value. For uplink transmissions, the threshold can be configured using RRC signaling to meet any regulatory requirement.

---

[e] The description is slightly simplified; the specifications describe in detail how to handle different cases of bundled acknowledgments, see [63] for details.

[f] The threshold scales with the channel bandwidth, thus in essence being a threshold for the power spectral density rather than the power.

The procedure above is described with FR1, in particular the 5 GHz and 6 GHz bands, in mind. For FR2-2, the same general procedure is used albeit with some changes, the main one being

- 8 μs and 5 μs are used for the defer period and the sensing slot duration, respectively, instead of 16 μs and 9 μs; and
- there is a single priority class with the contention window fixed to 3; and
- the energy threshold is $-60 + (P_{max}-P_{out})$ dBm in a 2 GHz channel, where P_{max} and P_{out} are the RF output power and the maximum EIRP, respectively, for the intended transmission.

The reason for the overall simplified handling of priority classes and contention windows in FR2-2 is the extensive use of beamforming at these frequencies, which results in a high degree of spatial isolation compared to FR1.

20.3.1.2 Channel-access procedure type 2 and COT sharing

Type 1 is, as stated, the procedure used to initiate transmissions within one COT. There could of course be a single transmission within the COT, for example data transmission from the gNB to the device. However, once the initiator, which can be the gNB but also a device, has obtained access to the radio channel, it may actually be used for multiple transmissions from different nodes, immediately following each other in which case a complete type 1 procedure is not necessary other than for the initial transmission.[g] This is known as COT sharing for which channel-access type 2 is used. There are different flavors of channel-access type 2 depending on the frequency range and the size of the gap between two transmission burst:

- Type 2A (also known as "LBT cat2"), used in FR1 when the COT gap is 25 μs or more, and for transmission of the discovery burst;
- Type 2B, used in FR1 when the gap is 16 μs;
- Type 2C (also, somewhat incorrectly, known as "LBT cat 1"), used in FR1 when the gap is 16 μs or less;
- Type 2, used in FR2-2

Channel-access types 2A/2B/2C are described in the following, assuming operation in FR1. The same overall procedure is used for FR2-2 but with different parameters as discussed at the end of this section.

If the next transmission follows at most 16 μs after the preceding one, no idle sensing is required between the transmission bursts as illustrated in Fig. 20.8. This is known as channel-access procedure type 2C, sometimes referred to as LBT cat 1 (incorrectly

[g] There are some limitations, for a gNB-initated COT, sharing DL-UL and DL-UL-DL are supported, while for a device-initiated COT only UL-DL is allowed (with restrictions).

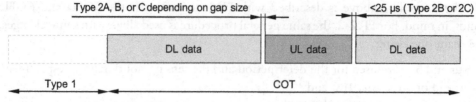

Fig. 20.8 An example of COT sharing.

as there is no LBT operation but only a constraint on the gap length). The transmission burst duration is limited to at most 584 μs. Such a short burst can carry small amounts of user data but also, more importantly, uplink control information such as hybrid-ARQ status reports and CSI reports. In essence, the uplink transmission can "piggyback" on the type 1 procedure performed for the downlink transmission (as long as the maximum COT is not exceeded).

The defer period of 25 μs for type 1 has been designed with type 2C and COT sharing for feedback information in mind – as long as the next transmission is within 16 μs, the defer period of at least 25 μs ensures that another transmitter trying to grab the channel using type 1 will not be able to interrupt the ongoing COT.

Longer gaps in the COT sharing are also possible but require channel sensing and channel access procedure type 2A or 2B. In essence, types 2A and 2B can be seen as type 1 but without the random back-off – if the channel is detected idle it is declared to be available, if it is detected busy the COT sharing has failed and the transmission cannot occur using COT sharing in this COT.

If the COT sharing gap is 16 μs, channel-access procedure type 2B is used and the channel must be detected to be idle in the 16 μs gap prior to the transmission.

If the COT sharing gap is 25 μs or longer (but the transmission is still within the COT), channel-access procedure type 2A is used. The channel must be detected idle during at least the 25 μs immediately preceding the next transmission burst. Type 2A is can also be used for non-unicast transmissions with a duty cycle of at most 1/20 and a channel occupancy of at most 1 ms, for example infrequent transmission of control information such as the SSB in case of stand-alone operation as discussed in Section 20.8.2.

To use COT sharing, the gaps between the transmission bursts typically need to be small, smaller than the duration of an OFDM symbol. Hence, the regular time-domain resource allocation, operating on OFDM symbol level, is not sufficient. This is addressed by the possibility to indicate an extension of the cyclic prefix such that the transmission starts earlier than the OFDM symbol boundary, see Fig. 20.9. In addition to no cyclic extension, three different alternatives of cyclic extension can be signaled as part of the uplink scheduling grant: $C_2 T_{symb} - T_{TA} - 16$ μs, $C_3 T_{symb} - T_{TA} - 25$ μs, and $T_{symb} - 25$ μs.

A cyclic extension of $C_2 T_{symb} - T_{TA} - 16$ μs is used when a gap of 16 μs between a downlink burst and an uplink burst sharing the same COT is desirable, typically in

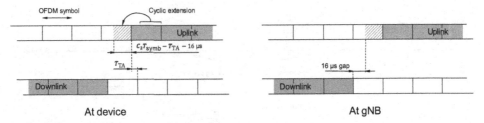

Fig. 20.9 Illustration of cyclic extension in the uplink in case of COT sharing between downlink and uplink (16 μs gap and $C_2 = 1$ assumed in this example).

combination with channel-access procedure type 2B. The reason for including the timing advance in the expression is to ensure a gap of 16 μs between downlink and uplink at the gNB, despite the amount of timing advance applied at the transmitter in the device, for example to achieve the downlink-to-uplink COT sharing illustrated in Fig. 20.8. Similarly, a cyclic extension of $C_3 T_{\text{symb}} - T_{\text{TA}} - 25$ μs can be used to create a gap of 25 μs, typically in combination with a channel-access procedure type 2A. A cyclic extension of $T_{\text{symb}} - 25$ μs, which does not compensate for the timing advance, is also possible.

The integers C_2 and C_3 in the expressions above are configured using RRC signaling. The reason is to handle scenarios where larger amounts of timing advance are expected. If C_2 and/or C_3 are not set to values larger than one, these expressions might result in a negative cyclic extension, which clearly is not possible.

In the uplink, the amount of extension to use is indicated in the scheduling grant as described in Section 20.6.4.2. Cyclic extension is useful in the downlink as well for similar reasons as in the uplink. However, as downlink transmission procedures in general are implementation choices, the specifications do not mention this case but leave it for implementation.

For FR2-2, the same overall procedure as above is used but with 8 μs and 5 μs for the defer period and the sensing slot duration, respectively. Furthermore, there is only a single type 2 procedure, without any subdivision into 2A, 2B, and 2C.

20.3.1.3 Channel-access procedure type 3

Channel-access procedure type 3 implies that the transmission takes place without a preceding sensing procedure. It is only available in FR2-2 and used for the discovery burst and msg1/msgA in the random-access procedure in regions where short control signaling transmissions do not require channel sensing.

20.3.2 Semi-static channel access procedures (FBE)

In deployments where it can be guaranteed that no other technology is using the spectra on a long-term basis, for example through regulation or deployments in a specific,

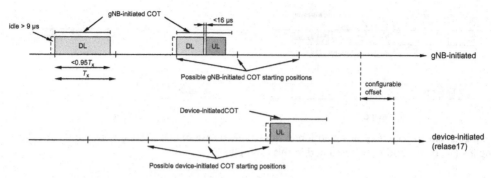

Fig. 20.10 Example of a semi-static channel-access procedure.

controlled area, semi-static channel access can be used as an alternative to dynamic channel access. In the uplink, the device can be configured to use either dynamic or semi-static channel access, while it is an implementation choice in the downlink.

In semi-static channel-access procedures, a COT can be initiated at regular time instants, that is, a COT can start once every T_x milliseconds subject to the channel being idle at least 9 µs before the start.[h] The interval T_x between two consecutive time instants is configurable from 1 ms to 10 ms. There is also a requirement that a gap of at least 5% of T_x (and at least 100 µs) must occur between each COT to give other transmitters a chance to acquire the radio channel. If sensing finds the channel busy at the start of a COT, the next attempt to start a COT Is at the next time instant as illustrated in Fig. 20.10. It is also possible to configure an offset for a device such that the time instants when a COT may start do not overlap between the gNB and the device. Note that, unlike dynamic channel access (LBE), there is *no* random back-off in semi-static channel access (FBE). Thus, transmissions in a system using semi-static channel access are often more predictable and semi-static channel access is therefore beneficial for, for example, time-critical industrial IoT applications (see Chapter 21).

A COT can be initiated by the gNB or, starting from release 17, by the device. COT sharing can be used in a similar manner as for dynamic channel-access if the gap between two channel uses within a COT is at most 16 µs. Cyclic prefix extension can be used to fill the gap but with different values of the extension compared to dynamic channel access. For longer gaps, the channel must have been sensed idle for at least 9 µs. In the examples in Fig. 20.10, the first COT is initiated by the gNB and used by a gNB transmission only, while the middle COT is initiated by the gNB but used by both the gNB and the device. The rightmost COT, finally, is initiated by the device and used by the device only.

---

[h] The variable T_u is used for devices instead of T_x. Values larger than 16 µs for device-initiated COT can be configured.

Uplink transmission may use configured grants. In this case, there are rules around which COT to use for the configured grant – if there is an ongoing gNB-initiated COT and the uplink transmission will complete prior to the end of the COT, the gNB-initiated COT is used, otherwise the device must initiate its own COT for this transmission, which is possible only at the COT starting positions. Detecting the presence of an gNB-initiated COT can be done by monitoring for gNB-transmissions at each gNB COT starting position. The background to this rule is to increase the likelihood of a configured grant being able to transmit. Always using a device-initiated COT would restrict the possible transmissions to the time instants where a COT may start. By allowing the device to reuse the gNB-initiated COT, the configured grant can start transmissions at almost any time, subject to the 9 μs idle sensing gap.

20.3.3 Carrier aggregation and wideband operation

The dynamic channel-access procedures above are described assuming a single channel. In FR2-2, the corresponding channel bandwidth is 2 GHz. A single channel thus provides sufficient bandwidth and no further arrangements are needed. In FR1, on the other hand, many regulatory requirements divide the overall spectra into 20 MHz channels which may not be sufficient. However, operation of NR in unlicensed spectra is not limited to a single 20 MHz channel in FR1 and transmission over wider bandwidths are supported.

Starting with discussing dynamic channel access, two approaches to access transmission bandwidth is larger than 20 MHz are defined, differing in how the channel, sometimes somewhat sloppily referred to as "LBT bandwidth", in the channel-access procedure is defined:

- carrier aggregation, where multiple carriers, each at most 20 MHz wide and corresponding to one channel, are aggregated to obtain the desired total bandwidth; and
- wideband carrier, where the carrier has a bandwidth wider than 20 MHz and is split into multiple channels for channel-access purposes. Carrier aggregation can be used in this case as well, but any carrier wider than 20 MHz needs to be split into multiple 20 MHz channels.

The carrier-aggregation approach is straightforward. Multiple component carriers, each at most 20 MHz wide, are used with each component carrier corresponding to a channel from a channel-access perspective. Each carrier is separately scheduled although there can be a dependency across the carriers in the channel-access procedure with either one back-off counter shared across multiple carriers or each carrier maintaining its own back-off counter.

If a single back-off counter is shared across multiple component carriers for dynamic channel access, a transmission burst can take place on multiple carriers once the back-off counter has expired and each of the carriers involved in the transmission

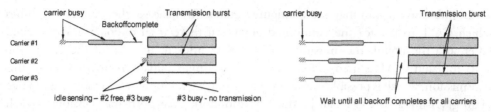

Fig. 20.11 LBT for multiple carriers, single back-off counter (left), individual back-off counters (right).

has been declared to be idle for at least 25 μs to the transmission. This is illustrated to the left in Fig. 20.11.

If multiple back-off counters are used for dynamic channel access, one per component carrier, transmission can take place on a carrier once its back-off counters have reached zero. The back-off counters for different carriers may reach zero at different time instants. In principle carriers may thus start to transmit at different time instants. However, in practice an "early" carrier cannot start to transmit until all carriers have completed their back-off procedures as transmission on one carrier would negatively impact the possibility for the listening on a neighboring carrier, see the right part of Fig. 20.11.

In the second approach with one (or more) carriers wider than 20 MHz, each of the carriers must be divided into multiple 20 MHz channels upon which the channel-access procedure is defined. This is achieved by splitting the overall carrier bandwidth into several sets of resource blocks with each resource block set corresponding to one channel (or "LBT bandwidth") for the purpose of channel access. Although the channel-access procedure operates per resource block set, the actual transmission is scheduled across the whole carrier, subject to the scheduled resource blocks being declared available by the channel access procedure. As an example, assume an 80 MHz carrier consisting of four sets of resource blocks, each corresponding to a 20 MHz channel. If three of these sets have been declared as available by their respective channel-access procedure but not the fourth set, the scheduling assignment must not schedule resource blocks corresponding to the fourth set. Since transmissions should start very shortly after a successful channel-access procedure, the time available for determining the scheduling assignment and assembling the corresponding transport block is very short, in the order of microseconds, for downlink carriers larger than 20 MHz.[i] The carrier aggregation approach is more relaxed in this aspect as the scheduling decision and transport block assembly can be performed in advance as there is limited dependency across carriers. For uplink transmissions on a wideband carrier, the device transmits only if all the scheduled resource blocks are subject to a successful channel-access procedure and are contiguous in frequency. Thus, if one or more of the scheduled resource

---

[i] One possibility to handle this is for the gNB to speculatively prepare PDSCH transmission for one or a few of the possible outcomes of the per-RB-set channel-access procedures, for example all resource-blocks sets being declared available. Obviously, this would be suboptimal but allow for a simpler implementation.

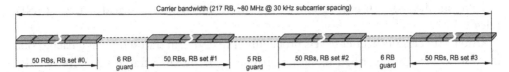

Fig. 20.12 Example of wideband operation using a single carrier with guard bands.

blocks belong to a resource-block set not subject to a successful channel access, no uplink transmission takes place on any of the scheduled resource-block sets.

Operating with wider carriers also requires guard bands between the resource block sets as shown in Fig. 20.12. The size of the guard bands has been chosen such that no filtering is needed to ensure that transmission on one resource block set does not cause significant interference to a neighboring resource block sets not available for transmission. The guard bands are either configured via RRC signaling or derived from the RF requirements.

For both carriers at most 20 MHz wide and carriers wider than 20 MHz, it is possible to indicate through group-common signaling to a device that some carriers or some resource block sets are not available for transmission during a downlink transmission burst. This can be useful for the device as a way to reduce power consumption by not monitoring for PDCCH on carriers or resource block sets not being used for downlink data transmission.

The description above assumes dynamic channel access in FR1 (the channel bandwidth in FR2-2 is, as described in the beginning of the section, sufficiently wide an no specific wideband handling is needed). For semi-static channel access in FR1, a similar approach is used with a wideband carrier divided into multiple resource block sets. For uplink transmissions, all resource block sets upon which transmissions are to take place must be available in order for the transmission to occur. Furthermore, PUSCH repetitions type B are dropped if they occur outside the COT. For downlink transmissions, the gNB may transmit upon any resource block set ("LBT bandwidth") found available, regardless of the other resource block sets, as long as guard bands are configured as illustrated in Fig. 20.12. Without guard bands configured, all resource block sets must be declared available prior to transmission on any of them.

20.4 Downlink data transmission

Downlink data transmission in unlicensed spectra is largely similar to data transmission in licensed spectra with the addition of the channel-access mechanisms described. Many of the requirements specific to unlicensed spectra can be handled in an implementation-specific manner. For example, spreading out the transmission over a larger distributed bandwidth to meet limitation on the power spectral density can be achieved by suitable scheduling decisions using resource allocation mechanisms already part of NR release 15.

After a successful channel-access procedure, the gNB can start scheduling data to one or more devices as described in earlier chapters. The flexible frame structure of NR, where data transmissions are not constrained to the slot boundaries, is beneficial as it reduces the delay from a successful channel access to transmission of the data burst (and between transmission burst in case of COT sharing). The front-loaded DM-RS design of NR, where for PDSCH mapping type B the reference signal is located at the beginning of the transmission, is also highly beneficial as it reduces the processing time in the device. To fully exploit these benefits, PDSCH mapping type B is extended to support any PDSCH length from 2 to 13 symbols (in release 15, only 2, 4, and 7 symbols are supported in the downlink as part of the device capabilities despite the general structure allowing any length).

20.4.1 Downlink hybrid ARQ

Hybrid-ARQ feedback in response to downlink data transmission, including the codebook used to multiplex multiple acknowledgments, is enhanced in release 16. In short, as discussed in Chapter 13, the release 15 design is based on the gNB controlling the time instant when the device should transmit the hybrid-ARQ acknowledgment, and the one-to-one mapping in the time domain between the PDSCH transmission and the corresponding feedback. These assumptions may not necessarily hold in unlicensed spectra as the exact timing of both downlink data transmission as well as uplink feedback information are subject to a successful channel-access procedure, calling for enhancements to the hybrid-ARQ design.

In licensed spectra, the gNB indicates when the device should transmit the hybrid-ARQ acknowledgment through the hybrid-ARQ timing field as described in Chapter 13. This provides flexibility on when to transmit the acknowledgment, which is needed also for unlicensed operation. However, once the gNB has informed the device when the acknowledgment is to be transmitted, there is no choice for the device but to transmit. This does not blend well with the channel-access procedures required when operating in unlicensed spectra where the device may not be able to transmit in case of an unsuccessful channel-access procedure. If all devices have very fast processing (see Section 13.1.4) and can generate the acknowledgment almost immediately after the PDSCH transmission, they could in principle feed back the result within the same COT. However, at least some devices have less aggressive decoding capabilities and cannot transmit the acknowledgment within the same COT but must defer it to a later point in time.

Furthermore, even if the device transmits a hybrid-ARQ acknowledgment, the gNB may not receive it properly. From a gNB perspective, a failed uplink channel-access procedure at the device or a transmitted but not received hybrid-ARQ message are indistinguishable. Due to the one-to-one mapping between PDSCH transmission and the corresponding feedback in the time domain, if the gNB fails to detect the feedback at the predefined time location, the gNB will have to assume NACK and retransmit all the corresponding PDSCHs. While missing a PUCCH transmission on a licensed carrier is unlikely, it is much more likely to happen on an unlicensed band due to collisions.

Finally, the device may miss the PDCCH transmission in which case the device and the gNB may have different understanding of the number of PDSCH transmissions to acknowledge. To handle this situation, the DAI is used as described in Chapter 13 to calculate the codebook size. For every PDSCH transmission, the cDAI signaled in the DCI is incremented by one and represents the number of scheduled PDSCHs up to the point the PDCCH was received. By comparing the cDAI value between received PDCCHs, the device can determine whether it missed a PDCCH or not and account for this when generating the hybrid-ARQ acknowledgment. In release 15, two bits are used for the DAI, that is, after reaching the highest DAI value the counter wraps around to zero again. The consequence of this is that if four or more PDCCHs are missed, the device will not be able to correctly calculate the codebook size. In licensed spectra, missing four or more consecutive PDCCHs is unlikely and the limited DAI size is not an issue. However, in unlicensed spectra, collisions between transmissions are more likely and the limited DAI size is a problem.

To handle these issues, release 16 introduces the possibility to postpone the transmission of the hybrid-ARQ acknowledgment to later, unspecified point in time. As described in Chapter 10, the hybrid-ARQ timing indicator fields points into an RRC-configured table from which the timing is obtained. By setting one of the entries in the table to "later", the gNB can instruct the device not to transmit the hybrid-ARQ acknowledgment but instead store it until a later point in time, see Fig. 20.13.

To handle the limited DAI field and the impact missing multiple sequential PDSCH transmissions has on the dynamic hybrid-ARQ codebook, the concept of PDSCH groups are introduced as part of an enhanced dynamic hybrid-ARQ codebook. Up to two PDSCH groups can be configured and the DAI operates

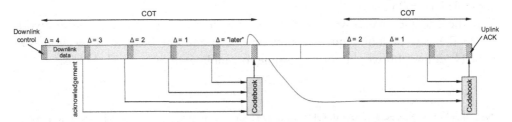

Fig. 20.13 Same-COT and cross-COT hybrid-ARQ acknowledgments.

independently between the two groups,[j] while the acknowledgments transmitted on PUCCH can include both groups. The downlink control signaling includes the group number when scheduling PDSCH transmissions in order to assist the device in determining the codebook and the resulting acknowledgment message. Additionally, the new feedback indicator (NFI) is introduced in the downlink control signaling to indicate whether the gNB has received the previous acknowledgment message for a group or not. The new feedback indicator for a group is toggled whenever the acknowledgment message is correctly received by the gNB. By using this indicator, the device can determine whether to include the feedback from previous downlink transmissions for the corresponding group or not.

An example of the operation is provided in Fig. 20.14. The first two downlink transmissions, to the left in the figure, belong to PDSCH group 0 and are received by the device (at least the control signaling, data decoding may or may not succeed), while the next two transmissions are not seen by the device, for example because of collisions with some other usage of the unlicensed spectra. As a result, when it is time to transmit the hybrid-ARQ report for PDSCH group 0 in the fifth slot, the gNB expects a report covering four PDSCH transmissions, while the device is only aware of two transmissions and hence only reports the outcome of those two. In other words, there is a mismatch between the device and the gNB about the size of the hybrid-ARQ report and the decoding of the PUCCH will fail.

In (the first part of) the fifth slot there is also downlink transmission which is to be acknowledged at a later point in time. If the device would have been fast enough to decode the downlink transmission in time for inclusion in the uplink hybrid-ARQ report in the fifth slot, the two missed transmissions could have been detected, but this is not the case in this example. Instead, the downlink transmission in the fifth slot is indicated to be part of PDSCH group 1 and the acknowledgment of this transmission is indicated to be part of a future acknowledgment report.

At a later point in time, two downlink transmissions in PDSCH group 1 take place. Since the gNB did not receive a proper hybrid ARQ report for PDSCH group 0, the gNB request feedback from both groups by not toggling the new feedback indicator for either of the groups, indicating to the device to include not only the acknowledgments from PDSCH group 1 but also the acknowledgments for the not-yet-acknowledged PDSCH group 0.

In the example, two downlink transmissions were lost but the scheme would work also in the case of four sequential downlink transmissions being lost. Without the PDSCH groups, this would not be possible to handle with only two DAI bits.

---

[j] Note that DAI is increased in size; the cDAI and tDAI for the current group is signaled, as well as tDAI for the other group.

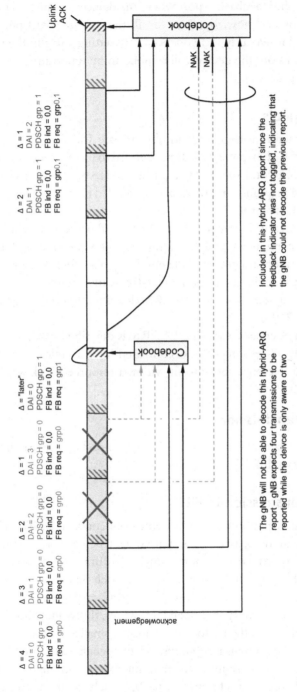

Fig. 20.14 Hybrid-ARQ feedback using multiple PDSCH groups.

Beside acknowledge reporting using the dynamic codebook as described, there is also the possibility for a one-shot feedback report where the device is requested to report the status, positive or negative acknowledgment, for all the hybrid-ARQ processes. This type of reporting is also known as type 3 codebook. By setting a flag in the DCI scheduling a downlink transmission, the device will respond to by transmitting a status report across all hybrid-ARQ processes.

20.4.2 Reference signals

The reference signal structure is largely identical to release 15.

The DM-RS for PDSCH mapping type B is extended to provide additional flexibility in scheduling only part of a slot. Instead of being restricted to transmissions being 2, 4, or 7 symbols long, PDSCH mapping type B has been extended to support any lengths from 2 to 13 as described in Chapter 9.

The CSI-RS configuration in release 15 is sufficiently flexible and a reasonable configuration would be to configure CSI-RS resource(s) to be confined within resource-block set(s). The specifications do not preclude configuring a CSI-RS spanning multiple resource-block sets, but a device assumes no CSI-RS is transmitted as soon as one or more of the resource-block sets over which the CSI-RS is configured is signaled as unavailable by DCI format 2_0.

The TRS, which in essence is a specific CSI-RS as described in Chapter 9, can be configured to cover only 48 resource-blocks compared to 52 in release 15. The reason is to ensure the tracking reference signal to fit within a resource-block set.

20.5 Uplink data transmission

Enhancements for uplink data transmission have a larger impact on the specifications than in the corresponding downlink enhancements.

20.5.1 Interlaced transmission in FR1

One aspect of unlicensed operation that calls for enhancements to the NR standard is the regulatory limit not only on the maximum output power a device may use but also limits on the maximum power spectral density, for example, 10 dBm/MHz in some regulatory regions in the 5 and 6 GHz bands. In principle, one approach would be to reuse the basic NR structure and set the transmission power such that regulatory limitations on both output power and power spectral density are fulfilled. However, this would limit the coverage in some cases, for example when the payload to transmit is small and only a fraction of the carrier bandwidth is required for transmission. Instead, it is beneficial to spread out the payload across a larger bandwidth to maximize the transmit power. Although this in principle could be achieved with resource allocation type 0, which is the approach used for non-contiguous frequency domain resource allocations in the downlink, only type 1 is supported for the uplink in release 15. Therefore, interlaces and resource allocation type 2 are introduced as part of the uplink resource allocation mechanism as a way to spread out the transmission across a larger bandwidth. Not only does this help resolving the coverage issue, it is also beneficial in terms of fulfilling requirements on the minimum occupied bandwidth defined in some regulatory regions. This is primarily of interest in FR1, which has been an underlying assumption when defining resource allocation type 2. In FR2-2, regulations and output powers are such that there is no need for interlaced transmissions.

To support interlaced transmission in FR1, the overall carrier bandwidth is divided into a number of interlaces for carriers wider than 10 MHz. The number of interlaces depends on the subcarrier spacing, 10 interlaces for 15 kHz subcarrier spacing and 5 interlaces for 30 kHz. Thus, for 15/30 kHz subcarrier spacing every 10th/5th resource block is part of the same interlace.

The interlaces are based on the common resource blocks, that is, are relative to point A. Having a common reference point results in a clean structure and simpler resource allocation. Otherwise, different bandwidth parts might need to use different interlace indices to refer to the same underlying interlace which would be an extra complication in scheduling and resource allocation for PUSCH and PUCCH among different users.

The interlaces are illustrated in Fig. 20.15, where interlace i consists of CRBs m, $m + M$, $m + 2M$, ... with M denoting the number of interlaces (5 or 10 depending on the subcarrier spacing). Note that the first resource block in a bandwidth part is not necessarily part of the first interlace.

If the device is configured to use interlaced transmission in the uplink, PUSCH resource allocation type 2 is used. The scheduling grant contains information on which interlace(s) and which resource block set(s) to use for the transmission (see Section 20.6.3 for details). PUCCH transmissions can also be configured to use the interlace structure with modifications to the PUCCH formats as outlined in Section 20.7.1.

If interlaces are not configured, the resource allocation operates as in release 15, that is, resource allocation type 0 is used for the PUSCH and the PUCCH structure is as described in Chapter 10.

20.5.2 Dynamic scheduling for uplink data transmission

Uplink data transmission in unlicensed spectra can, similar to release 15, either rely on dynamic scheduling or configured grants.

Dynamically scheduled uplink transmissions are relatively straightforward. The gNB provides the device with a scheduling grant which the device follows in line with the release 15 procedures. The enhancements are mainly limited to the support of interlaced transmission and scheduling of multiple transport blocks with a single grant as described in Section 20.6.4.2, and to the channel-access procedure required prior to

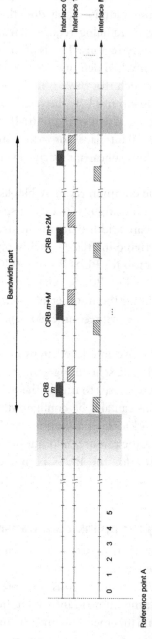

Fig. 20.15 Interlace definition.

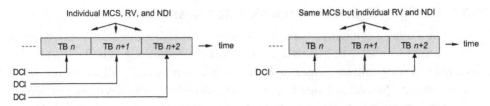

Fig. 20.16 Individually scheduled uplink transmissions (left) and one DCI scheduling multiple uplink transmissions (right).

an uplink transmission as described in Section 20.3. The scheduling grant is extended to include information necessary for the enhanced resource allocation, cyclic prefix extension, and the channel-access procedure the device should apply prior to transmission. Note that, as it is unknown at the point of transmitting the scheduling grant from the gNB whether the channel-access procedure will succeed or not, it is not guaranteed that the device will transmit in the uplink. This is different from when operating in licensed spectra where uplink transmission is guaranteed (assuming the device properly receives the grant).

In the first release of NR, a scheduling grant triggers transmission of one transport block. This is possible also in unlicensed spectra. However, since the transmissions typically are subject to a channel-access procedure with unpredictable outcome at the time of sending the scheduling grant, successful uplink transmission requires both the downlink and uplink channel-access procedures to be successful. To mitigate this cost, and to be able to transmit larger amounts of data once, one scheduling grant transmitted in the downlink can schedule multiple transport blocks in the uplink, see Fig. 20.16. The transport blocks are transmitted, one after each other, in separate slots (or "mini-slots"). Without these enhancements, a new scheduling grant would need to be transmitted in the downlink, subject to a channel-access procedure and (potentially) an associated back-off, for each uplink transmission.

The grant also includes a new-data indicator and a redundancy version for each of the transport blocks, while the modulation-and-coding scheme and the frequency-domain allocation is the same across the scheduled transport blocks.

A single grant scheduling multiple uplink transmission could in principle be beneficial as a way to reduce overhead also in licensed spectra, but the need is significantly pronounced when operating in unlicensed spectra for the reasons discussed.

The scheduling requests needed for dynamic scheduling are handled in the same way as in release 15, subject to a channel-access procedure. The handling of the scheduling request prohibit timer is slightly modified. If the scheduling request it is not transmitted as a result of a non-successful channel-access procedure, the device can transmit the scheduling request again even if the timer has expired.

20.5.3 Configured grants for uplink data transmission

Transmission using a configured grant, see Section 14.4, is supported for licensed spectra already in the first release of NR as a mean to reduce control signaling overhead. This reason holds also for unlicensed spectra, but more important is that it allows for data transmission without a preceding request–grant phase, something which is more problematic when each transmission requires a channel-access procedure which may fail. Consequently, configured grants, both type 1 and type 2, are supported with the transmission being subject to a successful channel-access procedure. Apart from the channel-access procedures required, two main enhancements have been added to better support operation in unlicensed spectra:

- Back-to-back uplink transmission within a single COT and at the same time allow the gNB to reserve slots for other purposes such as uplink and downlink control signaling.
- Decoupling of the hybrid-ARQ process identifier from the slot number. This requires uplink control information for configured grant operation, as well as downlink feedback information to indicate to the device whether the transmission was successfully received or not.

Similar to release 15, the device is configured with a periodicity for the configured grant. The starting time instant is provided either through configuration (type 1) or through the PDCCH (type 2). At those time instants, the device is allowed to perform a channel-access procedure and, if successful, transmits data in the uplink. Multiple consecutive transport blocks can be transmitted, following the same lines as for dynamic scheduling. If the channel-access is unsuccessful, the device must wait until next time instant before trying again.

To better exploit the transmission opportunity obtained by a device, configured grants also support transmission of multiple transport blocks back-to-back. With this, a single COT can be used to transmit multiple transport blocks over multiple slots, resulting in a longer COT than if only single transport blocks were allowed as is the case in release 15.

COT sharing as discussed in Section 20.3.1.2 is supported also when the COT is initiated by the device using a configured grant. This is done by the device signaling to the gNB using uplink control information on PUSCH the identity of the device and information necessary for COT sharing. Thus, even if the device initiated a COT for uplink data transmission, the gNB can benefit from it for downlink data transmission by using a quick type 2 channel-access procedure instead of type 1 including the random back-off. The rules for configured grants with semi-static channel access, described in Section 20.3.2, are such that the configured grant can have a very high likelihood of being able to transmit when needed. In principle the gNB could initiate a COT by performing a short transmission at the beginning of every COT period, which allows the devices to use the gNB-initiated COT for their configured grants.

The other main area affected by configured grants in unlicensed spectra relates to the hybrid-ARQ protocol. In release 15, the hybrid-ARQ process number is linked to the symbol number within the configured periodicity for configured grant transmission. This ensures that the device and gNB has the same understanding of the hybrid ARQ process used for the transmission but also assumes that transmission can take place at a specific time, an assumption that does not hold in unlicensed spectra where a channel-access procedure is used. Hence, the hybrid-ARQ process number needs to be signaled in the uplink and is therefore included in the UCI on the PUSCH. In addition to the process number, the UCI includes other hybrid-ARQ-related information required by the gNB for reception, more specifically the new-data indicator and the redundancy version.

Retransmissions in NR release 15 are dynamically scheduled, regardless of whether the initial transmission was dynamically scheduled or transmitted using a preconfigured grant, by using the new-data indicator in the DCI. This is possible also when operating in unlicensed spectra, but in addition retransmissions can take place using configured grants.

In licensed spectra, the initial transmission takes place at a predefined time instant for configured grants and the gNB knows when the uplink transmission is supposed to happen. Dynamically scheduling the retransmission is therefore straightforward. In unlicensed spectra, on the other hand, the initial transmission is subject to a channel-access procedure and the gNB cannot distinguish between a failed channel access and a successful channel access but failed reception of the initial transmission. Therefore, retransmissions can be autonomously initiated by the device, either by receiving a negative acknowledgment in the downlink, or, if no response has been received from the gNB, upon expiration of a timer. The timer is initialized whenever an initial transmission takes place and serves the purpose of ensuring retransmissions until the gNB indicates successful reception of the data.

The hybrid-ARQ feedback sent from the gNB in the downlink is known as the *downlink feedback information* (DFI) and is transmitted using the PDDCH, that is, no new physical channel is defined. The DFI consists of a bitmap with one acknowledgment bit per hybrid-ARQ process configured. The minimum time from PUSCH reception to transmission of the downlink feedback indicator can be configured. By observing the acknowledgment bit for a particular hybrid-ARQ process, the device can conclude whether the gNB has successfully received the uplink data and the process can be used for transmission of a new transport block, or if a retransmission is necessary. The new-data indicator in the UCI is toggled whenever a new transport block is transmitted on the related hybrid-ARQ process.

20.5.4 Uplink sounding reference signal

Uplink sounding can be useful in unlicensed spectra as well. In many cases it is preferable to perform uplink sounding in conjunction with other transmissions in order to avoid an extra channel-access procedure. Therefore, the SRS configuration is extended in release 16 such that any OFDM symbol in the slot can be used, not only the last 6 symbols in a slot.

20.6 Downlink control signaling

The downlink control signaling required for operation of NR in unlicensed spectra follows the same principles and structure as for licensed spectra with some additions motivated by the new features. The main enhancements are in the area of CORESET configuration, blind decoding, and DCI contents, while the PDCCH structure remains the same.

20.6.1 CORESET

The CORESET configuration in release 15 is flexible and it is in principle possible to configure CORESETs for PDCCH monitoring as frequent on every OFDM symbol although not supported by release 15 devices.[k] Frequent monitoring allows downlink data transmission to start in any OFDM symbol, which is useful for operation in unlicensed spectra, and devices capable of supporting unlicensed spectra therefore allows configurations with more frequent monitoring instants.

In case of aggregation of multiple 20 MHz carriers to exploit a large, contiguous bandwidth in FR1, no other enhancements are needed as the carrier aggregation framework already handles CORESET configuration per carrier. Thus, each carrier can be separately scheduled if needed. Similarly, for FR2-2, the baseline CORESET structure is sufficient.

In case of multiple resource blocks sets in FR1 as described in Section 20.3.3 to handle carrier bandwidths larger than 20 MHz, additional enhancements are needed. Since the availability of a certain resource-block set within a wide carrier is not known in advance, there must be at least one CORESET present per resource-block set in order to be able to send scheduling information in case that resource-block set is available. This is solved by extending the CORESET configuration[l] provided by the carrier such that the configured CORESET is repeated across all resource-block set in the frequency domain, see Fig. 20.17. This way, it is ensured that the device can monitor control channels individually for each resource-block set.

---

[k] In the specifications, the CORESET location in the time domain is given by the search-space configuration.

[l] In the specifications, the CORESET extension in the frequency domain is defined as being part of the time-domain search-space configuration and not the CORESET configuration.

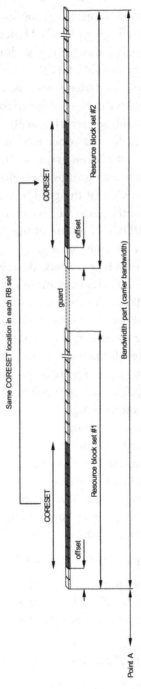

Fig. 20.17 Extending the CORESET in case of wideband operation using a single carrier in FR1.

20.6.2 Blind decoding and search space groups

Transmissions can occur at any time, subject to a successful channel-access procedure. Hence, the devices need to perform frequent PDCCH monitoring in case the gNB transmits a scheduling command. Ideally, monitoring is done every OFDM symbol for dynamic channel access and every possible COT start for semi-static channel access. To reduce the power consumption associated with this, especially for dynamic channel access, two groups of search space sets can be configured for device-specific search spaces (multiple groups are not used for common search spaces). This uses the same mechanism as described in Section 14.5.6 (actually, the concept of multiple search-space-set groups was originally developed for unlicensed access in release 16 and enhanced for NR in general in release 17). If two search space set groups are configured, each search space set is part of one or both of the groups. One of the groups is active and the device switches between the groups, either explicitly based on dynamic group-common signaling or implicitly based on detection of PDCCH in one of the groups. Both approaches make use of a timer to switch back to a "default" group.

By using the two groups of search space sets, device power consumption can be reduced. Prior to the start of a COT, frequent monitoring of the PDCCHs to determine when the COT starts and if the device is scheduled is beneficial. Monitoring as frequent as every OFDM symbol can be configured. Once the COT has started, less frequent monitoring can often be sufficient. For example, PDCCH can be monitored at the beginning of a slot only. Less frequent monitoring may also be sufficient at low traffic loads and when the latency requirements are less stringent. Search space groups can be used to achieve this flexibility with group 0 used for frequent monitoring and group 1 is used for less frequent monitoring.

Switching between the search space groups can be done through dynamic signaling using DCI format 2_0 (see later) with a bit indicating the group to activate. Which search space group to use is directly controlled by the gNB. There is also a timer mechanism defined, which is used as a complement to dynamic signaling of the search space groups. In this case, the device switches to group 1 (less frequent monitoring) whenever a valid DCI is detected and (re)starts a (configurable) timer. When the timer expires, the device returns to search space group 0 (Fig. 20.18).

20.6.3 Downlink scheduling assignments – DCI formats 1_0 to 1_3

To support the enhancements targeting unlicensed spectra, additional bits and information fields are needed in the DCI (see Table 10.2). For downlink scheduling assignments, DCI formats 1_0 and 1_1, and in later releases also formats 1_2 and 1_3, are extended with the following information:

- Hybrid-ARQ related information (the enhancements are primarily applicable to DCI format 1_1 only)
  - PDSCH group index (0 or 1 bit), used to indicate the PDSCH group and controlling the hybrid-ARQ codebook as described in Section 20.4.1.

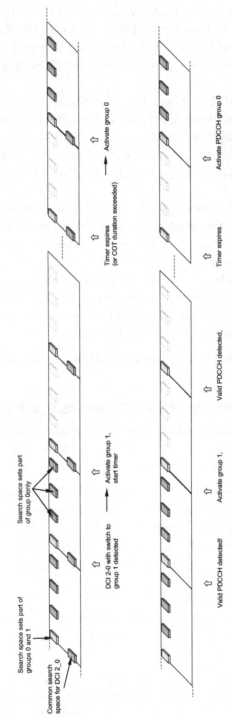

Fig. 20.18 Example of switching between search space groups using DCI format 2_0 (top) or purely timer-based (bottom).

o Downlink assignment index, DAI, is extended up to 6 bits to allow the tDAI to be transmitted also for the non-active PDSCH group as described in Section 20.4.1.

o One-shot hybrid-ARQ request (0 or 1 bit), used to trigger a hybrid-ARQ report for all hybrid-ARQ processes across all carriers and PDSCH groups as mentioned in Section 20.4.1.

o Number of requested PDSCH groups (0 or 1 bit), used to indicate whether hybrid-ARQ feedback should include only the current PDSCH group or also the other PDSCH group, see Section 20.4.1

o New feedback indicator (0–2 bit), used to indicate whether the gNB has received the hybrid-ARQ feedback (in which case the bit is toggled) or not, see Section 20.4.1.

• Channel access and CP extension (0–4 bit), used to indicate which type of channel-access procedure to use for uplink transmissions as described in Section 20.3.1. This field is present in DCI formats 1_0, 1_1, and 1_2; thus the size of the fallback format 1_0 is 2 bits larger when operating in unlicensed spectra compared to licensed spectra. Since a device knows whether it is operating in licensed or unlicensed spectrum, the difference in size is not a problem. The reason for a PUSCH-related information field in a PDSCH scheduling assignment is COT sharing where the scheduled PDSCH transmission may be followed by a PUSCH transmission.

Note that not all of the above fields are present in all DCI formats as shown in Table 10.2.

20.6.4 Uplink scheduling grants – DCI formats 0_0, 0_1, 0_2, and 0_3

Similar to the downlink formats, the uplink DCI formats are also extended to support features necessary for operating in unlicensed spectra. Most of the additions are due to the enhancements to the hybrid-ARQ protocol as summarized where:

• DFI flag (0 or 1 bit), present in format 0_1 only and serves as a header to indicate whether the DCI is an activation/release of a configured uplink grant or a request for downlink feedback information. If the flag is set for DCI format 0_1 with CS-RNTI, the remainder of the DCI content is interpreted downlink feedback information (see Section 20.6.5), otherwise it is a scheduling grant. For other RNTIs than the CS-RNTI, the bit is reserved.

• Hybrid-ARQ related information.

o New-data indicator is extended with additional bits; in case of multi-PUSCH scheduling there is one NDI bit per transport block scheduled by the DCI.[m]

---

[m] To avoid DCI size ambiguities, the number of bits for the new-data indicator field is given by the *maximum* number of transport blocks possible to schedule given the current configuration, not the *actual* number of scheduled transport blocks. The redundancy version field is handled similarly.

- o Redundancy version is extended with additional bits; in case of multi-PUSCH scheduling there is one RV value per PUSCH scheduled by the DCI.
- o Downlink assignment index, DAI, used for handling UCI on PUSCH and extended up to 6 bits to allow for the tDAI for both PDSCH groups to be transmitted.
- Channel access and CP extension (0–4 bit), used to indicate which type of channel-access procedure to use for uplink transmissions. This field is present in all formats, which impacts the fallback DCI size as discussed above.
- Resource allocation in time and frequency domains; these bitfields serve the same purpose as in licensed spectra but are extended to support interlaced resource allocation in the frequency domain and to support multi-PUSCH scheduling in the time domain, see below.

Similar to the downlink case, not all of the above fields are present in all DCI formats as shown in Table 10.3.

20.6.4.1 Signaling of frequency-domain resource allocation

Uplink resource allocation in the time and frequency domains follows the principles described in Chapter 10 for resource allocation type 1 with the addition of allocation type 2 for interlaced mapping in FR1. If interlaced mapping is not configured, resource allocation type 1 as described in Chapter 10 is used. If interlaced mapping is configured for FR1, and consequently type 2 resource allocation is used, the frequency-domain resource allocation bits in the DCI select one or more interlaces and the resource block sets within those interlaces. Note that RRC signaling is used to select interlaced transmission and hence resource allocation type 2 and there is thus no possibility for dynamic switching between type 2 and other types. This is not a problem as the preferable allocation type often is dictated by the regulatory requirements and thus does not change over time.

For resource allocation type 2, the overall number of bits for the frequency-domain resource allocation field is split into two parts – a first part indicting the interlace(s) and a second part indicating resource block sets. The resource blocks used for the actual transmission is then determined as the intersection of the resource blocks indicated by these two parts.

The first part is encoded using different approaches for 15 kHz and 30 kHz subcarrier spacing. For 30 kHz, a size-5 bitmap is used to indicate which of the five interlaces that are part of the scheduled resource. For 15 kHz, a bitmap is not used. Instead, the starting interlace number and the number of sequential interlaces is jointly coded using 6 bits. Out of the $2^6 = 64$ alternatives possible to signal with 6 bits, 56 of them are needed to cover all possible combinations of starting interlace number and the number of consecutive interlaces. The remaining 8 alternatives are used to encode a set of common non-contiguous interlace allocations.

The second part in encoded in the same way for both 15 kHz and 30 kHz subcarrier spacing. The starting resource block set (also known as starting LBT bandwidth) and the number of contiguous resource block sets in the active bandwidth part are jointly encoded.

Finally, the set of virtual resource blocks scheduled is determined as the resource blocks forming the intersection of the selected interlaces and the selected resource block sets. Since only non-interleaved mapping is supported, the virtual resource blocks scheduled directly correspond to physical resource blocks within the active uplink bandwidth part. Resource allocation type 2 is illustrated in Fig. 20.19.

20.6.4.2 Signaling of time-domain resource allocation

In the time domain, uplink resource allocation follows the same principles as described in Chapter 10 – the DCI used as an index into an RRC-configured table from which the set of OFDM symbol to transmit upon is obtained – enhanced such that one scheduling grant can schedule multiple transport blocks (see Fig. 20.20). The number of transport blocks to transmit is obtained from the RRC-configured table, extended with an extra column such that each row additionally contains time-domain allocation information for each of the transport blocks. The grant also includes a new-data indicator and a redundancy version for each of the transport blocks as described earlier, while the modulation-and-coding scheme and the frequency-domain allocation is the same across the scheduled transport blocks.

20.6.4.3 Signaling of cyclic extension and channel-access type

To use COT sharing, the gaps between the transmission bursts need to be small, smaller than the duration of an OFDM symbol. The time-domain resource signaling, with OFDM symbol resolution, is not sufficient to achieve this. Therefore, the possibility to indicate an extension of the cyclic prefix such that the transmission starts earlier than the OFDM symbol boundary is introduced in the uplink as mentioned in Section 20.3.1.2. The channel-access type to use, see Section 20.3, also needs to be indicated to the device. This is done by using the channel-access-and-CP-extension field as an index into a RRC-configured table, from which the channel-access type and the cyclic prefix extension are derived.

20.6.5 Downlink feedback information – DCI format 0_1

The downlink feedback information (DFI) is used for handling the hybrid-ARQ protocol in conjunction with configured grant transmission in the uplink. It is transmitted using the regular PDCCH structure and the CS-RNTI, that is, no new physical channel is defined. Rather, DCI format 0_1 is reused with the DFI flag indicating whether the rest of the DCI is to be interpreted as an uplink scheduling grant or downlink feedback information.

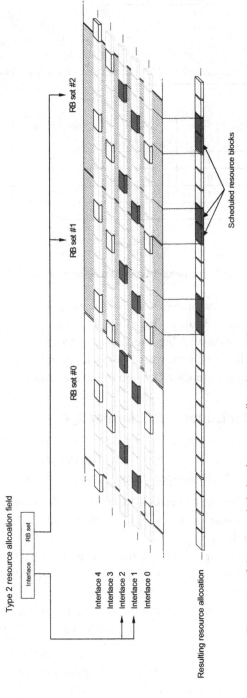

Fig. 20.19 Illustration of the principle behind resource allocation type 2.

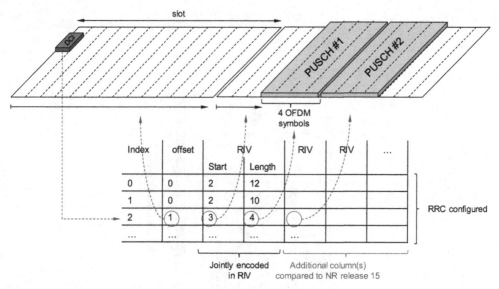

Fig. 20.20 Illustration of the principle behind time-domain resource allocation.

If the DFI flag is set, the rest of the DCI (except two bits for PUCCH power control) is interpreted as a bitmap to indicate positive or negative acknowledgment for each of the hybrid–ARQ processes. Reserved bits are included to ensure that the overall size is the same regardless of whether DCI format 0_1 carries an uplink grant or downlink feedback information, hence the number of blind decoding attempts is not increased.

20.6.6 Slot format indication – DCI format 2_0

The slot format indication can serve a wider purpose when operating in unlicensed spectra compared to the licensed counterpart. Apart from the slot format indication as described in Section 10.1.6, it has been extended to also include information about.

- COT duration; an index into an RRC-configured table where each entry represents the remaining COT duration expressed in OFDM symbols. In case of carrier aggregation there is one index per cell.
- RB set availability; a bitmap to indicate the availability of each resource block set (or LBT bandwidth) within a carrier as discussed in Section 20.3.3. In case of carrier aggregation the indication is per carrier.
- Search space group switching; a bit to indicate which search space group to activate as described in Section 20.6.2. In case of carrier aggregation there is one bit per cell group.

All these fields are optional, that is, it is possible to operate in unlicensed spectra without these fields (and without DCI format 2_0).

20.7 Uplink control signaling

Uplink control signaling in unsliced bands basically follow the same structure as in release 15 and can be carried on PUCCH or on PUSCH.

20.7.1 Uplink control signaling on PUCCH

PUCCH transmissions are subject to a successful channel-access procedure, unlike the licensed case. If the PUCCH is transmitted immediately after receiving a downlink transmission, COT sharing can be used while a successful type 1 channel-access procedure is required if there is no COT to share. A consequence of this is that the gNB cannot know when the PUCCH is transmitted and have to handle this uncertainty, for example by using energy detection to detect the presence of the PUCCH prior to decoding.

Apart from the channel-access procedure, the main enhancement to PUCCH for operation in unlicensed spectra is the changes to support interlaced transmission, an enhancement that it introduced for similar reasons as for uplink data transmission on PUSCH. The use of interlaced transmission is configured through RRC signaling. If interlaced transmission is not used the PUCCH formats are the same as in release 15. If interlaced transmission is configured, all resource blocks in one resource-block set in the interlace that are within the active bandwidth part are used for transmission. Frequency hopping is not supported in this case, which is reasonable as sufficient diversity is obtained through the interlacing mechanism itself. For carriers of 20 MHz or less, there is only a single resource-block set and hence the full interlace is used.

PUCCH format 0 with interlaced mapping supports transmissions over one interlace. This is achieved by repeating the single resource block resulting from the release 15 structure across all resource blocks in the interlace (and within the resource-block set). Repeating the same signal across all resource blocks would however result in an increase cubic metric, requiring a larger back-off in the power amplifier. To mitigate this, the phase rotation (corresponding to a cyclic shift in the time domain) is cycled through the 12 different possibilities across the 12 different values across the resource blocks in the interlace as illustrated in Fig. 20.21.

PUCCH format 1 is extended in a similar manner as format 0 to support interlaced mapping over one interlace. For each OFDM symbol, the content of the single resource block resulting from the release 15 structure is repeated across all resource blocks in the interlace. For the same reason as for PUCCH format 0, the phase rotation changes across resource blocks for PUCCH format 1 with interlaced mapping (Fig. 20.22, compare with the non-interlaced case in Fig. 10.17).

PUCCH format 2 is extended to also support interlaced mapping using one or two interlaces with higher-layer signaling configuring the interlace(s) a device should use. For smaller payloads, a single interlace is used, while for larger payloads, two interlaces are needed.

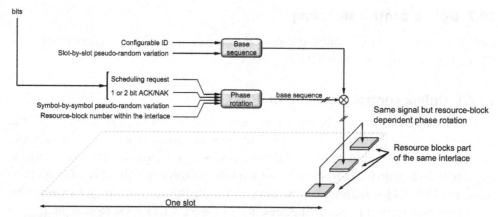

Fig. 20.21 Example of PUCCH format 0 with interlaced mapping.

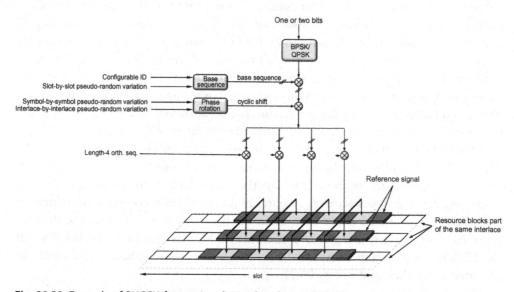

Fig. 20.22 Example of PUCCH format 1 with interlaced mapping.

The overall structure for interlaced PUCCH format 2 is similar to the non-interlaced case – coding, scrambling and QPSK modulation. However, before mapping to the resource blocks, the interlaced version using a single interlace adds the possibility to spread each QPSK symbol with an orthogonal code of length 2 or 4 as illustrated in Fig. 20.23. Since interlaced mapping implies a larger number or resource blocks being used for transmission than what is motivated by the payload size only, spreading useful as it allows multiple devices to transmit

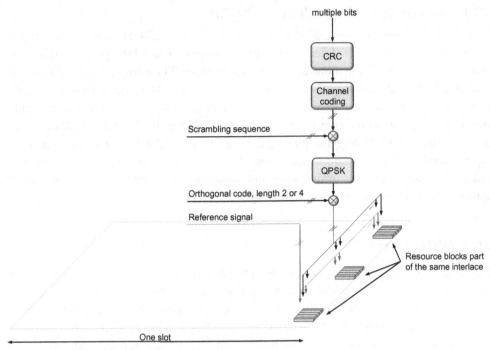

Fig. 20.23 Example of PUCCH format 2 with interlaced mapping (one interlace).

using the same resource blocks and separating the devices in the code domain. By using the orthogonal code, the resource efficiency for the PUCCH format 2 can be kept at a reasonable level despite the fact that the overall bandwidth is larger when using interlaced mapping compared to the non-interlaced case. Although the channel-access mechanism typically ensures that only one device at a time is transmitting, COT sharing and scheduling can be such that two devices transmit simultaneously in the uplink in which case the additional multiplexing capacity can be useful. To control the cubic metric of the transmitted signal, the orthogonal code to use varies between resource blocks in the interlace.

In case of a non-interlaced PUCCH format 2 or interlaced PUCCH format 2 with two interlaces, no spreading is used.

PUCCH format 3 is also extended to support interlaced mapping using one or two interlaces with the number of interlaces depending on the payload size. To increase the multiplexing capacity, spreading using an orthogonal code of length 2 or 4 is added for the single-interlace case, using a similar structure as for PUCCH format 4.

Deriving the resources to use for PUCCH for any of the formats discussed here follows the same principle as described in Section 10.2.7. In other words, the UCI payload determines the PUCCH resource set and the PUCCH resource indicator in the DCI determines the PUCCH resource configuration within the PUCCH resource set.

20.7.2 Uplink control signaling on PUSCH

Uplink control information can also be carried on the PUSCH in the same way as described in Section 10.2.8. There is one enhancement compared to operation in licensed spectra though, namely the transmission of UCI on PUSCH for configured grants. As discussed in Section 20.5.3, there is a need to transmit hybrid-ARQ related information for configured grants as a consequence of allowing retransmissions, and not only initial transmissions, to use configured grants. If configured grants are enabled, this UCI information is always present on PUSCH. Multiplexing the UCI on the PUSCH follows the same principle as in release 15, see Section 10.2.8, with the configured-grant related UCI being treated as the highest-priority information and consequently mapped in the first OFDM symbol after the demodulation reference signal.

20.8 Initial access

Initial access refers to the procedures where the device finds a cell, obtains the necessary system information, and performs a random access to connect to the cell. If license-assisted access is used to access the unlicensed spectra, almost all of these functions are handled by the primary cell on a licensed carrier. On the other hand, if NR is accessing the unlicensed spectra in a stand-alone manner, all these functions need to operate in the unlicensed spectra. Due to the specific requirements, for example channel-access procedures, enhancements are needed compared to the licensed case. In addition to the enhancements discussed in the following, the number of paging occasions has been increased to compensate for the risk that some occasions may not be useful due to the channel-access procedures. Radio-link failure procedures have also been updated to distinguish repeated channel-access failures from a radio link failure.

20.8.1 Dynamic frequency selection

The purpose of DFS is to determine the carrier frequencies for the carriers in order to find an available or at least lightly loaded carrier frequency. Since around 25 frequency channels, each 20 MHz wide, are part of the 5 GHz band (and an even larger number in the 6 GHz band), and the output power is fairly low, there is a reasonably high likelihood to find unused or lightly loaded channels. In the 60 GHz band, the range is limited and hence finding an available channel is not too difficult.

Dynamic frequency selection is performed at power-up of an NR cell in unlicensed spectra. In addition, the base station can periodically measure the interference or power level when not transmitting in order to detect whether the carrier frequency is used for other purposes and if a more suitable carrier frequency is available. If this is the case, the base station can reconfigure the carrier to a different frequency range (essentially an inter-frequency handover).

DFS is, as already mentioned, a regulatory requirement for some frequency bands in many regions. One example motivating DFS being mandated is radar systems, which often have priority over other usage of the spectra. If the NR base station detects radar usage, it must stop using this carrier frequency within a certain time (typically 10 s). The carrier frequency is not to be used again until at least 30 min has passed.

The details of dynamic frequency selection are up to the implementation of the base station and there is no need to mandate any particular solution in the specifications.

20.8.2 Cell search, discovery bursts, and stand-alone operation

Cell-search is the procedure to detect and find time synchronization to a new cell. In licensed spectra, periodically transmitted synchronization sequences, part of the SSB as described in Chapter 16, are used. Once the SSB is detected and the master information block (MIB) is properly received, the remaining system information, SIB1 and other SIBs, are scheduled and transmitted on the PDSCH.

In unlicensed spectra, a similar approach is used, but since channel-access procedures need to be supported, the transmission timing of the SSB cannot be guaranteed. Instead, a time window is defined within which the device could expect the SSB. Furthermore, it is beneficial if SIB1 is transmitted close in time with the SSB to enable a single channel access to serve both of them. The combination of SSB, the PDSCH carrying SIB1, and the associated PDCCH scheduling the PDSCH, is referred to as a *discovery burst* (DB). The DB has a short duration and is rather infrequent, hence channel-access type 2A can be used in FR1 and type 2 in FR2-2.

For the DB, the release 15 configurations for CORESET#0 and SSB are reused with some restrictions for the CORESET which can span at most 2 OFDM symbols in the time domain. In the frequency domain, CORESET#0 always occupies 48 resource blocks with 30 kHz subcarrier spacing, which is the subcarrier spacing assumed by the device for initial access in unlicensed 5 and 6 GHz bands. For secondary cells, the device can in addition be configured to search for SSBs using 15 kHz subcarrier spacing in which case 96 resource blocks are used for the SSB. When operating in FR2-2, the SSB uses 480 kHz or 960 kHz.

The SSB, PDSCH, and the associated PDCCH are located such that they are transmitted as one single, time-contiguous block. Consequently, not all combinations of SSB and CORESET#0 configurations provided by release 15 are relevant; configurations resulting in gaps in time between transmission of the different DB components would require multiple independent channel-access operations, which could result in the device receiving only part of the DB. Receiving only parts of the discovery burst is in itself not a problem (the device would simply continue to search for a complete DB) but it is an inefficient way to operate the system.

The configurable transmission window is known as the *discovery burst window* and can be up to 5 ms in length. The DB window starts at the first OFDM symbol in either of the

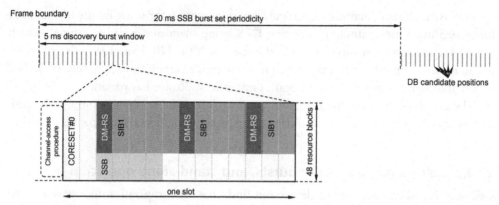

Fig. 20.24 Illustration of the discovery burst window and the discovery burst structure (30 kHz subcarrier spacing).

first or second half-frame and a DB could in principle be transmitted anywhere within this window. Up to 10 different candidate SSB positions can be monitored in a DB window for 15 kHz subcarrier spacing; for 30 kHz subcarrier spacing the upper limit is 20 and for 480/960 kHz in FR2-2 the upper limit is 32. The carrier raster is defined such that the SSB is located at the edge of the DB, see Fig. 20.24. Rate matching of the PDSCH around the SSB is not used, hence the SSB and SIB1 are frequency multiplexed.

Since the exact transmission time of a DB is unknown to the device, the transmission timing within the DB window needs to be included in the DB. This is done in a similar way as for operation in the licensed FR2 regime as described in Chapter 16. Three of the timing bits are implicitly encoded in the PBCH scrambling sequence and the remaining 1, 2 or 3 bits in the PBCH payload. This results in 4, 5, and 6 bits for 15 kHz subcarrier spacing, 30 kHz subcarrier spacing, and operation in FR2-2, respectively, which is sufficient to handle the 10, 20, or 32 candidate SSB positions. Once the SSB has been detected and decoded, the device can use these bits to determine the timing of the SSB within the frame.

In licensed spectra, the SSB time position and the QCL relation are equivalent – SSBs transmitted in different SS burst sets but at the same time instant within a set are quasi-colocated, that is, are transmitted using the same beam. When the SSBs are allowed to shift in time as the exact transmission timing within the DB window cannot be guaranteed, a new mechanism is needed for the QCL relations. Therefore, the QCL assumption is linked to the SSB candidate index.[n] DBs with the same index modulo the

---

[n] In the specifications, linking to the DM-RS sequence index is also mentioned as an alternative, but since the PBCH DM-RS sequence index is given by the SSB index the two methods are equivalent.

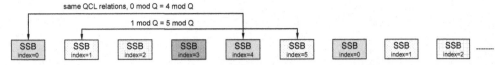

Fig. 20.25 Example of QCL relations for DRSs (Q=4 assumed in this example).

parameter Q are assumed to have the same QCL relation, see Fig. 20.25. The parameter Q is signaled to the device in the system information for the serving cell.[°]

Measurements for radio-resource management (RRM) are, as in the licensed case, based on the SSB and/or CSI-RS. To support neighbor cell RRM measurements in idle/inactive/connected states, Q is also signaled by broadcast and dedicated signaling at least per frequency.

20.8.3 Random access

Once the device has found a cell using the earlier cell search procedure, random access is used to initiate establishing a connection. The same mechanism as in licensed spectra is used and both the four-step procedure and the two-step procedure introduced in release 16 are supported, subject to a successful type-1 channel-access procedure prior to the transmission (if required by regulations). In many cases the two-step procedure is preferable in unlicensed spectra as it reduces the number of channel-access procedures needed and therefore has a smaller delay.

Two new, and longer, preamble sequences have been added for operation in unlicensed spectra: length 1151 for 15 kHz subcarrier spacing and length 571 for 30 kHz subcarrier spacing. The reason for this is to obtain a preamble covering a full 20 MHz channel in FR1 and thereby increase the transmitted energy while still meeting the power-spectral limitations set by regulations. It also reduces the likelihood of another device observing the channel to be available despite an ongoing random-access transmission, something which could happen with a more narrowband preamble. Which sequence length to use for the preamble, length 139 (or 839) as defined in release 15 or one of the new lengths is indicated as part of the system information. The new preamble lengths also use a different table for deriving the cyclic shifts with increased focus on system capacity in small cells (unlicensed spectra are unlikely to be used in wide-area deployments as the allowed transmission power is limited).

In both the two-step and four-step random access procedures, power ramping is used as described in Chapter 17 if the device does not obtain a response from the network. However, if a preamble transmission is dropped due to channel-access failure, preamble power ramping is not performed.

[°] For FR1, the LSB of the CRB grid offset and the SIB1 numerology bit in the MIB are reused to signal $Q \in \{1,2,4,8\}$ while for FR2-2, the SIB1 numerology bit is used to signal $Q \in \{32,64\}$.

CHAPTER 21

Industrial IoT and URLLC enhancements

Internet-of-things and machine-type communication are very wide terms and different applications may pose vastly different requirements, see Fig. 21.1 for a rough positioning of different technologies. For some scenarios, for example remote sensor reading, low cost and low power consumption are the most important aspects while data rates and latency are less of a concern. NB-IoT and eMTC are suitable technologies to handle these cases [26]. Although both NB-IoT and eMTC existed before NR was standardized, the spectrum-sharing mechanisms in NR enable a tight integration of these technologies into an NR carrier.

Other scenarios may stress reasonably high data rates with relaxed latency requirements at a modest cost. Video surveillance cameras is an example of this category and technology-wise, RedCap, which in essence are complexity-reduced NR devices and discussed in more detail in the next chapter, is an appropriate choice for such applications.

In yet other scenarios, high reliability and low latency are the most important requirements. *Industrial Internet of Things* is one example of such a scenario, referring to use cases such as factory automation, electrical power distribution, and transport industry. NR enhancements in this area is the focus of this chapter.

Already from the start, NR was designed to handle high reliability simultaneously with low latency. Several mechanisms and principles in NR contribute to this, for example

- transmissions are not restricted to start at slot boundaries, sometimes referred to as "mini-slot transmission" as discussed in Chapter 7;
- a front-loaded design, see Chapter 9, and requirements on fast processing time;
- downlink inter-device preemption, where an ongoing transmission to one device can be overridden by a latency-critical transmission to another device, see Chapter 14;
- robust MCS and CQI tables can be configured, increasing the robustness of transmission at the cost of a slight reduction in spectral efficiency as mentioned in Chapter 10; and
- data duplication and multi-site connectivity to increase reliability as outlined in Chapter 6.

In many cases, this set of mechanisms is sufficient. Nevertheless, several enhancements are introduced in release 16 and later releases to even further improve the support for industrial IoT. Many of the enhancements might seem small at first, but when used together they significantly enhance NR in the area of URLLC and industrial IoT. The main enhancements introduced in release 16 are:

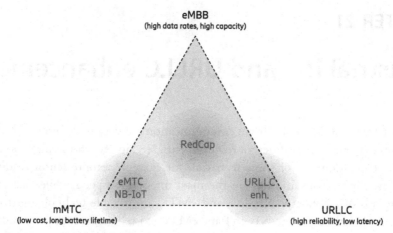

Fig. 21.1 Relative positioning of different IoT technologies.

- extending preemption to handle also uplink transmissions from different devices as described in Section 21.1;
- improved prioritization of uplink transmissions from the same device as discussed in Section 21.2;
- enhancements in the area of configured grants to allow multiple configurations and controlling the grants a certain traffic flow is allowed to exploit as discussed in Section 21.3;
- PUSCH enhancements, see Section 21.4, to lower the latency by providing better control of the time-domain resource allocation;
- PDCCH enhancements, see Section 21.5, to support some of the other enhancements;
- multi-connectivity and PDCP duplication enhancements to improve robustness as described in Section 21.7;
- time-sensitive networking, where tight time-synchronization across devices and small latency variations are as important. The tools to address this are described in Section 21.9.

Many of these areas are further refined in release 17, but two new major features are also added:

- better support for URLLC in unlicensed spectra when operating in a controlled environment (Section 21.8);
- enhancements to feedback of hybrid-ARQ acknowledgements and CSI reporting (Section 21.6).

These enhancements are described in more detail in the following sections.

21.1 Uplink preemption

Different services handled by a cellular system may have different priorities. Some services require very low latency, while other services are more relaxed in this respect. It is up to the scheduler to handle these differences as discussed in Chapter 14. In many cases, the available bandwidth is sufficient and latency-critical data for one device arriving while data transmission for another device is ongoing can be scheduled on resource blocks not used for the already ongoing transmission and thereby avoid collisions. However, at higher system load this may not be possible and the scheduler need to schedule the latency-critical data on resource blocks already used by an ongoing low-priority transmission.

In the downlink this is straightforward. The latency-critical downlink transmission can be scheduled on whatever resource blocks needed, regardless of their use for transmissions to other devices. The reception of the preempted transmissions is naturally impacted but given their less latency-critical nature this can be handled by the regular retransmission mechanisms, for example hybrid ARQ. There is also the possibility for a downlink preemption indicator as described in Chapter 14 to assist recovery of the preempted, low-priority traffic.

In the uplink, the situation is more complicated. If the high-priority and low-priority traffic originates from the same device, it is a multiplexing issue as discussed in Section 21.2. If, on the other hand, the high- and low-priority traffic originates from different devices, it is a matter of uplink preemption. Scheduling a high-priority uplink transmission on top of an already ongoing low-priority transmission from another device is possible, but due to the interference between the two transmissions it is likely that neither of them are correctly received. Therefore, release 16 adds support for uplink preemption between devices, that is, mechanisms for controlling this interference situation. Two mechanism are defined:

- cancellation, where the low-priority transmission is canceled (suppressed), and
- power boosting, where the high-priority transmission uses a higher power level than what would be used in absence of preemption.

21.1.1 Uplink cancellation

The main addition required to support the cancellation approach is reception of the cancellation indicator, which is transmitted using DCI format 2_4 scrambled with the CI-RNTI. The cancellation indicator uses a similar format as the downlink preemption indicator described in Section 14.1.1 and is thus a bitmap indicating a (set of) OFDM symbols and resource blocks upon which transmissions should be canceled. Upon reception of a cancellation indicator, a device should stop transmission of any PUSCH or SRS (but not PUCCH) that (partially) overlaps with any of the canceled resources.

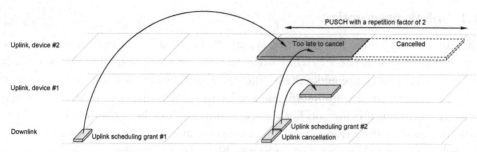

Fig. 21.2 Uplink cancellation in case of PUSCH repetition.

The cancellation must come a certain minimum time before the start of a PUSCH (or SRS) transmission in order for that transmission to be canceled. This is needed in order to allow the device to properly process and account for the cancellation indicator. It also means that an already ongoing PUSCH transmission cannot be stopped, with one exception – a PUSCH transmission using repetitions can be stopped between the repetitions. As an example, a PUSCH transmission configured with a repetition factor of two and receiving the cancellation indicator during the first transmission will cancel the second transmission (see Fig. 21.2, assuming the device is capable of simultaneous reception and transmission and there is sufficient time for canceling the repeated PUSCH).

Preemption using cancellation has the advantage of completely avoiding the interference from the canceled resources. This is useful for, for example, latency-critical transmission from cell-edge devices which may not have enough transmission power available to rely solely on the power-boosting approach. A drawback is the need for frequent monitoring of the cancellation indicator, typically several times per slot, by devices running the risk of being preempted. If this would not be the case, a device being preempted (like device #2 in Fig. 21.2) would not be able to cancel its uplink transmission and hence cause interference to the high-priority traffic. Devices not having this capability, for example release 15 devices, should therefore not be scheduled on time-frequency resources where high-priority traffic might be encountered. This can be seen as an incentive for a device to include support for frequent monitoring of DCI format 2_4; if it has this capability the network has greater scheduling flexibility in terms of resource assignment and the device may experience higher data rates.

21.1.2 Uplink power boosting for dynamic scheduling

The other approach to preemption, uplink power boosting, is relatively straight forward (Fig. 21.3). The device is configured with up to three values of the open-loop power-control parameter P_0 for PUSCH (see Section 15.1 for a discussion of uplink power control). One of the P_0 values correspond to the normal transmission power as would be the situation for a release 15 device while the other P_0 value(s) are configured such that the transmission power is increased compared to the normal case. Which of the configured P_0 values to use is indicated in the scheduling DCI. This way, the network can choose to

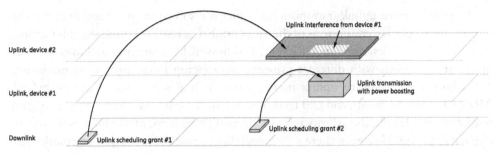

Fig. 21.3 Uplink preemption using power boosting.

dynamically boost the power of a high-priority device such that the interference from other devices with overlapping time-frequency allocation is less of a problem and the high-priority transmission is properly received. Reception of the data from the pre-empted device is likely not to succeed (unless some form of interference-canceling receiver is used in the gNB) but given the less critical nature of this traffic a hybrid-ARQ retransmission can easily be used to overcome this.

The power-boosting approach does not require reception of a cancellation indicator. Devices not supporting high-priority traffic do not need to implement any extra functionally, which is a benefit when introducing this feature in existing networks. Only devices supporting high-priority traffic need to implement the power-boosting functionality. On the other hand, dynamic power boosting cannot be applied to configured grants. It also assumes that the device has power available to afford the boosting, something which may not be the case in coverage-limited scenarios where the device already has reached its maximum power. Furthermore, interference from other devices is still present and result in an interference floor limiting the lowest achievable error probability.

21.2 Uplink collision resolution

Handling of uplink resource conflicts within a device is another critical aspect to ensure low latency. Multiple uplink channels and traffic flows can compete for the same resources within the device for several reasons, for example between dynamic grant and configured grant transmissions, conflicts involving multiple uplink configured grants, uplink control-vs-data collisions, and uplink control-vs-control collisions. Such conflict handling will primarily improve the resource efficiency when mixing different traffic flows in the same system, thereby enabling smooth introduction of URLLC into cellular networks. There are rules how to prioritize conflicts already in release 15, for example ignoring a configured grant if a dynamic grant is occurring at the same time, or "rerouting" uplink control information to PUSCH instead of using the PUCCH. However, the set of rules are extended in release 16, motivated by the simultaneous existence of control and data associated with higher priority and those associated with lower priority.

The intra-device uplink transmission conflict is resolved in two steps: first, the collision among uplink transmissions of the same priority is resolved using the rules defined in release 15 and described in Chapter 14, followed by resolving collisions between uplink transmissions with different priority by dropping lower priority transmissions. Accordingly, each type of uplink transmissions such as scheduling requests, hybrid-ARQ acknowledgements, and CSI feedback, and data are assigned a priority for collision resolution. By default, the priority is set to 'normal', which essentially implies the second step is transparent. However, there is the possibility to raise the priority of an uplink transmission to "high". If a high-priority uplink transmission would collide with an uplink transmission of normal priority, the normal priority uplink transmission is dropped (assuming there is sufficient time for cancellation or dropping). This way, it is possible to ensure that high priority uplink transmissions are prioritized in favor of low priority transmissions as exemplified in Fig. 21.4. For example, it ensures the hybrid-ARQ in response to a high-priority downlink transmission is not blocked by an uplink transmission with a lower priority as shown in the left part of the figure. Similarly, it also ensures that an ongoing normal priority PUSCH transmissions do not block high-priority scheduling requests as illustrated in the right part of the figure. Without this mechanism, the scheduling request would be delayed until the potentially long normal-priority PUSCH transmission is over, a PUSCH transmission that potentially can be very long given that a repetition factor of up to 16 can be configured in release 16. In either case, the support for low-latency traffic would be hampered.

The priority of dynamically scheduled uplink data or the uplink control information resulting from a downlink data transmission can be indicated by a priority indicator field in the scheduling grant/assignment, while the priority of uplink configured grant data is provided via the RRC configuration. The priority of a scheduling request is provided as part of the scheduling request configuration. Typically, high priority scheduling requests maps to high priority logical channels; see Chapter 14 for a discussion on logical channel multiplexing. Explicit configuration of the scheduling request priority avoids defining mapping from the two-level scheduling request priorities to the 16-level logical channel priorities.

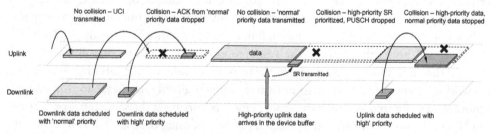

Fig. 21.4 Examples of uplink prioritization within a device.

21.3 Configured grants and semi-persistent scheduling

Uplink-configured grant transmission is a key mechanism to enable low-latency data transmission by pre-allocating resources to avoid the scheduling request/scheduling grant phase prior to uplink data transmission. In the downlink, semi-persistent scheduling is a useful tool to reduce control signaling, as is the use of configured grants in the uplink. Both of these mechanisms are available already in release 15 as discussed in Section 14.4. They are also useful tools from a robustness perspective as the need for a robust PDCCH for each data transaction is avoided.

To improve the support for high-reliability low-latency traffic in release 16, a number of improvements are made in the area of configured grants and semi-persistent scheduling.

For downlink semi-persistent scheduling, the periodicity can be as low as one slot, compared to the smallest value of 10 ms in release 15. For uplink configured grants, the periodicity can already in release 15 be as low as every second symbol, providing very short periodicities.

Multiple configurations can be active simultaneously, both for uplink configured grants and downlink semi-persistent scheduling. This can be used to support multiple services, where each service may have a different performance requirement in terms of latency and reliability. Multiple configurations differing only in the possible starting points can also be used to reduce latency. When a packet arrives, transmission can use the configuration that is closest in time.

To control which traffic flow that uses a certain grant, the logical channels which are allowed to use a certain configured grant can be restricted as part of the configuration. For example, as illustrated in Fig. 21.5, two configurations can be active. One with frequent transmission opportunities to meet the latency requirements for traffic flow #1 and one less frequent but larger resource allocation to allow larger amounts of less latency-critical data to be transmitted coming from flow #2. To avoid the low-priority data to use the frequent grant, it is possible to restrict a logical channel to use only some of the configured grant opportunities. Letting the high-priority logical channels to use the less frequent grant is typically fine and can be accounted for when configuring the logical channel restrictions. If grant opportunities from different configurations occur at the same time,

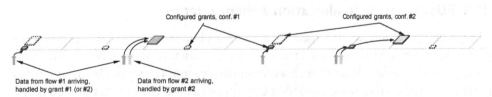

Configured grants, conf. #1   Configured grants, conf. #2

Data from flow #1 arriving,   Data from flow #2 arriving,
handled by grant #1 (or #2)   handled by grant #2

Fig. 21.5 Example of multiple uplink configured grants and different logical channel priorities.

for example for both normal and high priority traffic, the prioritization rules discussed in Section 21.2 resolve the conflicts, if any.

In addition to configuring which logical channel that is allowed to use a given configured grant, it is also possible to configure whether a dynamic grant is allowed to override a configured grant or not.

In release 15, if a device receives dynamic grant indicating transmission at the same time as a configured grant, the device always follows the dynamic grant. This is reasonable as dynamic grants can be used, for example, to grant the device a larger amount of resources than the configured grant when there is a large amount of data awaiting transmission. In many cases, MBB-like traffic is sporadic with large variations over time in the amount of needed resources where dynamic scheduling is required for efficient resource utilization. Critical traffic, on the other hand, is often dominated by periodic and deterministic traffic. Thus, in a mixed-traffic scenario, it can be less desirable to always prioritize the dynamic grant. As an example, assume that a dynamic grant instructs the device to transmit at the same time as a configured grant intended for critical traffic. If the uplink transmission from the dynamic grant is not properly received by the gNB, the delay until the data are retransmitted might be significant as the gNB is not aware of the presence of high-priority data.

The priority level – normal or high – can be used to address this problem. Given the periodic nature of the critical traffic, configured grants are well suited. By setting the priority of the configured grants to 'high' and using the normal priority for the dynamically scheduled, less critical traffic, the prioritization rules discussed earlier result in the desired behavior – if the transmission instance for a dynamic grant and a configured grant coincide and there is both normal-priority data and critical data to transmit, the high-priority configured grant will "win" and determine the uplink transmission. If there are no critical data to transmit, the device follows the dynamic grant. In case of the same priority for both the configured and dynamic grant, the dynamic grant will be followed, that is, the same behavior as in release 15.

Note that the gNB may need to blindly detect which uplink transmission that occurred – one triggered by the configured grant or one triggered by the dynamic grant. Alternatively, scheduling could be done such that the conflict never occurs. There are also other possibilities for the implementation to handle this potential ambiguity.

21.4 PUSCH resource allocation enhancements

The possibility to, already in release 15, transmit uplink data using only a part of a slot, sometimes referred to as "mini-slot transmission", is a useful feature to reduce the overall latency. However, in release 15 such a transmission cannot cross a slot boundary, meaning that transmissions sometimes need to be postponed until the next slot or use a shorter duration than what is motivated by the payload and required modulation-and-coding

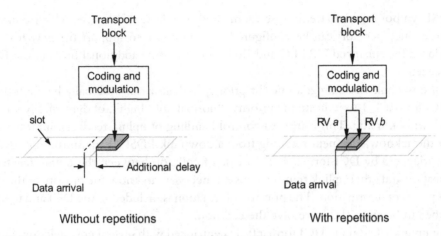

Fig. 21.6 Example of latency gain with "mini-slot" repetitions.

scheme, see the left part of Fig. 21.6. This restriction is in principle lifted in release 16 where repetitions can be dynamically indicated in the DCI. Assume, as an example, a four-symbol long transmission spanning the slot boundary is needed. The scheduling grant can, in this example, indicate a two-symbol transmission using the last two OFDM symbols of the first slot, together with one repetition as shown in the right part of Fig. 21.6. The transport block would in this case be coded, modulated, and transmitted in the last two OFDM symbol of the first slot and then repeated, typically using a different redundancy version, in the first two symbols of the next slot. In essence this would be a four-symbol transmission spanning a slot boundary. Without repetition, the transition would either have to use only the last two symbols of the first slot, resulting in a less robust transmission, or be postponed and use the first four symbols of the next slot, resulting in a larger delay.

To avoid having the repetitions colliding with other transmissions, it is possible to configure an invalid symbol pattern. This is a bitmap spanning one or two slots and indicating the OFDM symbols not allowed to be used for repetitions. Furthermore, it is possible to dynamically control as part of the DCI whether this bitmap should be used to indicate invalid OFDM symbols or not. Release 17 further enhances the handling by extending hybrid-ARQ feedback for these repeated resources to also support semi-static code books (release 16 supports dynamic codebooks only).

21.5 Downlink control channels

NR is fundamentally a scheduled system where each device monitors a set of downlink control channels to determine whether it is scheduled to transmit or receive. To reduce latency, more frequent control-channel monitoring, up to every second

OFDM symbol in the extreme case (compared to release 15 devices which typically monitors once per slot) can be configured. Furthermore, to support the new features introduced for enhanced URLLC and IIoT support, some additional fields in the DCI are needed.

For downlink scheduling, a one-bit priority indicator can optionally be configured for DCI format 1_1 to indicate the priority, "normal" or "high", of dynamically scheduled downlink traffic. This is used to control handling of uplink feedback information, either the acknowledgement resulting from a downlink PDSCH transmission or a CSI report triggered by DCI format 1_1. In case of collision between the uplink feedback information and other uplink transmissions it is necessary to know the priority of the different pieces of information. The priority information is included in the DCI and used as described in Section 21.2 to resolve the collision.

For uplink scheduling, DCI format 0_1 is enhanced with several new fields or extensions to existing fields:

- Open-loop power control to allow uplink power-boosting by selecting different pre-configured values of the open-loop power control parameter P_0 as described in Section 21.1.2;
- Priority indicator used to control the priority level of PUSCH as described in Section 21.2;
- The time-domain allocation field pointing into an extended time-domain allocation table where the number of repetitions, see Section 21.4, is obtained from a new column in the table. Thus, each table entry Fig. 10.11 indicates not only the start and length of the resource used but also the number of times the allocation should be repeated. This allows dynamic indication of the number of repetitions, unlike release 15 where it is semi-statically configured. Fully exploiting this additional flexibility may imply a larger allocation table and therefore the time-domain allocation field is increased in size to allow for up to 64 rows instead of 16
- Invalid symbol pattern indicator, controlling if the RRC-configured invalid symbol pattern should be applied or not when determining the symbols allocated for the PUSCH, see Section 21.4.

Two new DCI formats are also introduced in release 16, format 0_2 for uplink scheduling and format 1_2 for downlink scheduling, see Tables 10.2 and 10.3 in Chapter 10. They provide almost the same functionality as the formats 0_1 and 1_1, respectively, but allows for a greater configurability of the size of the different fields, including the possibility to configure a size of zero for several of them. Thus, these formats can allow for a smaller DCI size and hence more robust reception in cases where not all DCI information fields are needed or if a small number of bits is sufficient. For example, in DCI formats 0_1 and 1_1, the hybrid-ARQ process number always uses (at least) four bits and the redundancy version two bits while formats 0_2 and 1_2 allows for a smaller number of bits in situations where the full range of hybrid-ARQ processes and redundancy versions are not needed. As another example, the carrier indicator can be configured with zero bits in the new formats even in cases where carrier aggregation requires three bits for formats 0_1 and 1_1. A similar situation holds for several of the other fields as well. In release 17, DCI formats 0_2 and 1_2 are extended with information fields already present in 0_1 and 1_1 to support unlicensed spectra, as well as extensions of the HARQ-related signaling.

21.6 Feedback enhancements

Release 17 enhances feedback reporting from the devices to the gNB mainly in two areas: reducing the delay and increasing the reliability of hybrid-ARQ acknowledgements and improving the link adaptation by more detailed CSI reporting.

Hybrid-ARQ acknowledgements are transmitted on the PUCCH (unless there is a PUSCH transmission at the same time). As described in Chapter 10, either of the PCell or the PUCCH-SCell is used, depending on which of the downlink carriers the data was transmitted upon. In many cases this is a good and robust setup. However, when aggregating multiple TDD carriers with different uplink-downlink patterns, especially with a downlink-heavy pattern on the PCell or PUCCH-SCell, there can be a non-significant delay until there is an uplink slot upon which the PUCCH transmission can take place. To reduce this delay, release 17 allows the PUCCH to be switched to another cell, known as the *PUCCH switching cell* (PUCCH-sSCell). One PUCCH-sSCell can be configured in each of the two PUCCH groups. The switching can either be dynamically or semi-statically controlled. In the former case, the PUCCH cell indicator in the DCI selects which of two semi-statically configured cells – the PCell or the PUCCH-sScell – that should be used for the PUCCH transmission. In the latter case, a semi-static pattern is configured, indicating for each of the PUCCH groups which of the two carriers to use for PUCCH transmission.

A simple example of PUCCH cell switching is illustrated in Fig. 21.7. Note that sometimes the PCell is best from a latency perspective, while at other points in time the PUCCH-sSCell is better.

Another enhancement in release 17 is the possibility to retransmit the hybrid-ARQ acknowledgments. The reason for this is that, in release 16, acknowledgements might be dropped for several reasons. For example. Uplink prioritization rules may have prioritized another uplink transmission or the PUCCH was overlapping with invalid symbols. In this cases, the gNB would not know whether the PDSCH was correctly received or not and would schedule a potentially unnecessary retransmission. To handle this, the gNB can request the acknowledgement to be retransmitted by setting the HARQ retransmission indicator in the DCI.

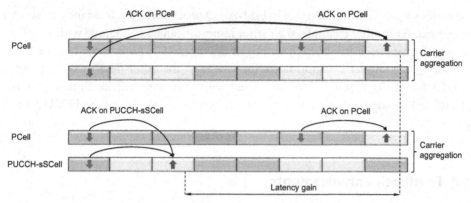

Fig. 21.7 Illustration of the latency gain obtained from switching to PUCCH-sSCell.

There is also possibility to defer the acknowledgements when using semi-persistent scheduling. To support low latency, very short periodicities for semi-persistent scheduling can be used. However, the periodicities necessary from a downlink latency perspective may not match the uplink-downlink pattern in a TDD carrier, causing the acknowledgement to be dropped in releases prior to release 17. To address this, the possibility to defer the acknowledgement until the next available uplink resource is introduced.

To improve the link adaptation, CSI reporting has been extended to full CSI reporting per sub-band using 4 bits instead of relying on 2-bit differential reporting. Release 17 also provides some enhancements in multiplexing low priority and high-priority UCI in the same transmission.

21.7 Multi-connectivity with PDCP duplication

Duplication, that is, transmitting the same downlink data more than once as a mean to increase reliability, is possible already from the first release of the NR where the duplicate detection mechanism in PDCP can remove the duplicates. In case of carrier aggregation, the RLC can also be configured to remove downlink duplicates. This is primarily an implementation aspect and the gNB can exploit duplication strategies within this framework as needed. The multi-TRP enhancements in release 16 and later, see Chapter 12, can also be used as a form of multi-connectivity to increase robustness.

In the uplink, the picture is more complex. Packet duplication cannot be left as an implementation-specific aspect as the network in that case would not be in control of the performance. Instead, the device behavior needs to be specified.

In release 15, PDCP duplication can be configured with up to two RLC entities for one radio bearer. Each PDCP PDU is duplicated with one copy mapped to the primary

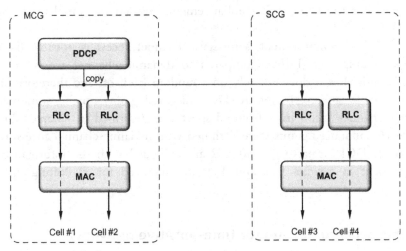

Fig. 21.8 Illustration of duplication across carriers and sites.

logical channel and one copy to the secondary logical channel. Since each logical channel has its own RLC entity, there will be two RLC entities associated with the radio bearer. These RLC entities can either belong to different cells from the same gNB (duplication using carrier aggregation) or different cells from different sites (duplication using dual connectivity). MAC control elements can be used to enable/disable duplication.

In release 16, this framework has been extended to support up to four RLC entities (Fig. 21.8). Thus, simultaneous duplication across carriers *and* sites can be handled, unlike earlier releases where either carrier aggregation or dual connectivity could be used in the uplink but not simultaneously. Similar to release 15, the enhanced PDCP duplication can be activated/deactivated by MAC control elements.

21.8 IIoT and URLLC in unlicensed spectra

Operation in unlicensed spectra, with initial support in release 16, can unleash significant amounts of spectrum without requiring a license. This can be advantageous in many scenarios although the in general unpredictable interference situation causes variations in latency and data rates and limits the performance compared to licensed spectra. However, in some scenarios the network owner can control which users that enter a specific area. In such controlled environment, for example a factory, the interference situation is more predictable and URLLC and IIoT type-of-services can often be offered also in unlicensed spectra. One important enhancement in release 17 is to enhance the support for these type of services in unlicensed spectra. Most of the mechanism required for unlicensed spectra are introduced in release 16 and described in

Chapter 21, but there are a few enhancements needed to complete the support of URLLC.

In a controlled environment, semi-static channel access is possible (known as frame-based equipment, FBE). Compared to dynamic channel access, semi-static access is beneficial as it does not rely on a random back-off and therefore provides more predictable latency. In release 17, configured grants can be combined with semi-static channel access in unlicensed spectra as described in Chapter 20, while in release 16 configured grants were designed with dynamic channel access in mind. Furthermore, DCI formats 0_2 and 1_2 are extended with the information fields already present in formats 0_1 and 1_1 and required for exploiting unlicensed spectra.

21.9 Time synchronization for time-sensitive networks

Time-sensitive networks (TSN) refer to a class of communication networks where very accurate timing is required. The nodes involved must have the same understanding of the time and communication packets need to be delivered within a time budget. This does not necessarily imply that the latency requirement is very low, although this can of course be the case, but rather that there is a common time reference and, in many cases, that the delay jitter is low. The mechanisms discussed in the previous sections, for example configured grants and the associated priority handling, can be used to provided latency-bounded low-jitter communication. However, in many scenarios this is not enough but here is also a need to have a common and very accurate time synchronization across several nodes.

Industrial automation is one typical TSN scenario targeted by NR where the requirements is to provide a common time reference with at most 1 µs inaccuracy across a 100 m × 100 m industrial site. In a wired environment, Ethernet is commonly used for time-sensitive networking, both for controlling the time references (there can be several) and for transmitting user data. Several protocols are available for time synchronization across a wired Ethernet network, for example the IEEE 1588 standard. However, these protocols are not developed with wireless connectivity in mind. To better support time-sensitive communication also over 5G networks, functionality is added to provide an accurate time reference over the 5G network. The time reference can be used at the device side to derive the different time references needed by the machines connected to the wireless receiver, see Fig. 21.9. Given the accuracy requirement, the accuracy of the radio network should be around 0.5 µs or less.

As a baseline, the device is not aware of the absolute time in the 5G network (at least not with sufficient accuracy). The timing advance value is known though from which the device can derive the propagation delay from the base station to the device assuming the timing advance has been set to twice the propagation delay. To obtain an absolute timing

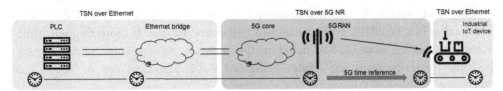

Fig. 21.9 Illustration of TSN over 5G.

reference in the device, the gNB transmits an RRC message with the absolute time at a future point in time corresponding to the end of a future system frame with a specific number. The content of the message is set is such that any delay in the transmitter-internal processing is compensated for, that is, the absolute time provided in the message represent the time at the antenna connector of the base station at the end of a future system frame number. The device can, upon receiving this message, compensate for the propagation delay and determine an absolute timing reference, see Fig. 21.10. Estimates of the propagation delay, necessary to perform the compensation, can be obtained either from the timing-advance value already available in the device or, if higher accuracy is required, from roundtrip-time measurements. Roundtrip-based measurements are based on the gNB transmitting a reference signal, either a TRS or a PRS. Upon reception of the reference signal, the device responds with an SRS. The gNB measures the time from the transmission of the TRS or PRS to the reception of the SRS and sends this value to the device. As the device knows its TRS/PRS-to-SRS delay, it can compute the propagation delay.

An alternative to compensating for the propagation delay in the device, the compensation can be handled in the gNB instead. The roundtrip measurements are done in the same way as previously describe, but with the device reporting the TRS/PRS-to-SRS delay to the gNB, which adjusts its transmitted timing reference accordingly.

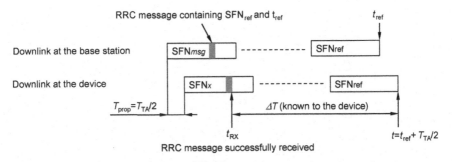

Fig. 21.10 Computation of the timing reference at the device.

The RRC messages with the absolute time are transmitted in one of the system information blocks, SIB9, typically transmitted every 160 ms. If more frequent time-reference signaling is needed to maintain the accuracy, or it is important to protect the time-reference using ciphering, dedicated RRC signaling can be used as an alternative.

Once the device in the 5G network has obtained an accurate absolute time in the 5G network, the clocks connected to the device can be synchronized with this 5G clock and thus also with the clocks in the overall time-sensitive network.

As already mentioned, Ethernet is commonly used for communication in wired TSN. Efficient support of Ethernet-based communication is therefore important and, in addition to time synchronization, NR is also enhanced with Ethernet header compression for efficient delivery of Ethernet frames. This is done by extended the PDCP protocol to support Ethernet header compression.

CHAPTER 22

RedCap and small data transmission

In the previous chapter, URLLC enhancements were discussed, primarily for industrial IoT. However, not all IoT use cases are latency critical. Instead, aspects such as reasonably high data rates, higher than eMTC/NB-IoT, at a modest cost can be more important. To address these use cases, *reduced capability* (RedCap) devices are introduced in release 17.

Release 17 also introduces *small data transmission* (SDT) to improve the handling of small amounts of data. This is a common scenario in many machine-type communication scenarios, although SDT is not limited to machine-type communication but can be used as a general tool.

22.1 RedCap devices

The first release of NR was designed focusing primarily on mobile broadband services and, to some extent, on URLLC services. For massive machine-type communication, the LTE-derived technologies eMTC and NB-IoT are used and NR provides mechanisms for supporting eMTC/NB-IoT on the same carrier as NR itself, thereby providing an integrated system supporting mobile broadband, critical IoT and massive IoT.

However, some IoT services have higher requirements than what eMTC/NB-IoT provides but lower than what can be provided by the full set of NR features. Industrial wireless sensor networks, video surveillance, and wearables are all examples requiring lower device cost and longer battery lifetime than NR, but higher data rates and lower latency than eMTC/NB-IoT [109]. The lower-end LTE device categories, cat 1 – cat 4, is one possibility but to address these needs within the NR family, release 17 introduces *reduced capability* (RedCap) devices. These devices are NR devices but with lower cost and complexity than regular NR by limiting some of the features, see Fig. 22.1. RedCap devices can operate in both paired and unpaired spectrum and in both FR1 and FR2. Downlink peak data rates in the order of 85 Mbit/s, 50 Mbit/s, and 240 Mbit/s supported in FR1 FDD, FR1 TDD, and FR2 TDD, respectively, for the simplest RedCap devices [110]. These numbers are significantly higher than eMTC/NB-IoT but not as high as "regular" NR.

The work on RedCap started with a study on potential technology options for reducing device complexity and cost [111]. Based on this study, it was concluded that significant cost reduction could be obtained by limiting the carrier bandwidth to 20 MHz in FR1 and 100 MHz in FR2, targeting single-antenna devices, removing 256QAM, and

5G/5G-Advanced
https://doi.org/10.1016/B978-0-443-13173-8.00013-X
495

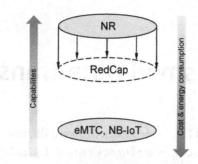

Fig. 22.1 Illustration of the relation between NR, RedCap, and eMTC/NB-IoT.

using half-duplex FDD. In total, 65% cost reduction for FR1 and 50% for FR2 can be achieved for the simplest RedCap devices although the exact numbers vary with the assumptions made.

22.1.1 Bandwidth reduction and initial access

RedCap devices support less bandwidth than ordinary NR devices. For RedCap in FR1, at most 20 MHz is supported, and for FR2 at most 100 MHz is supported. This is significantly smaller than regular NR, especially in FR1, and allows for a substantial cost reduction in the RF components. It also implicitly limits the data rates, thereby reducing cost also in the baseband processing.

The minimum bandwidth supported by RedCap is much larger than what is required by the SSB. Redcap devices are therefore capable of acquiring the same SSB as regular NR devices, in contrast to NB-IoT/eMTC devices which require separate system information delivery given their narrow bandwidth. Once the SSB has been acquired, the remaining SIBs can be scheduled by the gNB with the RedCap bandwidth limitation in mind. Given the reduced bandwidth of the device, the initial bandwidth part configured in the cell might be larger than what the device can handle. A separate initial bandwidth part, specific for RedCap devices, can therefore be configured through the system information.

Cell barring can be signaled in SIB1 separately for RedCap devices, allowing the network to separately restrict one-antenna and two-antenna RedCap devices from accessing the network. The background to this functionality is the concern of potential performance degradation of the system performance from the presence of less capable devices in the network. If this is a concern for an operator in specific cells, these cells can be configured not to allow some or all RedCap devices from accessing the cell.

Early indication whether a device is a regular device or a RedCap device, and hence may require special treatment, is provided as part of the random-access procedure. RedCap-specific preambles can be configured, otherwise, msg3 (or msgA for 2-step random access) indicates that random access is from a RedCap device by using one of two

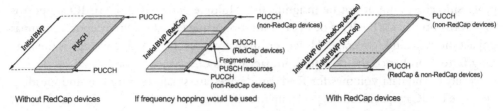

Fig. 22.2 Avoiding uplink resource fragmentation when introducing RedCap.

specific LCID values for the CCCH. Using a specific LCID value instead of explicit sig-naling was chosen as there was no additional bits available without changing the signaling structure.

The use of a smaller bandwidth part for RedCap devices than for regular devices can result in resource fragmentation due to RedCap PUCCH transmissions, see Fig. 22.2. If all devices support the same bandwidth, which is the case for baseline NR, the PUCCH can be transmitted at the edges of the initial bandwidth part as shown in the left of the figure. However, if some devices do not support the full bandwidth, as is the case for RedCap devices, their PUCCH transmissions will typically take place at a different fre-quency region than the baseline devices and cause fragmentation of the PUSCH resource. Resource fragmentation would limit the possibility to schedule wide band-widths for non-RedCap devices and negatively impact the data rates they may achieve.[a] To mitigate this, PUCCH frequency hopping can be disabled for msg4. By disabling PUCCH frequency hopping and placing the RedCap initial bandwidth part at the end of the regular initial bandwidth part, see the right part of Fig. 22.2, uplink resource fragmentation can be avoided.

One consequence of the limited bandwidth of RedCap devices is that they may all end up camping on the same part of the overall carrier bandwidth, namely the part where the SSB is transmitted. This may not be desirable as this part of the carrier may become congested. To alleviate this, multiple bandwidth parts can be configured in the Redcap devices with one SSB per bandwidth part. Non–cell-defining SSBs, see Chapter 16, are used for the bandwidth parts not covering the cell-defining SSB. RRM measurements can be configured to use the non–cell-defining SSBs.

22.1.2 Single-branch Rx antenna

As a baseline, although depending on the frequency bands supported, NR devices all have at least two and often four reception antennas and are therefore capable of receive diversity and typically also two-layer MIMO. This is beneficial from a performance per-spective. However, for low-end devices, the cost of two antennas and the associated

---

[a] If non-contiguous resource allocation in the uplink would be supported, fragmentation would not be an issue.

processing chains might not be insignificant. Multiple antennas and MIMO support is therefore optional for RedCap devices.[b] Not supporting (at least) two reception antennas impact the downlink coverage. However, in many cases the uplink coverage is the limiting factor and hence the downlink impact is less of a concern for a RedCap device, especially as the data volumes for RedCap devices are small. During the study leading up to the RedCap specifications [111], up to 3 dB loss[c] in downlink coverage was deemed acceptable and only for msg2 and msg4 during random access this needs to be mitigated. For msg2, a lower modulation-and-coding scheme can be used and for msg4 hybrid-ARQ retransmissions will helps to ensure sufficient coverage. Since RedCap devices are distinguishable from normal devices already during the random-access procedure, the gNB can select the appropriate transmission parameters for msg3 (or msgB).

In FR1, RedCap devices can support one or two antennas and the corresponding number of MIMO layers. The number of antennas is indicated as a device capability and the gNB takes this into account when performing scheduling.

In FR2, RedCap devices support one or two MIMO layers, independently of the number of antennas implemented (obviously the number of antennas need to larger than or equal to the number of MIMO layers). The number of MIMO layers supported is indicated as part of the device capabilities. There is also the possibility to support a new lower transmit power class to further reduce the device cost in FR2.

Another simplification made for RedCap devices is to not require support for 256-QAM (and obviously not for 1024-QAM either).

22.1.3 Half-duplex FDD

Typically, FDD devices support full duplex communication, that is, simultaneous transmission and reception in the uplink and downlink, respectively. Simultaneous transmission and reception is beneficial from, for example, a capacity and latency perspective. However, it requires a duplex filter which comes at a cost.

To reduce cost, an FDD RedCap device may support half-duplex operation only (the network can still operate using full duplex by scheduling different devices in uplink and downlink). Using half-duplex at the device implies that no duplex filter is needed in the device. Instead, a switch can be used to isolate the receiver from the transmitter, see Fig. 22.3. Such a switch has around 0.8 dB lower insertion loss compared to a duplex filter. Consequently, HD-FDD RedCap devices are required to have up to 0.8 dB improved receiver sensitivity compared to a full-duplex FDD device in some frequency bands.

---

[b] For RedCap devices in FR1, the number of MIMO layers is always identical to the number of supported reception antennas.

[c] During the actual specification work, no 3 dB loss was assumed and hence the coverage is better than what is suggested by the study in [111].

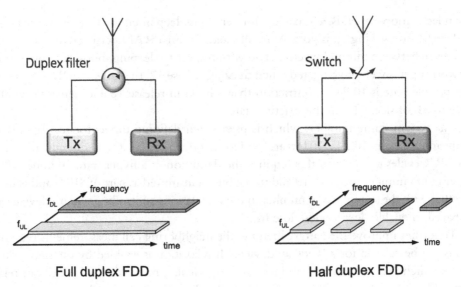

Fig. 22.3 Full duplex (left) and half duplex (right) devices.

The switching times for a half-duplex FDD device are the same as used for TDD. Similarly, the rules for prioritization between downlink and uplink in the device also follows the rules defined for TDD with some exceptions [112].

Whether a cell is supporting half-duplex FDD RedCap devices or not is indicated in SIB1. The purpose is to ensure that half-duplex RedCap devices refrain from accessing a cell not supporting the duplex characteristics required by the device.

22.1.4 Higher-layer complexity reductions

To reduce the higher-layer complexity, there are a set of smaller enhancements introduced as part of RedCap. The changes are mainly related to limitations of the parameter ranges, rather than changing the basic protocol structure. For example, a RedCap device does only have to support up to 8 radio bearers, unlike up to 16 as required for a regular device. Given the lower data rates, PDCP and RLC sequence numbers can be limited to 12 bits (as opposed to 18 bits for regular devices). Finally, automatic neighbor (ANR) functionality used to support mobility is optional for RedCap devices as the scenarios targeted by RedCap devices have no or limited mobility requirements.

22.1.5 Enhanced DRX and neighboring cell measurements

Device energy consumption is a concern, especially for wearables which is one of the device types targeted by RedCap. To reduce the energy consumption and thereby prolonging the battery lifetime for RedCap devices, the DRX mechanism has been extended – known as eDRX – to allow for very long sleep intervals. Close to three hours of sleep before waking up is possible in idle state. From a RAN perspective, there is no difference between idle and inactive state with respect to sleeping, but since not all core network aspects were investigated when freezing release 17, the longest eDRX cycle for the inactive state is 10.24 s, a restriction that is lifted in release 18 and hence providing close to 3 h of sleep also in the inactive state.

The system frame number, which is used when defining timers for regular DRX, wraps around every 10.24 s (10 bits are used to encode the SFN). Consequently, handling of eDRX cycles longer than this requires the definition of a hyper frame number. The hyper frame number provides 10 additional bits, transmitted as part of SIB1, and is used together with the system frame number in the calculations controlling when a device can be paged to enable sleep cycles close to 3 h.

To further improve the battery lifetime, the neighboring cell measurements requirements can be relaxed for a RedCap device.[d] If relaxation is enabled by the network, a device which is judged to be stationary and, optionally, not at the cell edge, can relax its neighboring cell measurements. Whether the device is considered stationary or not is based on RSRP/RSRQ measurements.

22.2 Small data transmission

The RRC state machine in NR has been designed with different RRC states as discussed in Chapter 6. Most of the time the device is in the idle state with no possibility for data transfer. When the need for data transfer arises, a connection is established with the device moved to the connected state and data transfer subsequently takes place. Once the data has been transferred, the device, at some point in time, returns to the idle or inactive state. Limiting data transfer to the connected state only is a good approach for larger amounts of data as transmission can make use of the device's unique capabilities, something which is possible in connected state only.

Moving from idle to connected state requires signaling of configuration parameters and establishing a connection with the core network. The amount of signaling for this may be non-negligible and the inactive state was therefore introduced in NR from the first release. Inactive state is a good step towards reducing the overhead as the configuration information and the connection to the core network are kept when transitioning back and forth between inactive and connected states, unlike the case when moving from idle state. However, for transmission of small payloads, which is common in, but not limited to, IoT scenarios, the signaling overhead related to moving to connected state might be non-negligible compared to the amount of user data, despite the use of the inactive state. Therefore, release 17 introduces support for *small data transmission* (SDT), which in essence is a procedure allowing (uplink) data and/or signaling transmission while

---

[d] Relaxation was introduced already in release 16 but further refined in release 17.

remaining in the inactive state, that is, without transitioning to from the inactive to the connected state. Release 17 focused on mobile-originated data transmission; enhancements for small amounts of mobile-terminated data are part of release 18.

22.2.1 Triggering of SDT

Small-data transmission is configured on a radio-bearer basis. Upon data arrival in a device in the inactive state and with SDT enabled, the device takes a decision whether to use the SDT procedure for data transmission or whether to initiate a transfer to the connected state and subsequent data transfer. If the payload is below a threshold and the downlink RSRP is above another threshold, the device initiates the SDT procedure, otherwise it uses random access to establish a connection in the same way as in earlier releases. The intention behind these rules is to use SDT in situations when the channel conditions are good and the amount of data is small. For larger payloads, connected state is more efficient and SDT is not used.

Two schemes for uplink small data transmission are defined: SDT with random access and SDT with configured grants.

22.2.2 SDT with random access

For the random-access-based PUSCH transmission, user data is transmitted as part of msg 3 or msg A for four-step and two-step random access, respectively (Fig. 22.4). Compared to regular random access, which does not allow for data transmission as part of msg 3 as configuration of, for example, security has not been completed, uplink data can be transmitted in an SDT-initiated random access. To handle early data transmission, which includes resuming the DRB and security configuration as well as providing a sufficiently large scheduling grant in msg3, it is necessary for the network to distinguish between SDT-imitated random access and regular random access. This is achieved by configuring a subset of the random-access preambles for SDT only, while the remaining set of preambles are used for regular random access.

In case of SDT with four-step random access, upon transmitting the selected preamble, the network responds with msg 2 (also known as random-access response, RAR) as outlined in Chapter 17. Compared to a regular random access, the size of the uplink resources allocated by msg2 and the associated transport-block size are larger for SDT to accommodate (some of) the user data as part of msg 3.

Once msg 3, including the user data, has been transmitted, the device does not know whether there was an uplink collision. Therefore, similar to regular random access, the device needs to receive msg 4 for contention resolution. If the contention resolution was successful, the device knows that the (first part of) the user data was properly received. The device will now start monitoring the PDCCH using the C-RNTI such that subsequent data can be scheduled on the uplink despite the device still being in the inactive state.

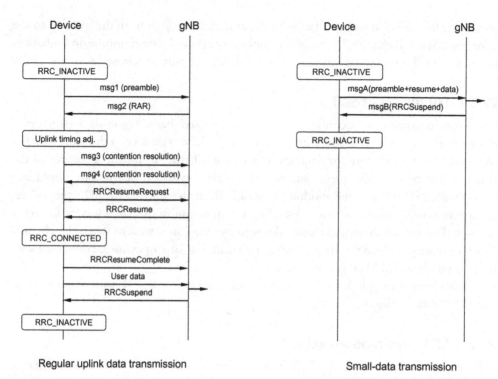

Fig. 22.4 Example of the random-access-based SDT procedure vs regular random-access followed by data transmission.

The motivation for this is to handle situations when the amount of data to transmit is larger than what would fit into msg3 but still small enough not to motivate the transition to connected state. Although large amounts of data can be transmitted from the device using this mechanism without moving to the connected state, it is inefficient to do so as the advanced features available in the connected state cannot be used. Therefore, msg4 may also contain RRC signaling to transfer the device to the connected state to handle transmission of additional data using regular uplink scheduling, or to release it to the inactive (or even idle) state if there is no need for further data transfer at this point in time.

The two-step SDT procedure is similar to the four-step procedure. The set of preambles associated with SDT and used in msgA are in this case linked to a larger PUSCH allocation, sufficient to handle (the first segment of) the user data. Contention resolution is necessary also in this case. This is handled as described in Chapter 17 by msgB, which, similar to the four-step procedure, may contain RRC signaling to change the RRC state of the device.

The SDT attempt can fail for several reasons, for example if the maximum number of random-access attempts have been reached, the SDT failure detection timer expires, or if the device performs a cell reselection. In case of an unsuccessful SDT attempt, the device moves to the idle state.

22.2.3 SDT with configured grants

Configured grants can be used in combination with SDT. In a scenario with stationary devices and periodic traffic, it could be an efficient alternative to random-access-based SDT if longer periodicities than in connected state would have been supported which unfortunately is not the case in release 17.

The device is configured with a type 1 configured grant. The grant is only valid in the PCell from which the device was moved to the inactive state, that is, mobility between cells is not supported and random-access-based SDT or conventional connection establishment has to be used in these cases.

One challenge with configured grants is that the device needs a valid timing advance for successful PUSCH transmissions. In connected state, the gNB can measure on the uplink transmissions and send timing advance commands to the device, but unlike random-access-based SDT there is not feedback from the network. Hence, the device needs to determine by itself whether the timing advance value still is valid. For this purpose, a configurable timer is used. Upon expiry of this timer, the uplink timing is considered invalid and the configured grant resources are released. For the initial configured-grant transmission the downlink RSRP is also taken into account; if the RSRP of a configurable set of SSBs falls below a threshold, the uplink timing is not considered valid and the configured-grant resources are released. The reason for this mechanism is to have the configured grant valid only in some beams and to drop it when the device moves outside the area covered by these beams.

CHAPTER 23

Multicast-broadcast services

Data transmission in the first releases of NR focuses of data intended for a single user only and transmission protocols and radio-resource management procedures are all designed with this in mind.[a] However, there are use cases where *multiple* users are interested in the *same* information. Radio and TV broadcasting networks are two well-known examples of networks focusing on covering very large areas with the same content, with no or limited possibilities for transmission of data intended for a single user. Other examples, potentially more relevant to cellular systems, when delivery the same data to multiple devices is of interest are first responders, where a dispatcher need to communicate with a group of responders, and software upgrades of multiple devices.

To better address these types of scenarios, release 17 introduces broadcast and multicast functionality to NR. The provision of broadcast/multicast services in a mobile-communication system implies that the same information is simultaneously provided to multiple devices, which in many cases reduces the amount of network resources required compared to individual transmissions to each of the devices. The remainder of this chapter will describe the details behind broadcast/multicast in release 17 and some of the enhanceents in release 18.

23.1 Unicast, multicast and broadcast

Unicast communication is the baseline in NR, that is, data is intended for (or coming from) a *single* device. IP packets enter (or leave) the core network via the *user plane function* (UPF) with a *PDU session* handling the end-to-end connectivity between the UPF and the device. In the radio-access network, data are transmitted over one or more *data radio bearers* (DRBs). This was outlined in Chapter 6 and is the background to many of the subsequent chapters.

Multicast/broadcast services (MBS), on the other hand, refer to services where the data can be (and typically are) intended for *multiple* devices. In analogy with unicast transmissions, IP packets arrive to the MB-UPF in the core network and the *MBS session* provides the end-to-end connectivity between the MB-UPF and the device. Two types of services are provided, *broadcast* services and *multicast* services.

---

[a] System information is an exception as the data is relevant for all users in a particular cell.

5G/5G-Advanced
https://doi.org/10.1016/B978-0-443-13173-8.00014-1

Broadcast services allows any MBS-capable device in the broadcast area, irrespective of the RRC state, to receive the data without contacting the network to join an MBS session. Devices may start and stop receiving a broadcast transmission whenever they want without informing the network and no uplink transmissions are required. Consequently, the network is not aware of which devices that are receiving what content. This is similar to receiving a TV program at home without information the TV broadcast operator about which channel you are watching. Retransmissions based on feedback from the devices cannot be used, neither can other features relying on device feedback. Thus, transmission parameters have to be chosen such that all positions within the broadcast area are sufficiently well covered. Since no feedback signaling is used, broadcast transmissions are highly scalable with no upper limit on the number of users possible to accommodate.

Multicast services, on the other hand, implies that the data is transmitted to a group of devices. Prior to receiving multicast data, the device needs to contact the network to join an MBS session and be added to the corresponding multicast group. Once the device has joined the MBS session, the device is provided with the relevant RRC configuration information and is ready for the session to be activated and data to be transmitted. Multicast inherits most of the unicast functionality, including the possibility for feedback from the devices. This can result in highly efficient transmissions but is not as scalable as broadcast transmissions as the need for feedback signaling eventually will limit the capacity. Furthermore, MBS data reception is only possible in the connected state. Therefore, when an MBS session is activated, group paging is used to ensure all devices in the corresponding group are in the connected state before any data transfer may take place. In release 18, multicast reception is possible also in inactive state and the overhead from paging can thus be avoided.

For both broadcast and multicast services, MBS data enters the MB-UPF and is transmitted to the MBS-capable gNBs using shared transport, that is, each gNB receives a single copy of each MBS data packet. The involved gNBs in turn uses the *MBS radio bearer* (MRB) to deliver the packets to all interested devices using *point-to-multipoint* (PTM) or *point-to-point* (PTP) transmission as illustrated in Fig. 23.1. Broadcast services can use PTM only, while multicast services can use PTM or PTP for conveying the data to the devices.

In addition, in order to support gNBs not capable of handling the MBS functionality added in release 17 but still involved in the multicast transmission, MBS data can be delivered individually to each of the devices using unicast transmission, see the rightmost part of Fig. 23.1. In this case, a DRB session is used for each of the devices and separate copies of the data, one for each devices involved, is delivered to the gNB.

23.2 Channel structure

Multicast/broadcast services use MRBs to carry the MBS data in an MBS session. Each MBS session has an associated identifier, the *Temporary Mobile Group Identity* (TMGI) and the MRB data is transmitted to the devices using PTP ("PTP-MRB"), PTM

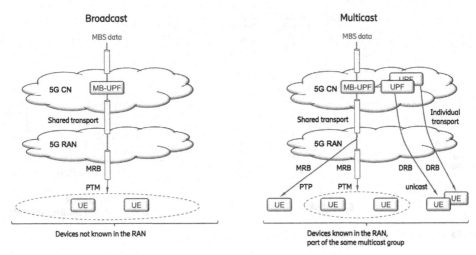

Fig. 23.1 Illustration of broadcast and multicast.

("PTM–MRB"), or a combination of the two ("split MRB"). Broadcast transmissions can only use PTM, while multicast can use any of the three alternatives. The same general user plane protocol architecture as described in Chapter 6 is reused for multicast/broadcast service although not all the functionality is used:

- SDAP; each QoS flow is mapped to one MRB, either the same MRB for multiple QoS flow or individual MRBs, one for each QoS flow. If both multicast and broadcast transmissions are used in a cell separate MRBs are used for the corresponding MBS sessions.
- PDCP; for multicast services the PDCP handles header compression and, in case of a split MRB, selection between PTM or PTP delivery. Ciphering and integrity protection are not used for MBS data. For PTM links, PDCP retransmissions are not used.
- RLC; for broadcast and multicast delivery with PTM-MRB, or with the PTM leg of a split-MRB, UM is used as there is no UL-SCH and hence no possibility for RLC status reports in this case. For PTP-MRB, or the PTP leg of the split MRB, AM or UM can be used.
- MAC; hybrid-ARQ retransmissions are possible for both PTM and PTP delivery.

Two new logical channels are introduced to support PTM delivery for multicast/broadcast services, the *multicast traffic channel* (MTCH) and the *multicast control channel* (MCCH), both of which are mapped to the DL-SCH transport channel (Fig. 23.2).

The MTCH is the logical channel type used to carry MBS data transmitted using PTM from one MRB, either from an PTM-MRB or from the PTM leg of a split MRB. A device may receive multiple MTCHs, for example if it is interested in multiple MBS services. Scheduling and transmission of MTCH uses the G-RNTI or, in case of

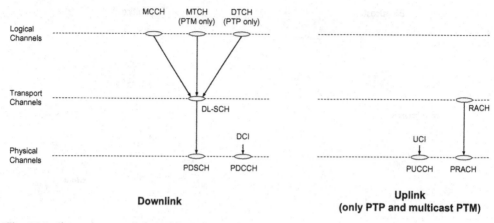

Fig. 23.2 Channel structure for MBS.

semi-persistent scheduling, the G-CS-RNTI. Multiple G-RNTIs (and G-CS-RNTIs) can be configured in a device in case the device is supposed to receive multiple MRBs.[b]

The MCCH is the logical channel type used to carry control information necessary for reception of MBS data transmitted using broadcast. The MCCH uses RLC-UM and the MCCH-RNTI is used when scheduling and transmitting data from the MCCH.

For MBS data transmitted using PTP, either from an PTP-MRB or the PTP leg of a split MRB, the DTCH is the logical channel and the C-RNTI is used as for any unicast transmission.

Some examples of different possibilities for MBS data delivery are illustrated in Fig. 23.3:

- broadcast session using RLC-UM and PTM delivery,
- multicast session using RLC-UM and PTM delivery, and
- multicast session using RLC-UM or RLC-AM and PTP delivery.

23.3 Downlink data transmission

Point-to-point transmissions of MBS data on a DTCH is straightforward from a DL-SCH processing point-of-view as these transmissions are no different from other unicast transmission targeting a single device. The parameters required to receive the data, including the scrambling sequence and time-frequency resources used, are conveyed in the same way as for other unicast transmissions using the PDCCH.

Point-to-multipoint transmissions, on the other hand, require careful attention as a single transmission is to be received by more than one device. The MTCH and MCCH

---

[b] It is possible to map multiple MRBs to a single G-RNTI (not all devices support multiple G-RNTIs).

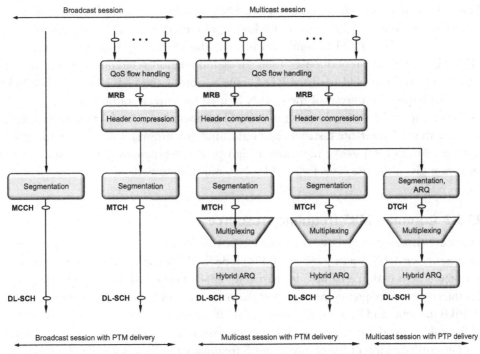

Fig. 23.3 Illustration of different possibilities for broadcast and multicast services.

are mapped to the DL-SCH and use the transport-channel processing described in Chapter 9 with some restrictions. First, to limit the amount of specification changes needed to support multicast/broadcast and to increase the likelihood of devices implementing this new feature, no new numerologies are introduced. In other words, the same subcarrier spacing and cyclic prefix as supported in release 15 are used also for multicast/ broadcast in release 17. Furthermore, as all involved deceives must have the same understanding of the parameters used for the transmission, including scrambling sequences and time-frequency resources, there are some enhancements in this area as described in the following sections.

23.3.1 Single frequency networks

The baseline for downlink transmissions in NR is to transmit the signal to be received from one site without coordinating the transmission activity across sites. The device receives the signa, subject to interference due to the transmission activity in neighboring sites. However, if the transmissions from the different sites are *truly identical* and *mutually time/frequency-aligned*, the resulting signal at the device will appear as a single transmission, subject to (severe) multi-path propagation. This can easily be handled by the OFDM receiver "for free" without knowing which sites that are involved in the transmission.

This is known as *single-frequency network* (SFN[c]) and can improve the performance as transmissions from neighboring sites no longer add interference but useful signal.

To support SFN PTM transmission in NR, the scrambling sequence used for the PDSCH, as well as the sequences used for demodulation reference signals, are both dependent on the RNTI used for PTM transmissions – the G-RNTI, G-CS-RNTI, or MCCH-RNTI – instead of the C-RNTI used for unicast transmissions.

Given the cyclic prefix duration in NR, SFN operation is useful mainly in deployments with small inter-site distances and tight time synchronization between the gNBs, and less relevant for TV-like high-tower high-power deployments. This is not a major limitation given the use cases targeted by NR multicast/broadcast.

23.3.2 Common MBS frequency resource

To ensure that all devices have the same understanding of the time-frequency resources used, a *common MBS frequency resource* (CFR) is defined as a contiguous subset of common resource blocks within a bandwidth part (Fig. 23.4). Point-to-multipoint transmissions use the common frequency resource for the resource-block mapping, including VRB-to-PRB mapping and PRB-to-CRB mapping. It can thus to some extent be viewed as an additional bandwidth part used for PTM transmissions although it is not defined as such.

The configuration of the common MBS frequency resource is handled by RRC signaling but differs slightly between multicast and broadcast as the latter is supported entirely in the inactive and idle states.

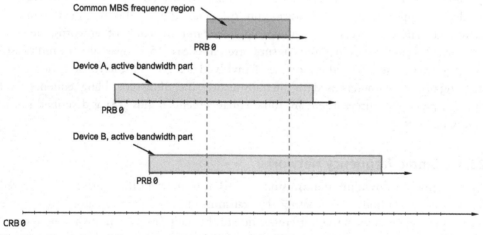

Fig. 23.4 Illustration of the common MBS frequency resource for multicast PTM.

[c] Not to be confused with *System Frame Number*, which also is abbreviated SFN.

For multicast MRBs, irrespective of PTP or PTM as the delivery method, all the MBS configuration is provided using dedicated RRC signaling, including configuration related to PDSCH, PDCCH, and semi-persistent scheduling. The common MBS frequency resource must be within the active bandwidth part as a device is not supposed to receive anything outside the active bandwidth part. It is up to the network to ensure that any changes of the active bandwidth part in a device are such that this assumption is not violated.

For broadcast MRBs, the common frequency resource by default covers the same frequency as CORESET#0, but it is possible to configure different values as part of SIB20. The MTCH and MCCH use the same common frequency resource.

23.4 Hybrid ARQ retransmissions

Retransmissions, either hybrid-ARQ or RLC retransmissions, serve an important role in unicast transmissions to ensure error-free data delivery. For MBS services delivered using PTP transmissions, these mechanisms can obviously be used in the same way as for any other unicast transmission.

For PTM delivery, the situation is different. As no UL-SCH upon which RLC status reports can be transmitted is present in this case, neither for multicast nor for broadcast, RLC retransmissions cannot be used and the RLC operates in unacknowledged mode. However, hybrid-ARQ retransmissions can be used for multicast transmissions with acknowledgements handled according to one of three different alternatives, selected as part of the RRC configuration:

- ACK/NAK,
- NAK only,
- No ACK/NAK.

The first two alternatives are applicable for multicast only as they require the possibility to transmit feedback information while the third alternative, no ACK/NAK, can be used for both multicast and broadcast services.

ACK/NAK operates in the same way as for unicast transmissions. Preferably, the configuration and signaling is such that different devices use different PUCCH resources to avoid collisions. This also allows the gNB to identify which device(s) that could not decode the transport block and allows the retransmission to use transmission parameters optimized for the particular device(s), for example by using PTP to a single device. Statistics on the hybrid-ARQ feedback can also be used to adjust the transmission parameters, for example the modulation-and-coding scheme or the bandwidth used, thereby optimizing the resource usage in the system. The same hybrid-ARQ codebook mechanism as for unicast are reused for multicast. Multicast and unicast may also be multiplexed using the same codebook and PUCCH resources, alternatively they may use separate codebooks and PUCCH resources.

NAK only, where only negative acknowledgements are transmitted and nothing is transmitted in case of successful reception of a transport block, is another alternative. In this case, the configuration is typically such that multiple devices are configured with the same PUCCH resource. Only PUCCH formats 0 and 1 are supported in this case and the gNB can use energy detection to determine whether one or more devices requested a retransmission. The gNB cannot identify which device(s) that could not decode but can conclude that at least one device was not able to do so. PTM is used for the retransmissions in this case. NAK-only feedback can be indicated for up to four transport blocks by selecting one out of 15 resources for transmitting the NAK.

No ACK/NAK transmissions, that is, no feedback from the device, is the third alternative and applicable not only to multicast but also broadcast. Obviously the gNB does not know whether devices successfully decoded the data or not, but it allows for "blind" retransmissions. The gNB can choose to transmit the same transport block multiple times and a device not successful in decoding the first attempt could perform soft combining of multiple retransmissions to increase the likelihood of successful reception. This is useful to improve the coverage of an MBS session.[d]

Different devices may use different hybrid-ARQ feedback signaling as they are individually configured. It is also possible to enable/disable the transmission of acknowledgements for all devices receiving a certain G-RNTI (assuming that devices are configured appropriately to transmit acknowledgements when enabled to do so).

23.5 Downlink control signaling

To support the scheduling of multicast and broadcast transmissions, the regular PDCCH structure is used. The CORESET used for monitoring of multicast/broadcast-related PDCCH transmissions must be within the common MBS frequency resource, otherwise not all devices may be able to receive the PDCCH.

Three new DCI formats are introduced in release 17 to support multicast and broadcast transmissions:

- DCI format 4_0, used for broadcast transmission of MBS data (with the G-RNTI) or MCCH transmissions (with MCCH-RNTI). The format is based on, and size-aligned with, the fallback DCI format 1_0.
- DCI format 4_1, used for multicast transmission of MBS data (using the G-RNTI or CS-G-RNTI). The format is based on, and size-aligned with, the fallback DCI format 1_0.

---

[d] For multicast services, slot aggregation is also possible, see Section 23.6.1.

- DCI format 4_2, used for multicast transmission of MBS data (using the G-RNTI or CS-G-RNTI). The format is based on the non–fallback DCI format 1_1. The size is configured by RRC signaling, which allows the network to align the size with the size of, for example, DCI size 1_1 to meet the DCI size budget.

The content of the three multicast/broadcast related DCI formats are summarized in Table 23.1. Comparing with the unicast formats in Table 10.2, many information fields are the same between unicast and broadcast/multicast, but there are also some differences.

Table 23.1 DCI formats 4_1, 4_1, and 4_2 used for broadcast/multicast downlink scheduling.

	Field	DCI format 4_0	DCI format 4_1	DCI format 4_2
Format identifier				
Resource information	CFI			
	BWP indicator			
	Frequency domain allocation	•[17]	•[17]	•[17]
	Time-domain allocation	•[17]	•[17]	•[17]
	VRB-to-PRB mapping	•[17]	•[17]	•[17]
	PRB bundling size indicator			•[17]
	Reserved resources			•[17]
	Zero-power CSI-RS trigger			•[17]
	Scheduling offset			
	Channel-access type			
	Dormancy indication			
Transport-block related	MCS	•[17]	•[17]	•[17]
	NDI		•[17]	•[17]
	RV	•[17]	•[17]	•[17]
	MCS, 2nd TB			•[17]
	NDI, 2nd TB			•[17]
	RV, 2nd TB			•[17]
	Priority indication			•[17]
Hybrid-ARQ related	Process number		•[17]	•[17]
	DAI		•[17]	•[17]
	PDSCH-to-HARQ feedback timing		•[17]	•[17]
	CBGTI			
	CBGFI			
	PDSCH group index			
	One-shot HARQ request			
	Number of PDSCH groups			
	New feedback indicator			
	Enhanced type 3 codebook indicator			
	HARQ retransmission indicator			
	Enable/disable HARQ feedback			•[17]

Continued

Table 23.1 DCI formats 4_1, 4_1, and 4_2 used for broadcast/multicast downlink scheduling—cont'd

	Field	DCI format 4_0	4_1	4_2
Multi-antenna related	Antenna ports			•[17]
	TCI			•[17]
	SRS request			
	SRS offset indicator			
	DM-RS sequence initialization			•[17]
PUCCH-related information	PUCCH power control			
	PUCCH resource indicator		•[17]	•[17]
	PUCCH cell indicator			
PDCCH-related information	PDCCH monitoring adaptation			
MCCH change indicator		•[17]		

For example, there is no need for the BWP indicator as multicast and broadcast transmissions always use the common MBS frequency resource, neither is there any need for the SRS request field as there is no uplink data transmission associated with the downlink-only multicast and broadcast transmission. Some of the HARQ-related fields are also omitted. For example, CBG retransmissions are not supported and hence there is no need for these information fields.

There is also a new field in DCI format 4_0, the MCCH change indicator, used when indicating an update of the MCCH information as discussed in Section 23.6.3.

As seen in Table 23.1, there are no power control commands in the new DCI formats and an interesting question is how the PUCCH is power controlled. Including power control commands in DCI formats 4_1/4-2 would not solve the problem as the required PUCCH transmission power is different for different devices. Therefore, uplink power control has to be handled using individual signaling to each of the devices, either by using DCI format 2_2 for power control or, if unicast transmissions are also taking place, the power control commands present in downlink scheduling assignments (DCI formats 1_x).

23.6 Scheduling

Scheduling of MBS transmissions controls when and on what resources the transmission takes place. For PTP transmissions, no special treatment with respect to scheduling is required and the same mechanisms as for any other unicast transmission can be used. (C-RNTI, SPS, etc.). For PTM transmissions – broadcast or multicast – some enhancements are needed as the targeted devices may not necessarily be in the connected state. The mechanisms used are somewhat different for multicast and broadcast transmissions as discussed below.

23.6.1 Scheduling of multicast services

Multicast transmissions on the MTCH can only be received in the connected state in release 17 while release 18 also enables reception in inactive state. However, a device subscribing to a multicast service may not be in the connected state when data arrive and an MBS session is activated. To cater for devices not in connected state, the paging mechanism is used to ensure that all devices in the multicast group transition to connected state. In all paging occasions where there is at least one device subscribing to the MBS service, that is, has joined the MBS session, a paging transmission takes place, but using the TMGI of the MBS session to be activated instead of the device identity. A device subscribing to this particular MBS session will, upon detecting the corresponding TMGI, resume the connection when in inactive state or establish the connection when in idle state using the random-access mechanism. Once all the devices are in connected state, the multicast transmission can be scheduled using DCI format 4_1 or 4_2.

Semi-persistent scheduling can be used for multicast transmissions. In this case, the device is configured in advance with the periodicity of the transmissions using RRC signaling. Activation (and deactivation) of semi-persistent scheduling is done using the PDCCH as for dynamic scheduling but with the G-CS-RNTI instead of the G-RNTI normally used for multicast scheduling. The parameters necessary to receive the multicast transmission, for example frequency allocation and modulation-and-coding scheme, are provided on the PDCCH when the semi-persistent scheduling is activated. This follows the same principles as for conventional unicast transmissions, see Section 14.4.

To conserve device power, it is possible to configure DRX patterns for multicast reception, one per G-RNTI configured in the device. These DRX patterns operate in a similar way as, and independent from, the unicast DRX patterns (see Section 14.5.1).

Slot aggregation, that is, transmitting one transport block across multiple slots to improve coverage, is possible. The same mechanism as for unicast transmissions are used with the time-domain allocation field of the DCI pointing into a table from which (among other information) the number of aggregated slots is obtained (see Section 10.1.15).

23.6.2 Scheduling of broadcast services

Scheduling of broadcast transmissions on the MTCH differs from scheduling of multicast transmissions as broadcast data can be received by a device irrespective of the RRC state. For idle and inactive devices, the network has no knowledge of which devices that are interested in a particular broadcast service. The paging mechanism described above can therefore not be used. Instead, the MCCH provides a list of all broadcast services transmitted on the MTCHs, their TMGIs, and the associated G-RNTI used when scheduling data transmissions. A device interested in MBS reception will monitor the PDCCH for potential scheduling of broadcast transmissions using the relevant G-RNTIs.

Upon detecting a DCI with the relevant G-RNTI, the device will receive the data scheduled. Note that the use of DCI format 4_0 implies that broadcast transmissions are dynamically scheduled – a PDCCH monitoring occasion does not imply that a broadcast transmission will occur at a certain point in time, only that a DCI scheduling a broadcast transmission *may* occur at that point in time. Dynamic scheduling allows the amount of resources used in the frequency domain to be dynamically adjusted to match the amount of data to transmit. It is thus a more flexible scheme compared to the fixed MBMS scheduling used in LTE [26].

Monitoring for G-RNTIs by default uses the same common search space as used for SIB1 transmissions, but it is possible to configure a different search space using the MCCH. MCCH may also provide additional MTCH scheduling information, which in essence is a DRX configuration informing the device about when transmission of a certain MBS session may occur. Together with the MTCH search space, the device thus knows when to monitor for a certain G-RNTI. Multiple G-RNTIs can be monitored in case the device is interested in more than one MBS session.

For MBS broadcast, there is no support for hybrid-ARQ retransmissions as there is no possibility for feedback from the devices. Furthermore, the network is not aware of which devices that receive what services. Nevertheless, the hybrid-ARQ buffers and soft combining mechanism can be used to mimic slot aggregation to improve the coverage of a service. The gNB can transmit the same transport block multiple times, potentially with different redundancy versions, and the device can perform soft combining of these attempts. Using one of the hybrid-ARQ processes, including the soft combining buffer, reduces the device complexity in situations when the device is receiving both unicast and broadcast transmissions. In this case the slot aggregation buffer may already be occupied by the unicast transmissions and it is preferable to use one of the hybrid-ARQ buffers instead of implementing a separate slot aggregation buffer for the broadcast service.

23.6.3 Scheduling of broadcast control information

The broadcast control information is carried on the MCCH and includes a list of ongoing MBS broadcast sessions and, for each such session, the associated G-RNTI. It may also, for each ongoing MBS session, include information about the PDCP, RLC, and PDSCH used for the MTCH.

The MCCH content is transmitted on the PDSCH within periodically occurring transmission windows as illustrated in Fig. 23.5. Within each transmission window, the device monitors the PDCCH for DCI format 4_0 with MCCH-RNTI. If a valid DCI is detected, the PDSCH is received to obtain the MCCH content. The monitoring

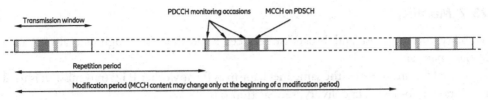

Fig. 23.5 Monitoring for MCCH transmissions.

occasions, that is, when the MCCH can be transmitted, is configurable but defaults to the same occasions as used for SIB1.

The content of the MCCH may change over time, for example if an MBS service is added or removed from the set of services broadcasted in the area or if the potential scheduling frequency of a service need to change. Such changes are only allowed to occur at a so-called modification period boundary. To indicate MCCH changes, two bits in DCI format 4_0 used to schedule the PDSCH carrying the MCCH are used. One of the bits indicate that a new MBS service has been added, the other bit indicate other changes (for example configuration changes for an ongoing MBS service).

23.6.4 Multiplexing of unicast, multicast, and broadcast transmissions

Unicast, multicast, and broadcast transmissions can be multiplexed, both in time and frequency, within the limitations of the device capabilities.

For unicast and multicast services, the device capabilities, for example the amount of spatial multiplexing supported by the device or whether carrier aggregation is supported, are known to the network and can be taken into account in the scheduling.

For broadcast services, on the other hand, the network is not aware of the device capabilities as a device can choose to receive broadcast transmissions at its own discretion. However, for the case of simultaneous broadcast and unicast reception, there might be limitations in what the device is capable of. Therefore, there is a possibility for the device to send an MBS interest indication to the network to indicate that it is interested in receiving certain broadcast services. The scheduler in the gNB can take this into account to schedule unicast transmissions to this device such that they do not interfere with the possibility to receive broadcast transmissions. Release 18 also provides a mechanism where a device receiving broadcast temporarily reduces its capabilities for unicast reception. The background to this feature is that the device may need to use some of its capabilities, for example receiver processing otherwise used for carrier aggregation, when receiving broadcast transmission. Expressed differently, without informing the network which broadcast services the device is receiving, it would not be possible for unicast services to benefit from the processing otherwise used for broadcast reception.

23.7 Mobility

Mobility for multicast and broadcast services is possible, although not always in a lossless manner.

For PTP transmissions, the same mechanism as for any other PTP transmission is used to guarantee lossless data delivery, see Section 6.6.

For multicast transmissions, PTM-to-PTM lossless mobility is possible if blind PDCP retransmissions are used and the PDCP sequence numbers are synchronized across the involved cells. In this case, the device may receive duplicate copies of the same packet, but the duplicate removal mechanism in the PDCP will discard those packets.

For broadcast transmissions, mobility is based on cell reselection and lossless mobility cannot be ensured. However, the source cell can indicate on the MCCH whether neighboring cells provide the same MBS session. This information can be used by the device in the cell reselection decision.

CHAPTER 24

Integrated access backhaul

NR *integrated access backhaul* (IAB) was introduced in 3GPP release 16. It provides functionality that allows for the use of the NR radio-access technology not only for the link between base stations and devices, sometimes referred to as the *access link*, but also for wireless backhaul links, see Fig. 24.1.

Wireless backhaul, that is, the use of wireless technology for backhaul links, has been used for many years. However, this has then been based on radio technologies different from those used for the access links. More specifically, wireless backhaul has typically been based on some proprietary, that is, non-standardized, radio technology operating in mmwave spectrum above 10 GHz and constrained to line-of-sight propagation conditions.

However, there are at least two factors that now make it more relevant to consider an *integrated* access/backhaul solution, that is, reusing the standardized cellular technology, used by devices to access the network, also for wireless-backhaul links.

- With the emergence of 5G NR, the cellular technology is anyway extending into the mmwave spectrum, that is, the spectrum range historically used for wireless backhaul
- With the emergence of small-cell deployments with base stations located, for example, on street level, there is a demand for a wireless-backhaul solution that allows for backhaul links to operate also under non-line-of-sight conditions, that is, the kind of propagation scenarios for which the cellular radio-access technologies have been designed.

In other words, the radio-related characteristics and requirements of the access and backhaul links are becoming more similar. Using the same radio technology for these links then has several benefits.

- A single technology used for both the access and backhaul links allows for reuse of technology development and larger equipment volumes, enabling lower cost
- An integrated access/backhaul solution improves the possibilities for pooling of spectrum where it can be up to the operator to exactly decide what spectrum resources to use for access and backhaul respectively, rather than having this decided on in an essentially static manner by spectrum regulators.

5G/5G-Advanced
https://doi.org/10.1016/B978-0-443-13173-8.00033-5

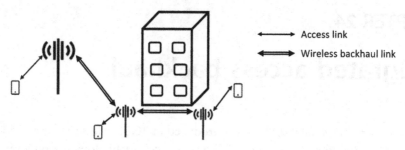

Fig. 24.1 Integrated access backhaul.

24.1 IAB architecture

The overall architecture for IAB is based on the *CU/DU split* of the gNB, introduced already in 3GPP release 15. According to the CU/DU split, a gNB consists of two functionally different parts with a standardized interface in between.

- A *Centralized Unit* (CU) including the PDCP and RRC protocols
- One or several *Distributed Units* (DUs) including the RLC, MAC, and physical-layer protocols

The standardized interface between the CU and a DU is referred to as the *F1 interface* [105]. The specification of the F1 interface only defines the higher-layer protocols, for example, the signaling messages between the CU and DU, but is agnostic to the lower-layer protocols. In other words, it is possible to use different lower lower-layer mechanisms to convey the F1 messages and data. As we will see below, with IAB, the NR radio-access technology (the RLC, MAC, and physical layer protocols,) together with some IAB specific protocols provides the lower-layer functionality on top of which the F1 interface is implemented.

IAB specifies two types of network nodes, see Fig. 24.2:

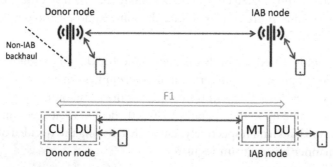

Fig. 24.2 Overall IAB architecture with IAB donor node including CU and DU functionality and IAB node including MT and DU functionality.

- The *IAB donor node* consists of CU functionality and DU functionality and connects to the remaining network via non-IAB backhaul, for example fiber-based backhaul. A donor-node DU may, and typically will, serve devices, like a conventional gNB, but will also serve wirelessly connected IAB nodes.
- The *IAB node* is the node relying on IAB for backhaul. It consists of DU functionality serving UEs as well as, potentially, additional IAB nodes in case of multi-hop IAB (see below). At its other side, an IAB node includes an *MT* ("mobile terminal") *functionality* (formally referred to as *IAB-MT*) that connects to the DU of the next higher node, referred to as the *parent node* of the IAB node.[a] Note that the parent node could either be an IAB donor node or another IAB node in case of multi-hop backhauling, see below.

The MT connects to the DU of the parent node essentially as a normal device. The link between the parent node DU and the MT of the IAB node then provides the lower-layer functionality on top of which the F1 messages are carried between the donor node CU and the IAB-node DU.

IAB also supports *multi-hop backhauling* where an IAB node is backhauled to the donor node via one or multiple intermediate IAB nodes, see Fig. 24.3. Note how, in the multi-hop case, the F1 interfaces of the DUs of all the cascaded IAB nodes terminate at the same donor-node CU.[b]

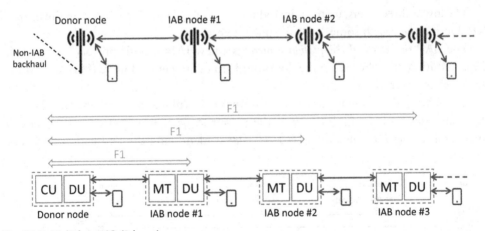

Fig. 24.3 Multi-hop IAB (3 hops).

[a] The term "mobile terminal functionality" is somewhat misleading as the MT is not a terminal/device. The term was selected to indicate the relatively high degree of commonality in functionality between the MT and a device.
[b] In release 17 this may not be the case as discussed in Section 24.6 and Fig. 24.15.

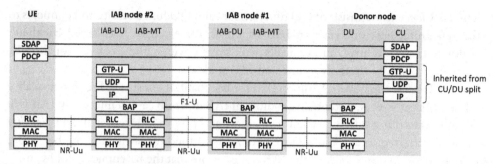

Fig. 24.4 IAB protocol stack – U-plane.

In principle, IAB supports multi-hop backhauling with a large number of cascaded IAB nodes. In practice though, one hop (no multi-hop as in Fig. 24.2) or two hops (one intermediate IAB node) can be expected to be the most common cases, at least in early IAB deployments.

Figs. 24.4 and 24.5 illustrate the IAB protocol stack for the U-plane and C-plane respectively.

As indicated in Figs. 24.4 and 24.5, the upper part of the protocol stacks is directly inherited from the general CU/DU split. The layers below are then providing the channel on top of which the F1 interface (F1-U and F1-C for the user plane and control plane respectively) is implemented.

The lower three layers, up to an including the RLC protocol, are based on the normal NR Uu interface with some IAB-specific extensions.

The BAP or *Backhaul Adaptation Protocol* is a new IAB-specific protocol responsible for the routing of packets from the donor node to the target IAB node (IAB node #2 in this case) and vice versa.

Each IAB node is assigned a *BAP address*. In the downlink direction, the BAP layer of the donor node adds a *BAP header* to the packet to be transmitted. The BAP header includes a *routing ID* consisting of a *BAP address* identifying the destination node and

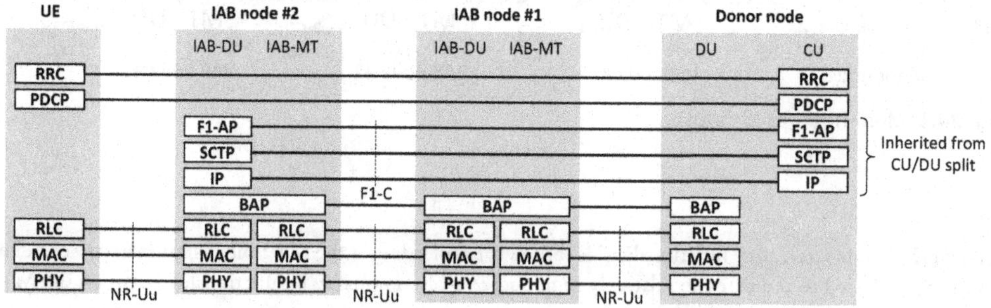

Fig. 24.5 IAB protocol stack – C-plane.

a *BAP path ID* identifying the path to the destination (in case there are multiple possible paths). The BAP header also includes a flag indicating if the packet is a user-plane packet or a control-plane packet.

When a packet arrives at an IAB node the BAP layer reads the BAP header. If the BAP address in the BAP header corresponds to the node itself, either because the packet is intended for a device directly under the node or is a control-plane packet for the IAB node itself, the packet is elevated to higher layers (GTP-U and F1-AP for user-plane packets and control-plane packets respectively). If the BAP address of the packet does not correspond to the IAB node itself, the BAP layer compares the packet routing ID with entries in a *routing table* configured by the donor node. The routing table indicates the next node for a given BAP address and BAP path ID. Assuming the routing table includes an entry for the routing ID of the packet, the packet is delivered to the appropriate DU for transmission to the next node indicated by the routing table.

The BAP protocol also operates in the uplink direction, in which case the first IAB node, that is, the IAB node to which the device is connected, attach the BAP header with the routing ID. Note that each IAB node has separate routing tables for the downlink and uplink directions.

24.2 Spectrum for IAB

IAB supports the full range of NR spectrum, from sub-GHz to mmwave frequencies. However, for several reasons, the mmwave spectrum is most relevant for IAB.

- The potentially large amount of mmwave spectrum makes it more justifiable to use part of the spectrum resources for wireless backhaul. In contrast, the more limited lower-frequency spectrum is often seen as too valuable to use for wireless backhaul
- Massive beam forming enabled at higher frequencies is especially beneficial for the wireless-backhaul scenario with stationary nodes at both ends of the radio link

As indicated in Chapter 5, higher-frequency spectrum is mainly organized as unpaired spectrum. Thus, operation in unpaired spectrum, with time-domain separation between downlink and uplink, has been the main focus for the 3GPP discussions on IAB. However, specification-wise, IAB can equally well be applied to paired spectrum at lower frequencies.

IAB supports both outband and inband backhauling

- Outband backhauling: The wireless backhaul links operate in a different frequency band, compared to the access links
- Inband backhauling. The wireless backhaul links operate on the same carrier frequency, or at least within the same frequency band, as the access links

In case of outband backhauling there is no interference between the access and backhaul links and these can be operated essentially independent of each other.

In contrast, in case of inband backhauling there could be significant interference between the access and backhaul links. Especially, means must be taken to handle/avoid the potentially very strong intra-node interference between the DU and MT parts of an IAB node as will be further discussed in Section 24.5.[c]

24.3 Initial access of an IAB node

When an IAB node, or more exactly the MT part of an IAB node, is initially connecting to the network (either directly to an IAB donor node or via another already up-and-running IAB node in case of multi-hop backhauling) it does so essentially as a normal device.

- The MT carries out cell search exactly as a device (see Chapter 16).
- From the system information of the found cell the MT determines if it can connect to the cell (the cell could, for example, be a release-15 cell not capable of supporting IAB nodes).
- If the MT may connect to the found cell it carries out an initial access in essentially the same way as a device (see Chapter 17) and a connection to the cell in established.

When IAB is used to extend the coverage of the cellular network, an IAB node may have to access a parent node from larger distance, compared to the distance from which devices may access the cell. As a consequence, IAB-node access may require a RACH configuration that allows for larger range, for example a RACH configuration with preambles of longer duration and larger guard space.

A RACH configuration supporting a longer round-trip time can always be used also for normal devices accessing from shorter distance. However, such a RACH configuration may imply extensive overhead, especially if the RACH occasions are to occur relatively frequently which may be needed if a low RACH latency is to be enabled for devices. In contrast, initial access by IAB nodes, which typically occur only rarely, in the extreme case only once when the IAB node is initially deployed, may be much less delay sensitive and thus a RACH configuration with less frequent RACH occasions may be sufficient.

Thus, in some cases it may be beneficial to provide two different RACH configurations within a cell

- One RACH configuration supporting the expected maximum round-trip time for devices and with RACH occasions occurring relatively frequently
- Another RACH configuration specifically targeting IAB nodes, supporting longer round-trip time but with less frequent RACH occasions.

[c] As discussed in Section 24.5, in case of multihop IAB such intra-node interference needs to be handled/avoided also in case of outband backhauling.

To enable this, NR allows for the cell system information to provide a separate IAB-specific RACH configuration. This would then typically be associated with preambles supporting larger range in combination with less frequently occurring RACH occasions. If no such IAB-specific RACH configuration is provided, an IAB node should carry out initial access according to the normal RACH configuration.

Once the MT has a connection to the network, the F1 interface between the donor-node CU and the IAB-node DU is established, the DU is configured for operation and the cells of the DU are established. The IAB node is then fully operational.

24.4 IAB-node transmission timing

An IAB node carries out two types of transmissions

- MT uplink transmissions toward the parent node
- DU downlink transmissions toward devices and, potentially, child IAB nodes in case of multi-hop IAB

24.4.1 MT transmission timing

As described in Section 15.2 the transmission timing of a device or, more specifically, the time offset T_{TA} between the uplink transmission timing and the downlink reception timing is given by

$$T_{TA} = \left(N_{TA} + N_{TA,offset}\right) \cdot T_c \qquad (24.1)$$

where

- $N_{TA,offset}$ is a cell-specific parameter
- N_{TA} is a device-specific parameter which can be regularly updated by the network, typically with the aim to align the reception timing of different uplink transmissions

The parameters $N_{TA,offset}$ and N_{TA} exist also for the MT of an IAB node and they are provided in the same way as for a device. However, for an IAB-node MT, the actual transmission timing can be set in different ways depending on the *timing mode* which can be configured independently for different slots.

In case of *timing-mode case 1* (illustrated in Fig. 24.6), the MT transmission timing is controlled exactly as for a device. In other words, the time offset between the MT transmission timing and the MT reception timing is given by

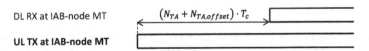

DL RX at IAB-node MT $\xleftarrow{\left(N_{TA} + N_{TA,offset}\right) \cdot T_c}$

UL TX at IAB-node MT

Fig. 24.6 Timing-mode case 1 – MT transmission timing controlled similar to devices.

$$T_{TA} = \left(N_{TA} + N_{TA,offset}\right) \cdot T_c \tag{24.2}$$

By controlling the MT transmission timing in the same way as the transmission timing of devices, the parent node can make the transmissions from MTs of different IAB nodes, as well as transmission directly from devices, arrive time aligned at the receiver, thereby retaining receiver-side orthogonality between different uplink transmissions.

In release 16, that is, the first 3GPP release to support IAB, case 1 was the only supported timing mode. However, two additional timing modes, *timing-mode case 6* and *timing-mode case 7* were introduced as part of release 17.[d] The reason for introducing these additional timing modes was to allow for the alignment in time between MT and DU transmissions within an IAB node (for timing-mode case 6), and the alignment in time between MT and DU reception (for timing-mode case 7). This, in turn, improves the operation of so-called FDM and SDM between the MT and DU of an IAB node, see Section 24.5 below.

In case of timing mode 6 (illustrated in Fig. 24.7), the IAB node will ignore expression (24.1) and will instead directly align its MT transmission timing with the transmission timing of its DU (how the transmission timing of the DU is determined is described in the next section).

In case of timing mode case 7 (illustrated in Fig. 24.8), the transmission timing of the MT is adjusted, compared to timing-mode case 1, by the introduction of an additional cell-specific parameter $N_{TA,offset.2}$. In other words, the time offset T_{TA} between the MT transmission timing and the MT reception timing, for case-7 timing, given by

$$T_{TA} = \left(N_{TA} + N_{TA,offset} + N_{TA,offset.2}\right) \cdot T_c \tag{24.3}$$

The basic idea is that, by setting the parameter $N_{TA,offset.2}$ properly, the parent node can control the MT transmission timing in such a way that the MT transmission is

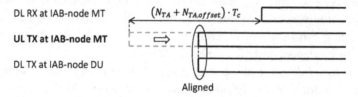

Fig. 24.7 Timing-mode case 6 – MT transmission timing aligned with DU transmission timing.

---

[d] The names of the timing modes are for historical reasons. Seven different schemes, or "cases, for MT transmission timing was initially discussed arelady for release 16. Out of these, only case 1 was eventually adopted for release 16. Two other cases (case 6 and case 7) was then included in release 17.

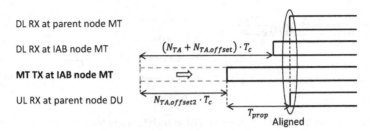

DL RX at parent node MT

DL RX at IAB node MT $(N_{TA} + N_{TA,offset}) \cdot T_c$

MT TX at IAB node MT

UL RX at parent node DU $N_{TA,offset2} \cdot T_c$

T_{prop} Aligned

Fig. 24.8 Timing-mode case 7 – DU reception timing aligned with MT reception timing at parent node.

received at the parent node DU with a timing that is aligned with the parent node MT reception, that is, the reception of transmissions from a "grand-parent" node. Thus, timing-mode case 7 provides a tool to align the MT reception and DU reception at the parent node.

Note that timing-mode case 7 is only relevant if the parent node is also an IAB node with its own parent, that is, in case of multi-hop IAB.

24.4.2 DU transmission timing and OTA timing alignment

Regarding the timing of DU transmissions, there is a general requirement in the 3GPP specifications that, in case of operation in unpaired spectrum, the downlink transmission timing of all cells should be mutually aligned within a window of 3 μs [104]. As, from a device point-of-view, the cells created by an IAB node should be indistinguishable from any other cell, this requirement on mutually aligned downlink transmissions between cells also applies for cells created by an IAB-node DU, at least in case of operation in unpaired spectrum.

Such alignment of the downlink transmission timing between nodes can be achieved in several ways, including, for example, the use of GPS reception at the IAB node together with an agreed absolute transmission timing. However, IAB also supports *over-the-air* (OTA) based transmission-timing alignment where an IAB node can derive its DU transmission timing solely from signals received from the parent node.

The basic principle of OTA-based transmission-timing alignment is that the IAB node should set its DU transmission timing an amount T_{prop} ahead of the timing of signals received from the parent node, where T_{prop} is the propagation time from the parent node to the IAB node. In this way, the DU transmission timing of the IAB node is aligned with the DU transmission timing of its parent node, in line with the general requirement of aligned downlink transmission timing between cells. The task of OTA-based transmission-timing alignment is thus equivalent to estimating the propagation time between the IAB node and its parent node.

Fig. 24.9 illustrates the timing relation (transmit and receive timing) between the DU and MT on each side of an IAB backhaul link (DU at the parent node and MT

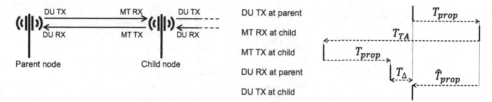

Fig. 24.9 Timing relations between parent-node DU and child-node MT.

at the child node). Given an offset T_{TA} between the downlink reception timing and uplink transmission timing at the child-node MT and an offset T_Δ between the uplink reception timing and downlink transmission timing at the parent-node DU, one can see that the propagation time can be estimated as $\hat{T}_{prop} = (T_{TA} - T_\Delta)/2$. Note that T_{TA} is the true offset between the transmission timing and reception timing at the MT, that is, T_{TA} will depend on the MT timing mode.

T_{TA} is inherently known at the child node while T_Δ is known at the parent node. To enable the child node to estimate the parent-to-child propagation time, and thus to enable OTA-based timing alignment, the parent node can provide T_Δ to the child node by means of MAC-CE signaling, that is, the same type of signaling that is providing, for example, uplink time-alignment commands, see Section 15.2. The reason for providing T_Δ by means of MAC-CE signaling instead of more reliable RRC signaling, is that the RRC protocol is terminated at the donor node while T_Δ originates at the parent node. In case of multi-hop IAB, where the parent node may not be the same as the donor node, the parent node would then need to first provide T_Δ to the donor node after which the donor node would provide T_Δ to the child node by means of RRC signaling. By using MAC-CE signaling, which terminates at the parent-node DU, the parent node can directly provide T_Δ to the child node, leading to less signaling overhead and enabling faster updates of T_Δ.

24.5 DU/MT interaction

When the DU and MT of an IAB node are operating in the same frequency band (inband backhauling) there is typically a need for some level of coordination, in terms of resource usage, between the DU and the MT.

Especially, one typically needs coordination to avoid the "full-duplex" situation of Fig. 24.10 in which transmissions to be received by the MT are severely interfered by simultaneous DU transmissions, or vice versa.

It should be noted though that in some deployment scenarios there could be sufficient antenna isolation between the IAB-node DU and MT to actually allow for the kind of "full-duplex" DU/MT operation illustrated in Fig. 24.10. One such scenario could be

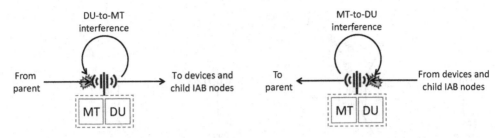

Fig. 24.10 MT reception being severely interfered by simultaneous DU transmission (left) and DU reception being severely interfered by simultaneous MT transmission (right).

when an IAB node is used to provide outdoor-to-indoor coverage, with the MT part of the IAB node located on the outside and the DU part located on the inside.

Fig. 24.10 illustrated a situation with simultaneous DU and MT operation in *the same transmission direction*, that is, in the downlink direction (MT reception interfered by DU transmission), or in the uplink direction (DU reception interfered by MT transmission).

However, as illustrated in Fig. 24.11, one can also envision simultaneous DU and MT operation in *opposite* transmission directions, that is, simultaneous DU and MT transmission in the downlink and uplink transmission directions respectively (left part of Fig. 24.11) or simultaneous DU and MT reception in the uplink and downlink transmission directions respectively (right part of Fig. 24.11).

In the transmission scenarios illustrated in Fig. 24.11 there would not be the same kind of extreme intra-node interference between the DU and the MT as in the "full-duplex" scenario of Fig. 24.10. Rather, from an interference point-of-view one would just need to ensure that the simultaneous DU and MT transmissions or receptions take place in different resources. This could, for example, imply a separation in the frequency domain, that is, the use of different resource blocks for the DU and MT, in 3GPP referred to as *FDM* operation. Alternatively, there could be a separation in the spatial domain between the DU and the MT, in 3GPP referred to as *SDM* operation. The DU and MT could, for example, transmit or receive simultaneously on the same frequency resource but with different antenna panels pointing in different directions, as illustrated in Fig. 24.12.

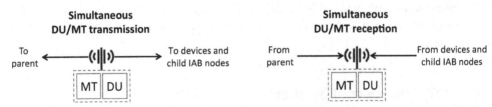

Fig. 24.11 Simultaneous DU and MT transmission (left) and simultaneous DU and MT reception (right).

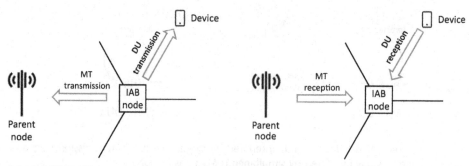

Fig. 24.12 DU/MT SDM on the transmitter side (left) and receiver side (right).

Note that, in some sense, this can be seen as multi-user MIMO (MU-MIMO) between MT and DU transmissions (left part of Fig. 24.12) or between MT and DU reception (right part of Fig. 24.12).

This kind of FDM/SDM between the DU and MT parts of an IAB node would benefit if the DU and MT would operate in a time-aligned fashion, that is, with aligned transmission timing in case of simultaneous DU and MT transmission, or aligned reception timing in case of simultaneous DU and MT reception. This was the reason for the release-17 introduction of the case-6 and case-7 timing modes as described in Section 24.4. With the case-1 timing, which was the only supported timing mode in release 16, the DU and MT transmission timings, as well as the DU and MT reception timings, are inherently not aligned. However, the case-6 timing allows for an IAB node to directly align the timing of its MT transmissions with the timing of the DU transmissions, thereby providing better support for FDM/SDM between DU and MT transmissions. Likewise, the case-7 timing allows for an IAB node to adjust the MT transmission timing of a child node to align the corresponding uplink reception (at the IAB nodes DU) with its MT reception, thereby providing better support for FDM/SDM between DU and MT reception at the IAB node.

To conclude there is a need to provide means to coordinate the DU and MT operation to be able to avoid collision between DU and MT transmissions/receptions. At the same time, simultaneous DU and MT operation on different, and in some case even on the same, resources should be possible when conditions so allow. Within IAB, this is ensured by being able to separately configure the DU and MT for specific transmission direction(s) and providing means to, for a given resource, prioritize either the MT or DU.

24.5.1 MT resource configuration

In Chapter 7 it was described how, from a device point-of-view, time-domain resources (OFDM symbols) can be configured/indicated as *downlink, uplink* or *flexible,* limiting the direction in which the resource will be used by the network. The same is true for the MT of an IAB node, that is, an MT time-domain resource can be configured/indicated as

- *Downlink* (D), implying that the resource will only be used by the parent node in the downlink direction (MT reception)
- *Uplink* (U), implying that the resource will only be used by the parent node in the uplink direction (MT transmission)
- *Flexible* (F), implying that the resource may be used in both the downlink and uplink directions (MT reception and transmission). The instantaneous transmission direction is then determined by the parent-node scheduler in the same way as for devices, see Section 7.8.3.

In the same way as for devices, this is done by a combination of common configuration by means of system information (same configuration for devices and IAB-node MTs within a cell), dedicated configuration on a per-MT basis by means of RRC signaling, and semi-dynamic indication by means of SFI. One difference is that, while the dedicated D/F/U configuration for devices only allows for patterns in the order D-F-U, the MT dedicated configuration is extended to also allow for patterns in the order U-F-D. The same is true for the semi-dynamic configuration by means of SFI where, for MTs, the set of predefined slot patterns has been extended to also include a set of formats with the order U-F-D.

The reason for these extensions is the possible use of simultaneous DU/MT operation by means of SDM/FDM as described above. In that case the direction of the child link of an IAB node, that is, downwards from the DU, will be the opposite of that of the parent link of the same IAB node (upwards from the MT). Thus, if one link is operating in the uplink (U) direction the other link should operate in the downlink (D) direction, and vice versa. In other words, to match a D-F-U pattern on one link, a corresponding U-F-D pattern is needed on the other link.

Note though that the MT D/U/F common configuration is the same as the common configuration for devices, that is, limited to the order D-F-U. Furthermore, similar to the dedicated configuration for devices, the dedicated MT configuration can only restrict flexible resources in the common configuration but cannot change the configuration for downlink and uplink resources. Consequently, the only way to have a true U-F-D order of the MT resources is to provide an all-flexible common configuration. As the common configuration is the same for IAB-node MTs and devices, this means that also devices would need to operate with an all-flexible common configuration. Any restrictions in the device configuration must then be provided by means of the dedicated per-device configuration.

24.5.2 DU/MT flexible coordination

During the initial specification of IAB in release 16, it was typically assumed that DU and MT operation was separated in time. Thus, there were no special considerations to the possibility for DU and MT separation by means of FDM or SDM. Rather, the focus was on coordination tools by which time-domain resources could be flexibly shared between the DU and MT.

However, for release-17, there was a renewed focus also on the possibility for FDM or SDM between the DU and MT. One result of this was the introduction of timing-mode case 6 and case 7 as described above. However, it also led to an extension of the coordination tools to also cover flexible resource sharing between the DU and MT in the frequency domain.

24.5.2.1 Time-domain coordination

Similar to the MT, DU time-domain resources (symbols) can be configured as

- *downlink* (D), implying that the DU can only use the resource in the downlink direction (DU transmission),
- *uplink* (U), implying that the DU can only use the resource in the uplink direction (DU reception), or
- *flexible* (F), implying that the DU can use the resource in both the downlink (DU transmission) and uplink (DU reception) direction.

Alternatively, a DU time-domain resource can be configured as *Not Available*, implying that the DU should not use the resource at all.

In parallel to the D/U/F configuration and as a tool for more flexible resource sharing between the DU and MT of an IAB node, release-16 allowed for DU time-domain resources to be configured as *Hard* or *Soft*. In case of a hard configuration, the DU can use the resource in the transmission direction or direction(s) allowed by the D/U/F configuration without having to take into account the impact on the MTs ability to transmit/receive according to its configuration and scheduling. In practice this implies that, if a certain DU time-domain resource is configured as hard, the parent node must assume that the IAB-node MT may not be able to receive or transmit. Consequently, the parent node should not schedule transmissions to/from the MT in this resource.

In contrast, in case of a DU time-domain resource configured as soft, the DU can use the resource if and only if this does not impact the MTs ability to transmit/receive according to its configuration and scheduling. This means that the parent node can schedule a downlink transmission to the MT in the corresponding MT resource and assume that the MT is able to receive the transmission. Similarly, the parent node can schedule MT uplink transmission in the resource and assume that the MT can carry out the transmission.

The configuration of DU resources as hard or soft is done on a slot basis and per resource type (D, U, and F). In other words, for each slot, the sets of DU resources configured as downlink, uplink and flexible can independently be configured as hard or soft. As an example, within a slot, resources configured as downlink (D) can be configured as hard while resources configured as uplink (U) and flexible (F) can be configured as soft.

The possibility to configure soft DU resources allows for more dynamic utilization of DU resources. Take as an example a soft DU resource corresponding to an MT resource configured as uplink (U). If the MT does not have a scheduling grant for that resource, the IAB node knows that the MT will not have to transmit within the resource. Consequently, the DU can dynamically use the resource, for example, for downlink transmission, even if the IAB node is not capable of simultaneous DU and MT operation.

The possibility to configure soft DU resources also provides a possibility for an IAB node to benefit from being able of simultaneous DU and MT operation. As described above, whether or not a specific IAB node is capable of simultaneous DU and MT operation may depend on the IAB-node implementation and may also depend on the exact deployment scenario. Thus, an IAB node designed or deployed so that it can support simultaneous DU and MT operation can use a soft DU resource without the parent node even knowing about it.

These situations, when an IAB node, by itself, can conclude that it can use a soft DU resource has, in the 3GPP discussions, been referred to as *implicit indication of availability* of soft DU resources. The parent node can also provide an *explicit indication of availability* of a soft DU resource.

The explicit indication of availability of soft DU resources is done on a slot basis. It is also done per DU resource type (D, U, or F). There are thus in total eight possible indications for a given slot, see Table 24.1.

Note that, even if a certain set of soft symbols are not explicitly indicated as available, they can still be available according to implicit indication as described above. In some sense one can see the explicit indication of availability of a soft DU resource as a way for the parent node to indirectly inform the IAB node that it will not use the MT resource corresponding to (overlapping with) the soft DU resource. From this, the IAB node can conclude that the soft DU resource can be used without impacting the MT operation.

The indication of availability is provided to the IAB node in a semi-dynamic way by the IAB node first being configured with a table of *availability combinations*. The parent node then indicates a certain availability combination by means of DCI.

Table 24.1 Different types of availability of soft resources of a slot.

Availability indication	Availability
0	No soft symbols indicated as available
1	Only downlink (D) soft symbols indicated as available
2	Only uplink (U) soft symbols indicated as available
3	Only flexible (F) symbols indicated as available
4	Only downlink (D) and uplink (U) soft symbols indicated as available
5	Only downlink (D) and flexible (F) soft symbols indicated as available
6	Only uplink (U) and flexible (F) soft symbols indicated as available
7	All soft symbols (D, U, and F)) indicated as available

Each availability combination within the table of availability combinations corresponds to a sequence of availabilities, where each availability (taking one of the eight values of Table 24.1) corresponds to a given slot in a sequence of slots. Furthermore, each availability combination within the table of availability combinations is associated with an *availability-combination index*. The semi-dynamic indication of availability is then done by the parent providing the index of one of the availability combinations, that is, one of the rows of the table of availability combinations, by means of a new DCI format 2_5.

The DCI format 2_5 has the same size as, for example, DCI format 2_0, implying that no additional blind decodings are needed to detect DCI format 2_5. It is encoded with a special AI-RNTI which is provided to the IAB node as part of the IAB-node initial configuration.

Fig. 24.13 shows an example of explicit indication where each availability combination covers eight slots. A DCI of format 2_5 indicating index 3 points to the fourth row of the table of availability combinations. This provides the explicit indication of availability over eight slots as indicated in the lower part of the figure. Note that the figure ignores that availability may also be implicitly indicated, that is, it assumes that the availability of a soft resource is only determined by the explicit indication of availability provided by the parent by means of DCI Format 2_5.

The IAB specifications allow for the table of availability combinations to consist of up to 512 availability combinations. Each availability combination can then correspond to the resource availability of up to 256 consecutive slots (eight slots assumed in Fig. 24.13).

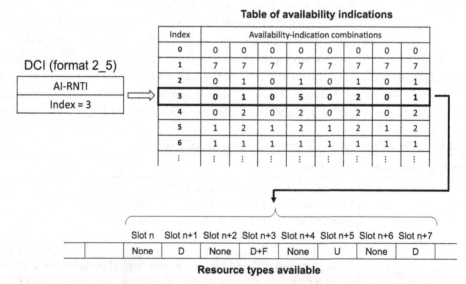

Fig. 24.13 Explicit indication of availability of soft resources of based on a configured table of availability combinations together with an availability-combination index provided in DCI.

Also note that, in practice, a DU may create more than cell. The DCI format 2_5 will then provide multiple indices or pointers, one for each DU cell.

24.5.2.2 Frequency-domain coordination

As part of release 17, the hard/soft configuration of DU resources was extended to the frequency domain in order to allow for flexible resource sharing also in the case of FDM between the DU and MT. This was done by dividing the carrier into *resource-block sets* (RB sets) each consisting of a set of consecutive resource blocks. The bandwidth of an RB set is configurable from as small as 360 kHz (two resource blocks at a sub-carrier spacing of 15 kHz) to as large as ≈184 MHz (64 resource blocks with a subcarrier spacing of 240 kHz).

For a given slot, each RB set can be separately configured as *hard*, *soft*, or *not-available*, over-riding the per-slot configuration. Similar to the per-slot configuration, the per-RB configuration is done separately for the different resource types, that is, separately for downlink, uplink, or flexible symbols. If an RB set is not provided with a hard/soft configuration, the per-slot configuration is assumed for the RB set.

The interpretation of hard and soft RB sets is the same as for slots, that is,

- Resource blocks in RB sets configured as hard can be used by the DU without it having to take into account the impact to the MT operation
- Resource blocks in RB sets configured as soft can be used by the DU only if this does not impact the MTs ability to transmit/receive according to its configuration and scheduling.

Similar to slots, the availability of RB sets configured as soft can be *implicitly* indicated, that is, the DU by itself concludes that it can use the soft resource, or *explicitly* indicated. The explicit indication is enabled by a straightforward extension of the release-16 availability indication described above. For release 16, an indication of availability for a given slot implied that soft symbols within the slot were available. With the release-17 extensions, an indication of availability for a given slot implies that RB sets configured as soft are available. Similar to the per-slot indication in release 16, the availability indication is done per DU resource type (D, U, or F), compare Table 24.1. Thus, as an example, an indication of availability could indicate that, in a given slot, the RB sets configured as soft within symbols configured as D (downlink) are available (this would correspond the Availability Indicator = 1 in Table 24.1).

24.6 IAB mobility

In most cases there is no need for IAB-node mobility, that is, an IAB node can be assumed to be stationary and remain under the same parent node which could be either a donor node or another IAB node in case of multi-hop IAB. However, one may envision scenarios where it could be beneficial with non-stationary IAB nodes, for example, to allow for IAB nodes onboard moving vehicles such as buses or trains.

The MT part of an IAB node very much behaves like a device, that is, an MT can move within a cell of a parent node and handover to a cell of a different parent node essentially as a "normal" device. IAB-node mobility is rather limited by the IAB architecture based on the CU/DU split and with the DU of an IAB node connecting to the CU of the donor node by means of the F1 interface, possibly via one or multiple IAB nodes in case of multi-hop.

In general, the F1 interface, which was specified in release 15 in a non–IAB context, does not support the change of CU-DU relation, that is, a DU can be activated under a given CU but will then remain under that CU until it is de-activated. For this reason, IAB-node mobility in release 16 was limited to intra-CU mobility. In other words, an IAB node could move between DUs of the same CU but could not move into an area covered by a different CU, see Fig. 24.14. The change of DU under the same CU would consist of a handover of the MT to the new DU, essentially like a device handover, together with an update of the BAP routing tables in any intermediate IAB nodes if present (not illustrated in Fig. 24.14).

To enable an IAB node to connect via a parent node DU under a different CU, release 17 introduced the possibility for tunneling of the F1 interface as illustrated in Fig. 24.15. In this case, the MT of the migrating IAB node is handover to the new CU/DU implying, for example, that the RRC protocol of the IAB-node MT is terminated in CU_2. However, the DU of the IAB node remains connected to CU_1, with

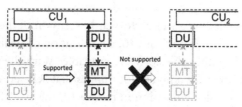

Fig. 24.14 Release 16 support for intra-CU mobility only.

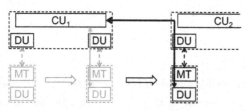

Fig. 24.15 Inter-CU mobility by means of tunneling (release 17).

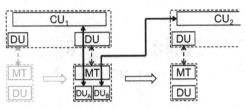

Fig. 24.16 Release-18 inter-CU mobility.

the F1 interface being tunneled via the donor node of CU_2. As a consequence, devices and potential child IAB nodes that connect via the migrating IAB node will also logically remain under CU_1.

Although allowing for a certain degree of inter-CU mobility, the release-17 solution of Fig. 24.15 is limited in the sense that it requires that the two CUs have a direct Xn interface between them. Release 18 therefore introduced a more general solution for inter-CU mobility as part of a work item on *mobile IAB* [132].

As already mentioned, the fundamental problem of inter-CU mobility for an IAB node is that the F1 specification does not support the change of CU for an active DU. The release-18 solution goes around this limitation by activating a new DU within the mobile IAB node, with the new DU being under the new/target CU, see Fig. 24.16. The new DU (DU_B in Fig. 24.16) creates new cells and the devices under the cell(s) of the old DU (DU_A) are handed over to the corresponding cell(s) of DU_B by means of intra-frequency handover. Once the handovers have been executed, DU_A is de-activated and the change of CU is complete.

24.7 Network-controlled repeaters

An IAB node is a type of *decode-and-forward* relay, that is, data received on the backhaul link is demodulated/decoded and re-encoded/remodulate before forwarded on the access link and vice versa. A less complex alternative to decode-and-forwarding is *amplify-and-forwarding*, often referred to as a *repeater*. As the name suggests, amplify-and-forwarding implies that the received signal is just amplified, that is, without any demodulation/decoding and corresponding re-encoding/remodulation, before forwarding.

A classical repeater is "dumb" in the sense that it does not have any knowledge about the signal to be forwarded, including the intended ultimate receiver. Thus, it typically applies a certain gain, alternatively a certain fixed transmit power, without considering the transmit power actually needed to reach the target receiver. It also forwards whatever signal it receives, regardless of if the target device is really within the coverage of the repeater. Finally, there is no way to apply beam forming to the signal to be forwarded. The latter is true also in the uplink direction, that is, there is no way to apply receiver side beam forming to focus the repeater reception in the direction of a transmitting device.

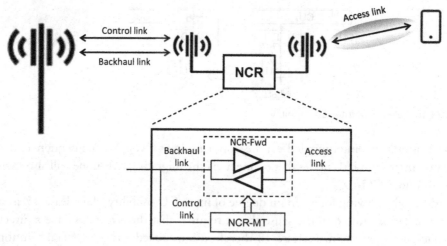

Fig. 24.17 Network-controlled repeater.

For these reasons, and as a less complex complement to IAB, 3GPP release 18 introduced support for so-called *network-controlled repeaters* (NCRs).

A network-controlled repeater consists of two parts, see also Fig. 24.17,

- A bi-directional forwarding part, in the specifications referred to as *NCR-Fwd*. This is, essentially, an amplify-and-forward relay with possibility for selective forwarding and controllable beam forming on the access link.
- A controller part, in the specifications referred to as *NCR-MT*, by means of which the network can control the forwarding part and, especially, control the beam forming applied on the access link.

Although in the more general case, a repeater can also apply frequency-translation, 3GPP NCR is currently limited to inband repeating where the backhaul link and the access link operate on the same carrier frequency. Furthermore, release-18 NCR only supports single-hop repetition, that is, not the cascading of multiple network-controlled repeaters. Finally, there is an assumption that the control link operates on the same frequency as the backhaul link or, if the NCR supports forwarding of multiple carriers, on one of these carriers.

24.7.1 NCR transmission timing

In terms of transmission timing, there are no specific timing relations for the forwarding part of the NCR. Rather, the NCR simply forwards the received signal with a minimum delay in both the downlink and uplink directions and will, to the base station, just appear as a small additional propagation delay. The uplink transmission timing of the NCR-MT is controlled by the network in the same way as the uplink transmission timing of a device, see Section 15.2.

24.7.2 NCR beam management and access-link beam indication

In general, an NCR is transparent to a device. This implies that, in terms of beam forming, a device sees a cell with a set of beams without being able to distinguish between beams generated at the cell site and beams generated by the NCR, see Fig. 24.18. In practice, a subset of these beams originates at the actual cell site (beam 0-11 in Fig. 24.18) while another subset originates at the NCR (beam 12-15). The base station communicates via the NCR forwarding part, as well with the NCR-MT, using one of the base-station-originating beams with the NCR typically using a corresponding receiver beam.

Beam-management, that is, the selection of a suitable beam pair, for the backhaul/control links is done in the same way as beam management for device, see Chapter 12. Note that it can typically be assumed that the same beam pair can be used for the control and backhaul links.

The key feature of an NCR is the *access-link beam management*, that is, the means by which the network, via the NCR-MT, controls what beam should be used on the access link for a given time-domain resource. That beam will then be used for downlink access-link transmission if the time resource has been configured as a downlink resource or for uplink access-link reception if the time resource has been configured as an uplink resource.

The access-link beam indication can be done by three different means that can be used in parallel, *periodic beam indication*, *semi-persistent beam indication*, and *aperiodic beam indication*.[e]

24.7.2.1 Periodic beam indication

In case of periodic beam indication, the NCR can be configured with one or multiple periodic patterns where each pattern consists of a set of time-domain resources and each time-domain resource is associated with a specific beam index referring to one of the beams on the NCR access link, see Fig. 24.19. The periodicity of the pattern can range

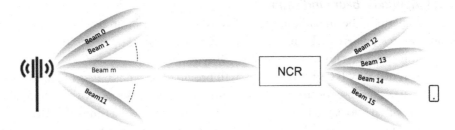

Fig. 24.18 NCR beam forming.

[e] Note that the NCR "beam indication" discussed here is different from the "beam indication" by means of TCI states discussed in Chapter 12.

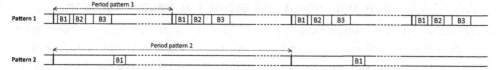

Fig. 24.19 Periodic bean indication.

from as short as 1 ms to as long as 10 s and each beam index can indicate one out of up to 64 different beams. Note that, in most cases, an NCR will support significantly less than 64 beams on the access link.

Each time-domain resource can start at an arbitrary symbol within the period (configured as a certain slot offset and a certain symbol offset within the slot) and can have a length of up to 112 symbols.

One use of periodic beam indication is for common signals with a specific periodicity, for example, SSBs to be broadcast in different beams in the downlink and RACH occasions with corresponding receiver-side beam forming in the uplink.

In principle, different simultaneously configured periodic patterns could be partly overlapping, that is, include overlapping time-domain resources. However, the specification states that, in such cases, those time-domain resource should correspond to the same beam index.

24.7.2.2 Semi-persistent beam indication

Semi-persistent beam indication is very similar to periodic beam indication in that a periodic pattern of time-domain resources with corresponding beam indices are configured exactly as for periodic beam indication. The only difference is that semi-persistent beam indication can be activated and de-activated by means of MAC CE signaling to the NCR-MT, see Fig. 24.20 as an illustration.

24.7.2.3 Aperiodic beam indication

In case of aperiodic beam indication, an NCR is first configured with a set of time-domain resources, where each time-domain resource is defined by a relative symbol-level start position and a certain duration (up to 28 symbols).

By means of a new DCI format (DCI 5_0) provided to the NCR-MT, the network can then indicate a subset of the configured time-domain resources, with each indicated resource having an associated beam index. DCI 5_0 also includes a time offset that defines the reference point for time-domain resources identified by the DCI.

Fig. 24.20 Semi-persistent beam indication.

If there is an overlap between time-domain resources of configured periodic indication/active semi-persistent beam indication and time-domain resources indicated by DCI 5_0 the configured indications (periodic/semi-persistent) can be explicitly configured to have priority over the dynamic indications. If not explicitly indicated, the aperiodic indication has priority.

24.7.3 Selective forwarding

Forwarding by the NCR-Fwd is by default disabled and only carried out if the corresponding time-domain resource is explicitly associated with a beam by means of the periodic, semi-periodic or aperiodic beam indication described above. This means that selective forwarding, that is, forwarding only when there is something to forward, is inherently supported by means of the NCR beam-indication functionality.

CHAPTER 25

Non-terrestrial NR access

A *non-terrestrial network* or NTN is a wireless-communication network with air/space-borne network nodes carrying equipment, also referred to *payloads*, providing wireless access to devices on the ground.[a] As illustrated in Fig. 25.1, the air/space-borne nodes could, for example, be satellites, high-altitude platforms (HAPs) or lower-altitude drones.[b] However, within 3GPP the use of the term NTN has been limited to satellites and HAPs.

The main reason for introducing support for non-terrestrial network nodes complementing the ground-based/terrestrial network infra-structure is that it may be a more cost-efficient way, and in some cases the only way, to provide wireless coverage in some scenarios, for example, in extreme rural areas or at sea. One could of course use completely different wireless-access technologies for the terrestrial and non-terrestrial access. However, there are obvious benefits in terms of economy-of-scale, especially on the device side, if one can extend the terrestrial radio-access technology, in this case NR, to also support access via non-terrestrial network nodes.

3GPP work on NR-based non-terrestrial access started in 2017 with a study item on *NTN scenarios and channel models* [80]. This was followed by a second study item on *Solutions for NTN* [130]. Actual specification to enable NR support for non-terrestrial access was then included as part of 3GPP release 17. Although the 3GPP work on NTN has targeted both satellites and HAPs, the main focus has been on satellite-based NTN. The reason for this is that support for satellite-based NTN was assumed to require more extensive extensions to the NR specifications and that, with such extensions, HAP-based NTN would be inherently supported as well. Thus, below we will mainly discuss NTN from the point-of-view of satellite-based payloads.

25.1 Satellite basics

Before discussing the NR extensions targeting non-terrestrial networks, we will first discuss some more general aspects of satellites and satellite-based communication.

---

[a] Note that, in NTN terminology, the term "payload" is used for the communication equipment/functionality within the NTN node, for example, within a satellite.

[b] Instead of HAP, one sometimes also use the term HAPS (*high-altitude platform station/system*). Another term, specifically in the context of IMT technology, is HIBS (*high-altitude IMT base station*).

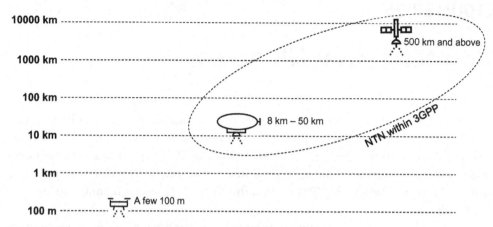

Fig. 25.1 Different types of NTN nodes and their rough altitudes.

25.1.1 Satellite orbits and their characteristics

Satellites can orbit the earth at vastly different altitudes, ranging from a couple of 100 km to several 10,000 km. Furthermore, satellite orbits may be circular or elliptical and may also have different *inclinations*, that is, different angles relative to the equator, see Fig. 25.2.

In terms of altitude, the 3GPP work on NTN has focused on satellites in *geo-stationary orbit* (GSO satellites) and satellites in *low-earth orbit* (LEO satellites), see also Fig. 25.3.

A GSO satellite, sometimes also referred to as a GEO (*Geo-stationary Equatorial Orbit*) satellite, has a circular orbit aligned with the equator and with an altitude of 35,786 km. In such an orbit, the speed of the satellite matches the rotation of the earth around its own axis. A GSO satellite will therefore be stationary relative to the surface of the earth, hence the term *"geo-stationary"*.

A LEO satellite orbits the earth at a significantly lower altitude, normally assumed to be between 500 km and 2000 km. The orbit of a LEO satellite may be circular or elliptical. It may also have different inclinations, that is, the orbit of a LEO satellite does not have to be aligned with the equator.

Fig. 25.2 Circular and elliptical satellite orbits and satellite orbits not aligned with the equator.

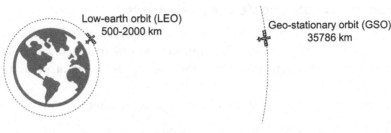

Fig. 25.3 Geo-stationary orbit vs low-earth orbit.

The benefit of a GSO satellite is that it is stationary relative to the surface of the earth. Thus, as seen from the surface, the position of a GSO satellite and its coverage area will appear stationary, that is, will not vary in time. However, this comes at the expense of a large propagation distance and corresponding very large propagation time. The maximum satellite-to-device distance for a GSO satellite is in the order of 40,000 km, corresponding to a one-way (satellite-to-device) propagation time of more than 130 ms.[c] The large propagation distance also implies a relatively high path loss in case of GSO satellites.

In contrast, a LEO satellite is associated with a much shorter satellite-to-device distance leading to smaller propagation time and less path loss. Note though that the propagation time will still by much larger than what will be experienced in conventional terrestrial networks. As an example, assuming a satellite altitude of 700 km, the maximum satellite-to-device distance may exceed 2000 km, corresponding to a one-way propagation time of around 7 ms.

However, the main drawback of LEO satellites is that the satellite will move rapidly relative to the surface of the earth. This means that, even for stationary devices, there will be a very high, and rapidly varying, Doppler shift of the communication signal as well as a need to regularly switch the satellite via which the device is communicating. Also, even if one is only to cover a specific limited surface area, a large number of LEO satellites is still needed as each satellite will only cover the target area for a limited time.

It should be noted that the satellite channel is typically line-of-sight and thus the movement of the satellite will mainly cause a Doppler *shift* of the communication signal. This is in contrast to terrestrial networks for which the channel is typically non-line-of-sight and movements of the device or within the environment mainly cause a Doppler *spread* of the signal.

---

[c] Note that the maximum satellite-to-device distance exceeds the altitude of the satellite orbit as the satellite is normally not located directly above the device. This difference between altitude and maximum satellite-to-device distance is even more pronounced for lower-orbit satellites.

25.1.2 Ephemeris data and Keplerian elements

The traditional way to describe a satellite orbit, also referred to as *Ephemeris data*, is by means of so called *Keplerian elements*. The set of Keplerian elements consists of six parameters that jointly provide information about the position and speed of an orbiting body, see also Fig. 25.4.

Two parameters describe the shape of the, in general, elliptical orbit

- The *eccentricity* of the orbit (zero for circular orbits)
- The length of the *semi-major axis* of the orbit (equals the radius for a circular orbit)

Three additional parameters specify the orientation of the orbit relative to a reference plane which, in case of satellites orbiting the earth, corresponds to a plane through the equator with a well-defined reference direction.

- The *inclination* describes the angle between the reference plane and the plane of the orbit
- The *longitude of the ascending node* (denoted Ω in Fig. 25.4) describes the orientation of the orbit around an axis perpendicular to the reference plane
- The *argument of periapsis* (denoted ω in Fig. 25.4) describes the orientation of the orbit in the plane of the orbit, in practice the direction (in the plane of the orbit) towards which the semi-major axis is pointing.

Jointly these five parameters fully describe the orbit of the satellite. The sixth parameter, the angular parameter υ in Fig. 25.4, gives the instantaneous position of the satellite within this orbit at a given time instant often referred to as the "*epoch*". In principle, knowledge about the Keplerian elements makes it possible to determine the satellite position at any future time.

25.1.3 Transparent vs regenerative payloads

Fig. 25.5 illustrates the different types of links in a non-terrestrial network.

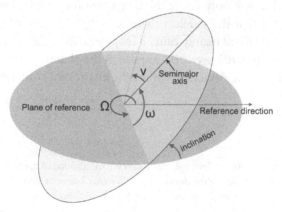

Fig. 25.4 Keplerian elements.

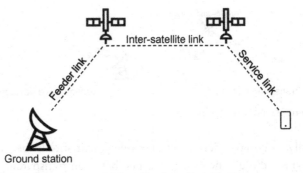

Fig. 25.5 Different links in a non-terrestrial (satellite-based) network.

- The *service link* is the link between a satellite and a device
- The *feeder link* is the link between a satellite and a ground station

In addition, there may be *inter-satellite links* that provide connectivity between satellites. Such inter-satellite connectivity is needed, for example, when a satellite does not have direct line-of-sight to its ground station.

For communication via satellites, one distinguishes between *transparent payloads*, sometimes also referred to as a *bent pipe*, and *regenerative payloads*.

In case of a transparent payload, the satellite essentially operates as an amplify-and-forward repeater, that is, it amplifies and, in some cases, frequency translates the signal received on the feeder link before transmitting it on the service link, and similarly in the opposite direction. Note that, in case of transparent payload, the feeder link inherently relies on the same basic air-interface technology as the service link, except that it may operate on a special feeder-link frequency.

On the other hand, in case of a regenerative payload the satellite operates more like a decode-and-forward relay. In the 3GPP context this has often be described as having "the gNB onboard the satellite" although one can clearly envision alternative splits where parts of the gNB functionality remain on the ground.

At least in principle, in case of a regenerative payload the feeder and service links can be based on completely different air-interface technologies. In the 3GPP context the service link would be based on the normal gNB-to-UE interface (the Uu interface) while the feeder link could be based on the same technology or a different technology. Note that this is similar to conventional wireless-backhaul solutions where the backhaul link can be based on the same technology as the access link, as is the case, for example, for IAB (Chapter 24), or a different backhaul-specific technology such as *mini-link* [131]. Note that, although Figs. 25.5 and 25.6 assume a satellite-based NTN, the same basic figures and related discussions are applicable also for HAP-based NTN.

3GPP initially considered both transparent and regenerative payloads for NTN. However, release-17 eventually focused on transparent payload with regenerative

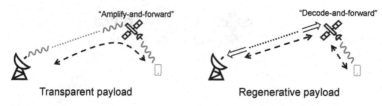

Fig. 25.6 Transparent vs. regenerative payload.

payload potentially to be considered for later releases. It should be noted that, from a device point-of-view, there is no real difference between transparent and regenerative payload.

25.1.4 Fixed beams vs steerable beams

A satellite typically covers its target surface area with a relatively large number of beams. For GSO satellites these beams are obviously stationary, that is, each beam, covers a specific area. In case of LEO satellites and, more generally, satellites that are non-stationary relative to the surface of the earth, there are two alternatives, see also Fig. 25.7.

- The satellite has a set of fixed beams in which case the beam footprint (the area covered by each beam) continuously moves in line with the movement of the satellite
- The satellite has a set of steerable beams that are controlled in such a way that the beam footprint remains (approximately) fixed on the surface as long as the satellite remains within range. When the satellite beam can no longer reach its target area, it is adjusted to cover a new area. At the same time, a new emerging satellite must take over the coverage of the area previously covered by the "disappearing" satellite.

Note that Fig. 25.7 illustrates only a single beam out of, typically, a large number of beams generated by each satellite.

Although both fixed beams and steerable beams are possible, the second alternative, that is, steerable beams, is typically assumed for satellite-based NTN, as this makes mobility within an NTN (see further Section 25.2.5) somewhat less complex.

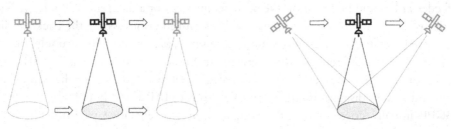

Fixed beam, moving footprint   Steerable beam, fixed footprint

Fig. 25.7 Fixed beam with moving footprint vs steerable beam with fixed footprint.

25.2 NR-based NTN

The extensions needed to the NR air-interface specifications to also support NTN-based access are relatively minor and primarily deals with functionality to handle

- The large propagation time, especially for GSO satellites
- The rapid movement of the satellite relative to the device (for LEO satellites)

25.2.1 Spectrum for NTN

Like 3GPP specifications in general, the NTN-related features in 3GPP are mainly spectrum agnostic. At the same time, assumptions on potential spectrum can often impact specific detailed solutions.

During the study items, 3GPP considered both frequencies below 6 GHz and mm-wave frequencies for the satellite-device link. However, due to time limitations, 3GPP eventually decided to focus release 17 on lower frequency bands and postponing any features specifically targeting operation in higher bands to later releases (see also Section 25.3). Eventually, release 15 specified two frequency bands for the service (satellite-device) link:

- 1626.5–1660.5 MHz (uplink)/1525–1559 MHz (downlink)
- 1980–2010 MHz (uplink)/2170–2200 MHz (downlink)

Note that the large propagation time associated with satellite-based NTN makes the use of time-division duplex essentially impossible due to the need for extreme guard times. Therefore, FDD operation, that is, frequency separation between the satellite-to-device and device-to-satellite links, is generally assumed for satellite-based NTN.

25.2.2 Extensions to uplink/downlink time alignment

As described in Section 15.2, uplink/downlink time alignment, that is, the timing offset T_{TA} between the start of an uplink slot n and the start of the corresponding downlink slot n is given by

$$T_{TA} = \left(N_{TA} + N_{TA,offset}\right) \cdot T_c \qquad (25.1)$$

where

- $N_{TA,offset}$ is a cell-specific offset provided by the network
- N_{TA} is a device-specific offset based on timing-advance commands provided by the network
- T_c is the basic time unit, see Section 7.1

Especially, for initial random access $N_{TA}=0$ and the timing of the uplink PRACH/preamble transmission has a fixed offset $N_{TA,offset}$ relative to the downlink reception timing, with no considerations to the propagation time between the network and the device. As a consequence, the network receiver must "search" for preamble transmissions within a window corresponding to twice the maximum base-station-to-device propagation time. There must also be a sufficiently large guard time to ensure that preamble transmissions do not collide with other transmissions.

In case of satellite communication, the propagation time is significantly larger than for any terrestrial link. The propagation time will also vary significantly depending on the exact position of the satellite relative to the device. To handle this, the device is assumed to estimate and pre-compensate for the propagation time between the device and the satellite.

In addition to the satellite-to-device propagation time there will, in case of transparent payload, be an additional component of the overall propagation time due to the satellite-to-ground-station link. This propagation time is the same for all devices connected via the satellite but is constantly changing due to the satellite movement. To enable the device to compensate also for this propagation time, an additional term is added to the overall time-alignment expression.

Thus, for NTN, the uplink/downlink time alignment, that is, the timing offset between uplink transmission and downlink reception at the device, is given by

$$T_{TA} = \left(N_{TA} + N_{TA,offset} + N_{TA,adj}^{common} + N_{TA,adj}^{UE} \right) \cdot T_c \qquad (25.2)$$

where

- N_{TA} and $N_{TA,offset}$ are the same as for expression (25.1) above
- $N_{TA,adj}^{common}$ is the compensation for the satellite-to-ground-station propagation time
- $N_{TA,adj}^{UE}$ is the device-calculated compensation based on estimates of the satellite-to-device propagation time

As the satellite-to-ground-station distance is continuously changing, $N_{TA,adj}^{common}$ is not provided explicitly to the device. Rather, it assumed that the one-way ground-station-to-satellite propagation time T_{gs} varies in time according to a second-degree polynomial, that is,

$$T_{gs} = A + B \cdot (t - t_o) + C \cdot (t - t_o)^2 \qquad (25.3)$$

Instead of directly providing $N_{TA,adj}^{common}$, which varies continuously and thus continuously have to be updated, the parameters A, B and C (in the specifications referred to as TA_{Common}, $TA_{CommonDrift}$ and $TA_{CommonDriftVariant}$ respectively) are provided to the device together with the time t_0. The compensation $N_{TA,adj}^{common}$ is then calculated as

$$N_{TA,adj}^{common} = T_{gs}/T_c$$

For regenerative payload, where the reference point for the timing is at the satellite rather than at the ground station, $N_{TA,adj}^{common}$ should be set to zero.

Regarding $N_{TA,adj}^{UE}$, to enable estimation of the satellite-to-device propagation time, information about the instantaneous satellite orbit, that is, Ephemeris data, is broadcast as part of NTN-specific system information provided within a new system-information block (SIB19). The specification allows for this information to be provided in two different formats

- As Keplerian elements, see Section 25.1.2
- As standard cartesian (x, y, z) coordinates and corresponding velocity components $(v_x, v_y, v_z)^{d}$

This broadcast Ephemeris data is updated regularly and SIB19 also includes the exact time instant for which the information is valid. Based on this, the device can, at least in principle, estimate the exact satellite position at any future time.

To estimate the satellite-to-device propagation distance/time, a device also needs to know its own position. Thus, at least for current 3GPP releases, devices that support access via satellite-based NTN are assumed to support self-positioning based on GNSS.

In addition to estimating and compensating for the satellite-to-device propagation time, the device is also assumed to estimate and compensate for the varying Doppler-shift due to the satellite movement relative to the device. The Doppler shift can also be estimated from the Ephemeris data broadcast by the satellite as this does not only enable the device to calculate the instantaneous satellite location but also the instantaneous satellite velocity.

In contrast, any Doppler shift due to the satellite movement relative to the ground station is assumed to be handled internally within the network with no visibility to the device.

25.2.3 Timing relations between downlink and uplink transmission

For NR there are several specified timing relations that determine in which slot a certain uplink transmission should take place, given the slot in which a corresponding downlink transmission is received. This includes, for example,

- the DCI-to-PUSCH timing relation that determines the uplink slot within which a dynamically scheduled PUSCH is to be transmitted, given the downlink slot in which the corresponding scheduling grant (on DCI) is received. As described in Section 10.1.5, this is given by a slot offset, in the specifications referred to as K_2, in the range 0 to 32 slots provided as part of the scheduling grant.
- the PDSCH-to-HARQ-feedback timing relation that determines the uplink slot within which Hybrid-ARQ feedback is to be transmitted, given the downlink slot in which the corresponding PDSCH is received. As described in Section 13.1.4, this is given by another slot offset, in this case referred to as K_1, provided as part of the DCI carrying the scheduling assignment for the PDSCH transmission.
- The timing of aperiodic SRS transmissions that determines in which uplink slot the SRS transmission is to take place, given the slot in which the triggering DCI is received

---

[d] Note that, in contrast to the Keplerian elements, (x,y,z) coordinates can also be used to describe the position of, for example, a HAPs.

In general, these relations identify an uplink slot within which transmission is to take place relative to a corresponding downlink slot carrying information about the uplink slot. There is then obviously an implicit assumption that the downlink slot is received by the device *before* the corresponding uplink slot is to be transmitted by the device.

For terrestrial networks, the propagation time between the base station and a UE is typically limited to a few 100 µs at most (and even less for higher numerologies at higher frequencies, due to shorter range at such frequencies). As a consequence, the DL/UL timing dis-alignment, that is, the difference in transmission timing of a specific uplink slot and the corresponding downlink slot, is limited to, at most, a few slots and is, in most cases, less than a slot, see upper part of Fig. 25.8.

However, in case of a satellite-based NTN, the propagation time is much larger, leading to a very large timing dis-alignment between an uplink slot n and the corresponding downlink slot n, see lower part of Fig. 25.8.[e] As a consequence, the tools provided in earlier releases to specify a certain uplink/downlink timing relation are not sufficient for the case of NTN. As an example, a scheduling grant received in downlink slot n can schedule uplink transmissions in an uplink slot in the range n to $n+32$ depending on the slot offset (K_2) provided in the scheduling grant. This is obviously insufficient in the case when the dis-alignment between the uplink and downlink exceeds 32 slots. Taking into account the required processing time in the device, it is clear that this is in many cases insufficient for satellite-based NTN.

For this reason, release 17 introduced an additional offset K_{offset} to the above timing relations. Thus, as an example, instead of the HARQ feedback being transmitted in slot number $n + K_1$, where K_1 is the slot offset signaled in the DCI, the HARQ feedback is to

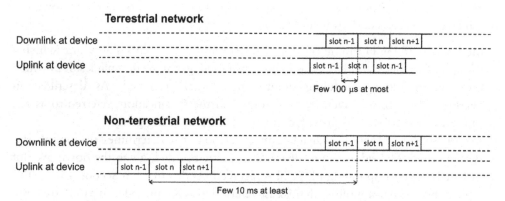

Fig. 25.8 DL/UL time alignment for terrestrial network (upper part) vs satellite-based non-terrestrial network (lower part).

[e] Note that there is an implicit assumption that downlink slot n is essentially time aligned with uplink slot n at the gNB.

be transmitted in slot $n + K_{offset} + K_1$, and similar for the other timing relations. K_{offset} is provided as part of the NTN-related system information (SIB19) and its maximum value is 1024 slots. Assuming 15 kHz numerology, this is sufficient to cover total round-trip time of roughly 1 s. K_{offset} should typically be set so that it matches the maximum propagation time within the NTN cell, including the propagation over the feeder link in case of transparent payload. There is also a possibility to configure K_{offset} on a per-device basis.

The parameter K_{offset} is introduced to handle the large dis-alignment between uplink and downlink slots at the device side in case of satellite-based NTN. More specifically, it ensures that the downlink/uplink timing relations are such that a device can receive a downlink slot before the corresponding uplink slot is to be transmitted despite a large dis-alignment between uplink and downlink slots. In addition to K_{offset} there is a second parameter introduced in the NR specification to handle the large propagation time in case of satellite-based NTN, more specifically its impact on the activation of configurations provided by a MAC Control Element (MAC CE).

Before the network carries out downlink transmission according to a new configuration provided by means of MAC CE it needs a confirmation that the MAC CE has been properly received by the device and that the corresponding configuration has been activated. Such confirmation is assumed to be provided by the reception of HARQ feedback from the device indicating correct detection of the MAC CE.

According to release 15/16, if the device transmits HARQ feedback, related to the reception of a MAC-CE, in an uplink slot n the device should activate the configuration provided by the MAC CE in downlink slot $n + 4$.[f] It is then critical that the gNB receives the HARQ feedback, that is, receives uplink slot n, before downlink slot $n + 4$ is to be transmitted. Otherwise, the gNB will not know if the MAC CE has been properly received by the device, that is, will not know if the device has activated the configuration provided by the MAC CE or not.

The timing relation between downlink and uplink slots at the gNB (the ground station) will be determined by two parameters (see Eq. 25.2)

- The device-calculated parameter $N_{TA,adj}^{UE}$ compensating for the satellite-to-device propagation time
- The parameter $N_{TA,adj}^{common}$ compensating for the ground-station-to-satellite propagation time

If $N_{TA,adj}^{common}$ equals two times the ground-station-to-satellite propagation time, this propagation time will be fully compensated for and uplink slot n and downlink slot n will be

[f] This assumes the same numerology for uplink and downlink, which can typically be assumed. In case of the downlink numerology being different from that of the uplink, the expression will be more complicated.

essentially aligned at the gNB. In other words, uplink slot n will be received by the gNB well before downlink slot $n+4$ occurs at the gNB.

However, there may be cases when, for some reason, the network is operating with a different timing advance, in practice with a different value for $N_{TA,adj}^{common}$. In the extreme case, with $N_{TA,adj}^{common}=0$, the timing advance will only compensate for the satellite-to-device propagation time in which case uplink slot n and downlink slot n will be aligned *at the satellite*. Assuming a transparent payload with the gNB at the ground station, downlink slot $n+4$ may then occur at the gNB long before uplink slot n is received.

To handle this situation, the parameter K_{mac} has been introduced. K_{mac} simply changes the time instant when the configuration provided by the MAC CE is to be activated from slot $n+4$ to slot $n+4+K_{mac}$, where K_{mac} should be set so that uplink slot n is received at the gNB well before downlink slot $n+4+K_{mac}$ occurs. Similar to K_{offset}, K_{mac} is provided as part of the NTN-related system information in SIB19.

In the "worst" case ($N_{TA,adj}^{common}=0$), the value of K_{mac} should be set to twice the ground-station-to-satellite propagation time. This should be compared to K_{offset} which should be set to twice the sum of the ground-station-to-satellite propagation time and the satellite-to-device propagation time. For this reason, the maximum value of K_{mac} is 512 slots, that is, half that of K_{offset}.

Fig. 25.9 illustrates the two cases above, that is the case when the downlink and uplink are aligned at the gNB (at the ground station for transparent payload) and K_{mac} is not needed (upper part of Fig. 25.9) vs. the case when the downlink and uplink are not aligned at the gNB and K_{mac} is needed (lower part of Fig. 25.9). The figure also illustrates how the parameter K_{offset} ensures that downlink data is received before the HARQ feedback is to be transmitted by "pushing" the time of HARQ feedback an additional amount K_{offset} forward.

Note that K_{mac} is only needed for transparent payload and then only in the case when the parameter $N_{TA,adj}^{common}$ *is not used* to fully compensate for the ground-station-to-satellite propagation time, that is, when uplink slot n and downlink slot n are not aligned at the ground station. For regenerative payloads, $N_{TA,adj}^{common}$ should be set to zero and K_{mac} is not needed.

25.2.4 HARQ operation and number of HARQ processes

Hybrid-ARQ is a key feature for good link performance, especially in situations with unpredictable channel conditions. As described in Chapter 13, release 15/16 was limited to 16 HARQ processes. This means that, if the HARQ round trip time exceeds 16 slots, including processing time in the network and device, one cannot sustain continuous transmission to or from a UE as one will run out of available HARQ processes. Obviously, for a satellite-based NTN, the round-trip time may far exceed this limit.

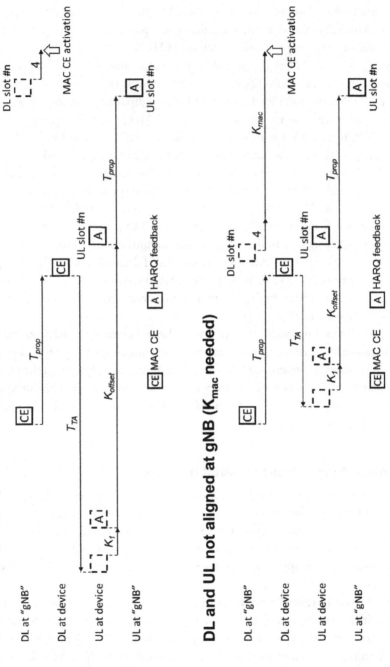

Fig. 25.9 Situation when K_{mac} is not needed (upper figure) and when K_{mac} is needed (upper figure).

To somewhat enhance the situation, release 17 introduced support for up to 32 HARQ processes by extending the HARQ-process indication from 4 bits to 5 bits for both downlink and uplink. Assuming a slot length of 1 ms (15 kHz numerology), this allows for an overall round-trip time, including any processing in network and device, of up 32 ms without running the risk of stalling HARQ.

In many cases of NTN, and especially for geo-stationary satellites, the overall round-trip time is much larger than 32 ms. One alternative is then to simply accept that the HARQ will stall with a reduced overall link throughput as a consequence.

Another alternative is to operate without HARQ, something which is possible already in NR release 15/16. In the downlink, this can be done by simply scheduling new downlink data with the same HARQ process and a toggled NDI flag indicating "*new data*". However, release 17 NTN takes this further by introducing the possibility to explicitly disable HARQ feedback, thereby avoiding the overhead of this feedback. The possibility to disable HARQ feedback is configured on a HARQ-process basis, that is, HARQ feedback can be disabled for certain HARQ processes while remain enabled for other HARQ processes. In this way, full throughput can be retained while still ensuring high reliability based on HARQ for more critical data. Also note that, for uplink data transmission there is, in general, no explicit HARQ feedback as the request for retransmissions are implicit in the scheduling grant. Thus, operating without HARQ is straightforward already with release 15/16.

Obviously, operating without HARQ will have a negative impact on the basic link performance as additional margins are needed in the link adaptation. However, the NTN radio link is typically experiencing line-of-sight condition implying that the channel conditions, even if not good, are still typically much more predictable, compared to the terrestrial case. This makes the need for HARQ less critical for NTN, compared to conventional terrestrial networks.

25.2.5 Mobility in a non-terrestrial network

In Section 25.1.4 it was described how a satellite covers the earth with, typically, a large set of beams. It was also described how these beams can be fixed, implying that the beam footprint continuously moves in line with the movement of the satellite. However, a more common case is steerable beams that are controlled in such a way that the beam footprint remains at least approximately fixed on the surface as long as the satellite remains within range.

Similar to the terrestrial infrastructure, a satellite creates one or multiple cells by means of which devices access the network. Each such cell may correspond to a single satellite beam. Alternatively, each cell may correspond to an area jointly covered by multiple beams of the same satellite. In the extreme case, there may be a single cell per satellite.

Regardless of whether the satellite uses fixed beams or steerable beams, the movement of satellites in non-geostationary orbit implies that even a stationary device needs to regularly handover to a cell of a new satellite. The speed by which a satellite may disappear in combination with the high latency in any network-to-device signaling implies problems for conventional network-triggered handover. At the same time, in contrast to terrestrial networks where mobility is due to the more or less random movement of the device, handover within an NTN is much more predictable as it, to a large extent, is due to the known movement of the satellite. For these reasons, *conditional handover* is expected to be extensively used for NTN. As described in Chapter 6, conditional handover implies that a device is configured in advance with a set of candidate cells and associated triggering conditions. Once a combination of the trigger conditions associated with a specific candidate cell is fulfilled, the device should execute a handover to that cell without the need for the reporting and signaling that need to be carried out prior to conventional network-controlled handover.

To utilize the predictability of handover in the NTN scenario, release 17 extended the triggering conditions for conditional handover with a complementary *timing condition*. Each timing condition defines a time interval within which handover to the corresponding candidate cell can take place. The timing condition is always combined with a conventional measurement-based condition, that is, the handover is executed when a measurement condition is fulfilled during the time interval defined by the timing condition.

There is also the possibility for a *location-based condition* for condition handover. Once again, this is used in combination with a conventional measurement-based condition, that is, the handover is executed when the measurement condition is fulfilled while the device is within an area defined by the location-based condition.

25.3 NTN extensions on release 18

At the time of this writing, the 3GPP work on release-18 extensions to NTN is still ongoing. At this stage, the work has focused on three topics.

25.3.1 NTN in mm-wave bands

As already discussed, due to time limitations, release 17 was limited to operation in the lower-frequency spectrum, more specifically spectrum around 1.6 GHz and 2 GHz. As part of release 18, 3GPP is extending NTN to support operation also in mm-wave spectrum. Currently, the following spectrum been agreed upon.

- Uplink: 27.5–30 GHz
- Downlink: 17.3–20.2 GHz

Note that, despite being mm-wave bands which have until now been solely specified for TDD operation in 3GPP, this is paired spectrum for FDD operation. As already mentioned in Section 25.2.1, TDD-based operation is essentially impossible for satellite-based NTN due to the large propagation time.

25.3.2 Coverage enhancements

With regards to coverage enhancements, 3GPP is currently only focusing on one specific case, namely the transmission of HARQ feedback (on PUCCH) for downlink Message 4 of the random-access procedure (see Chapter 17). The proposed solution/extension is to allow for PUCCH repetition a factor two, four, or eight for the HARQ acknowledgement where the repetition factor can be dynamically decided on by the network.

25.3.3 Network-verified UE (device) location

In general, an NTN cell will typically be very large, something which makes network-based positioning of a device difficult. At the same time, for legal reasons the network must be able to determine at least the rough location of a device (for emergency calls, lawful intercept, etc.).

Obviously an NTN device knows its own location from GNSS (note that this is currently anyway an assumption for NTN devices, see Section 25.2.2) and could report its coarse location to the network.[g] However, GNSS measurements cannot be fully trusted as they can relatively easily be jammed and/or spoofed. The work on network-verified UE location aims at developing network-based solutions to verify such coarse GNSS-based UE locationing.

---

[g] Privacy requirements forbid a device to report its precise GNSS-based location.

CHAPTER 26

Sidelink communication

The possibility for 3GPP *device-to-device* (D2D) communication, that is, direct communication between devices, also referred to as *sidelink communication*, was first introduced for LTE as part of 3GPP release 12 [98]. Later releases then extended the LTE sidelink communication with specific focus on the *vehicle-to-vehicle* (V2V) use case, that is, direct communication between vehicles [101].

The first release of the NR specifications did not include support for sidelink communication. However, support for NR sidelink communication was introduced in 3GPP release 16 as part of a work item on V2X (Vehicle-to-Anything) [102]. The aim of the V2X work item was to ensure that NR could provide the connectivity required for advanced V2X services, with focus on the following more specific use cases (see, for example, Ref. [103] for more details):

- Vehicle Platooning
- Extended sensors
- Advanced driving
- Remote driving

Although the scope of the release-16 V2X work item was not limited to vehicle-to-vehicle communication but also included, for example, the required vehicle-to-infrastructure communication for these use cases, the absolute main part of the work-item activities focused on the introduction of NR sidelink communication targeting the vehicle-to-vehicle use case.

In principle, there is nothing that prevented the use of the release-16 sidelink also for other scenarios and use cases for which there were a need for direct device-to-device communication. However, the focus on vehicular-related use cases led to some design choices that was partly in conflict with requirements of other use cases. For example, the focus on the vehicular-related use cases led to a limited focus on low device energy consumption as a vehicular-mounted device can be assumed to have access to abundant amount of energy. For this reason, 3GPP release 17 introduced extensions to NR sidelink communication, making it more applicable to other use cases, for example, those for which low device energy consumption is important [133].

Separately from these sidelink enhancements, 3GPP release 17 also introduced support for *sidelink relaying* [134], that is, the use of the sidelink functionality as a tool to extend network coverage by using devices as relay nodes.

5G/5G-Advanced
https://doi.org/10.1016/B978-0-443-13173-8.00025-6

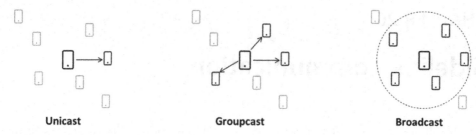

Fig. 26.1 Sidelink transmission scenarios.

26.1 NR sidelink – Transmission and deployment scenarios

NR sidelink supports three basic transmission scenarios, see also Fig. 26.1.

- *Unicast*, in which case the sidelink transmission targets a specific receiving device
- *Groupcast*, in which case the sidelink transmission targets a specific group of receiving devices
- *Broadcast*, in which case the sidelink transmission can be received by any device that is within the range of the transmission

There are two basic deployment scenarios for NR sidelink communication in terms of the relation between the sidelink communication and an overlaid cellular network, see also Fig. 26.2.

- *In-coverage operation*, in which case the devices involved in the sidelink communication are within the coverage of an overlaid cellular network. The network can then, to a smaller or larger extent depending on the exact mode-of-operation, control the sidelink communication.
- *Out-of-coverage operation*, in which case the devices involved in the sidelink communication are not within the coverage of an overlaid cellular network

Note that there is also a *"partial-coverage"* scenario where only a subset of the devices involved in the device-to-device communication are within the coverage of an overlaid network.

In case of in-coverage operation, the sidelink communication may share carrier frequency with the overlaid cellular network. Alternatively, sidelink communication

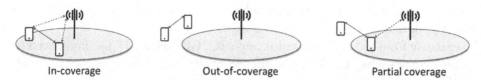

Fig. 26.2 Sidelink deployment scenarios.

may take place on a sidelink-specific carrier frequency which is different from the carrier frequency of the overlaid network.

In general, before a device can engage in NR-based sidelink communication it needs to be properly configured. For devices under network coverage such configuration is provided by means of a sidelink-specific system-information block referred to as SIB12. In addition to the common configuration provided by SIB12, there may be additional device-specific sidelink-related configurations provided by means of dedicated signaling.

For devices outside network coverage the sidelink-related configuration may, for example, be hard-wired into the device itself or stored on the device SIM card. In 3GPP terminology this is referred to as "pre-configuration", to differentiate from the more conventional configuration taking place for devices under network coverage. Here we will avoid these details and use the term "configuration" also for the case when a device is outside network coverage and sidelink-related parameters are provided by means of pre-configuration.

As already mentioned, sidelink communication, including the vehicle-to-vehicle use case, is supported already in LTE. There may be situations where one would like to have LTE-based sidelink communication under the coverage of, and controlled by, an overlaid NR network. Similarly, there may be situations where one would like to have NR sidelink communication together with an overlaid LTE network. Support for such sidelink scenarios is included in the 3GPP specifications. However, we will here not dwell more into this but assume NR sidelink operating under the coverage of an NR network, alternatively operating out-of-coverage.

26.2 Resources for sidelink communication

NR sidelink transmission is based on conventional OFDM with the same 15 kHz-based numerology as is used for network-to-device communication, see Chapter 7. This is a difference compared to LTE sidelink which was based on DFTS-OFDM, that is, the uplink transmission scheme of LTE.[a]

In case of NR sidelink communication sharing carrier frequency with a conventional cellular network, sidelink transmission takes place in uplink resources, that is, on an uplink carrier in case of paired spectrum and in slots configured as uplink slots in unpaired spectrum. In other words, sidelink transmissions, which are inherently device transmissions, use the same set of resources that is used for device transmission also in the overlaid network.

When an NR device is configured for sidelink transmission it is configured with a sidelink *resource pool* which, among other things, defines the overall time/frequency resource that can be used for sidelink communication within a carrier, see also Fig. 26.3.

---

[a] Note that NR supports both conventional OFDM and DFTS-OFDM for uplink transmission.

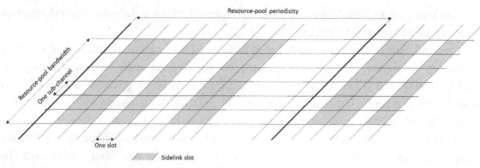

Fig. 26.3 Example structure of sidelink resource pool.

- In the time–domain the resource pool consists of a set of slots repeated over a *resource-pool period*.
- In the frequency domain the resource pool consists of a set of consecutive *subchannels*, where a subchannel consists of a number of consecutive resource blocks. The subchannel is the smallest unit of sidelink data transmission in the frequency domain.

Overall, the time/frequency structure of a sidelink resource pool is thus defined by

- A configurable resource-pool period
- A configurable set of *sidelink slots* within the resource-pool period
- A configurable subchannel bandwidth which can be of a size corresponding to 10, 12, 15, 20, 25, 50, 75 or 100 resource blocks
- A resource-pool bandwidth corresponding to a configurable set of consecutive sub-channels
- The frequency-domain location of the first subchannel of the resource pool

It should be pointed out that a device may be configured with multiple resource pools for sidelink transmission. Furthermore, in addition to resource pool(s) for sidelink transmission, a device will also be independently configured with one or multiple resource pools for sidelink reception. Note though that, typically, the transmit and receive resource pools are identical.

Although the resource-pool configuration has a slot-based granularity in the time domain, this does not mean that all symbols of a sidelink slot are necessarily available for sidelink transmission. Rather, the network can impose limitations so that only a limited set of consecutive symbols within a sidelink slot is actually available for sidelink communication, see Fig. 26.4.[b] This is done by configuring

---

[b] This is actually not a property of the resource pool but a property of the bandwidth part within which the resource pool is configured. The same limitations are thus valid for all resource pools configured within the same bandwidth part.

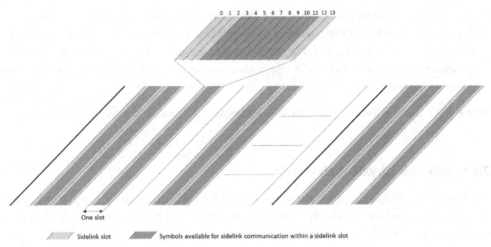

Fig. 26.4 Limiting the sets of available symbols within the sidelink slots.

- the first symbol of the set of consecutive symbols available for sidelink communication, ranging from symbol number 0 to symbol number 7 (symbol number 3 assumed in Fig. 26.4)
- the number of consecutive symbols available for sidelink communication, ranging from 7 symbols to 14 symbols (9 symbols assumed in Fig. 26.4)

In this way one can, for example, ensure that the first and/or last few symbols of a slot are available for downlink or uplink control signaling for the case when sidelink communication shares carrier frequency with conventional downlink and uplink communication. Note that, in case of a sidelink-specific carrier frequency, it can typically be assumed that all symbols within a sidelink slot are available for sidelink communication.

As will be further discussed below, some of the available symbols of a sidelink slot will/may also be used

- for Hybrid ARQ feedback
- for AGC (Automatic Gain Control)
- as guard symbol(s)

This will further limit the number of symbols available for actual sidelink data transmission within a slot.

There are two basic modes for sidelink transmission in terms of how the exact set of resources to use for the sidelink transmissions is decided on.

- In case of *resource-allocation mode 1*, an overlaid network schedules the sidelink transmissions in a way that, at least on a high level, is similar to the scheduling of uplink PUSCH transmissions (see for example Chapter 14). Resource-allocation mode 1 is obviously only applicable when the transmitting device is within network coverage.

- In case of *resource-allocation mode 2*, a decision on sidelink transmission, including decision on the exact set of resources to use for the transmission, is made by the transmitting device itself based on a *sensing and resource-selection/reservation procedure*. Resource-allocation mode 2 is applicable to both the in-coverage and out-of-coverage deployment scenarios.

The two resource-allocation modes will be discussed in more details in Section 26.4.

26.3 Sidelink physical channels

Similar to downlink and uplink transmissions, sidelink transmission takes place over a set of physical channels on to which a transport channel is mapped and/or which carry different types of L1/L2 control signaling. This includes

- the *physical sidelink control channel* (PSCCH) which carries *sidelink control information* (SCI), more specifically the *1st-stage SCI* or *SCI-1*. SCI-1 includes information needed by receiving devices for proper demodulation/detection of the PSSCH. The SCI-1 information is also used as part of the sensing and resource-selection procedure of resource-allocation mode 2, see further Section 26.4.2.
- the *physical sidelink shared channel* (PSSCH) on to which the *sidelink shared channel* (SL-SCH) transport channel is mapped. In other words, the PSSCH carries the actual sidelink data between devices. The PSSCH also carries some additional control signaling referred to as *2nd-stage SCI* or *SCI-2*.
- the *physical sidelink feedback channel* (PSFCH) which carries Hybrid-ARQ feedback from a receiving device to the transmitting device.

As we will see later, there is also a *physical sidelink broadcast channel* (PSBCH) which is part of the so-called *sidelink SSB* (SL-SSB) The SL-SSB is used for sidelink synchronization, with the PSBCH carrying a small amount of "system information" (the sidelink *master information block* or sidelink *MIB*) needed to be shared between devices. Sidelink synchronization and the SL-SSB is discussed in more details in Section 26.7.

Fig. 26.5 summarizes the sidelink physical channels (including the PSBCH) and the information they carry.

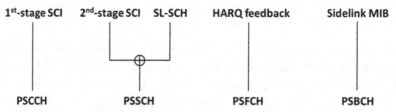

Fig. 26.5 Sidelink physical channels and corresponding information carried by each channel.

We should already now comment on the split of the sidelink control information (SCI) into two parts, SCI-1 and SCI-2. As mentioned above, in addition to information needed for the demodulation and detection of the PSSCH, the SCI-1 also includes information related to sensing and resource reservation. As will be discussed further in Section 26.4.2, this information is relevant for multiple devices, in principle for all devices operating under resource-allocation mode 2 within range of the transmitting device. Thus, even if the sidelink data transmission is unicast, SCI-1 has to be broadcast with a known format.

In contrast, SCI-2 only contains information relevant for the device, or devices in case of multicast/broadcast, for which the actual sidelink data transmission is intended. This includes, for example, the *destination ID*, that is, the identity of the device, or group of devices, for which the sidelink data transmission is intended, and information related to Hybrid ARQ. Thus, SCI-2 can, for example, be beam-formed, especially if there is only a single target device. There are also different SCI-2 formats, depending on, for example, the exact Hybrid-ARQ mode-of-operation, with the actual SCI-2 format being signaled within SCI-1. This also provides a certain degree of future-proofness in the sense that new SCI-2 formats supporting new functionality can be introduced in later releases without impacting legacy devices. Already with release 16 there were two different SCI-2 formats, SCI-2A and SCI-2B, that differed somewhat in terms of the Hybrid-ARQ-related information, see also Section 26.5. In release 17, an additional format (SCI-2C) was introduced to support signaling related to *inter-UE coordination*, see Section 26.4.2.3.

26.3.1 PSSCH/PSCCH

The PSSCH and PSCCH are jointly transmitted within the time/frequency resource, consisting of one slot over a number of contiguous subchannels, either scheduled for the sidelink transmission by the network in case of resource-allocation mode 1 or autonomously selected by the transmitting device itself in case of resource-allocation mode 2. Fig. 26.6 illustrates the time/frequency structure of a PSSCH/PSCCH

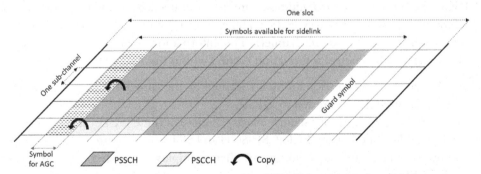

Fig. 26.6 Structure for PSSCH/PSCCH assuming eleven symbols available for sidelink transmission, including symbol for AGC and guard symbol, and assuming PSCCH extending over three symbols.

transmission. As already mentioned, only a subset of the symbols of a sidelink slot may be available for PSSCH/PSCCH transmission. The actual PSSCH/PSCCH transmission always starts at the second of these available symbols, with the first available symbol being a copy of the second symbol. The reason for this is to provide a time interval during which a receiving device can carry out automatic gain control (AGC), that is, adjust the gain of the receiver amplifier to fit the power of the received signal. There is also a guard symbol at the end of the PSSCH/PSCCH transmission. The guard symbol is needed, for example, to provide time for switching from sidelink transmission to sidelink reception and vice versa, as well as for switching between sidelink transmission/reception and regular downlink or uplink transmissions.

The exact number of symbols over which the PSSCH/PSCCH transmission occurs depends on the number of symbols available for sidelink transmission within a slot. However, it also depends on if there are resources assigned for PSFCH transmission within the slot, as will be further described below (no PSFCH resources assumed in Fig. 26.6).

In the time domain, PSCCH is transmitted within the first two or three symbols of the resource assigned for the PSSCH/PSCCH transmission, not including the AGC symbol (three symbols assumed in Fig. 26.6). In the frequency domain the PSCCH is transmitted starting at the lowest resource block of the PSSCH/PSCCH resource and with a bandwidth up to one sub-channel. The PSSCH is then mapped to the remaining resource elements of the overall PSSCH/PSCCH resource. The bandwidth and duration (two or three symbols) of the PSCCH are part of the resource-pool configuration and are thus known to a receiving device in advance.

The location of the PSCCH at a fixed position within the overall PSSCH/PSCCH resource implies that

- A receiving device only needs to search for PSCCH at the lower end of each sub-channel.
- Once a receiving device has found the PSCCH and decoded the corresponding SCI-1, it has also found the frequency-domain starting position of the overall PSSCH/PSCCH resource. The only additional information needed for the receiving device to completely know the overall PSSCH/PSCCH resource is the PSSCH/PSCCH bandwidth in number of subchannels. This information is provided as part of the SCI-1, that is, the control information carried within the PSCCH.

Channel coding and modulation for SCI-1 is based on the same Polar code as is used for DCI and with modulation limited to QPSK.

For PSSCH, the situation is slightly more elaborate as the PSSCH carries transport-channel data (SL-SCH) but also SCI-2. SL-SCH and SCI-2 are separately channel coded and modulated. The modulated symbols are then multiplexed together before mapping to the PSSCH time/frequency resource.

- Channel coding and modulation for SCI-2 is the same as for SCI-1, that is, the same Polar code as is used for DCI and with modulation limited to QPSK
- Channel coding and modulation for SL-SCH is the same as is used for the downlink and uplink shared channels (see Chapter 9), that is, LDPC codes and with modulation up to 256QAM.

SCI-1 contains information about the transmission format for the SCI-2, including information that allows a receiving device to determine the set of resource elements used for SCI-2 and SL-SCH respectively. Once a device has decoded SCI-1, it can thus properly extract SCI-2 and the SL-SCH from the PSSCH.

SL-SCH supports the transmission of one transport block over up to two layers. In case of two-layer transmission, for SCI-2, which relies on the same DM-RS as SL-SCH, the same symbol is mapped to both antenna ports.

26.3.2 PSFCH

The PSFCH carries Hybrid-ARQ feedback for sidelink transmissions received on the PSSCH.

The basic structure of the PSFCH is the same as PUCCH format 0 (Section 10.2.2), that is, the feedback information (ACK or NACK) is conveyed by applying different phase rotations to a frequency-domain base sequence of length twelve. The phase-rotated sequence is then mapped to a single resource block (twelve sub-carriers) assigned for the PSFCH transmission.

As illustrated in Fig. 26.7, the PSFCH is transmitted in the second last available symbol of a sidelink slot (the last available symbol is always used as a guard symbol). Furthermore, to allow for AGC, the PSFCH symbol is copied to the immediately prior symbol in the same way as for PSSCH/PSCCH, see above. The guard symbol between the PSSCH/PSCCH and the PSFCH is needed to provide a switching time between PSSCH/PSCCH reception and PSFCH transmission.

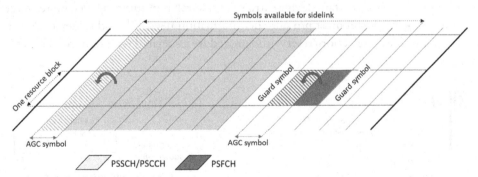

Fig. 26.7 Joint PSSCH/PSCCH and PSFCH structure assuming eleven symbols available for sidelink transmission (including AGC and guard symbols).

This implies that, if PSFCH resources are configured for a sidelink slot, this will require of a total of three symbols for the PSFCH itself, the AGC symbol and the extra guard symbol, with a corresponding reduction in the number of symbols available for PSSCH transmission.

There does not have to be PSFCH resources in every slot. Rather, a resource pool can be configured to have PSFCH resources in every slot, in every second slot, or in every fourth slot. This means that each PSFCH occasion may have to carry HARQ feedback corresponding to multiple PSSCH transmissions. A resource pool can also be configured without any PSFCH resources in which case Hybrid ARQ will not be used for the side-link transmission within the resource pool.

More details on sidelink Hybrid-ARQ and the related feedback signaling are given in Section 26.5.

26.4 Resource allocation

As already mentioned in the introduction, there are two different modes in terms of how the exact set of resources to use for a specific sidelink transmission is decided on, see also Fig. 26.8.

- *Resource-allocation mode 1*, in which case an overlaid network schedules sidelink transmissions
- *Resource-allocation mode 2*, in which case the device autonomously decides on the resource to use for sidelink transmission by means of a sensing and resource-selection procedure

It is important to understand that the resource-allocation mode is relevant only from a transmitter point-of-view and a receiving device does need to not know under what resource-allocation mode the corresponding transmitting device is operating. Also, different devices within the same area may operate under different resource-allocation modes.

Therefore, even if a certain device carries out sidelink transmission based on resource-allocation mode 1, it still must provide the information needed by other devices for the sensing and resource-selection/reservation procedure associated with resource-allocation mode 2.

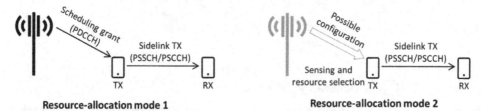

Fig. 26.8 Resource-allocation mode 1 and 2.

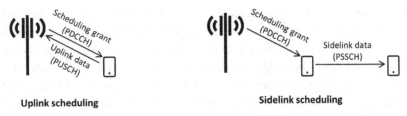

Fig. 26.9 Scheduling grant for uplink transmission (left) vs. sidelink transmission (right).

26.4.1 Resource-allocation mode 1

In case of resource-allocation mode 1, sidelink (PSSCH/PSCCH) transmissions can only be carried out by a device if the network has provided it with a scheduling grant that indicates the exact set of resources used for the transmission. This is in many respects similar to the scheduling of uplink transmissions (Chapter 14) with the important difference that the grant is for a sidelink (PSSCH/PSCCH) transmission rather than an uplink (PUSCH) transmission, see Fig. 26.9.

Similar to uplink scheduling, sidelink scheduling can be done by means of both dynamic and configured grants.

A dynamic grant for sidelink transmission is provided by means of *DCI format 3_0* introduced specifically for sidelink scheduling grants.[c] Each dynamic grant can provide resources for sidelink transmission of the same SL-SCH transport block in up to three slots (one initial transmission and up to two retransmissions) within a window of 32 slots, see Fig. 26.10. The first scheduled resource occurs a time offset ΔT after the slot within which the DCI carrying the scheduling grant is received. The remaining up to two

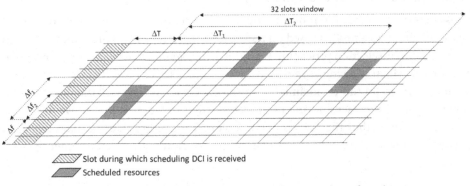

Fig. 26.10 Scheduling of up to three sidelink resources within a window of 32 slots.

[c] There is also a DCI format 3_1 that is used to when an overlaid NR network schedules LTE-based sidelink transmissions.

scheduled resources have time offsets ΔT_1 and ΔT_2 relative to the first scheduled resource. The up to three resources have the same bandwidth (four subchannels assumed in Fig. 26.10) but may have different frequency-domain locations given by the frequency offsets Δf, Δf_1 and Δf_2, where Δf is the frequency offset of the first scheduled resource relative to the start of the resource pool and Δf_1 and Δf_2 are the frequency offsets of the second and third scheduled resources relative to the first scheduled resource. Information about the parameters ΔT, ΔT_1 and ΔT_2, and Δf, Δf_1 and Δf_2, as well as the bandwidth of the scheduled resource, is provided within the scheduling grant.

Note that the possibility for a scheduling grant to include resources for up to two different retransmissions does not mean that there can only be up to two retransmissions of a transport block. However, if additional retransmissions are needed, the device has to request, and be explicitly scheduled, new resources for these retransmissions.

A configured grant provides periodically occurring resources for sidelink transmission. Similar to configured grants for uplink transmission, see Section 14.4, there are two types of configured grant for sidelink transmission.

- *Configured grant type 1* for which the entire grant, including the resources to use for sidelink transmission, is configured by means of RRC signaling
- *Configured grant type 2* for which the periodicity is configured by means of RRC signaling while the activation of the grant, as well as the periodic resources to use for sidelink transmission, is provided by DCI format 3_0 using an RNTI different from the one used for dynamic grants.

For each period, the configured grant (both type 1 and type 2) may provide resources in up to three slots similar to a dynamic grant.

26.4.2 Resource-allocation mode 2

In case of resource allocation mode 2, a device autonomously decides on the resources to use for sidelink transmissions based on a *sensing and resource-selection* procedure. The sensing and resource-selection procedure is assisted by *resource-reservation* announcements, the intention of which is to inform other devices about a set of resources that a device has selected/reserved for future sidelink transmissions. Other devices will then use this information as part of their sensing and resource-selection procedure, that is, when selecting the set of resources they themselves will reserve/use for future sidelink transmissions. As already mentioned, the resource-reservation information is provided to other devices as part of SCI-1.

26.4.2.1 Resource reservation/indication

In addition to the PSSCH/PSCCH transmission of the current slot, that is, the slot in which SCI-1 is transmitted, a device can reserve/indicate resources for up to two additional transmissions of the same transport block within a time window of 32 slots, see Fig. 26.11. Each of these two resources has the same bandwidth as the transmission of

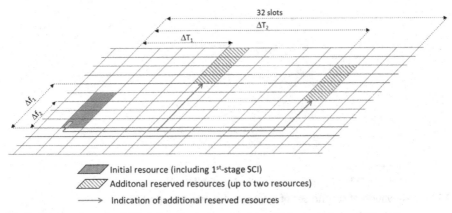

Initial resource (including 1st-stage SCI)

Additonal reserved resources (up to two resources)

Indication of additional reserved resources

Fig. 26.11 Reservation of additional up to two resources within a window of 32 slots.

the current slot but can have different frequency-domain locations. Information about these reserved resources, defined by the time offsets ΔT_1 and ΔT_2 and the frequency shifts Δf_1 and Δf_2 is provided within SCI-1. Note that these additional resources are assumed to be used for the retransmission of the same transport block, that is, not for the transmission of a new transport block.

It can be noted that the structure of these up to three resources (current resource plus up to two additional resources) is the same as the up to three scheduled resources that can be provided by means of a dynamic or configured grant in case of resource-allocation mode 1 (compare Figs. 26.10 and 26.11). Information about the parameters ΔT_1, ΔT_2, Δf_1 and Δf_2 in Fig. 26.10 are included in SCI-1 also when the transmitting device operates under resource-allocation mode 1. A receiving device operating under resource-allocation mode 2 will then interpret the corresponding resources as "reserved" when carrying out its sensing and resource selection. This enables simultaneous operation of resource-allocation mode 1 and mode 2 by different devices within the same area.

In addition to one or two additional resources within a time window of 32 slots as described above, a device operating under resource-allocation mode 2 may also reserve an identical set of up to three resources for the transmission of another transport block. As illustrated in Fig. 26.12, this additional set of resources are time shifted an amount T_p relative to the first set of resource, where the offset T_p can range from as small as 1 ms to as large as 1000 ms.[d] As part of the resource-pool configuration, devices are provided with up to 15 possible values for T_P. A device then selects one of these values and announces it in form of a four-bit parameter that is part of the resource reservation within SCI-1. The remaining (all-zero) parameter value is used to indicate that no additional resources are reserved.

---

[d] Note that his additional set of resources should not overlap with the first set of resources within the 32 ms window, something which may impose additional restrictions on the lower-limit of T_p.

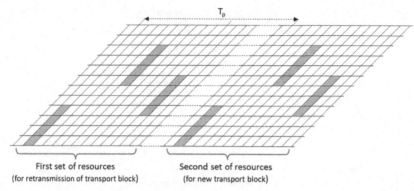

Fig. 26.12 Reservation of second set of resources for new transport block.

The possibility to reserve additional time-shifted resources is typically used in scenarios when a device is to carry out periodic sidelink transmissions. Note though that the device does not reserve periodic resources but only a single set of additional resources. At the time of transmission on these additional reserved resources, the device could/ should reserve yet another set of additional resources for the transmission of yet another transport block and so on.

26.4.2.2 Sensing and resource selection

The sensing and resource selection is the procedure by which a device operating under resource-allocation mode 2 selects the set of resources to use for sidelink transmission based on, among other things, resource reservations announced by other devices.

The data to be transmitted is assumed to require a certain amount of frequency-domain resources, that is, a certain number of sub-channels. It is also assumed to have a certain delay budget, in practice implying that the data should be transmitted within a certain time window, as well as a certain priority.

Release 16 only supported so-called *full sensing*. In release 17 this was extended with the possibility for *partial sensing* as well as the possibility for resource selection without any prior sensing. Below we will first describe the release-16 full sensing and then describe the release-17 extensions.

The resource selection starts by the device listing all *potential candidate resources*, which is the same as all N-subchannel resources within the resource-pool bandwidth for all sidelink slots within a *selection window*, see Fig. 26.13, where N is the required amount of frequency-domain resources.[e] The minimum length of the selection window is a

---

[e] The number of N-subchannel resources within a slot equals $(M - N + 1)$, where M is the overall resource-pool bandwidth measured in subchannels.

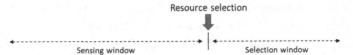

Fig. 26.13 Resource selection with sensing window and selection window.

configurable parameter, while the maximum length is limited by the delay budget for the data to be transmitted.

Based on SCI-1 transmissions of other devices received during a preceding *sensing window*, the device is assumed to have acquired knowledge about resource reservations made by such devices. For full sensing, the length of the sensing window can be configured to either 100 ms or 1000 ms.

If a resource in the list of potential candidate resources partly overlaps with a resource reserved by another device, based on information acquired from received SCI-1 transmissions, and the SCI-1 transmission was received with a signal strength that exceeds a configured threshold, the resource is removed from the list of potential candidate resources. The threshold with which the received signal strength is compared depends on

- The priority of the transmission to be made by the sensing device (the higher priority, the lower threshold)
- The priority of the announced resource reservation (the higher priority, the higher threshold), information about which is provided as part of the resource-reservation information within SCI-1

If, at the end of this procedure, the remaining list of potential candidate resources contains less than 20% of the original list, that is, more than 80% of the original potential candidate resources have been removed from the list, the procedure is restarted with the signal-strength threshold increased by 3 dB. This will reduce the probability that a resource will be removed from the list of potential candidate resources, that is, increase the number of remaining candidate resources. This is repeated, with further increased thresholds, until the remaining list of candidate resources includes at least 20% of the resources of the original list. From the remaining list of candidate resources, the device selects up to three N-channel resources in different slots for the transmission of a transport block and, potentially, an additional set of up to three resources for the transmission of a second transport block as illustrated in Fig. 26.12.

On a high level, what the sensing procedure does is thus to

- Prioritize resources not reserved by other devices
- Prioritize resources reserved by devices received with lower signal strength, with the aim to reduce the impact of any collision due to the use of the same resource for sidelink transmissions by nearby devices while allowing for spatial reuse of the resources by devices at larger distance.

- Prioritize resources reserved with a lower relative priority by other devices, that is, prioritize resources for which a collision may be less critical
- Guarantee that there are at least 20% of the original potential candidate resources within the delay-budget time window to do the final random resource selection from

In some cases, a sensing device may carry out transmissions of its own, and thus not be able to monitor for SCI-1 transmissions, during some slots of the sensing window. For such slots, the device should make a worst-case assumption and remove, from the list of potential candidate resources, all resources that could hypothetically have been indicated by a potentially missed SCI-1 transmission. However, this is only true for the reservation of resources for a second transport block (the T_p-shifted resources in Fig. 26.12) and not for the reservation of resources for retransmissions within the 32-slots window of Fig. 26.11.

Release-17 extensions – Partial sensing and random-resource allocation

The drawback of the release-16 full sensing is that a device has to carry out essentially continuous reception in order to have sensing information available for resource selection if a packet arrives for sidelink transmission. Alternatively, the device has to delay the sidelink transmission for a time corresponding to the sensing window, something which is only possible for non-time-critical transmissions.

For vehicle-mounted devices, which was the focus for release-16 sidelink communication, continuous sensing is typically not a serious problem as such devices can be assumed to have an abundant amount of energy available. However, for use cases where low device energy consumption is more important, full sensing may lead to an unacceptable drain of the device battery. Release 17 therefore introduced two additional schemes for sensing under the umbrella of resource-allocation mode 2, *periodic-based partial sensing* and *contiguous partial sensing*. These two schemes target somewhat different scenarios, with periodic-based partial sensing more suitable for detecting and avoiding periodic transmissions and contiguous partial sensing targeting the detection/avoidance of aperiodic transmissions. Thus, contiguous partial sensing is always used if partial sensing is configured for a resource pool, as a resource pool may always be used for aperiodic sidelink transmissions. In contrast, periodic-based partial sensing is only used when the resource pool allows for the reservation of resources for transmission of an additional transport blocks based on the parameter T_p as described above.

The basic principle of resource selection based on partial sensing is the same as for the case of full sensing.

- There is a set of candidate resources which, in case of full sensing, corresponds to resources in all slots within the selection window
- There is a set of slots, in case of full-sensing all slots of the sensing window, within which the device monitors for resource reservations within SCI-1 transmissions of

other devices. Based on this, resources may be removed from the set of candidate resources.

The difference between the different sensing schemes lies in

- the exact set of slots within which candidate resources are located (all slots within the selection window for full sensing)
- the exact set of slots within which sensing is carried out (all slots within the sensing window for full sensing)

Common to both schemes of partial sensing is that, instead of considering resources within *all the available slots* of the selection window as potential candidate resources, partial sensing starts by selecting a limited set of *candidate slots* within the selection window and limits the potential candidate resources to be within these candidate slots. The minimum number of candidate slots can be configured by the network from as few as a single candidate slot up to a maximum of 32 candidate slots. Note that the set of selected candidate slots does not have to be contiguous.

Contiguous and periodic-based partial sensing differ in terms of within what slots sensing is carried out. In case of contiguous sensing, sensing is carried out within a limited set of *consecutive* slots ending just before the first candidate slot, see upper part of Fig. 26.14. In contrast, in case of periodic-based partial sensing (lower part of Fig. 26.14), for a given candidate slot sensing is carried out within slot(s) $m - k \cdot P$, where

- m is the index of the candidate slot
- P is the set of the possible periods for periodic reservation as resources as described above.
- k is either a limited to single value $k = k_0$, determined so that the corresponding sensing slot occurs just before the first candidate slot, or takes two values $k = k_0$ and $k = k_0 - 1$ (this alternative is assumed in Fig. 26.14)

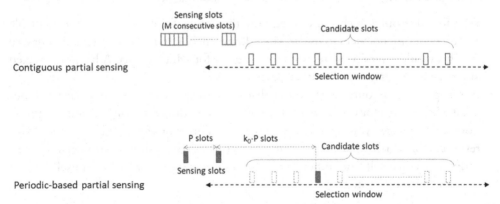

Fig. 26.14 Contiguous partial sensing vs periodic-based partial sensing. Two sensing slots assumed in the case of partial sensing.

Note that the lower part of Fig. 26.14 only shows the sensing slots for one specific candidate slot and for one specific periodicity. In the general case there will be multiple periodicities with each periodicity corresponding to one or two sensing slots for each candidate slot.

Once a set of slots within which sensing should take place has been determined according to above, the actual sensing is done in the same way for the two schemes of partial sensing and in the same way as for full sensing. That is, the device monitors for resource reservations of other devices within the slots identified for sensing. Based on this, together with signal-strength measurements and priority information, resources are potentially removed from the set of candidate resources.

In addition to periodic-based and contiguous partial sensing, release 17 also introduced the possibility for *random resource selection*. In that case, a device selects resources for transmission without any prior sensing.

26.4.2.3 Inter-UE coordination
Another extension of resource-allocation mode 2 in release 17 is the possibility for *inter-UE coordination*. This includes the possibility for devices to

- provide information to other devices about *preferred* and/or *non-preferred* resources, information that can then be taken into account in the resource selection of these other devices (referred to as *coordination-scheme 1*),
- provide information about collisions between resource reservations, information that can be used to trigger reselection of resources by other devices (referred to as *coordination-scheme 2*)

In case of coordination-scheme 1, a certain device A provides information of preferred and/or non-preferred resources to other devices This information is based on sensing carried out by device A, that is, it reflects the situation as seen by device A.

- A preferred resource is a resource that, from the point-of-view of device A, is a suitable resource for other devices to use if/when doing sidelink transmissions targeting device A. This would typically correspond to resources for which device A has not detected any resource reservations by other devices.
- A non-preferred resource is a resource that, from the point-of-view of device A, is not a suitable resource for other device B to use if/when they carry out sidelink transmissions with device A as an intended receiver. This would typically correspond to resources for which device A has detected resources reservations (with signal strength above a certain level) from other devices or resources on which device A itself intends to carry out transmissions.

Device reporting of preferred/non-preferred resources could be triggered by another device requesting this information. In addition to indicating if the request is for preferred or non-preferred resources, the request will then also include information about

- The time window within which the preferred/non-preferred resource is to be located
- The size of the preferred/non-preferred resource in number of subchannels
- A priority value which will impact the signal-strength threshold by means of which device A will characterize a resource as preferred/non-preferred

Alternatively, a device may broadcast information about preferred/non-preferred resources without having received any prior request. Note that, even if reporting of preferred/non-preferred resources is triggered by requests from other devices, the report itself can be received and utilized by any device.

Note that the parameters included in the above request are similar to the parameters that are input to the sensing and resource selection carried out by a device operating according to the conventional/non-coordinated resource-allocation mode 2. In some sense the inter-UE coordination could be seen as an extension of conventional sensing described above. With conventional sensing a device selects suitable resources for transmission based on resource-reservation information within SCI-1 received by the device itself. With inter-UE coordination, the resource selection is augmented with additional resource-reservation information received by other devices, in practice devices that are the targets of the sidelink transmission. While the conventional sensing is based on the situation at the transmitter itself, the inter-UE-coordination information is instead based in the situation at an intended receiver.

The coordination information, that is, the indication of preferred/non-preferred resources is provided by means of a new SCI-2 format (SCI-2C). The SCI-2C format is also used for the request of coordination information.

In case of inter-UE-coordination scheme 2, that is, collision detection, a device A can inform another device B that the resource reservation of device B, targeting sidelink transmission to device A, overlaps with a resource reservation of another device. The criterion for detecting a collision could either be based on the absolute received power of the reservations or on the relative difference in received power. The device A can also report a collision if the resource reservation of device B overlaps with a resource in which device A is not expecting to do sidelink reception, for example, due to half-duplex limitations.

In contrast to scheme 1, which can be seen as pro-active, that is, indicating referred and/or non-preferred resources to avoid collisions in the resource reservations of different devices, scheme 2 is more of a reactive scheme, that is, indicating that a collision in the resource reservation has been detected. On the reception of such an indication, a device may then decide to re-select resources as a means to avoid a collision in the actual sidelink transmission.

Similar to the indication of preferred/non-preferred resources, the collision indication is provided to device B by means of the new SCI-2C format.

26.5 Sidelink hybrid-ARQ

Depending on the configuration, receiving devices may provide Hybrid-ARQ feedback to the transmitting device using the PSFCH physical channel. Based on such feedback, the transmitting device may then carry out retransmissions, possible by first requesting resources for such retransmissions from an overlaid network.

26.5.1 Hybrid-ARQ feedback

Hybrid-ARQ feedback from a receiving device to the transmitting device is supported for both unicast and groupcast sidelink transmission.

In case of unicast transmission, both ACK and NACK feedback can be provided, encoded as different phase rotations of the PSFCH base sequence.

- ACK is provided if the receiving device has correctly decoded an SL-SCH transport block
- NACK is provided if the receiving device has detected the presence of an SL-SCH transport block from the decoding of SCI-1 and SCI-2 but has not been able to correctly decode the SL-SCH transport block

Note that a receiving device needs to correctly decode both SCI-1 and SCI-2 in order to determine the presence of an SL-SCH transport block aimed for the device.

In case of groupcast transmission, that is, a sidelink transmission targeting a group of receiving devices, there are two options for Hybrid-ARQ feedback.

- *ACK/NACK feedback*: A receiving device provides ACK feedback if it has correctly decoded an SL-SCH transport block. It provides NACK feedback if it has detected the presence of the SL-SCH transport block from SCI-1/SCI-2 but has not been able to decode the transport block. This is essentially the same as the Hybrid-ARQ feedback for unicast sidelink transmissions
- *NACK-only feedback*: A receiving device provides NACK feedback if it has detected the presence of an SL-SCH transport block from SCI-1/SCI-2 but has not been able to decode the transport block. If the device has correctly decoded the SL-SCH transport block it does not provide any Hybrid-ARQ feedback.

In case of NACK-only feedback, multiple devices can share the same PSFCH resource, that is, the same resource block and phase rotation, see Section 26.3.2. If any device is not able to decode the SL-SCH and thus provides NACK feedback, the transmitting device can detect a NACK and initiate a retransmission of the SL-SCH transport block. In contrast, in case of ACK/NACK feedback, each receiving device must be assigned its own PSFCH resource. This can be done by assigning different sets of PSFCH phase rotations and/or different resource blocks for PSFCH transmission from different devices.

In case of groupcast transmission with NACK-only feedback there is also a mechanism to limit NACK transmission so that only devices within a certain physical range from the transmitting device will provide Hybrid-ARQ feedback. This is enabled by the division of the geographical area in which sidelink communication takes place into a number of *zones*. To enable the distance-dependent restriction of Hybrid-ARQ feedback, a transmitting device will provide:

- information about the zone within which the device is located
- the range limit within which Hybrid-ARQ feedback should be provided

This information is included within SCI-2, more specifically using the SCI-2B format. A receiving device is assumed to know its own physical location, for example, by means of GNSS, and can then, based on the information within the received SCI-2, determine the distance to the center of the zone indicated by the transmitting device. By comparing this distance to the range limit provided within SCI-2, the device can determine whether or not Hybrid-ARQ feedback should be provided. Note that the calculated distance is not really the distance from the receiving device to the transmitting device but rather the distance from the receiving device to the center of the zone in which the transmitting device is located.

26.5.2 Hybrid-ARQ retransmissions

To enable sidelink Hybrid ARQ, a Hybrid-ARQ process number, a new data indicator and a redundancy-version indictor are all included within the SCI-2. The function of these parameters is essentially the same as for downlink and uplink Hybrid-ARQ.

For resource-allocation mode 2, the transmitting device can, by itself, decide on a retransmission based on the received sidelink Hybrid-ARQ feedback. There is, however, a possibility to configure a maximum number of retransmissions that can be carried out.

For resource-allocation mode 1, Hybrid-ARQ retransmissions are slightly more elaborate as any sidelink transmission, including a retransmission, can only take place on a resource granted by the network. In some cases, a device may already have a grant for retransmissions (remember that a device may be assigned resources for up to three transmissions, that is, two retransmissions, in the same grant). Otherwise, the device must explicitly request a grant for retransmission from the network. It does so by means of "Hybrid-ARQ feedback" to the network on a normal PUCCH physical channel. This device-to-network "Hybrid-ARQ feedback" can, in essence, be seen as a special scheduling request that is requesting resources for retransmission of a specific transport block. This is true also for the case of sidelink transmissions based on configured grants, that is, if resources are needed for additional retransmissions, beyond those provided for by the configured grant itself, the device must explicitly make a request for such resources. The resources are then provided by the network by means of a dynamic grant.

To enable this kind of TX-device-to-network retransmission request the sidelink scheduling grant with DCI format 3_0 includes a Hybrid-ARQ process number and a new data indicator, similar to scheduling grants for PUSCH, see Section 10.1.5. Note that this Hybrid-ARQ process number and new data indicator are only relevant for the TX-device/network "Hybrid-ARQ" loop. The process number and new data indicator conveyed from the transmitting device to the receiving device(s) within the SCI-2 may or may not the same. Also note that DCI format 3_0 does not include any redundancy-version indicator, that is, the transmitting device can, by itself, decide on the redundancy version to use for the sidelink transmission.

26.6 Other sidelink procedures
26.6.1 Sidelink power control

In addition to being assigned or selecting a resource for sidelink transmission, the device must also determine the transmit power for the sidelink transmission. There are several mechanisms for this.

First, a maximum sidelink transmit power $P_{max, config}$ can be configured as part of the resource-pool configuration.

Secondly, for devices under network coverage the transmit power for a sidelink transmission can be further limited in order to limit any interference to uplink transmissions within the cell. This is done by the device estimating the path loss to the cell site and, if needed, further reducing its maximum sidelink transmit power according to

$$P_{TX,max} = min\left\{P_{max,config}, P_0 + \alpha \cdot P_L + 10 \cdot log_{10}(M_{RB})\right\}$$

where

- P_0 is a network-provided parameter that, somewhat simplified, corresponds to the target maximum received power/interference level per subchannel at the cell site
- P_L is the estimate of the device-ot-cell-site path-loss
- α is a network-provided parameter ($\leq$1) for fractional path-loss compensation
- M_{RB} corresponds to the bandwidth of the sidelink transmission

This additional restriction of the maximum sidelink transmit power can be applied to all sidelink transmission scenarios (unicast, groupcast, and broadcast).

Note that the expression for the maximum sidelink transmit power is essentially a simplified version of the open-loop power-control expression for PUSCH (Section 15.1.1).

Finally, in case of unicast sidelink transmissions, a receiving device can provide the transmitting device with Layer-3-filtered RSRP reporting. The RSRP can then be used by the transmitter to estimate the transmitter-to-receiver path loss and, by means of this, further match the transmit power to the sidelink channel conditions. Note that this is only applicable to PSCCH/PSSCH transmissions, that is, not for the transmission of PSFCH and PSBCH.

26.6.2 Sidelink channel sounding and CSI reporting

NR sidelink supports sidelink CSI reporting where a receiving device sounds the channel based on CSI-RS transmitted by another device and reports CSI to the transmitting device. The reported CSI can then be used, for example, for selection of precoding for subsequent transmissions to the reporting device.

The sidelink CSI-RS structure re-uses the structure of the downlink CSI-RS, see Chapter 8, with the following restrictions:

• The number of CSI-RS ports is limited to one or two
• The CSI-RS density is limited to one, that is, CSI-RS is transmitted within every resource block within the sidelink transmission bandwidth.

Sidelink CSI-RS is only transmitted together with PSSCH/PSCCH, that is, there are no stand-alone CSI-RSs transmissions, and the presence of CSI-RS within the PSSCH/PSCCH is indicated within the SCI-2.

An indication of CSI-RS transmission within the SCI-2 also triggers the reporting of CSI. As there is no sidelink physical channel corresponding to the uplink PUCCH, reporting of sidelink CSI is done by means of MAC-CE signaling within a PSSCH. The signaling is limited to rank indication (rank one or two) and four-bit CQI. Thus, explicit sidelink PMI reporting is not supported.

26.7 Sidelink synchronization

Before devices can engage in sidelink communication, they should acquire a common timing reference.

For devices within coverage of an overlaid cellular network, such a common timing reference can be acquired by synchronizing to the cells of the network.[f] Furthermore, a device not under direct network coverage can indirectly acquire the common timing reference by synchronizing to another device that is under network coverage and has already acquired the common timing reference. The common timing reference can then be further propagated along a chain of devices as illustrated in Fig. 26.15.

Instead of acquiring the timing reference, either directly or indirectly, from an overlaid cellular network, a device may alternatively acquire the timing reference from GNSS. Similar to the case of network-based timing acquisition, acquiring the timing reference from GNSS may either be direct, if the device is under GNSS coverage, or indirect via another device, see Fig. 26.16.

---

[f] Note that this assumes a synchronized network deployment where different cells are operating with a common timing.

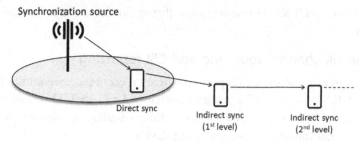

Fig. 26.15 Direct and indirect (via another device) network-based timing acquisition.

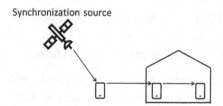

Fig. 26.16 Direct and indirect GNSS-based timing acquisition.

If there is no network available for synchronization, GNSS-based timing acquisition is the only alternative. However, even if sidelink communication takes place under network coverage, acquiring the timing reference from GNSS may still be preferred. One such situation could be if the cells of the overlaid network are not mutually synchronized. The sidelink-related system information in SIB12 provides information if devices should prioritize network-based or GNSS-based timing acquisition.

Although multi-level synchronization along a chain of devices as illustrated in Figs. 26.15 and 26.16 is possible, it is always preferred for a device to be, synchronization-wise, as close as possible to the ultimate source of the timing reference (a network cell or GNSS). In other words, acquiring the timing reference directly from the network/GNSS, if possible, is preferred over indirectly acquiring it from the network/GNSS via another device. Likewise, indirectly acquiring the timing reference via a device that is directly synchronized to the network or GNSS is preferred over acquiring it via a device that is, in itself, only indirectly synchronized to the network/GNSS. The sidelink synchronization procedure (Section 26.7.2) is designed to ensure this.

Note though that all devices within the same area should have the same ultimate timing reference. Thus, if network-based timing-acquisition is prioritized, even indirect network-based timing acquisition is prioritized over direct GNSS-based acquisition, and vice versa.

26.7.1 The sidelink SS/PSBCH block

It should be clear from above that an important component of sidelink synchronization is the possibility for a device to acquire its timing reference from another device. Such

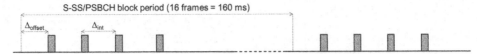

Fig. 26.17 Time-domain structure ("time domain allocations") for SL-SSB transmission assuming four SL-SSB within the SL-SSB period.

inter-device synchronization is enabled by the device transmission of a so-called *sidelink SS/PSBCH block* or SL-SSB.

As illustrated in Fig. 26.17, the SL-SSB block is transmitted with a periodicity of 16 frames (160 ms). As can also be seen in the figure, there can be multiple SL-SSB transmissions within each 160 ms period (four SL-SSB transmissions assumed in Fig. 26.17). The overall SL-SSB transmission is thus given by

- The number of SL-SSB transmissions within the 160 ms SL-SSB period
- The offset Δ_{offset} (in number of slots) from the start of the SL-SSB period to the start of the first SL-SSB transmission within the SL-SSB period
- The interval Δ_{int} (in number of slots) between consecutive SL-SSB transmissions

Note that the SL-SSB transmissions are not necessarily distributed uniformly over the SL-SSB period.

It should be pointed out that there are at least two, and in some cases three, different time-domain structures, or *time-domain allocations*, for the SL-SSB, which differ in terms of the offset Δ_{offset}.[g] The reason for this is to enable a device to both transmit and receive SL-SSB without collision between transmission and reception, see further Section 26.7.2.

The basic structure of each SL-SSB is similar to that of the cell SSB (Chapter 16) in the sense that it consists of

- a *sidelink primary synchronization signal* (S-PSS)
- a *sidelink secondary synchronization signal* (S-SSS)
- the *physical sidelink broadcast channel* (PSBCH) which carries a limited amount of information (the *sidelink MIB*) relevant for the inter-device synchronization.

The specific sequences used for the S-PSS and S-SSS are also the same as for the SSB, that is, length-127 m-sequences for S-PSS and length-127 Gold sequences for S-SSS.[h]

However, the time/frequency structure of the SL-SSB, as illustrated in Fig. 26.18, is somewhat different compared to the structure of the SSB.

As described in Chapter 16, the SSB covers 20 resource blocks (240 sub-carrier spacings) in the frequency domain. However, for the sidelink case, a minimum bandwidth of

---

[g] In principle, the different time allocations may also have different periodicity and different Δ_{int}.
[h] In Chapter 16, the SSS was described as a combination of two m-sequences which is the same as a Gold sequence.

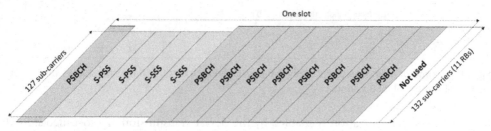

Fig. 26.18 Structure of S-SS/PSBCH.

20 resource blocks was concluded to be too large. Thus, the SL-SSB is limited to a bandwidth of 11 resource blocks or 132 subcarriers.

In the time-domain, the SL-SSB block covers 13 symbols consisting of

- Two symbols for S-PSS, with the same m-sequence being used in the two symbols
- Two symbols for S-SSS, with the same Gold sequence in the two symbols
- Nine symbols for PSBCH

It should be pointed out that the possible limitations in terms of actually available symbols within a sidelink slot, as discussed in Section 26.2, is not valid for sidelink slots in which SL-SSB is transmitted.

26.7.2 Synchronization procedure

As described above, similar to the cell SSB, the sidelink synchronization signal consists of an S-PSS and an S-SSS. In the sidelink case there are 2 different S-PSS and 336 different S-SSS jointly providing $2 \cdot 336 = 672$ different S-PSS/S-SSS combinations corresponding to 672 different *sidelink identities*. These 672 sidelink identities are divided into two groups, sometimes referred to as the *in-coverage group* and *out-of-coverage group* respectively:

- Group 1 (in-coverage group) corresponding to sidelink identities 0–335
- Group 2 (out-of-coverage group) corresponding to sidelink identities 336–671

A device that acquires its timing reference directly from an overlaid cellular network or GNSS uses a sidelink identity from the first group of sidelink identities. More specifically, if the timing is acquired directly from GNSS, the device uses a sidelink identity of 0 (zero). On the other hand, if timing is acquired from an overlaid network, the device uses a sidelink identity in the range 2–335 with information about the exact sidelink identity to use being provided as part of the sidelink-related system information in SIB 12.

In addition, a device that acquires its timing reference directly from an overlaid cellular network or GNSS should set an *in-coverage* flag within the sidelink MIB to TRUE, indicating that it has acquired timing directly from a network or GNSS.

A device not within network/GNSS coverage should prioritize synchronization to a device that is, in itself, within network coverage, that is, prioritize synchronization to a device for which the in-coverage flag is set to TRUE. If the device finds and synchronize to such device, it should

- Set its own sidelink identity to the same value as the sidelink identity of the device from which the timing is acquired
- Set its in-coverage flag to FALSE, indicating that it has not acquired timing directly from a network or GNSS.

On the other hand, if the device does not find any such device but instead acquires timing from a device that has, in itself, only indirectly acquired timing, the device should

- Set its own sidelink identity to the corresponding group-2 identity
- Set its in-coverage flag to FALSE

To allow for a device to both receive and transmit SL-SSB, there are, as earlier mentioned, two different time allocations for the SL-SSB transmissions. Devices synchronized directly to gNB or GNSS uses time-allocation 1. The time allocation is then switched for each step in the synchronization chain.

The overall structure of the synchronization chain, including the selection of sidelink identity and time allocation, as well as the value of the in-coverage flag, is summarized in Fig. 26.19.

There is an alternative procedure for GNSS-based synchronization when a device is configured with a third SL-SSB time allocation ("time-allocation 3"), see Fig. 26.20. In

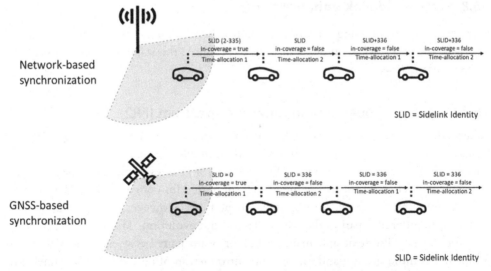

Fig. 26.19 Direct and indirect synchronization based on overlaid network and GNSS respectively.

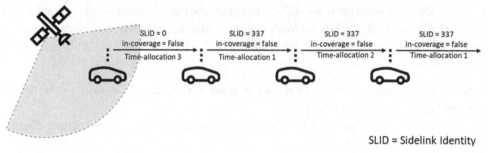

SLID = Sidelink Identity

Fig. 26.20 Alternative GNSS-based synchronization using time-allocation 3.

this case, a device synchronizing directly with GNSS sets its sidelink identity to 0 (as for the regular GNSS-based synchronization of Fig. 26.19) but sets its in-coverage flag to FALSE (despite being within coverage of GNSS) and transmits the SL-SSB using time allocation 3. Devices that are indirectly synchronized to GNSS will then use a sidelink identity of 337 (not used in case of the regular GNSS-based synchronization of Fig. 26.19) and will switch between time-allocation 1 and time-allocation 2. Thus, in this case it is the use of time-allocation, rather than the in-coverage flag, that indicates that a device is directly synchronized to GNSS.

This alternative procedure, which was inherited from LTE sidelink, ensures that devices that synchronize directly with the network or with GNSS transmit SL-SSB using different time allocations.

26.8 Further sidelink enhancements

Currently, work is ongoing on further evolution/extension of NR sidelink communication as part of the work on 3GPP release 18 [135]. At this stage, this evolution covers three different areas.

26.8.1 Enhanced operation in mmwave spectrum (FR2)

Release-16/17 sidelink does not take into account that devices operating in higher frequency bands are typically equipped with multiple antenna panels, each with a relatively high degree of directivity. Thus, release 16/17 does not include specific functionality for sidelink beam management, as is supported for conventional NR network-device communication (see, for example, Chapter 12). However, such functionality is currently considered as part of the release-18 sidelink evolution. This includes functionality for the establishment and maintenance of beam pairs between devices involved in sidelink communication and the possible introduction of new signal(s)/channel(s) to support such functionality.

26.8.2 Support for operation in unlicensed spectrum

As mentioned already in Chapter 5 and described in detail in Chapter 20, release 16 introduced support for NR operation in unlicensed spectrum. A key part of this was the introduction of channel-access procedures and more flexible timing relations, to enable operation aligned with the listen–before–talk (LBT) principle.

Release 18 aims to extend the support of operation in unlicensed spectrum to also include sidelink communication. Similar to the conventional network-device communication, this includes extending the sidelink channel access, including both mode of resource allocation (see Section 26.4), and possibly more flexible timing relations, for example for SL-SSB transmissions.

26.8.3 Carrier aggregation

In addition to the two areas above, release 18 also considers introducing support for carrier aggregation for sidelink communication. However, at the time of this writing, this activity is currently on-hold and it is unclear if sidelink carrier aggregation will be part of 3GPP release 18.

CHAPTER 27

Positioning

*Global navigation satellite system*s (GNSS), assisted by cellular networks, have for many years been used for positioning with GPS probably being the most well-known GNSS system. Each of the satellites has tightly synchronized clocks and by observing multiple satellites and measuring the time difference between them it is possible to determine the geographical position. However, satellite-based systems have very limited coverage in indoor scenarios. There is a range of applications, for example logistics and manufacturing, that require accurate positioning, not only outdoors but also indoors. NR is therefore extended in release 16 to provide better positioning support.

Architecture-wise, NR positioning is based on the use of a location server, similar to LTE. The location server collects and distributes information related to positioning (device capabilities, assistance data, measurements, position estimates, and so forth) to the other entities involved in the positioning procedures. A range of (proprietary) positioning methods, both downlink-based and uplink-based, can be implemented in the location server and used separately or in combination, see Fig. 27.1.

Common to all methods, at least those providing very accurate estimates of the position, is that tight time synchronization across the sites involved in the positioning of the device is required. By measuring the time difference between downlink signals transmitted from multiple, time-synchronized sites, the position can be derived using triangulation. This method is often referred to as *observed time-difference of arrival* (OTDOA) and the same method can be applied in the uplink. Angle-or-arrival (AoA), roundtrip time (RTT), cell identity, and received power are other quantities that can be used as input to a positioning algorithm.

Discussing different positioning algorithms is beyond the scope of this book, but there is a wide range of methods and result available in the literature, see for example [89,91]. In the following, some of the enhancements introduced in release 16 to support accurate positioning will be discussed, both downlink-based and uplink-based positioning. Positioning support is further enhanced in release 17, although large parts of release 17 focused on understanding error sources and how to handle those, aspects which to a large extent are not visible in the specifications. Carrier-phase measurements to improve the positioning accuracy is one topic in release 18 although for such measurements to be useful good phase coherency and line-of-sight propagation conditions from multiple transmitters are required. Sidelink-based positioning is another example of specification work in release 18.

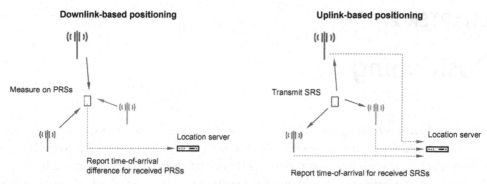

Fig. 27.1 Examples of downlink-based (left) and uplink-based (right) positioning.

27.1 Downlink-based positioning

Downlink-based positioning is supported by providing a new reference signal – the *positioning reference signal* (PRS) – and the associated measurement and reporting mechanism. To position a device, it is configured to measure on multiple positioning reference signals originating from different sites and report these measurements to the network for further processing and estimation of the location.

The positioning reference signal is a downlink signal intended for time-of-arrival measurements. Typically, measurements are carried out on multiple positioning reference signals originating from multiple as it otherwise would be difficult to determine the position of the device using, for example, triangulation. Multiple, non–colliding positioning reference signals are therefore needed, potentially transmitted on different carrier frequencies. NR uses a hierarchy with positioning frequency layers, PRS resource sets, and PRS resources to define the structure, see Fig. 27.2.

One positioning frequency layer consists of one or more PRS resource sets across one or more sites, all with the same carrier frequency and OFDM numerology. A PRS resource set contains PRS resources originating from the same site (one site may have more than one PRS resource set). Each PRS resources typically corresponds to a beam from that site. Thus, by configuring a device to measure on a certain PRS resource in a

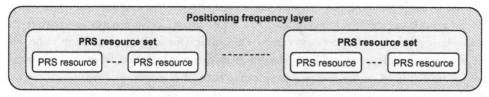

Fig. 27.2 Positioning frequency layer, PRS resource set, and PRS resource.

PRS resource set, the location server obtains knowledge not only about which site the reported measurements for this PRS resource set corresponds to, but also the particular beam the from that site.

A positioning reference signal is transmitted using one PRS resource and it is upon this reference signal the device performs the positioning-related measurements. The basis for the PRS resource is a so-called permuted staggered comb. Permuted in this context means that the comb in the different OFDM symbols has a different and not necessarily monotonically increasing offset in the frequency domain. This provides benefits compared to a non-permuted pattern in case coherent combining cannot be done across all OFDM symbols in the PRS resource but only a subset of them. An example is provided in Fig. 27.3 where every sixth subcarrier in 12 consecutive OFDM symbols is used for the PRS resource. The frequency offset from the edge of a resource block is 0, 3, 1, 4, 2, 5 for the six first OFDM symbols (a non-permuted comb would have offsets the corresponding offset set to 0, 1, 2, 3, 4, 5). The comb factor can be set to 2, 4, 6, or 12 subcarriers, that is, every 2nd to every 12th subcarrier is used for the comb. The use of a comb allows several simultaneous PRSs to be multiplexed by using different combs. In the time domain 2, 4, 6, or 12 OFDM symbols are used for one PRS resource.

In the frequency domain, a PRS resource can be configured to have a bandwidth up to 272 resource blocks with all PRS resources in a PRS resource set having the same bandwidth and location in the frequency domain (defined relative to point A). Note that the PRS resource is defined independent of any bandwidth parts and may have a bandwidth larger than the active bandwidth part. The bandwidth upon which the device measures is left for implementation as long as the measurement accuracy requirements are fulfilled.

In the time domain, a PRS resource occurs periodically in each cell with the periodicity configurable from a few milliseconds up to 10 s. Different PRS resources in the same PRS resource set may have different starting points and periodicities.

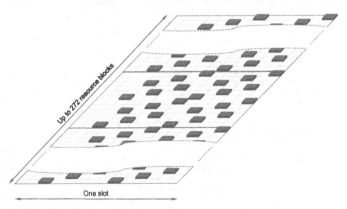

Fig. 27.3 Example of a PRS resource with a permuted staggered comb.

The actual positioning reference signal transmitted in a PRS resource is a QPSK-modulated PN sequence with configurable seed, punctured into the data transmission. Puncturing, as opposed to rate matching the PDSCH around the PRS, is used as devices not supporting positioning, and hence not being aware of the PRS resource configuration, cannot perform rate matching around the PRS resources. If the PRS-to-PDSCH interference is an issue, the gNB implementation can always schedule such that collisions between PRS and PDSCH are avoided.

Compared to LTE-based positioning, the NR PRS design is significantly better. First and foremost, the wider bandwidth possible in NR will give more accurate measurements. The structure where different PRS resources (with different identities) are used in different beams can be used to estimate the angle-of-arrival relative to the device.

Note that some of the positioning reference signal configurations are identical to the TRS. This allows the TRS to be reused for positioning purposes, which can be useful as a way to reduce overhead, albeit with a reduced performance compared to an unrestricted PRS configuration.

The PRS is transmitted on antenna ports in the 5000 series. Quasi-colocation can be configured and a PRS can be quasi-collocated with other PRSs or with the SSB.

A single PRS occasion may not result in sufficiently accurate measurements. Therefore, a PRS resource can be repeated in time as illustrated in Fig. 27.4. Up to 32 repetitions can be configured. The repetition pattern is specified not only by the repetition factor (4 in the example in the figure) but also by a time gap (8 and 1 in the figure). Together with different starting points in time, both sweep-and-repeat and repeat-and-sweep patterns can be configured in a multi-antenna system using beam sweeping. This provide flexibility to handle different beamforming strategies.

The use of orthogonal combs as described here allows for multiplexing of multiple positioning reference signals by using different, orthogonal combs. However, since the device needs to listen to positioning reference signals also from more distant sites, the near-far issue must be accounted for. Receiving a relatively weak signal from a distant

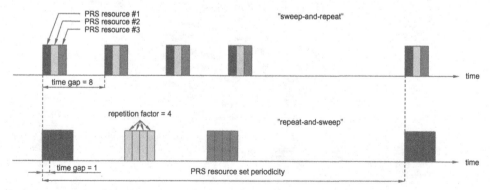

Fig. 27.4 Example of PRS resource configuration.

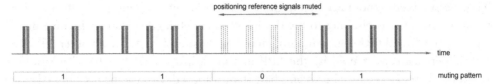

Fig. 27.5 Muting of positioning reference signals.

base station simultaneously with a more closely located base station is transmitting may not be possible, regardless of whether different frequency-domain resources are used – there is not sufficient dynamic range in the receiver to handle both of the signals. Therefore, a mechanism to ensure that the near base station is silent while measuring on the distant base station is required. This is achieved by specifying different muting patterns using a bitmap. There are multiple possibilities of specifying the muting pattern, but one example is provided in Fig. 27.5. When a bit in the bitmap is zero, the positioning reference signals at the corresponding time instant are not transmitted. Combined with not transmitting data from that site at the same time instant, the net effect is a silence gap from a particular site which allows the device to measure on a positing reference signal from a more distant base station.

Together with the introduction of the positioning reference signals, new measurements are introduced as well to support downlink-based positioning. Three measurements on the positioning reference signals are defined: positioning reference signal received power (PRS-RSRP), relative-signal-time-difference measurement (RSTD), and Rx-Tx time difference. The measurements can be performed in connected state and, from release 17, also in inactive state.

The positioning reference signal received power (PRS-RSRP) is, as the name indicated, the received power of the positioning reference signal. The primary usage for the PRS-RSRP is in combination with other positioning techniques. For example, it can be used as part of fingerprinting or as complementary input when assessing the accuracy of other PRS-related measurements. In release 17, it is possible to report not just the composite PRS-RSRP but also the received power for the individual propagation paths. It is also possible to report whether a measurement is, based on a device estimate, obtained from a line-of-sight path or not. This can be used by the network to distinguish line-of-sight propagation from non-line-of-sight propagation and thereby provide better positioning estimates. Furthermore, the network can provide beam-related angle-of-departure information to the device to assist in more accurate RSRP measurements.

The relative-signal-time-difference measurement is a measure of the difference in reception time for two positioning reference signals transmitted by two different nodes. It is a highly useful signal for positioning purposes, for example when doing triangulation.

The Rx-Tx time difference is a report from the device about the time difference between the start of a downlink frame and the start of the corresponding uplink frame.

This measurement is not restricted to the serving cell only – in which case the measurement in principle corresponds to the timing advance – but can measure the time difference relative to and PRS configuration, including those transmitted from other cells. The time difference can be used by the LPP in roundtrip-time (RTT) based positioning schemes where the distance between a base station and a device can be determined based on the estimated RTT. By combining several such RTT measurements, involving different base stations, the position can be determined.

There are also other measurements, originally defined for other purposes such as handover, that can be used as input to the positioning algorithm. The different variants of the RSRP measurement based on the SSB or CSI-RS is one example of a measurement that can be used in a similar way as the PRS-RSRP.

27.2 Uplink-based positioning

Uplink-based positioning is based on sounding reference signals (SRSs) transmitted from the devices. To better support positioning, the SRS structure is extended in several ways.

First, the duration in time is extended and can be to up to 12 OFDM symbols instead of 4 OFDM symbols. The reason for this is to ensure a sufficiently good signal-to-noise ratio for accurate measurements in the gNB. The starting point is also more flexible to account for the increased duration; an SRS can be transmitted anywhere within a slot and not only in the last six OFDM symbols as is the case in release 15. Note that the added flexibility in the time domain is needed as part of supporting unlicensed spectra as well as mentioned in Chapter 20. This is not the only example where an extension is useful for more than one purpose and underlines the fact that the NR specifications is a set of generic "tools" rather than a description of which "tool" to use for what scenario.

Second, in the frequency domain the largest comb size of the SRS is increased to up to 8. This allows multiplexing of a larger number of devices. Similar to the downlink PRS, a "permuted" comb is used for positioning purposes, see Fig. 27.6 for an example. Frequency hopping is not supported for SRS-based positioning.

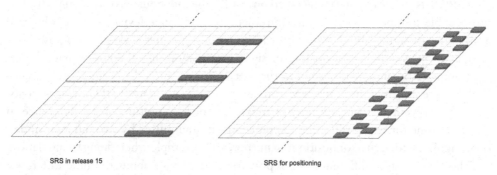

SRS in release 15   SRS for positioning

Fig. 27.6 Example of SRS configurations.

For the same reasons as in the downlink, additional measurements are specified to support uplink-based positioning. Four new gNB measurements are defined: relative time-of-arrival, Rx–Tx time difference, angle of arrival, and SRS reference signal received power.

The relative time-of-arrival measures the arrival time of the SRS relative to a configurable time reference. Rx–Tx time difference is similar but with the subframe boundary as the reference. Hence, it reports the arrival time of the SRS relative to the nearest downlink subframe boundary.

Angle-of-arrival is the angle of arrival for the signal transmitted by the device relative to either a global reference or the geographical North pole and the zenith, or relative to a local coordinate system. In a fixed-beam system this in practice corresponds to the direction of the beam receiving the signal.

The sounding reference signal received power (SRS-RSRP) is similar to its downlink counterpart, namely the received power of the sounding reference signal. It can for example be used for fingerprinting schemes. In release 17, the SRS-RSRP can be reported per detected propagation path, similar to the case for the downlink PRS-RSRP.

CHAPTER 28

RF characteristics

The RF characteristics of NR are strongly tied to the spectrum available for 5G as described in Chapter 3 and the spectrum flexibility required to operate in those spectrum allocations. While spectrum flexibility has been a cornerstone for previous generations of mobile systems, this becomes even more accentuated for NR. It consists of several components, including deployment in different-sized spectrum allocations and diverse frequency ranges, both in paired and unpaired frequency bands and with aggregation of different frequency allocations within and between different bands. NR also has the capability to operate with mixed numerology on the same RF carrier and has an even higher flexibility than LTE in terms of frequency domain scheduling and multiplexing of devices within a base station RF carrier. It is the use of OFDM in NR that gives flexibility both in terms of the size of the spectrum allocation needed and in the instantaneous transmission bandwidth used, and that enables frequency-domain scheduling.

Implementation of Active Antenna Systems (AAS) and multiple antennas in devices has been in use for LTE, but is taken one step further in NR, which supports massive MIMO and beam-forming applications both in existing bands and in the new mm-wave bands. Beyond the physical layer implications, this impacts the RF implementation in terms of filters, amplifiers, and all other RF components that are used to transmit and receive the signal and must be defined taking also the spectrum flexibility into account. These are further discussed in Chapter 29.

Note that for the purpose of defining RF characteristics, the physical representation of the gNB is called a base station (BS). A base station is defined with interfaces where the RF requirements are specified, either as conducted requirements at one or more antenna port(s) or as radiated requirements over-the-air (OTA).

28.1 Spectrum flexibility implications

Spectrum flexibility was a fundamental requirement for LTE and it had major implications for how LTE was specified. The need for spectrum flexibility is even higher for NR, because of the diverse spectrum where NR needs to operate and the way the physical layer is designed to meet the key characteristics required for 5G. The following are some important aspects impacting how the RF characteristics are defined:

- *Diverse spectrum allocations*: The spectrum used for 3G and 4G is already very diverse in terms of the sizes of the frequency of operation, bandwidth allocations, how they are arranged (paired and unpaired), and what the related regulation is. For NR it is even more diverse, with the fundamental frequency varying from 600 MHz up to 71 GHz in Rel-16, which is the maximum frequency presently identified for IMT in ITU-R. The size of allocated bands where NR is to be deployed varies from 5 MHz to 14 GHz, with both paired and unpaired allocations, where the intention is to use some allocations as supplementary downlinks or uplinks together with other paired bands. The spectrum that is planned and under investigation to be used for 5G and the related operating bands defined for NR is described in Chapter 3.

- *Various spectrum block definitions*: Within the diverse spectrum allocations, spectrum blocks are assigned for NR deployment, usually through operator licenses. The exact frequency boundaries of the blocks can vary between countries and regions and it must be possible to place the RF carriers in positions where the blocks are used efficiently without wasting spectrum. This puts specific requirements on the channel raster to use for placing carriers.

- *LTE-NR coexistence*: The LTE/NR coexistence in the same spectrum makes it possible to deploy NR with in-carrier coexistence in both uplink and downlink of existing LTE deployments. This is further described in Chapter 18. Since the coexisting NR and LTE carriers need to be subcarrier-aligned, this poses restrictions on the NR channel raster in order to align the placing of NR and LTE carriers.

- *Multiple and mixed numerologies:* As described in Section 7.1, the transmission scheme for NR has a high flexibility and supports multiple numerologies with subcarrier spacings ranging from 15 to 960 kHz, with direct implications for the time and frequency domain structure. The subcarrier spacing has implications for the RF in terms of the roll-off of the transmitted spectrum, which impact the resulting guard bands that are needed between the transmitted resource blocks. The guard bands define the RF carrier edges that are used as reference points for RF requirements (see Section 28.3). NR also supports mixed numerologies on the same carrier, which has further RF impacts since the guard bands may need to be different at the two edges of the RF carrier.

- *Independent channel bandwidth definitions.* NR devices do in general not receive or transmit using the full channel bandwidth of the BS but can be assigned what is called a bandwidth part (see Section 7.4). While the concept does not have any direct RF implications, it is important to note that BS and device channel bandwidth are defined independently, and that the device bandwidth capability does not have to match the BS channel bandwidth.

- *Variation of duplex schemes:* As shown in Section 7.2, a single frame structure is defined in NR that supports TDD, FDD, and half-duplex FDD. The duplex method is specifically defined for each operating band defined for NR as shown in Chapter 3. Some bands are also defined as supplementary downlink (SDL) or supplementary uplink (SUL) to be used in FDD operation. This is further described in Section 7.7.

Many of the frequency bands identified for deployment of NR are existing bands identified for IMT (see Chapter 3) and they may already have 2G, 3G, and/or 4G systems deployed. Many bands are also in some regions defined and regulated in a "technology-neutral" manner, which means that coexistence between different technologies is a requirement. The capability to operate in this wide range of bands for any mobile system, including NR, has direct implications for the RF requirements and how those are defined, in order to support the following:

- *Coexistence between operators in the same geographical area*: Operators may deploy NR or other IMT technologies, such as LTE, UTRA, or GSM/EDGE in different bands. There may in some cases also be non-IMT technologies. Coexistence requirements between technologies in both the same and different bands are to a large extent developed within 3GPP, but there may also be regional requirements defined by regulatory bodies in certain cases.
- *Co-location of base station equipment between operators*: There are in many cases limitations to where base-station equipment can be deployed. Often, sites must be shared between operators, or an operator will deploy multiple technologies in one site. This puts additional requirements on both base station receivers and transmitters to operate in close proximity to other base stations, even if they operate in different bands.
- *Coexistence with services in adjacent frequency bands and across country borders*: The use of the RF spectrum is regulated through complex international agreements, involving many interests. There are therefore requirements for coordination between operators in different countries and for coexistence with services in adjacent frequency bands. Most of these are defined in different regulatory bodies. In some cases, the regulators request that 3GPP includes such coexistence limits in the 3GPP specifications.
- *Coexistence between operators of TDD systems* in different parts of the same band is in general provided by inter-operator synchronization, in order to avoid interference between downlink and uplink transmissions of different operators. This means that all operators need to have the same downlink/uplink configurations and frame synchronization, which is not in itself an RF requirement, but it is implicitly assumed in the 3GPP specifications. RF requirements for unsynchronized systems become much stricter, but are not defined by 3GPP.
- *Release-independent frequency-band principles*: Frequency bands are defined regionally, and new bands are added continuously for each generation of mobile systems. This means that every new release of 3GPP specifications will have new bands added. Through the "release independence" principle, it is possible to design devices based on an early release of 3GPP specifications that support a frequency band added in a later release. The first set of NR bands (see Chapter 3) is defined in release 15 and additional bands are added in a release-independent way for subsequent releases.
- *Aggregation of spectrum allocations*: Operators of mobile systems have quite diverse spectrum allocations, which in many cases do not consist of a block that easily fits exactly

within one carrier. The allocation may even be noncontiguous, consisting of multiple blocks spread out in a band or in multiple bands. For these scenarios, the NR specifications supports *carrier aggregation*, where multiple carriers within a band, or in multiple bands, can be combined to create larger transmission bandwidths.

28.2 RF requirements in different Frequency Ranges

As discussed above and in Chapter 3, there is a very wide range of diverse spectrum allocations where NR can operate. The allocations vary in block size, channel bandwidth and duplex spacing supported, but what really differentiates NR from previous generations is the wide frequency range over which requirements need to be defined, where not only the requirement limits but also the definitions and conformance testing aspects may be quite different at different frequencies. Measurement equipment, such as spectrum analyzers, become more complex and expensive at higher frequencies and for the highest frequencies considered, including the harmonics of the highest possible carrier frequencies, requirements may not even be possible to test in a reasonable way.

For this reason, the RF requirements for both devices and base stations are divided into *frequency ranges* (FRs), where presently three are defined in 3GPP release 16 as shown in Table 28.1. The frequency range concept is not intended to be static. If new NR band(s) are added that are outside the existing frequency ranges, one of them could be extended to cover the new band(s) if the requirements align well with that range. If there are large differences compared to existing FR, a new frequency range could be defined for the new band. 3GPP Rel-16 had two frequency ranges defined (FR1 and FR2). The extension to higher frequencies in Rel-17 was done by extending FR2, but at the same time introducing a separate identification FR2-2 for the new range.

The frequency ranges are also illustrated in Fig. 28.1 on a logarithmic scale, where the related bands identified for IMT (in at least one region) are shown. FR1 starts at 410 MHz at the first IMT allocation and ends at 7.125 GHz. FR1 was originally defined from 450 to 6000 MHz, but this was later extended to cover neighboring ranges that were considered for the specifications.

FR2 in Rel-16 covered a subset of the new bands in the mm-wave range up to 52.6 GHz that are identified for IMT by the ITU-R (see Section 3.1). With the introduction of higher bands in Rel-17, FR2 was extended to 71 GHz and was at the same

Table 28.1 Frequency ranges defined in 3GPP release 15.

Frequency range designation	Corresponding frequency range
Frequency range 1 (FR1)	410–7125 MHz
Frequency range 2-1 (FR2-1)	24,250 MHz–52,600 MHz
Frequency range 2-2 (FR2-2)	52,600–71,000 MHz

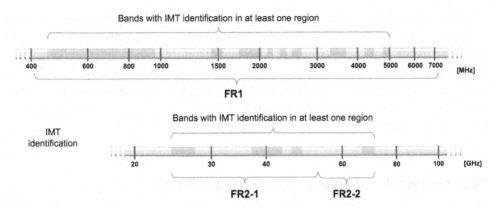

Fig. 28.1 Frequency ranges FR1 and FR2 and corresponding IMT identifications (in *blue—gray color in print version*). Note that the frequency scales are logarithmic.

time split into FR2-1 and FR2-2. The reason is that some aspects related to the new higher bands needed separate descriptions in the specification for the new higher range, while many other aspects still remain common. The term FR2 is therefore used for any aspects common to the full range 24.25 GHz to 71 GHz.

For the frequency range between FR1 and FR2-1 (7125 to 24,250 MHz), 3GPP studied the feasibility of NR operation and how it would impact specifications [106]. It could either result in further extensions of FR1 and FR2 within the new range, or the definition of a new FR3. The latter option has more far-reaching impacts on the specifications, not only for RF specifications, but also in specifications for Physical layer, Radio resource control and Radio resource management. Within ITU-R, there is an agenda item created for the coming WRC-23 to consider the band 10.0–10.5 GHz for IMT identification. There may also be regional or national allocations introduced for IMT in the range between FR1 and FR2.

For frequencies above 52.6 GHz in FR2-2, feasibility was studied by 3GPP and a work item completed in Rel-17 for NR operation in the range 52.6–71 GHz. Note that WRC-19 identified the frequency range 66–71 GHz for IMT operation in certain regions. The outcome of WRC-19, as well as the new bands to be considered for IMT at WRC-23 are further discussed in Section 3.1.1.

All existing LTE bands are within FR1 and NR thus needs to coexist with LTE and previous generations of systems in many of the FR1 bands. It is only in what is often referred to as the "mid bands" around 3.5 GHz (in fact spanning 3.3 to 5 GHz) that NR to a larger extent can be deployed in a "new" spectrum, that is spectrum previously not exploited for mobile services. FR2 covers a part of what is often referred to as the mm-wave band (strictly, mm-wave starts at 30 GHz with 10 mm wavelength). At such high frequencies compared to FR1, propagation properties are different, with less diffraction, higher penetration losses, and in general higher path losses. This can be compensated for by having more antenna elements both at the transmitter and receiver, to be used for narrower antenna beams with higher gain and for massive MIMO. This gives overall different coexistence properties and therefore leads to different RF requirements for coexistence. mm-wave RF implementation for FR2 bands also have different complexity and performance compared to FR1 bands, impacting all components including A/D and D/A converters, LO generation, PA efficiency, filtering, etc. This is further discussed in Chapter 29.

28.3 Channel bandwidth and spectrum utilization

The operating bands defined for NR have a very large variation in bandwidth, as shown in Chapter 3. The spectrum available for uplink or downlink can be as small as 5 MHz in some LTE re-farming bands, while it is up to 900 MHz in "new" bands for NR in frequency range 1, and up to several GHz in frequency range 2. The spectrum blocks available for a single operator are often smaller than this. Furthermore, the migration to NR in operating bands currently used for other radio-access technologies such as LTE, must often take place gradually to ensure that a sufficient amount of spectrum remains to support the existing users. Thus, the amount of spectrum that can initially be migrated to NR can be relatively small but may then gradually increase. The variation of the size of spectrum blocks and possible spectrum scenarios implies a requirement for very high spectrum flexibility for NR in terms of the transmission bandwidths supported.

The fundamental bandwidth of an NR carrier is called the channel bandwidth ($BW_{Channel}$) and is a fundamental parameter for defining most of the NR RF requirements. The spectrum flexibility requirement points out the need for NR to be scalable in the frequency domain over a large range. In order to limit implementation complexity, only a limited set of bandwidths is defined in the RF specifications. Channel bandwidths from 5 to 100 MHz in FR1 and from 100 to 2000 MHz inFR2 are supported.

The bandwidth of a carrier is related to the spectrum utilization, which is the fraction of a channel bandwidth occupied by the physical resource blocks. In LTE, the maximum spectrum utilization was 90%, but a higher number has been targeted for NR to achieve a higher spectrum efficiency. Considerations however must be taken for the numerology (subcarrier spacing), which impacts the OFDM waveform roll-off, and for the implementation of filtering and windowing solutions. In addition, spectrum utilization is related to the achievable error vector magnitude (EVM) and transmitter unwanted emissions, and also to receiver performance including adjacent channel selectivity (ACS). The spectrum utilization is specified as a maximum number of physical resource blocks, NRB, which is the maximum possible transmission bandwidth configuration, defined separately for each possible channel bandwidth.

What the spectrum utilization ultimately defines is a guard band at each edge of the RF carrier, as shown in Fig. 28.2. Outside of the guard band and thereby outside the RF channel bandwidth, the "external" RF requirements such as unwanted emissions are defined, while only requirements on the actual RF carrier such as EVM are defined inside. For a channel bandwidth $BW_{Channel}$, the guard band will be

$$W_{Guard} = \frac{BW_{Channel} - N_{RB} \cdot 12 \cdot \Delta f - \Delta f}{2} \qquad (28.1)$$

where N_{RB} is the maximum number of resource blocks possible and Δf is the subcarrier spacing. The extra $\Delta f/2$ guard applied on each side of the carrier is due to the relation to the RF channel raster, which has a subcarrier-based granularity and is defined independent of the actual spectrum blocks. It may therefore not be possible to place a carrier exactly in the center of a spectrum block and an extra guard of $\Delta f/2$ on each side of the transmission bandwidth will be required to make sure RF requirements can be met.

As shown in Eq. (28.1), the guard band and thereby the spectrum utilization depends on the numerology applied. As described in Section 7.3, different bandwidths are possible depending on the subcarrier spacing of the numerology, since the maximum value for N_{RB} is 275. In order to have reasonable spectrum utilization, values of N_{RB} below 11 are not used either. The result is a range of possible channel bandwidths and corresponding spectrum utilization numbers defined for NR, as shown in Table 28.2. Note that the subcarrier spacing used differs between frequency ranges 1 and 2. The spectrum

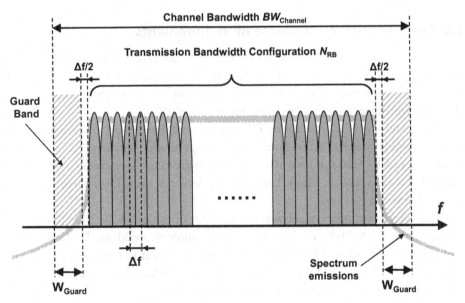

Fig. 28.2 The channel bandwidth for one RF carrier and the corresponding transmission bandwidth configuration.

Table 28.2 Range of channel bandwidths and spectrum utilization numbers defined for the different numerologies and frequency ranges in Rel-17.

Frequency range	Set of BW$_{Channel}$ used in frequency range (MHz)	SCS (kHz)	Range of possible BW$_{Channel}$ per SCS (MHz)	Corresponding range for spectrum utilization (N$_{RB}$)
FR1	5, 10, 15, 20, 25, 30, 40, 50, 60, 70, 80, 90, 100	15	5–50	25–270
		30	5–100	11–273
		60	10–100	11–135
FR2-1	50, 100, 200, 400	60	50–200	66–264
		120	50–400	32–264
FR2-2	100, 400, 800, 1600, 2000	120	100–400	66–264
		480	400–1600	66–248
		960	400–2000	33–148

utilization expressed as a fraction is up to 98% for the widest channel bandwidths and it is above 90% for all cases, except for the smaller bandwidths, where $N_{RB} \leq 25$.

Since the channel bandwidth is defined independently for base stations and devices (see earlier and in Section 7.4), the actual channel bandwidths that are supported by the base station and device specifications are also different. For a specific bandwidth, the supported spectrum utilization is however the same for base station and device, if the combination of bandwidth and subcarrier spacing is supported by both.

28.4 Overall structure of device RF requirements

The differences in coexistence properties and implementation between FR1 and FR2 mean that device RF requirements for NR are defined separately for FR1 and FR2. For a more detailed discussion of the implementation aspects in FR2 using mm-wave technology for devices and base stations, see Chapter 29.

For LTE and previous generations, RF requirements have in general been specified as conducted requirements that are defined and measured at an antenna connector. This is also the way much of the fundamental regulation is written. Since antennas are normally not detachable on a device, this is done at an antenna test port. Device requirements in FR1 are defined in this way.

With the higher number of antenna elements for operation in FR2 and the high level of integration expected when using mm-wave technology, conducted requirements are no longer seen as feasible. FR2 are therefore specified with radiated requirements and testing is done OTA. While this is an extra challenge when defining requirements, in particular for testing, it is seen as a necessity for FR2.

There is also a set of device requirements for interworking with other radios within the same device. This concerns primarily interworking with LTE for non-standalone (NSA) operation and interworking between FR1 and FR2 radios for carrier aggregation.

Finally, there is a set of device performance requirements, which set the baseband demodulation performance of physical channels of the device receiver across a range of conditions, including propagation in different environments.

Because of the differences between the different types of requirements, the specification for device RF characteristics is separated into four different parts, where the device is called *user equipment* (UE) in 3GPP specifications:

- TS 38.101-1 [5]: UE radio transmission and reception, FR1;
- TS 38.101-2 [6]: UE radio transmission and reception, FR2;
- TS 38.101-3 [7]: UE radio transmission and reception, interworking with other radios;
- TS 38.101-4 [8]: UE radio transmission and reception, performance requirements.

The conducted RF requirements for FR1 are described in Sections 28.6–28.11.

28.5 Overall structure of base station RF requirements

28.5.1 Conducted and radiated RF requirements for NR BS

For the continuing evolution of mobile systems, AAS have an increasing importance. While there were several attempts to develop and deploy base stations with passive antenna arrays of different kinds for many years, there have been no specific RF requirements associated with such antenna systems. With RF requirements in general defined at the base station RF antenna connector, the antennas have also not been seen as part of the base station, at least not from a standardization point of view.

Requirements specified at an antenna connector are referred to as *conducted requirements*, usually defined as a power level (absolute or relative) measured at the antenna connector. Most emission limits in regulation are defined as conducted requirements. An alternative way is to define a radiated requirement, which is assessed including the antenna, often accounting for the antenna gain in a specific direction. Radiated requirements demand more complex OTA test procedures, using for example an anechoic chamber. With OTA testing, the spatial characteristics of the whole BS, including the antenna system, can be assessed.

For base stations with AAS, where the active parts of the transmitter and receiver may be an integral part of the antenna system, it is not always suitable to maintain the traditional definition of requirements at the antenna connector. For this purpose, 3GPP developed RF requirements in release 13 for AAS base stations in a set of separate RF specifications that were applicable to both LTE and UTRA equipment.

For NR, radiated RF requirements and OTA testing are part of the specifications from the start, both in FR1 and FR2. Much of the work from AAS has therefore been taken directly into the NR specifications. The term AAS as such is not used within the NR base-station RF specification [4]; however requirements are instead defined for different BS types. NR is also included together with LTE and UTRA in the original AAS specifications in release 15.

The AAS BS requirements are based on a generalized AAS BS radio architecture, as shown in Fig. 28.3. The architecture consists of a *transceiver unit array* that is connected to a *composite antenna* that contains a *radio distribution network* and an *antenna array*. The transceiver unit array contains multiple transmitter and receiver units. These are connected to the composite antenna through a number of connectors on the *transceiver array boundary* (TAB). These TAB connectors correspond to the antenna connectors on a non-AAS base station and serve as a reference point for conducted requirements. The radio distribution network is passive and distributes the transmitter outputs to the corresponding antenna elements and vice versa for the receiver inputs. Note that the actual implementation of an AAS BS may look different in terms of physical location of the different parts, array geometry, type of antenna elements used, etc.

Based on the architecture in Fig. 28.3, there are two types of requirements defined:

- *Conducted requirements* are defined for each RF characteristic at an individual or a group of TAB connectors. The conducted requirements are defined in such a way that they are in a sense "equivalent" to the corresponding conducted requirement for a non-AAS base station, that is, the performance of the system or the impact on other systems is expected to be the same.

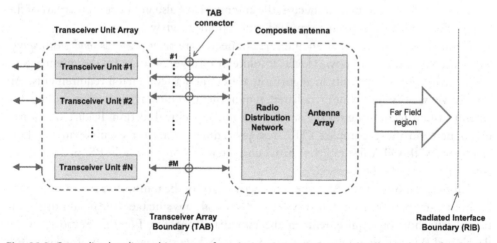

Fig. 28.3 Generalized radio architecture of an Active Antenna System (AAS), used also for the NR radiated requirements.

- *Radiated requirements* are defined OTA in the far field of the antenna system. Since the spatial direction becomes relevant in this case, it is detailed for each requirement how it applies. Radiated requirements are defined with reference to a *radiated interface boundary* (RIB), somewhere in the far-field region.

28.5.2 BS types in different frequency ranges for NR

A number of different base-station design possibilities have to be considered for the RF requirements. First in FR1, there are base stations built in a way similar to "classical" 3G and 4G base stations with antenna connectors through which external antennas are connected. Then we have base stations with AAS, but where antenna connectors can still be accessed for definition and testing of some RF requirements. Finally, we have base stations with highly integrated antenna systems where all requirements must be assessed OTA, since there are no antenna connectors. It is assumed that in FR2 where mm-wave technology is used for implementation of the antenna systems, only the latter type of base station needs to be specified.

3GPP has defined four base station types based on these assumptions, with reference to the architecture defined above in Fig. 28.3:

- *BS type 1-C*: NR base station operating in FR1, specified only with conducted requirements defined at individual antenna connectors.
- *BS type 1-O*: NR base station operating in FR1, specified only with conducted (OTA) requirements defined at the RIB.
- *BS type 1-H*: NR base station operating at FR1, specified with a "hybrid" set of requirements consisting of both conducted requirements defined at individual TAB connectors and some OTA requirements defined at the RIB.
- *BS type 2-O*: NR base station operating in FR2, specified only with conducted (OTA) requirements defined at the RIB.

BS type 1-C has requirements defined in the same way as for UTRA or LTE conducted requirements. These are described in Sections 28.6–28.11.

BS type 1-H corresponds to the first type of AAS base stations specified for LTE/UTRA in 3GPP Release 13, where two radiated requirements are defined (radiated transmit power and OTA sensitivity), while all others are defined as conducted requirements, as described in Sections 28.6–28.11. Many conducted requirements, such as unwanted emission limits, are for BS type 1-H defined in two steps. First a basic limit is defined, which is identical to the conducted limit at an individual antenna connector for BS type 1-C and thereby equivalent to the limit at a TAB connector for BS type 1-H. In a second step, the basic limit is converted to a radiated limit at the RIB through a scaling factor based on the number of active transmitter units. The scaling is capped at a maximum of 8 (9 dB), which is the maximum number of antenna elements used in defining certain regulatory limits. Note that the maximum scaling may vary depending on regional regulation.

BS type 1-O and BS type 2-O have all requirements defined as radiated. BS type 1-O has many requirements defined with reference to the corresponding FR1 conducted requirements, where unwanted emission limits also have a scaling applied as for BS type 1-H. The overall differences in coexistence properties and implementation between FR1 and FR2 mean that BS type 2-O has separate FR2 requirements defined that in many cases are different from the FR1 requirements for BS type 1-O.

An overview of the radiated requirements used for BS types 1-O and 2-O, and to some extent for BS type 1-H, is given in Section 28.12.

28.6 Overview of conducted RF requirements for NR

The RF requirements define the receiver and transmitter RF characteristics of a base station or device. The base station is the physical node that transmits and receives RF signals on one or more antenna connectors. Note that an NR base station is not the same thing as a gNB, which is the corresponding logical node in the radio-access network (see Chapter 6). The device is denoted UE in all RF specifications. Conducted RF requirements are defined for operating bands in FR1, while only radiated (OTA) requirements are defined for operating bands in FR2 (see Section 28.12).

The set of conducted RF requirements defined for NR is fundamentally the same as those defined for LTE or any other radio system. Some requirements are also based on regulatory requirements and are more concerned with the frequency band of operation and/or the place where the system is deployed, than with the type of system.

What is particular to NR is the flexible channel bandwidths and multiple numerologies of the system, which makes some requirements more complex to define. These properties have special implications for the transmitter requirements on unwanted emissions, where the definition of the limits in international regulation depends on the channel bandwidth. Such limits are harder to define for a system where the base station may operate with multiple channel bandwidths and where the device may vary its channel bandwidth of operation. The properties of the flexible OFDM-based physical layer also have implications for specifying the transmitter modulation quality and how to define the receiver selectivity and blocking requirements. Note that the channel bandwidth in general is different for the BS and the device as discussed in Section 28.3.

The type of transmitter requirements defined for the device is very similar to what is defined for the base station, and the definitions of the requirements are often similar. The output power levels are, however, considerably lower for a device, while the restrictions on the device implementation are much higher. There is tight pressure on cost and complexity for all telecommunications equipment, but this is much more pronounced for devices, due to the scale of the total market, being close to *two billion* devices per year. In cases where there are differences in how requirements are defined between device and base station, they are treated separately in this chapter.

The detailed background of the conducted RF requirements for NR is described in [70,71]. The conducted RF requirements for the base station are specified in [4] and for the device in [5]. The RF requirements are divided into transmitter and receiver characteristics. There are also performance characteristics for base stations and devices that define the receiver baseband performance for all physical channels under different propagation conditions. These are not strictly RF requirements, though the performance will also depend on the RF to some extent.

Each RF requirement has a corresponding test defined in the NR test specifications for the base station and the device. These specifications define the test setup, test procedure, test signals, test tolerances, etc. needed to show compliance with the RF and performance requirements.

28.6.1 Conducted transmitter characteristics

The transmitter characteristics define RF requirements for the wanted signal transmitted from the device and the base station, but also for the unavoidable unwanted emissions outside the transmitted carrier(s). The requirements are fundamentally specified in three parts:

- **Output power level** requirements set limits for the maximum allowed transmitted power, for the dynamic variation of the power level, and in some cases for the transmitter OFF state;
- **Transmitted signal quality** requirements define the "purity" of the transmitted signal and also the relation between multiple transmitter branches;
- **Unwanted emissions** requirements set limits to all emissions outside the transmitted carrier(s) and are tightly coupled to regulatory requirements and coexistence with other systems.

A list of the device and base-station transmitter characteristics arranged according to the three parts defined here is shown in Table 28.3. A more detailed description of the specific requirements can be found later in this chapter.

28.6.2 Conducted receiver characteristics

The set of receiver requirements for NR is quite similar to what is defined for other systems such as LTE and UTRA. The receiver characteristics are fundamentally specified in three parts:

- **Sensitivity and dynamic range** requirements for receiving the wanted signal;
- **Receiver susceptibility to interfering signals** defines receivers' susceptibility to different types of interfering signals at different frequency offsets;
- **Unwanted emissions** limits are also defined for the receiver.

A list of the device and base-station receiver characteristics arranged according to the three parts defined here is shown in Table 28.4. A more detailed description of each requirement can be found later in this chapter.

Table 28.3 Overview of conducted NR transmitter characteristics.

	Base station requirement	Device requirement
Output power level	Maximum output power Output power dynamics ON/OFF power (TDD only)	Transmit power Output power dynamics Power control
Transmitted signal quality	Frequency error Error Vector Magnitude (EVM) Time alignment between transmitter branches	Frequency error Transmit modulation quality In-band emissions
Unwanted emissions	Operating band unwanted emissions Adjacent Channel Leakage Ratio (ACLR and CACLR) Spurious emissions Occupied bandwidth Transmitter intermodulation	Spectrum emission mask Adjacent Channel Leakage Ratio (ACLR and CACLR) Spurious emissions Occupied bandwidth Transmit intermodulation

Table 28.4 Overview of conducted NR receiver characteristics.

	Base station requirement	Device requirement
Sensitivity and dynamic range	Reference sensitivity Dynamic range In-channel selectivity	Reference sensitivity power level Maximum input level
Receiver susceptibility to interfering signals	Out-of-band blocking In-band blocking Narrowband blocking Adjacent channel selectivity Receiver intermodulation	Out-of-band blocking Spurious response In-band blocking Narrowband blocking Adjacent Channel Selectivity Intermodulation characteristics
Unwanted emissions from the receiver	Receiver spurious emissions	Receiver spurious emissions

28.6.3 Regional requirements

There are a number of regional variations to the RF requirements and their application. The variations originate in different regional and local regulations of the spectrum and its use. The most obvious regional variation is the different frequency bands and their use, as discussed in Chapter 3. Many of the regional RF requirements are also tied to specific frequency bands.

When there is a regional requirement on, for example, spurious emissions, this requirement should be reflected in the 3GPP specifications. For the base station it is entered as an optional requirement and is marked as "regional." For the device, the same procedure is not possible, since a device may roam between different regions and will therefore have to fulfill all regional requirements that are tied to the operating bands in the regions where the band is used. For NR (and also for LTE), this becomes more complex than for UTRA, since there is an additional variation in the transmitter (and receiver) bandwidth used, making some regional requirements difficult to meet as a mandatory requirement. The concept of *network signaling* of RF requirements is therefore introduced for NR, where a device can be informed at call setup of whether some specific RF requirements apply when the device is connected to a network.

28.6.4 Band-specific device requirements through network signaling

For the device, the channel bandwidths supported are a function of the NR operating band, and also have a relation to the transmitter and receiver RF requirements. The reason is that some RF requirements may be difficult to meet under conditions with a combination of maximum power and high number of transmitted and/or received resource blocks.

In both NR and LTE, some additional RF requirements apply for the device when a specific network signaling value (NS_x) is signaled to the device as part of the cell handover or broadcast message. For implementation reasons, these requirements are associated with restrictions and variations to RF parameters such as device output power, maximum channel bandwidth, and number of transmitted resource blocks. The variations of the requirements are defined together with the NS_x in the device RF specification, where each value corresponds to a specific condition. The default value for all bands is NS_01. NS_x values are connected to an allowed power reduction called *additional maximum power reduction* (A-MPR) and may apply for transmission using a certain minimum number of resource blocks, depending also on the channel bandwidth.

28.6.5 Base-station classes for BS type 1-C and 1-H

In order to accommodate different deployment scenarios for base stations, there are multiple sets of RF requirements for NR base stations, each applicable to a *base station class*. When the RF requirements were derived for NR, base-station classes were introduced that were intended for macro-cell, micro-cell and pico-cell scenarios. The terms macro, micro, and pico relate to the deployment scenario and are not used in 3GPP to identify the base-station classes, instead the following terminology is used:

- *Wide area base stations*: This type of base station is intended for macro-cell scenarios, defined with a minimum coupling loss between base station and device of 70 dB. This is the typical large cell deployment with high-tower or above-rooftop installations, giving wide area outdoor coverage, but also indoor coverage.

- *Medium range base stations*: This type of base station is intended for micro-cell scenarios, defined with a minimum coupling loss between base station and device of 53 dB. Typical deployments are outdoor below-rooftop installations, giving both outdoor hot spot coverage and outdoor-to-indoor coverage through walls.
- *Local area base stations*: This type of base station is intended for pico-cell scenarios, defined with a minimum coupling loss between base station and device of 45 dB. Typical deployments are indoor offices and indoor/outdoor hotspots, with the BS mounted on walls or ceilings.

There is in Addition a BS class for High Altitude Platform (HAPS) scenarios, defined with a BS to ground UE minimum distance of typically around 20 m. Requirements for HAPS BS are in general the same as for Wide Area BS.

The local area and medium range base station classes have modifications to a number of requirements compared to wide area base stations, mainly due to the assumption of a lower minimum coupling loss:

- Maximum base station power is limited to 38 dBm output power for medium range base stations and 24 dBm output power for local area base stations. This power is defined per antenna and carrier. There is no maximum base station power defined for wide area base stations.
- The frequency error requirement is more relaxed for medium range and local area base stations.
- The spectrum mask (operating band unwanted emissions) has lower limits for medium range and local area, in line with the lower maximum power levels.
- Receiver reference sensitivity limits are higher (more relaxed) for medium range and local area. Receiver dynamic range and in-channel selectivity (ICS) are also adjusted accordingly.
- Limits for co-location for medium range and local area are relaxed compared to wide area BS, corresponding to the relaxed reference sensitivity for the base station.
- All medium range and local area limits for receiver susceptibility to interfering signals are adjusted to take the higher receiver sensitivity limit and the lower assumed minimum coupling loss (base station-to-device) into account.

28.7 Conducted output power level requirements

28.7.1 Base-station output power and dynamic range

There is no general maximum output power requirement for base stations. As mentioned in the discussion of base-station classes, there is, however, a maximum output power limit of 38 dBm for medium range base stations and 24 dBm for local area base stations.

In addition to this, there is a tolerance specified, defining how much the actual maximum power may deviate from the power level declared by the manufacturer.

The base station also has a specification of the total power control dynamic range for a resource element, defining the power range over which it should be possible to configure. There is also a dynamic range requirement for the total base-station power.

For TDD operation, a power mask is defined for the base-station output power, defining the off power level during the uplink subframes and the maximum time for the *transmitter transient period* between the transmitter on and off states.

28.7.2 Device output power and dynamic range

The device output power level is defined in three steps:

- *UE power class* defines a *nominal* maximum output power for QPSK modulation. It may be different in different operating bands, but the main device power class is today set at 23 dBm for all bands.
- *Maximum power reduction (MPR)* defines an allowed reduction of maximum power level for certain combinations of modulation used and resource block allocation. There are also MPR values defined for Carrier Aggregation, SUL, Uplink MIMO, V2X and Shared spectrum access.
- *Additional maximum power reduction (A-MPR)* may be applied in some regions and is usually connected to specific transmitter requirements such as regional emission limits and to certain carrier configurations. For each such set of requirements, there is an associated network signaling value NS_x that identifies the allowed A-MPR and the associated conditions, as explained in Section 28.6.4.

A minimum output power level setting defines the device dynamic range. There is a definition of the transmitter off power level, applicable to conditions when the device is not allowed to transmit. There is also a general on/off time mask specified, plus specific time masks for PRACH, PUCCH, SRS, and for PUCCH/PUSCH/SRS transitions. Dynamic range requirements are specified separately for CA, NR-DC, SUL, UL MIMO, V2X, Shared spectrum access, and CA with UL MIMO.

The device transmit power control is specified through requirements for the *absolute power tolerance* for the initial power setting, the *relative power tolerance* between two subframes, and the *aggregated power tolerance* for a sequence of power-control commands.

28.8 Transmitted signal quality

The requirements for transmitted signal quality specify how much the transmitted base station or device signal deviates from an "ideal" modulated signal in the signal and frequency domains. Impairments on the transmitted signal are introduced by the transmitter RF parts, with the nonlinear properties of the power amplifier being a major contributor.

The signal quality is assessed for base station and device through requirements on *EVM* and *frequency error*. An additional device requirement is device in-band emissions.

28.8.1 EVM and frequency error

While the theoretical definitions of the signal quality measures are quite straightforward, the actual assessment is a very elaborate procedure, described in great detail in the 3GPP specification. The reason is that it becomes a multidimensional optimization problem, where the best match for the timing, the frequency, and the signal constellation is found.

The EVM is a measure of the error in the modulated signal constellation, taken as the root mean square of the error vectors over the active subcarriers, considering all symbols of the modulation scheme. It is expressed as a percentage value in relation to the power of the ideal signal. The EVM fundamentally defines the maximum SINR that can be achieved at the receiver, if there are no additional impairments to the signal between transmitter and receiver.

Since a receiver can remove some impairments of the transmitted signal such as time dispersion, the EVM is assessed after cyclic prefix removal and equalization. In this way, the EVM evaluation includes a standardized model of the receiver. The frequency offset resulting from the EVM evaluation is averaged and used as a measure of the *frequency error* of the transmitted signal.

28.8.2 Device in-band emissions

In-band emissions are emissions within the channel bandwidth. The requirement limits how much a device can transmit into non-allocated resource blocks within the channel bandwidth. Unlike the out-of-band (OOB) emissions, the in-band emissions are measured after cyclic prefix removal and FFT, since this is how a device transmitter affects a real base-station receiver.

28.8.3 Base-station time alignment

Several NR features require the base station to transmit from two or more antennas, such as transmitter diversity and MIMO. For carrier aggregation, the carriers may also be transmitted from different antennas. In order for the device to properly receive the signals from multiple antennas, the timing relation between any two transmitter branches is specified in terms of a maximum time alignment error between transmitter branches. The maximum allowed error depends on the feature or combination of features in the transmitter branches.

28.9 Conducted unwanted emissions requirements

Unwanted emissions from the transmitter are divided into *OOB emissions* and *spurious emissions* in ITU-R recommendations [40]. OOB emissions are defined as emissions on a frequency close to the RF carrier, which results from the modulation process. Spurious emissions are emissions outside the RF carrier that may be reduced without affecting the corresponding transmission of information. Examples of spurious emissions are harmonic emissions, intermodulation products, and frequency conversion products. The frequency range where OOB emissions are normally defined is called the *OOB domain*, whereas spurious emission limits are normally defined in the *spurious domain*.

ITU-R also defines the boundary between the OOB and spurious domains at a frequency separation from the carrier center of 2.5 times the necessary bandwidth, which corresponds to 2.5 times the channel bandwidth for NR. This division of the requirements is easily applied for systems that have a fixed channel bandwidth. It does, however, become more difficult for NR, which is a flexible bandwidth system, implying that the frequency range where requirements apply would then vary with the channel bandwidth. The approach taken for defining the boundary in 3GPP is slightly different for base-station and device requirements.

With the recommended boundary between OOB emissions and spurious emissions set at 2.5 times the channel bandwidth, third- and fifth-order intermodulation products from the carrier will fall inside the OOB domain, which will cover a frequency range of twice the channel bandwidth on each side of the carrier. For the OOB domain, two overlapping requirements are defined for both base station and device: *spectrum emissions mask* (SEM) and *adjacent channel leakage ratio* (ACLR). The details of these are further explained here.

28.9.1 Implementation aspects

The spectrum of an OFDM signal decays rather slowly outside of the transmission bandwidth configuration. Since the transmitted signal for NR occupies up to 98% of the channel bandwidth, it is not possible to meet the unwanted emission limits directly outside the channel bandwidth with a "pure" OFDM signal. The techniques used for achieving the transmitter requirements are, however, not specified or mandated in NR specifications. Time-domain windowing is one method commonly used in OFDM-based transmission systems to control spectrum emissions. Filtering is always used, both time-domain digital filtering of the baseband signal and analog filtering of the RF signal.

The non-linear characteristics of the *power amplifier* (PA) used to amplify the RF signal must also be taken into account, since it is the source of intermodulation products outside the channel bandwidth. Power back-off to give a more linear operation of the PA can be used, but at the cost of a lower power efficiency. The power back-off should therefore be kept to a minimum. For this reason, additional linearization schemes can be employed. These are especially important for the base station, where there are fewer restrictions on implementation complexity and use of advanced linearization schemes is an essential part of controlling spectrum emissions. Examples of such techniques are feed–forward, feedback, predistortion, and postdistortion.

28.9.2 Emission mask in the OOB domain

The emission mask defines the permissible OOB spectrum emissions outside the necessary bandwidth. As explained above, how to take the flexible channel bandwidth into account when defining the frequency boundary between OOB emissions and spurious emissions is done differently for the NR base station and device. Consequently, the emission masks are also based on different principles.

28.9.2.1 Base-station operating band unwanted emission limits

For the NR base station, the problem of the implicit variation of the boundary between OOB and spurious domain with the varying channel bandwidth is handled by not defining an explicit boundary. The solution is a unified concept of *operating band unwanted emissions* (OBUE) for the NR base station instead of the spectrum mask usually defined for OOB emissions. The operating band unwanted emissions requirement applies over the whole base-station transmitter operating band, plus an additional 10–100 MHz on each side, as shown in Fig. 28.4. All requirements outside of that range are set by the regulatory spurious emission limits, based on the ITU-R recommendations [40]. As seen in Fig. 28.4, a large part of the operating band unwanted emissions is defined over a frequency range that for smaller channel bandwidths can be both in spurious and OOB domains. This means that

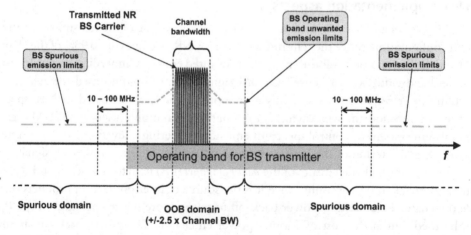

Fig. 28.4 Frequency ranges for operating band unwanted emissions and spurious emissions applicable to NR base station (FR1).

the limits for the frequency ranges that may be in the spurious domain also have to align with the regulatory limits from the ITU-R. The shape of the mask is generic for all channel bandwidths, with a mask that consequently has to align with the ITU-R limits starting 10–100 MHz from the channel edges, depending on the BS type and the width of the operating band. The operating band unwanted emissions are defined with a 100 kHz measurement bandwidth and align to a large extent with the corresponding masks for LTE.

In the case of carrier aggregation for a base station, the OBUE requirement (as other RF requirements) applies as for any multicarrier transmission, where the OBUE will be defined relative to the carriers on the edges of the RF bandwidth. In the case of non-contiguous carrier aggregation, the OBUE within a sub-block gap is partly calculated as the cumulative sum of contributions from each subblock.

There are also special limits defined to meet a specific regional regulation. These are for example set by the FCC (Federal Communications Commission, Title 47) for the operating bands used in the USA and by the ECC for some European bands. They are specified as separate limits in addition to the operating band unwanted emission limits.

28.9.2.2 Device spectrum emission mask

For implementation reasons, it is not possible to define a generic device spectrum mask that does not vary with the channel bandwidth, so the frequency ranges for OOB limits and spurious emissions limits do not follow the same principle as for the base station. The SEM extends out to a separation Δf_{OOB} from the channel edges, as illustrated in Fig. 28.5. For 5 MHz channel bandwidth, this point corresponds to 250% of the necessary bandwidth as recommended by the ITU-R, but for higher channel bandwidths it is set closer than 250%.

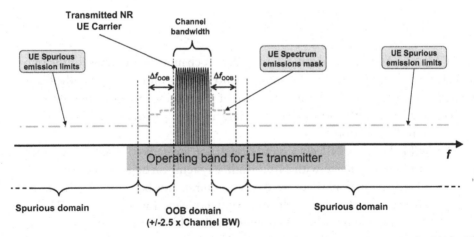

Fig. 28.5 Frequency ranges for spectrum emission mask and spurious emissions applicable to NR device.

The SEM is defined as a general mask and a set of additional masks that can be applied to reflect different regional requirements. Each additional regional mask is associated with a specific network signaling value NS_x. There are also variations of the SEM and its applicability defined for CA, V2X, shared spectrum channel access and CA with UL MIMO.

28.9.3 Adjacent channel leakage ratio

In addition to a spectrum emissions mask, the OOB emissions are defined by an ACLR requirement. The ACLR concept is very useful for analysis of coexistence between two systems that operate on adjacent frequencies. The ACLR defines the ratio of the power transmitted within the assigned channel bandwidth to the power of the unwanted emissions transmitted on an adjacent channel. There is a corresponding receiver requirement called Adjacent Channel Selectivity (ACS), which defines a receiver's ability to suppress a signal on an adjacent channel.

The definitions of ACLR and ACS are illustrated in Fig. 28.6 for a wanted and an interfering signal received in adjacent channels. The interfering signal's leakage of unwanted emissions at the wanted signal receiver is given by the ACLR and the ability of the receiver of the wanted signal to suppress the interfering signal in the adjacent channel is defined by the ACS. The two parameters when combined define the total leakage between two transmissions on adjacent channels. That ratio is called the *Adjacent Channel Interference Ratio* (ACIR) and is defined as the ratio of the power transmitted on one channel to the total interference received by a receiver on the adjacent channel, due to both transmitter (ACLR) and receiver (ACS) imperfections.

This relation between the adjacent channel parameters is [11]:

$$\text{ACIR} = \frac{1}{\frac{1}{\text{ACLR}} + \frac{1}{\text{ACS}}} \tag{28.2}$$

ACLR and ACS can be defined with different channel bandwidths for the two adjacent channels, which is the case for some requirements set for NR due to the bandwidth flexibility. Eq. (28.2) will also apply for different channel bandwidths, but only if the same two channel bandwidths are used for defining all three parameters ACIR, ACLR, and ACS used in the equation.

The ACLR limits for NR device and base station are derived based on extensive analysis [11] of NR coexistence with NR or other systems on adjacent carriers.

For an NR base station, there are ACLR requirements both for an adjacent channel with an NR receiver of the same channel bandwidth and for an adjacent LTE receiver. The ACLR requirement for NR BS is set to 45 dB. This is considerably more strict than the ACS requirement for the device, which according to (28.2) implies that in the downlink, the device receiver performance will be the limiting factor for ACIR and

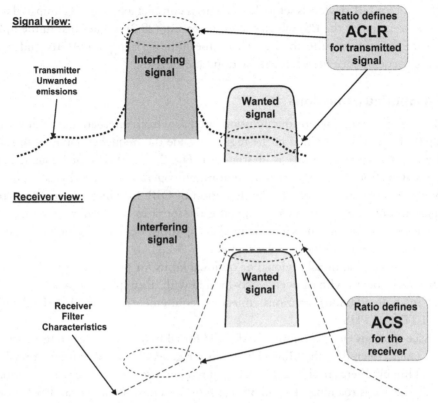

Fig. 28.6 Illustration of ACLR and ACS, with example characteristics for an "aggressor" interferer and a receiver for a "victim" wanted signal.

consequently for coexistence between base stations and devices. From a system-point-of-view, this choice is cost-efficient since it moves implementation complexity to the BS, instead of requiring all devices to have high-performance RF.

In the case of carrier aggregation for a base station, the ACLR (as other RF requirements) apply as for any multicarrier transmission, where the ACLR requirement will be defined for the carriers on the edges of the RF bandwidth. In the case of non-contiguous carrier aggregation where the sub-block gap is so small that the ACLR requirements at the edges of the gap will "overlap", a special *cumulative ACLR* requirement (CACLR) is defined for the gap. For CACLR, contributions from carriers on both sides of the sub-block gap are accounted for in the CACLR limit. The CACLR limit is the same as the ACLR for the base station at 45 dB.

ACLR limits for the device are set both with assumed NR and an assumed UTRA receiver on the adjacent channel. In the case of carrier aggregation, the device ACLR requirement applies to the aggregated channel bandwidth instead of per carrier. The

ACLR limit for NR devices is set to 30 dB. This is considerably relaxed compared to the ACS requirement for the BS, which according to Eq. (28.2) implies that in the uplink, the device transmitter performance will be the limiting factor for ACIR and consequently for coexistence between base stations and devices.

28.9.4 Spurious emissions

The limits for base station spurious emissions are taken from international recommendations [40], but are only defined in the region outside the frequency range of operating band unwanted emission limits as illustrated in Fig. 28.4 – that is, at frequencies that are separated from the base-station transmitter operating band by at least 10–100 MHz. This means that due to the definition of OBUE (which covers also parts of the spurious domain), the frequency range for the spurious emissions requirement does not coincide exactly with the range defined as "spurious domain" in international regulation. The limits are however fully aligned.

There are also additional regional or optional limits for protection of other systems that NR may coexist with or even be co-located with. Examples of other systems considered in those additional spurious emissions requirements are GSM, UTRA FDD/TDD, LTE, and PHS.

Device spurious emission limits are defined for all frequency ranges outside the frequency range covered by the SEM. The limits are generally based on international regulations [40], but there are also additional requirements for coexistence with other bands when the device is roaming. The additional spurious emission limits can have an associated network signaling value.

In addition, there are base-station and device emission limits defined for the receiver. Since receiver emissions are dominated by the transmitted signal, the receiver spurious emission limits are only applicable when the transmitter is not active, and also when the transmitter is active for an NR FDD base station that has a separate receiver antenna connector.

28.9.5 Occupied bandwidth

Occupied bandwidth is a regulatory requirement that is specified for equipment in some regions, such as Japan and the United States. It was originally defined by the ITU-R as a maximum bandwidth, outside of which emissions do not exceed a certain percentage of the total emissions. The occupied bandwidth is for NR equal to the channel bandwidth, outside of which a maximum of 1% of the emissions are allowed (0.5% on each side).

28.9.6 Transmitter intermodulation

An additional implementation aspect of an RF transmitter is the possibility of intermodulation between the transmitted signal and another strong signal transmitted in the proximity of the base station or device. For this reason, there is a requirement for *transmitter intermodulation*.

For the base station, the requirement is based on a stationary scenario with a co-located other base-station transmitter, with its transmitted signal appearing at the antenna connector of the base station being specified but attenuated by 30 dB. Since it is a stationary scenario, there are no additional unwanted emissions allowed, implying that all unwanted emission limits also have to be met with the interferer present.

For the device, there is a similar requirement based on a scenario with another device-transmitted signal appearing at the antenna connector of the device being specified but attenuated by 40 dB. The requirement specifies the minimum attenuation of the resulting intermodulation product below the transmitted signal.

28.10 Conducted sensitivity and dynamic range

The primary purpose of the *reference sensitivity requirement* is to verify the receiver *noise figure*, which is a measure of how much the receiver's RF signal chain degrades the SNR of the received signal. For this reason, a low-SNR transmission scheme using QPSK is chosen as a reference channel for the reference sensitivity test. The reference sensitivity is defined at a receiver input level where the throughput is 95% of the maximum throughput for the reference channel.

For the device, reference sensitivity is defined for the full channel bandwidth signals and with all resource blocks allocated for the wanted signal.

The intention of the *dynamic range requirement* is to ensure that the receiver can also operate at received signal levels considerably higher than the reference sensitivity. The scenario assumed for base-station dynamic range is the presence of increased interference and corresponding higher wanted signal levels, thereby testing the effects of different receiver impairments. In order to stress the receiver, a higher SNR transmission scheme using 16QAM is applied for the test. To further stress the receiver to higher signal levels, an interfering AWGN signal at a level 20 dB above the assumed noise floor is added to the received signal. The dynamic range requirement for the device is specified as a *maximum signal level* at which the throughput requirement is met.

28.11 Receiver susceptibility to interfering signals

There is a set of requirements for base station and device, defining the receiver's ability to receive a wanted signal in the presence of a stronger interfering signal. The reason for the multiple requirements is that, depending on the frequency offset of the interferer from the wanted signal, the interference scenario may look very different and different types of receiver impairments will affect the performance. The intention of the different combinations of interfering signals is to model as far as possible the range of possible scenarios with interfering signals of different bandwidths that may be encountered inside and outside the base-station and device receiver operating band.

While the types of requirements are very similar between base station and device, the signal levels are different, since the interference scenarios for the base station and device are very different. There is also no device requirement corresponding to the base-station ICS requirement.

The following requirements are defined for NR base station and device, starting from interferers with large frequency separation and going close in (see also Fig. 28.7). In all cases where the interfering signal is an NR signal, it has the same or smaller bandwidth than the wanted signal, but at most 20 MHz.

- *Blocking*: This corresponds to the scenario with strong interfering signals received outside the operating band (out-of-band blocking) or inside the operating band (in-band blocking), but not adjacent to the wanted signal. In-band blocking includes interferers in the first 20–100 MHz outside the operating band for the base station and the first 15 MHz for the device. The scenarios are modeled with a *continuous wave* (CW) signal for the out-of-band case and an NR signal for the in-band case. There are additional (optional) base-station blocking requirements for the scenario when the base station is co-located with another base station in a different operating band. For the device, a fixed number of *exceptions* are allowed from the out-of-band blocking requirement, for each assigned frequency channel and at the respective *spurious response frequencies*. At those frequencies, the device must comply with the more relaxed spurious response requirement.

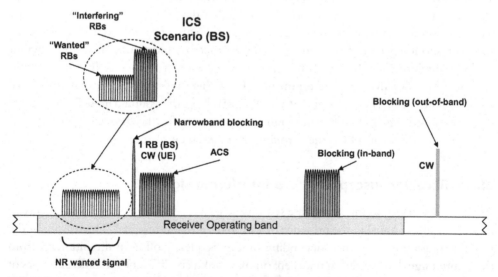

Fig. 28.7 Base-station and device requirements for receiver susceptibility to interfering signals in terms of blocking, ACS, narrowband blocking, and in-channel selectivity (BS only).

- *Adjacent channel selectivity*: The ACS scenario is a strong signal in the channel adjacent to the wanted signal and is closely related to the corresponding ACLR requirement (see also the discussion in Section 28.9.3). The adjacent interferer is an NR signal. For the device, the ACS is specified for two cases with a lower and a higher signal level.
- *Narrowband blocking*. The scenario is an adjacent strong narrowband interferer, which in the requirement is modeled as a single resource block NR signal for the base station and a CW signal for the device.
- *In-channel selectivity* (ICS): The scenario is multiple received signals of different received power levels inside the channel bandwidth, where the performance of the weaker "wanted" signal is verified in the presence of the stronger "interfering" signal. ICS is only specified for the base station.
- *Receiver intermodulation*: The scenario is *two* interfering signals near to the wanted signal, where the interferers are one CW and one NR signal (not shown in Fig. 28.7). The purpose of the requirement is to test receiver linearity. The interferers are placed in frequency in such a way that the main intermodulation product falls inside the wanted signal's channel bandwidth. There is also a *narrowband intermodulation* requirement for the base station where the CW signal is very close to the wanted signal and the NR interferer is a single RB signal.

For all requirements except ICS, the wanted signal uses the same reference channel as in the corresponding reference sensitivity requirement. With the interference added, the same 95% relative throughput is met as for the reference sensitivity, but at a "desensitized" higher wanted signal level.

28.12 Radiated RF requirements for NR

Many of the radiated RF requirements defined for devices and base stations are derived directly from the corresponding conducted RF requirements. Unlike conducted requirements, the radiated requirements will account also for the antenna. When defining emission levels such as base station output power and unwanted emissions, this can be done either by incorporating the antenna gain as a directional requirement using an *Effective Isotropic Radiated Power* (EIRP) or by definition of limits using *Total Radiated Power* (TRP). Two new radiated requirements are defined as directional for the base station (see Section 28.12.3), but most radiated device and base station requirements for NR are defined with limits expressed as TRP.

TRP and EIRP are directly related through the number of radiating antenna elements and depend also on specific base station implementation, considering the geometry of the antenna array and the correlation between unwanted emission signals from different antenna ports. The implication is that an EIRP limit could result in different levels of total radiated unwanted emission power depending on the implementation. An EIRP limit will thus not give control of the total amount of interference in the network, while a TRP requirement limits the total amount of interference injected in the network regardless of the specific BS implementation.

Another relevant element behind the 3GPP choice of defining unwanted emission as TRP is the different behavior between passive and active antenna systems. In the case of passive systems, the antenna gain does not vary much between the wanted signal and unwanted emissions. Thus, EIRP is directly proportional to TRP and can be used as a substitute. For an active system such as NR, the EIRP may vary between the wanted signal and unwanted emissions and also between implementations, all depending on how strong correlation there is between wanted and unwanted emissions in terms of the source of unwanted emissions, frequency separation between emissions etc. The implication is that EIRP in general is not proportional to TRP and using EIRP to substitute TRP would be incorrect.

The radiated RF requirements for device and base station are described below.

28.12.1 Base-station classes for BS type 1-O and 2-O

Since the radiated requirements are defined without assuming an antenna connector as reference, it is also not possible to define BS classes based on coupling loss as was done for conducted requirements (see Section 28.6.5). The BS classes are instead defined based on BS-to-device minimum distance, but are otherwise the same:

- *Wide area base stations*: This type of base station is intended for macro-cell scenarios, with a BS-to-device minimum distance along the ground equal to 35 m.
- *Medium range base stations*: This type of base station is intended for micro-cell scenarios, with a BS-to-device minimum distance along the ground equal to 5 m.
- *Local area base stations*: This type of base station is intended for pico-cell scenarios, defined with a BS-to-device minimum distance along the ground equal to 2 m.

The local area and medium range base station classes have modifications to a number of requirements compared to wide area base stations, mainly due to the assumption of a lower minimum base station to device distance, giving a lower minimum coupling loss. The requirements are modified in the same way as for conducted requirements for BS type 1-C and 1-H (see Section 28.6.5), with the following difference:

- Maximum base station rated carrier TRP is limited to 47 dBm for medium range base stations and 33 dBm for local area base stations. This power is defined per antenna and carrier. There is no maximum base station TRP defined for wide area base stations.

28.12.2 Radiated device requirements in FR2

As described in Section 28.4, the RF requirements in FR2 operating bands are described in a separate specification [6] for devices, because of the higher number of antenna elements for operation in FR2 and the high level of integration used for mm-wave technology. There are no radiated device requirements defined for FR1 operating bands. The set of requirements for FR2 mostly correspond to the conducted RF requirements defined for FR1 operating bands. The limits for many requirements are however different. The difference in coexistence at mm-wave frequencies leads to lower requirements on, for example, ACLR and spectrum mask. This is demonstrated through coexistence studies performed in 3GPP and documented in [11]. The possibility for different limits has also been demonstrated in academia [69].

The implementation using mm-wave technologies is more challenging than using the more mature technologies in the frequency bands below 6 GHz (FR1). The mm-wave RF implementation aspects are further discussed in Chapter 29.

It should also be noted that the channel bandwidths and numerologies defined for FR2 are in general different from FR1, making it not possible to compare requirement levels, especially for receiver requirements.

The following is an overview of the radiated RF requirements in FR2, as compared to the ones in FR1:

- *Output power level requirements:* Maximum output power is of the same order as in FR1 but is expressed both as TRP and EIRP. The minimum output power and transmitter OFF power levels are higher than in FR1. Radiated transmit power is an additional radiated requirement, which unlike the maximum output power is directional. There is also a *Spherical coverage* requirement for the variation of EIRP in different directions. It is based on the minimum EIRP at the 85th percentile of the distribution of radiated power measured over the full sphere around the UE.
- *Transmitted signal quality:* Frequency error and EVM requirements are defined similar to what is done in FR1 and mostly with the same limits.
- *Radiated unwanted emissions requirements:* Occupied bandwidth, ACLR, spectrum mask and spurious emissions are defined in the same way as for FR1. The latter two are based on TRP. Many limits are less strict than in FR1. ACLR is on the order of 10 dB relaxed compared to FR1, due to more favorable coexistence at higher frequencies.
- *Beam correspondence,* briefly mentioned already in Section 12.2 in connection with the need for uplink beam sweeping, is a new requirement that does not exist for FR1 UEs. The UE minimum Peak EIRP and Spherical coverage form the basis for the beam correspondence requirement. UEs can either support those requirements with autonomously chosen uplink beams or by using beam sweeping. For UEs declared to support the Peak EIRP and Spherical coverage requirements with uplink beam sweeping, there is a *beam correspondence tolerance* requirement defined.
- *Reference sensitivity and dynamic range:* Defined as a radiated requirements based on *equivalent isotropic sensitivity* (EIS), which can be seen as the receiver measure corresponding to EIRP. Levels are not comparable to FR1
- *Receiver Susceptibility to Interfering Signals:* ACS, in-band and out-of-band blocking are defined similar to FR1, but are defined as radiated requirements based on EIS as test

metric. There is no narrow-band blocking scenario defined since there are only wide-band systems in FR2. ACS is in the order of 10 dB relaxed compared to FR1, due to more favorable coexistence.

28.12.3 Radiated base station requirements in FR1

As described in Section 28.5, the RF requirements for BS type 1-O consisted of only radiated (OTA) requirements. These are in general based on the corresponding conducted requirements, either directly or through scaling. Two additional radiated requirements defined are *radiated transmit power* and *OTA sensitivity*, described further.

BS type 1-H is defined with a "hybrid" set of requirements consisting mostly of conducted requirements and in addition two radiated requirements, which are the same as for BS type 1-O:

- *Radiated transmit power* is defined accounting for the antenna array beamforming pattern in a specific direction as EIRP for each beam that the base station is declared to transmit. In a way similar to BS output power, the actual requirement is on the accuracy of the declared EIRP level.
- *OTA sensitivity* is a directional requirement based on a quite elaborate declaration by the manufacturer of one or more *OTA sensitivity direction declarations* (OSDDs). The sensitivity is in this way defined accounting for the antenna array beam-forming pattern in a specific direction as declared *equivalent isotropic sensitivity* (EIS) level toward a receiver target. The EIS limit is to be met not only in a single direction but within a *range of angle of arrival* (RoAoA) in the direction of the receiver target. Depending on the level of adaptivity for the AAS BS, two alternative declarations are made:
- If the receiver is adaptive to direction, so that the receiver target can be redirected, the declaration contains a *receiver target redirection range* in a specified *receiver target direction*. The EIS limit should be met within the redirection range, which is tested at five declared sensitivity RoAoA within that range.
- If the receiver is not adaptive to direction and thus cannot redirect the receiver target, the declaration consists of a single sensitivity RoAoA in a specified receiver target direction, in which the EIS limit should be met.

Note that the OTA sensitivity is defined in addition to the reference sensitivity requirement, which exists both as conducted (for BS type 1-H) and radiated (for BS type 1-O).

28.12.4 Radiated base station requirements in FR2

As described in Section 28.5, the RF requirements for BS type 2-O are radiated requirements for base stations in FR2 operating bands. These are described separately, together with the radiated requirements for BS type 1-O, but in the same specification [4] as the conducted base-station RF requirements.

The set of requirements is identical to the radiated RF requirements defined for FR1 operating bands described, but the limits for many requirements are different. As for the device, the difference in coexistence at mm-wave frequencies leads to lower requirements on, for example, ACLR, ACS as demonstrated through 3GPP coexistence studies [11]. The implementation using mmwave technologies is also more challenging than using the more mature technologies in the frequency bands below 7 GHz (FR1) as further discussed in Chapter 29.

The following is a brief overview of the radiated RF requirements in FR2:

- *Output Power Level Requirements:* Maximum output power is the same for FR1 and FR2, but is scaled from the conducted requirement and expressed as TRP. There is in addition a directional radiated transmit power requirement. The dynamic range requirement is defined similarly to FR1.
- *Transmitted signal quality:* Frequency error, EVM, and time-alignment requirements are defined similar to what is done in FR1 and mostly with the same limits.
- *Radiated unwanted emissions requirements:* Occupied bandwidth, spectrum mask, ACLR, and spurious emissions are defined in the same way as for FR1. The three latter are based on TRP and also have less strict limits than in FR1. ACLR is on the relaxed with more than 15 dB compared to FR1, due to more favorable coexistence.
- *Reference sensitivity and dynamic range:* Defined in the same way as in FR1, but levels are not comparable. There is in addition a directional OTA sensitivity requirement.
- *Receiver susceptibility to interfering signals:* ACS, in-band and out-of-band blocking are defined as for FR1, but there is no narrow-band blocking scenario defined since there are only wideband systems in FR2. ACS is relaxed compared to FR1, due to more favorable coexistence.

28.13 Multi-standard radio base stations

Traditionally the RF specifications have been developed separately for the different 3GPP radio-access technologies GSM/EDGE, UTRA, LTE, and NR. The rapid evolution of mobile radio and the need to deploy new technologies alongside the legacy deployments has, however, led to implementation of different radio-access technologies (RAT) at the same sites, sharing antennas and other parts of the installation. In addition, operation of multiple RATs is often done within the same base-station equipment. The evolution to multi-RAT base stations is fostered by the evolution of technology. While multiple RATs have traditionally shared parts of the site installation, such as antennas, feeders, backhaul, or power, the advance of both digital baseband and RF technologies enables a much tighter integration.

3GPP defines an MSR base station, as a base station where the receiver and the transmitter are capable of simultaneously processing multiple carriers of different RATs in common active RF components. The reason for this stricter definition is that the true potential of multi-RAT base stations, and the challenge in terms of implementation

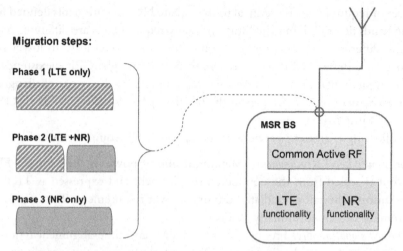

Fig. 28.8 Example of migration from LTE to NR using an MSR base station for all migration phases.

complexity, comes from having a common RF. This principle is illustrated in Fig. 28.8 with an example base station capable of both NR and LTE. Some of the NR and LTE baseband functionality may be separate in the base station but is possibly implemented in the same hardware. The RF must, however, be implemented in the same active components as shown in the figure.

MSR BS including NR is a part of 3GPP specifications from release 15 in FR1 bands (FR2 bands do not support multiple RATs). The main advantages of an MSR base station implementation for NR are twofold:

- Migration between RATs in a deployment, for example from previous mobile generations to NR, is possible using the same base station hardware. The operation of NR can then be introduced gradually over time when parts of the spectrum used for previous generations is freed up for NR.
- A single base station designed as an MSR base station can be deployed in various environments for single-RAT operation for each RAT supported, as well as for multi-RAT operation, where that is required by the deployment scenario. This is also in line with the recent technology trends seen in the market, with fewer and more generic base-station designs. Having fewer varieties of base station is an advantage both for the base-station vendor and for the operator, since a single solution can be developed and implemented for a variety of scenarios.

The MSR concept has a substantial impact for many requirements, while others remain completely unchanged. A fundamental concept introduced for MSR base stations is *RF bandwidth*, which is defined as the total bandwidth over the set of carriers transmitted and received. Many receiver and transmitter requirements are usually specified relative to the carrier center or the channel edges. For an MSR base station, they are instead specified relative to the *RF bandwidth edges*, in a way similar to carrier aggregation. By introducing the RF bandwidth concept and introducing generic limits, the requirements for MSR shift from being carrier centric toward being frequency block centric, thereby embracing technology neutrality by being independent of the access technology or operational mode.

While NR, LTE and UTRA carriers have quite similar RF properties in terms of bandwidth and power spectral density, GSM/EDGE carriers are quite different. The FR1 operating bands for which MSR base stations are defined are therefore divided into three *Band Categories* (BC):

- BC1 – All paired bands where NR, LTE and UTRA can be deployed with FDD operation.
- BC2 – All paired bands where in addition to NR, UTRA and LTE, GSM/EDGE can also be deployed with FDD operation.
- BC3 – All unpaired bands where NR, UTRA and E-UTRA can be deployed with TDD operation.

Since the carriers of different RATs are not transmitted and received independently, it is necessary to perform parts of the testing of the MSR base station with carriers of multiple RATs being activated. This is done through a set of multi-RAT *Test Configurations* defined in [2], specifically tailored to stress transmitter and receiver properties. These test configurations are of particular importance for the unwanted emission requirements for the transmitter and for testing of the receiver susceptibility to interfering signals (blocking, etc.). An advantage of the multi-RAT test configurations is that the RF performance of multiple RATs can be tested simultaneously, thereby avoiding repetition of test cases for each RAT. This is of essential for the very time-consuming tests of requirements over the complete frequency range outside the operating band.

The requirement with the largest impact from MSR is the spectrum mask, or the *operating band unwanted emissions* requirement, as it is called. The spectrum mask requirement for MSR base stations is applicable for multi-RAT operation where the carriers at the RF bandwidth edges are either GSM/EDGE, UTRA, LTE or NR carriers of different channel bandwidths. The mask is generic and applicable to all cases and covers the complete operating band of the base station.

Another important concept for MSR base stations is the supported *capability set* (CS), which is part of the declaration of base station capabilities by the vendor. Each capability set defines the RATs supported by the base station and how they can be combined. There are currently 19 capability sets CS1 to CS19 defined in the MSR BS test specification [2], where CS1 and CS2 define single-RAT support for UTRA and LTE respectively and CS3 to CS19 define different single- and multi-RAT combinations for all RATs. Capability sets that include NR in Release 16 are CS16 through CS19. The RAT combinations possible for CS16 to CS19 are listed in Table 28.5. Note the difference between the capability of a base station (as declared by the manufacturer) and the configurations in which a BS can operate. A BS has the capability to operate a number of RATs as defined by the Capability Set but is at any certain time only operating with one supported RAT configuration.

Table 28.5 Capability sets (CSx) defined for MSR base stations that include NR and the corresponding RAT configurations.

Capability set CSx supported by a base station	Applicable band categories	Supported RAT configurations
CS16	BC1, BC2 or BC3	Single-RAT: NR or LTE Multi-RAT: LTE + NR
CS17	BC1, BC2 or BC3	Single-RAT: NR, LTE or NB-IoT standalone Multi-RAT: LTE + NR LTE + NB-IoT standalone NR + NB-IoT standalone LTE + NR + NB-IoT standalone
CS18	BC2	Single-RAT: NR or LTE Multi-RAT: GSM + LTE GSM + NR LTE + NR GSM + LTE + NR
CS19	BC2	Single-RAT: NR, LTE or UTRA Multi-RAT: UTRA + LTE UTRA + NR LTE + NR UTRA + LTE + NR

Carrier aggregation is also applicable to MSR base stations. Since the MSR specification has most of the concepts and definitions in place for defining multicarrier RF requirements, whether aggregated or not, the differences for the MSR requirements compared to non-aggregated carriers are very minor.

28.14 Operation in non-contiguous spectrum

Some spectrum allocations consist of fragmented parts of spectrum for different reasons. The spectrum may be recycled 2G spectrum, where the original licensed spectrum was "interleaved" between operators. This was quite common for original GSM deployments, for implementation reasons (the original combiner filters used were not easily tuned when spectrum allocations were expanded). In some regions, operators have also purchased spectrum licenses on auctions and have for different reasons ended up with multiple allocations in the same band that are not adjacent.

For deployment of non-contiguous spectrum allocations there are a few implications:

• If the full spectrum allocation in a band is to be operated with a single base station, the base station has to be capable of operation in non-contiguous spectrum.

- If a larger transmission bandwidth is to be used than what is available in each of the spectrum fragments, both the device and the base station have to be capable of *intra-band non-contiguous carrier aggregation* in that band.

Note that the capability for the base station to operate in non-contiguous spectrum is not directly coupled to carrier aggregation as such. From an RF point-of-view, what is required by the base stations is to receive and transmit carriers over an RF bandwidth that is split in two (or more) separate sub-blocks, with a sub-block gap in-between as shown in Fig. 28.9. The spectrum in the sub-block gap can be deployed by any other operator, which means that the RF requirements for the base station in the sub-block gap are based on coexistence for uncoordinated operation. This has a few implications for some of the base station RF requirements within an operating band.

For base station operating in non-contiguous spectrum there are some additions and modifications to the requirements. The non-contiguous transmission consists of two or more *sub-blocks*, with *sub-block gaps* in-between. Transmitter requirements for Operating Band unwanted Emissions apply and ACLR apply as usual outside the RF-bandwidth edges, but also inside the sub-block gap as follows:

- **Operating band unwanted emissions mask (OBUE)**: The OBUE limit applies inside the gap calculated as a cumulative sum of contributions from the adjacent sub-blocks on each side of the gap.
- **ACLR**: For non-contiguous operation, ACLR applies inside the gap for the assumed first and second adjacent channels that do not overlap with the first or second channels from the other adjacent sub-block. For small sub-block gaps where adjacent channels overlap between the two sub-blocks, the Cumulative ACLR (CACLR)

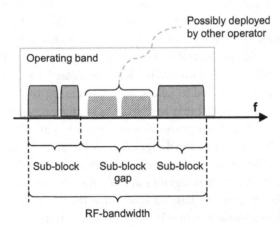

Fig. 28.9 Example of non-contiguous spectrum operation, illustrating the definitions of *RF bandwidth*, *sub-clock* and *sub-block gap*.

requirement will apply in the gap, with contributions counted from both bands (see also Section 28.9.3).

For the device, non-contiguous operation is tightly coupled to carrier aggregation, since multicarrier reception in the downlink or transmission in the uplink within a band does not occur unless carriers are aggregated. This also means that the definition of non-contiguous operation is different for the device than for the base station. For the device, intra-band non-contiguous carrier aggregation is therefore assumed to occur as soon as the spacing between two carriers is larger than the nominal channel spacing.

Compared to the base station, there are also additional implications and limitation to handle the simultaneously received and/or transmitted non-contiguous carriers. There is an allowed Maximum Power Reduction (MPR) already for transmission in a single component carrier, if the resource block allocation is non-contiguous within the carrier. For non-contiguous aggregated carriers, an allowed MPR is defined for sub-block gaps of up to 35 MHz between the aggregated carriers. The MPR depends on the number of allocated resource blocks.

28.15 Multiband-capable base stations

The 3GPP specifications have been continuously developed to support larger RF bandwidths for transmission and reception through multicarrier and multi-RAT operation and carrier aggregation within a band and with requirements defined for one band at a time. This has been made possible with the evolution of RF technology supporting larger bandwidths for both transmitters and receivers. From 3GPP release 11, there is support in the LTE and MSR base-station specifications for simultaneous transmission and/or reception in two bands through a common radio. Such a multiband base station covers multiple bands over a frequency range of a few 100 MHz. Support for more than two bands is given from 3GPP release 14. Support for multi-band operation is included in NR base station specifications from Release 15 for BS types 1-C, 1-H and 1-O. Multiband support for FR2 BS has been studied by 3GPP and may be added in a future release.

One obvious application for multiband base stations is for interband carrier aggregation. It should however be noted that base stations supporting multiple bands were in existence long before carrier aggregation was introduced in 3GPP Release 11. Already for GSM, dual-band base stations were designed to enable more compact deployments of equipment at base-station sites, but they were really two separate sets of transmitters and receivers for the bands that were integrated in the same equipment cabinet. The difference for "true" multiband-capable base stations is that the signals for the bands are transmitted and received in common active RF in the base station.

There are several scenarios envisioned for multi-band base station implementation and deployment. The possibilities for the multi-band capability are

- Multi-band transmitter + multi-band receiver
- Multi-band transmitter + single-band receiver
- Single-band transmitter + multi-band receiver

An example base station for the first case is illustrated in Fig. 28.10, which shows a base station with a common RF implementation of both transmitter and receiver for two operating bands X and Y. Through a duplex filter, the transmitter and receiver are connected to a common antenna connector and a common antenna. The example is also a multi-RAT capable MB-MSR base station, with NR + GSM configured in Band X and NR configured in Band Y. Note that the figure has only one diagram showing the frequency range for the two bands, which could either be the receiver or transmitter frequencies.

Fig. 28.10 also illustrates some parameters that are defined for multi-band base station.

- **_RF bandwidth_** has the same definition as for a multi-standard base station, but is defined individually for each band.
- **_Inter-RF-bandwidth gap_** is the gap between the RF bandwidths in the two bands. Note that the inter-RF bandwidth gap may span a frequency range where other mobile operators can be deployed in bands X and Y, as well as a frequency range between the two bands that may be used for other services.
- **_Radio bandwidth_** is the full bandwidth supported by the base station to cover the multiple carriers in both bands.

In principle, a multi-band base station can be capable of operating in more than two bands.

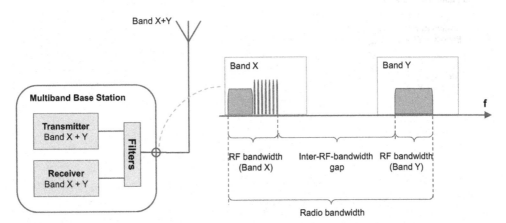

Fig. 28.10 Example of multiband base station with multiband transmitter and receiver for two bands with one common antenna connector.

While having only a single antenna connector and a common feeder that connects to a common antenna is desirable to reduce the amount of equipment needed in a site, it is not always possible. It may also be desirable to have separate antenna connectors, feeders and antennas for each band. An example of a multiband base station with separate connectors for two operating bands X and Y is shown in Fig. 28.11. Note that while the antenna connectors are separate for the two bands, the RF implementation for transmitter and receiver is in this case common for the bands. The RF for the two bands is separated into individual paths for band X and band Y before the antenna connectors through a filter. As for multiband base stations with a common antenna connector for the bands, it is also here possible to have either the transmitter or receiver be a single-band implementation, while the other is multiband.

Further possibilities are base station implementations with separate antenna connectors for receiver and transmitter, in order to give better isolation between the receiver and transmitter paths. This may be desirable for a multiband base station, considering the large total RF bandwidths, which will in fact also overlap between receiver and transmitter.

For a multi-band base station, with a possible capability to operate with multiple RATs and several alternative implementations with common or separate antenna connectors for the bands and/or for the transmitter and receiver, the declaration

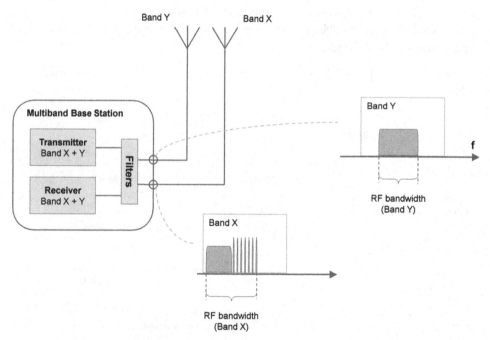

Fig. 28.11 Multiband base station with multiband transmitter and receiver for two bands with separate antenna connectors for each band.

of the base station capability becomes quite complex. What requirements that will apply to such a base station and how they are tested will depend on these declared capabilities.

Most RF requirements for a multi-band base station remain the same as for a single-band implementation. There are however some notable exceptions:

- **Transmitter spurious emissions**: For NR base stations, the requirements exclude frequencies in the operating band plus an additional 10–100 MHz "exclusion" on each side of the operating band, since this frequency range is covered by the OBUE limits. For a multi-band base station, the exclusion applies to both operating bands (plus 10–100 MHz on each side), and only the OBUE limits apply in those frequency ranges. This is called a "joint exclusion band".
- **Operating band unwanted emissions mask (OBUE)**: For multi-band operation, when the inter-RF bandwidth gap is less than two times the "exclusion" for spurious, the OBUE limit applies as a cumulative limit with contributions counted from both bands, in a way similar to operation in non-contiguous spectrum.
- **ACLR**: For multi-band operation, ACLR applies inside the *inter-RF bandwidth gap*. When the inter-RF bandwidth is so small that adjacent channels from the two RF bandwidths overlap, the Cumulative ACLR (CACLR) will apply in the gap with contributions counted from both bands, in the same way as for operation in non-contiguous spectrum.
- **Transmitter intermodulation**: For a multi-band base station, when the inter-RF bandwidth gap is small, the requirement only applies for the case when the interfering signals fit within the gap.
- **Blocking requirement**: For multi-band base station, the in-band blocking limits apply for the in-band frequency ranges of *both* operating bands. This can be seen as a "joint exclusion", similar to the one for spurious emissions. The blocking and receiver intermodulation requirements also apply inside the inter-RF bandwidth gap.
- **Receiver spurious emissions**: For a multi-band base station, a "joint exclusion band" similar to the one for transmitter spurious emissions will apply, covering both operating bands.

In the case where the two operating bands are mapped on separate antenna connectors as shown in Fig. 28.11, the above exceptions for transmitter/receiver spurious emissions, OBUE, ACLR and transmitter intermodulation do not apply. Those limits will instead be the same as for single-band operation for each antenna connector. In addition, if such a multiband base station with separate antenna connectors per band is operated in only one band with the other band (and other antenna connector) inactive, the base station will from a requirement point-of-view be seen as a single-band base station. In this case all requirements will apply as single-band requirements.

CHAPTER 29

RF technologies at mm-wave frequencies

The existing 3GPP specifications for 2G, 3G, and 4G mobile communications are applicable to frequency ranges below 6 GHz and the corresponding RF requirements consider the technology aspects related to below 6 GHz operation. NR also operates in those frequency ranges (identified as frequency range 1) but will in addition be defined for operation above 24.25 GHz (frequency range 2 or FR2), also referred to as mm-wave frequencies. A fundamental aspect for defining the RF performance and setting RF requirements for NR base stations and devices is the change in technologies used for RF implementation in order to support operation in those higher frequencies. In this chapter, some important and fundamental aspects related to mm-wave technologies are presented in order to better understand the performance that mm-wave technology can offer, but also what the limitations are.

In this chapter, Analog-to-Digital/Digital-to-Analog converters and power amplifiers are discussed, including aspects such as the achievable output power versus efficiency and linearity. In addition, some detailed insights are provided into receiver essential metrics such as noise figure, bandwidth, dynamic range, power dissipation, and the dependencies between metrics. The mechanism for frequency generation and the related phase noise aspects are also covered. Filters for mm-waves are another important part, indicating the achievable performance for various technologies and the feasibility of integrating filters into NR implementations.

The data sets used in this chapter indicate the current state-of-the-art capability and performance and are either published elsewhere or have been presented as part of the 3GPP study for developing NR [11]. Note that neither the 3GPP specifications nor the discussion here mandate any restrictions, specific models, or implementations for NR in frequency range 2. The discussion highlights and analyzes different possibilities for RF implementation of mm-wave receivers and transmitters.

An additional aspect is that essentially all operation in Frequency Range 2 will be with Active Antenna System base stations using large antenna array sizes and devices with multi-antenna implementations. While this is enabled by the smaller scale of antennas at mm-wave frequencies, it also drives complexity. The compact building practice needed for mm-wave systems with many transceivers and antennas requires careful and often complex consideration regarding the power efficiency and heat dissipation within a small area or volume. These considerations directly affect the achievable performance and possible RF requirements. The discussion here in many aspects applies for both NR base stations and NR devices, noting also that the mm-wave transceiver implementation between device and base station will have less differences compared to frequency bands below 6 GHz.

29.1 ADC and DAC considerations

The larger bandwidths available at mm-wave communication challenge the data conversion interfaces between analog and digital domains in both receivers and transmitters. The signal-to-noise-and-distortion ratio (SNDR)-based Schreier Figure-of-Merit (FoM) is a widely accepted power-efficiency metric for Analog-to-Digital Converters (ADCs) defined by [58]

$$\text{FoM} = \text{SNDR} + 10\log_{10}(f_s/2/P)$$

with SNDR in dB, power consumption P in W, and Nyquist sampling frequency f_s in Hz. Fig. 29.1 shows the Schreier FoM for a large number of ADCs vs. the Nyquist sampling frequency f_s (=2 × the signal bandwidth), published at the two most acknowledged

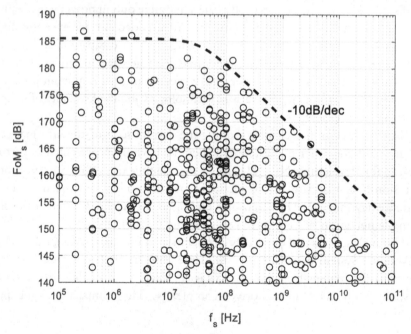

Fig. 29.1 Schreier figure-of-merit for published ADCs [62].

conferences [59] in this field of research. The dashed line indicates the FoM envelope which is constant at roughly 185 dB toward sampling frequencies below some 100 MHz. With constant FoM, the power consumption doubles for every doubling of bandwidth or 3 dB increase in SNDR. Above 100 MHz there is an additional 10 dB/decade penalty, and this means that a doubling of bandwidth will increase power consumption by a factor of 4.

Although the FoM envelope is expected to continue being slowly pushed toward higher frequencies by continued development of integrated circuit technology, RF bandwidths in the GHz range inevitably give poor power efficiency in the analog-to-digital conversion. The large bandwidths and array sizes assumed for NR at mm-wave will thus lead to a large ADC power footprint and it is important that specifications driving SNDR requirements are not unnecessarily high. This applies to devices as well as base stations.

Digital-to-Analog Converters (DACs) are typically less complex than their ADC counterparts for the same resolution and speed. Furthermore, while ADC operation commonly involves iterative processes, the DACs do not. DACs also attract substantially less interest in the research community. While structurally quite different from their ADC counterparts they can still be benchmarked using the same FoM and render similar numbers as for ADCs. In the same way as for ADC, a larger bandwidth and unnecessarily high SNDR requirement on the transmitter will result in a higher DAC power footprint.

29.2 LO generation and phase noise aspects

Local Oscillator (LO) is an essential component in radio communication systems for shifting carrier frequency up- or downwards in transceivers. One LO performance metric is the so-called phase noise (PN) of the signal generated by the LO. In plain words, PN is a measure of how stable the signal is in frequency domain. Numerically, it is defined as the single-side noise power spectral density at a frequency that is Δf Hz away from the desired LO frequency f_0, relative to the signal power. Therefore, PN is given in dBc/Hz for a specified offset frequency (Δf) and its value represents the likelihood that the LO oscillation frequency deviates by Δf Hz from the desired frequency.

LO phase noise may significantly degrade system performance; this is illustrated in Fig. 29.2, though somewhat exaggerated for a single-carrier example, where the constellation diagram for a 16-QAM signal is compared with and without phase noise, including in both cases Additive White Gaussian Noise (AWGN) modeling thermal noise. For a given symbol error rate, phase noise limits the highest modulation scheme that may be utilized. In other words, different modulation schemes pose different requirement on LO phase noise level.

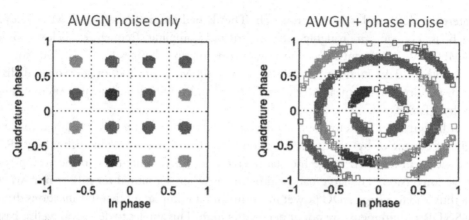

Fig. 29.2 Illustrative constellation diagram of a single-carrier 16-QAM signal without (left) and with (right) LO phase noise.

29.2.1 Phase noise characteristics of free-running oscillators and PLLs

A commonly used circuit solution for frequency generation is the Voltage Controlled Oscillator (VCO). Fig. 29.3 shows an empirical model describing the PN characteristic for a free-running VCO, where k is the Boltzmann constant, T is the absolute temperature, P_s is the signal strength, Q is the loaded quality factor of the resonator, F is a fitting parameter but has physical meaning of noise figure, and $\Delta f_{1/f}3$ is the 1/f-noise corner frequency of the active device in use [54].

The following can be concluded from the Leeson formula given in Fig. 29.3:

- PN increases by 6 dB per every doubling of the oscillation frequency f_0;
- PN is inversely proportional to signal strength, P_s;
- PN is inversely proportional to the square of the quality factor of the resonator, Q;
- 1/f noise up-conversion gives rise to close-to-carrier PN increase (i.e., at small offset).

Thus, several parameters may be used as design trade-offs in VCO development. For purpose of performance comparison of the VCOs made in various semiconductor

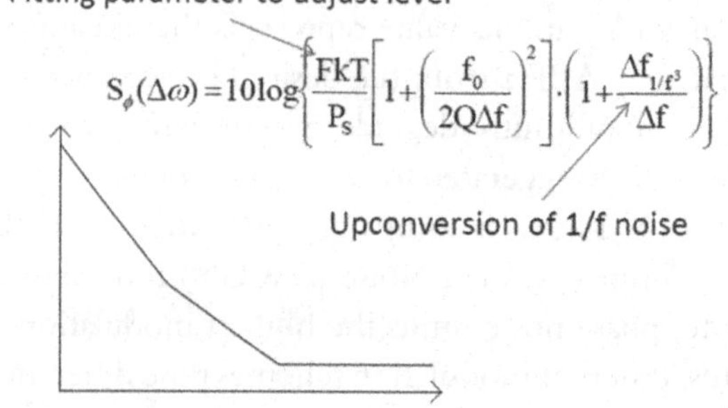

Fitting parameter to adjust level

$$S_\phi(\Delta\omega) = 10\log\left\{\frac{FkT}{P_s}\left[1+\left(\frac{f_0}{2Q\Delta f}\right)^2\right]\cdot\left(1+\frac{\Delta f_{1/f^3}}{\Delta f}\right)\right\}$$

Upconversion of 1/f noise

Fig. 29.3 Phase noise characteristic for a typical free-running VCO [57]: phase noise in dBc/Hz (y-axis) versus offset frequency in Hz (X-axis, logarithmic scale).

technologies and circuitry topologies, a Figure-of-Merit (FoM) is often used which takes into account power consumption in comparison [95]:

$$FoM = PN_{VCO}(\Delta f) - 20\log\left(\frac{f_o}{\Delta f}\right) + 10\log(P_{DC}/1mW)$$

Here $PN_{VCO}(\Delta f)$ is the phase noise of the VCO in dBc/Hz and P_{DC} is the power consumption in watt. One noticeable result of this expression is that both phase noise and power consumption in linear power are proportional to f_o^2. Thus, to maintain a phase noise level at a certain offset while increasing f_o by a factor N would require the power to be increased by N^2 (assuming a fixed FoM value).

A common way to suppress the phase noise is to apply a Phase Locked Loop (PLL) [18]. Basic PLL building blocks contain a VCO, frequency divider, phase detector, loop filter and a low-frequency reference source of high stability, such as a crystal oscillator. The total phase noise of the PLL output is composed of contributions from the VCO outside the loop bandwidth and the reference oscillator inside the loop. A significant noise contribution is also added by the phase detector and the divider.

As an example for the typical behavior of a mm-wave source, Fig. 29.4 shows the measured phase noise for a 28 GHz LO produced using a PLL at a lower frequency and multiplying up to 28 GHz. There are four distinctive offset ranges that show different characteristics:

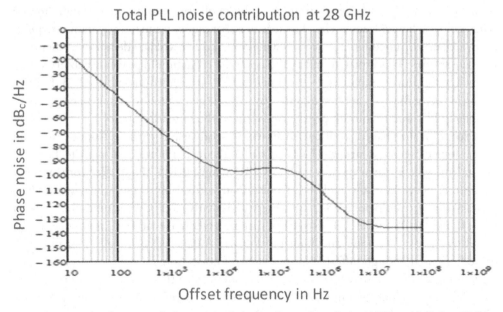

Fig. 29.4 Example of measured phase noise behavior for a phase locked VCO multiplied to 28 GHz. *(Source: Ericsson AB, used with permission.)*

- for offset <10 kHz: 30 dB/decade roll-off, due to 1/f noise up-conversion;
- for offset >10 kHz up to the PLL bandwidth (350 kHz): relatively flat and composed of contributions from several PLL blocks;
- for offset > PLL bandwidth up to 10 MHz: 20 dB/decade roll-off, dominant by VCO phase noise, and
- for offset >10 MHz: flat, white noise floor.

29.2.2 Challenges with mm-wave signal generation

As phase noise increases with frequency, increasing the oscillation frequency from 3 GHz to 30 GHz, for instance, will result in PN degradation of typically 20 dB. This will certainly limit the highest order of PN-sensitive modulation schemes applicable at mm-wave bands, thus poses a limitation on achievable spectrum efficiency for mm-wave communications.

Millimeter-wave LOs also suffer from the degradation in quality factor Q and the signal power P_s. Leeson's equation says that in order to achieve low phase noise, Q and P_s need to be maximized while the noise figure of the active device needs to be minimized. Unfortunately, all these three factors behave in an unfavorable manner with the increase of oscillation frequency. In monolithic VCO implementation, the Q-value of the on-chip resonator decreases rapidly when the frequency increases due mainly to (i) the increase of parasitic losses such as metal loss and substrate loss and (ii) the decrease of varactor Q. Meanwhile, the signal strength of the oscillator becomes increasingly limited when going to higher frequencies. The reason is that higher frequency operation requires more advanced semiconductor devices whose breakdown voltage decreases as their feature size shrinks. This is manifested by the observed reduction in power capability versus frequency for power amplifiers (-20 dB per decade) as detailed in Section 29.3. For this reason, a method widely applied in mm-wave LO implementation is to generate a lower frequency PLL and then multiply the signal up to the target frequency.

Except for the challenges discussed above, up-conversion of the 1/f noise creates an added slope close to the carrier. The 1/f noise, or the flicker noise, is strongly technology dependent, where planar devices such as CMOS and HEMT (High Electron Mobility Transistor) generally show higher 1/f noise than vertical bipolar-type devices such as SiGe HBTs. Technologies used in fully integrated MMIC/RFIC VCO and PLL solution range from CMOS and BiCMOS to III-V materials where InGaP HBT is popular due to its relatively low 1/f noise and high breakdown voltage. Occasionally also pHEMT devices are used, though suffering from severe 1/f noise. Some developments have been made using GaN FET structures in order to benefit from the high breakdown voltage, but 1/f is even higher than in GaAs FET devices and therefore seems to offset the gain from the high breakdown advantage. Fig. 29.5 summarizes phase noise performance at 100 kHz offset vs oscillation frequency for various semiconductor technologies.

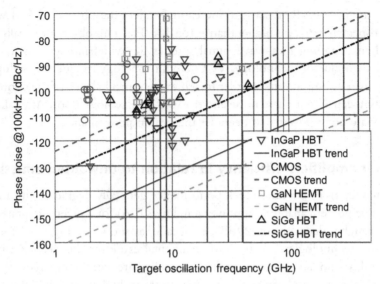

Fig. 29.5 Phase noise at 100 KHz offset versus oscillation frequency for oscillators in different semiconductor technologies [36].

Lastly, recent research has revealed the impact of the LO noise floor (Fig. 29.3) on the performance of wide bandwidth systems [21]. Fig. 29.6 shows the measured EVM from a transmitter using a 7.5 GHz carrier versus the LO noise floor level for different symbol rates. The impact of the flat LO noise floor is insignificant when the symbol bandwidth is low (<<100 MHz). However, when the bandwidth increases to beyond hundred MHz,

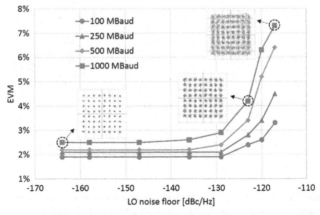

Fig. 29.6 Measured EVM of a 64-QAM signal at 7.5 GHz versus the LO noise floor level for different symbol rate [23].

such as the case in 5G NR, it starts to increasingly affect the observed EVM when the noise power spectral density is higher than −135 dBc/Hz. This observation calls for extra cautions when designing mm-wave signal sources for wideband systems in terms of choice of semiconductor technology, VCO circuit topology, frequency multiplication factor and power consumption, in order to achieve optimal balance of the contributions from the offset-dependent phase noise and the white phase noise floor. After all, it is the integrated phase noise power that degrades the system performance.

29.3 Power amplifiers efficiency in relation to unwanted emission

Radio Frequency (RF) building block performance generally degrades with increasing frequency. The power capability of power amplifiers (PA) for a given integrated circuit technology roughly decreases by 20 dB per decade, as shown in Fig. 29.7 for various semiconductor technologies. There is a fundamental cause for this decrease; increased power capability and increased frequency capability are conflicting requirements as observed from the so-called Johnson limit [52]. In short, higher operational frequencies require smaller geometries, which subsequently result in lower operational power in order to prevent dielectric breakdown from the increased field strengths. To uphold Moore's law, the gate geometries are constantly shrunk and hence the power capability per transistor is reduced.

A remedy is however found in the choice of integrated circuit material. mm-wave integrated circuits have traditionally been manufactured using so called III–V materials, that is a combination of elements from groups III and V of the periodic table, such as Gallium Arsenide (GaAs) and more recently Gallium Nitride (GaN). Integrated circuit technologies based on III–V materials are substantially more expensive than conventional

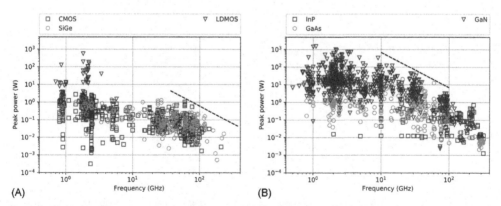

Fig. 29.7 Power amplifier output power versus frequency for various semiconductor technologies. (A) Silicon based and (B) III–V. The dashed line illustrates the observed reduction in power capability versus frequency (−20 dB per decade). The data points are from a survey of published microwave and mm-wave power amplifier circuits [96].

silicon-based technologies and they cannot handle the integration complexity of large scale digital circuits or radio modems for cellular handsets. Nevertheless, GaN-based technologies are now maturing rapidly and deliver power levels an order of magnitude higher compared to conventional silicon-based technologies.

There are mainly three semiconductor material parameters that affect the efficiency of an amplifier: maximum operating voltage, maximum operating current density and knee-voltage. Due to the knee-voltage, the maximum attainable efficiency is reduced by a factor that is proportional to:

$$\frac{1-k}{1+k}$$

where k is the ratio of knee-voltage to the maximum operating voltage. For most transistor technologies the ratio k is in the range of 0.05–0.01, resulting in an efficiency degradation of 10–20%.

The maximum operating voltage and current density limit the maximum output power from a single transistor cell. To further increase the output power, the output from multiple transistor cells must be combined. The most common combination techniques are stacking (voltage combining), paralleling (current combining), and corporate combiners (power combining). Either choice of combination technique will be associated with a certain combiner-efficiency. A technology with low power density requires more combination stages and will incur a lower overall combiner-efficiency. At mm-wave frequencies the voltage- and current-combining methods are limited due to the wavelength. The overall size of the transistor cell must be kept less than about 1/10th of the wavelength. Hence, paralleling and/or stacking are used in combination with corporate power combining to get the wanted output power. The maximum power density of CMOS is about 100 mW/mm compared to 4000 mW/mm for GaN. Thus, GaN technology will require less aggressive combining strategies and hence give higher efficiency.

Fig. 29.8 shows the saturated power-added efficiency (PAE) as a function of frequency. The maximum reported PAE is approximately 60% and 40%, at 30 GHz and 77 GHz, respectively.

PAE is expressed as $\mathrm{PAE} = 100^{*}\left\{[\mathrm{P_{OUT}}]_{\mathrm{RF}} - [\mathrm{P_{IN}}]_{\mathrm{RF}}\right\}/[\mathrm{P_{DC}}]_{\mathrm{TOTAL}}.$

At mm-wave frequencies, semiconductor technologies fundamentally limit the available output power. Furthermore, the efficiency is also degraded with higher frequency.

Considering the PAE characteristics in Fig. 29.8, and the non-linear behavior of the AM-AM/AM-PM characteristics of the power amplifier, significant power back-off may be necessary to reach linearity requirement such as the transmitter ACLR requirements (see Section 25.9). Considering the heat dissipation aspects and significantly reduced area/volume for mm-wave products, the complex interrelation between linearity, PAE and output power in the light of heat dissipation must be considered.

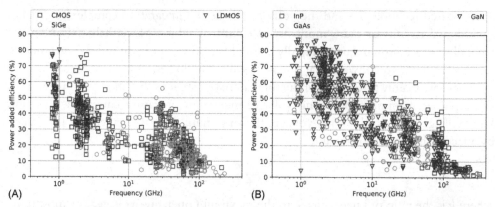

Fig. 29.8 Saturated power added efficiency versus frequency for various semiconductor technologies. (A) Silicon based and (B) III–V. Data from a survey of published microwave and mm-wave power amplifier circuits [96].

29.4 Filtering aspects

Using various types of filters in base station and device implementations is an essential component in meeting the overall RF requirements. This has been the case for all generations of mobile systems and is essential also for NR, both below 6 GHz and in the new mm-wave bands. Filters are used to mitigate unwanted emissions due to noise, LO-leakage, intermodulation, harmonics generation, and various unwanted mixing products. In the receiver chain, filters are used to handle either self-interference from own transmitter signal in paired bands, or to suppress interferers in adjacent bands or at other frequencies.

The RF-requirements are differentiated for different scenarios. For base station spurious emission, there are general requirements across a very wide frequency range, coexistence requirements in the same geographical areas, and co-location requirements for dense deployments. Similar requirements are defined for devices.

Considering the limited size (area/volume) and level of integrations needed for mm-wave frequencies, the filtering can be challenging where most discrete mm-wave filters are bulky and ill-suited for embedding in highly integrated structures for mm-wave products.

29.4.1 Possibilities of filtering at the analog front-end

Different implementations provide different possibilities for filtering. For the purpose of discussion, two extremes can be considered:

- Low-cost, monolithic integration with a few multi-chain CMOS/BiCMOS core-chips with built-in power amplifiers and built in down-converters. This case will give

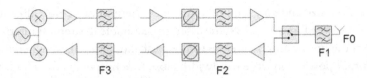

Fig. 29.9 Possible filter locations.

limited possibilities to include high-performance filters along the RF-chains since the Q-values for on–chip filter resonators will be poor.

- High-performance, heterogeneous integration with several CMOS/BiCMOS core chips, combined with external amplifiers and external mixers. This implementation allows the inclusion of external filters at several places along the RF-chains (at a higher complexity, size, and power consumption).

There are at least three places where it makes sense to put filters, depending on implementation, as shown in Fig. 29.9:

- Behind or inside the antenna element (F1 or F0), where loss, size, cost, and wide-band suppression is critical;
- Behind the first amplifiers (looking from the antenna side), where low loss is less critical (F2);
- On the high-frequency side of mixers (F3), where signals have been combined (in the case of analog and hybrid beam forming).

The main purpose of F1/F0 is typically to suppress interference and emissions far from the desired channel across a wide frequency range (for example, DC to 60 GHz). Ideally, there should not be any unintentional resonances or passbands in this wide frequency range (see Fig. 29.10). This filter will help relax the design challenge (bandwidth to consider, linearity requirements, etc.) of all following blocks. Insertion loss must be very low,

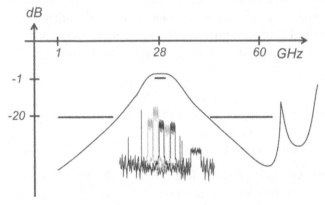

Fig. 29.10 Filter example for the 28 GHz band.

and there are strict size and cost requirements since there must be one filter at each element or sub-array (see Fig. 29.9). In some cases, it is desirable to suppress intermodulation products from the power amplifier close to the passband, particularly when transmitting with high output power and large bandwidth close to sensitive bands. This is very challenging for millimeter wave array antennas due to lack of suitable filters, and lack of isolators between amplifiers and filters.

The purpose of F2 is similar to that of F1/F0. There are still strict size requirements, but more loss can be accepted (behind the first amplifiers) and even unintentional passbands (assuming F1/F0 will handle that). This allows better discrimination (more poles), and better frequency precision (for example, using half-wave resonators).

The main purpose of F3 is typically suppression of LO-, sideband-, spurious- and noise-emission, and suppression of incoming interferers that accidentally fall in the IF-band after the mixer, and strong interferers that fall outside the IF-band that tend to block mixers and ADCs. For analog (or hybrid) beamforming it is enough to have just one (or a few) such filters. This relaxes requirements on size and cost, which opens the possibility to achieve sharp filters with multiple poles and zeroes, and with high Q-value and good frequency precision in the resonators.

The deeper into the RF-chain (starting from the antenna element), the better protected the circuits will get. For the monolithic integration case it is difficult to implement filters F2 and F3. One can expect performance penalties for this case, and output power per branch is lower. Furthermore, it is challenging to achieve good isolation across a wide frequency range, as microwaves tend to bypass filters by propagating in ground structures around them.

29.4.2 Insertion loss (IL) and bandwidth

Sharp filtering on each branch (at positions F1/F0) with narrow bandwidth leads to excessive loss at microwave and mm-wave frequencies. To get the insertion loss down to a reasonable level the passband can be made significantly larger than the signal bandwidth. A drawback of such an approach is that more unwanted signals will pass the filter. In choosing the best loss-bandwidth trade-off there are some basic dependencies to be aware of:

- IL decreases with increasing BW (for fixed fc);
- IL increases with increasing fc (for fixed BW);
- IL decreases with increasing Q-value;
- IL increases with increasing N.

To exemplify the trade-off, a three-pole LC-filter with $Q = 20$, 100, 500 and 5000, for 800 and 4×800 MHz 3 dB-bandwidth, tuned to 15 dB return loss (with $Q = 5000$), is examined as shown in Fig. 29.11.

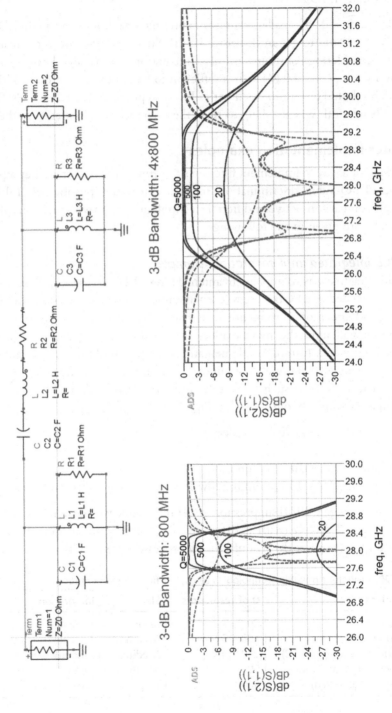

Fig. 29.11 Example 3-pole LC filter with 800 and 4 × 800 MHz bandwidth, for different Q values.

From this study it is observed that:

- 800 MHz bandwidth or smaller, requires exotic filter technologies, with a Q-value around 500 or better to get an IL below 1.5 dB. Such Q-values are very challenging to achieve considering constraints on size, integration aspects, and cost;
- For a bandwidth of 4×800 MHz, it is sufficient to have a Q-value around 100 to get 2 dB IL. This should be within reach with a low-loss printed circuit board (PCB). The increased bandwidth will also help to relax the tolerance requirements on the PCB.

29.4.3 Filter implementation examples

When looking for a way to implement filters in a 5G array antenna system, key aspects to consider include: Q-value, discrimination, size, and integration possibilities. Table 29.1 gives a rough comparison between different technologies, and two specific implementation examples are given in the following.

29.4.3.1 PCB integrated implementation example

A simple and attractive way to implement antenna filters (F1) is to use strip-line or micro-strip filters, embedded in a PCB close to each antenna element. This requires a low-loss PCB with good precision. Production tolerances (permittivity and patterning and via-positioning) will limit the performance, mainly through a shift in the passband and increased mismatch. In most implementations the passband must be set larger than the operating frequency band with a significant margin to account for this.

Typical characteristics of such filters can be illustrated by looking at the following design example, with the layout shown in Fig. 29.12:

- Five-pole, coupled line, strip-line filter;
- Dielectric permittivity: 3.4;
- Dielectric thickness: 500 μm (ground to ground);
- Unloaded resonator Q: 130 (assuming low loss microwave dielectrics).

The filter is tuned to give 20 dB suppression at 24 GHz, while passing as much as possible of the band 24.25–27.5 GHz (with 17 dB return loss). Significant margins are added to make room for variations in the manufacturing processes of the PCB.

Table 29.1 Different possible technologies to use for filter implementation.

Technology	Q of resonators	Size	Integration
On-chip (Si)	20	Small	Feasible
PCB (low-loss)	100–150	Medium	Feasible
Ceramic (thin film, LTCC)	200–300	Medium	Difficult
Advanced miniature filters	500	Medium	Difficult
Waveguide (air-filled)	5000	Large	Extremely difficult

Fig. 29.12 Layout of stripline filter on a PCB.

A Monte Carlo analysis was performed to study the impact of variations in the manufacturing process on filter performance, using the following quite aggressive tolerance assumptions for the PCB:

- Permittivity standard deviation: 0.02;
- Line width standard deviation: 8 μm;
- Thickness of dielectric standard deviation: 15 μm.

With these distribution assumptions, 1000 instances of the filter were generated and simulated. Fig. 29.13 shows the filter performance (S21) for these 1000 instances (blue traces; black in print version), together with the nominal performance (yellow trace).; white in print version). Red lines (dark gray in print version) in the graph (with hooks) indicate possible requirement levels that could be targeted with this filter.

From this design example, the following rough description of a PCB filter implementation is found:

- 3–4 dB insertion loss (IL);
- 20 dB suppression (17 dB if IL is subtracted);
- 1.5 GHz transition region with margins included;

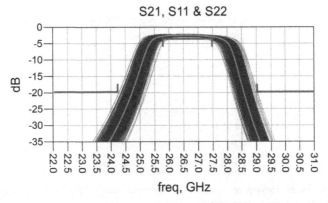

Fig. 29.13 Simulated impact of manufacturing tolerances on the filter characteristics of a strip line filter in PCB.

- Size: 25 mm^2, which can be difficult to fit in the case of individual feed and/or dual polarized elements;
- If a 3 dB IL is targeted, there would be significant yield loss with the suggested requirement, in particular for channels close to the pass-band edges.

29.4.3.2 LTCC filter implementation example

Another possibility to implement filters is to make components for Surface Mount Assembly (SMT), including both filters and antennas, for example based on Low-Temperature Cofired Ceramics (LTCC). One example of a prototype LTCC component was outlined in [29] and is also shown in Fig. 29.14.

The measured performance of a small batch of filter components, without antennas, based on LTCC is shown in Fig. 29.15. It is found that these filters add about 2.5 dB of insertion loss for a 3 GHz passband, while providing 20 dB of attenuation 0.75 GHz from the passband edge.

Additional margins relative to this example should be considered to account for large volume manufacturing tolerances, temperature sensitivity, improved S11 and degradation related to the integration with the antenna. Accounting for such margins, the LTCC-filter shown could be assumed to add approximately 3 dB of insertion loss, for 20–25 dB suppression (IL subtracted) 1–1.25 GHz from the pass-band edge.

In essence, considering the above two implementation examples and the challenges related to size, integration possibilities, Q-value, manufacturing tolerances, lack of tuning, radiation pattern degradation, impedance matching, cost, etc., the conclusion is that it is not realistic to rely on filtering closer than 1–1.25 GHz from the edge of the pass-band.

Fig. 29.14 Example of prototype of an LTCC-component containing both antenna elements and filters. *(Source: TDK Corporation, used with permission.)*

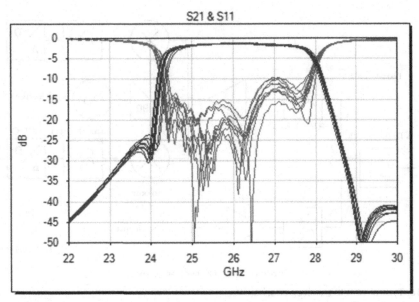

Fig. 29.15 Measured performance of an LTCC-based filter without antenna. *(Source: TDK Corporation, used with permission.)*

29.5 Receiver noise figure, dynamic range and bandwidth dependencies

29.5.1 Receiver and noise figure model

A receiver model as shown in Fig. 29.16 is assumed here. The dynamic range (DR) of the receiver will in general be limited by the front-end insertion loss (IL), the receiver (RX) low-noise amplifier (LNA) and the ADC noise and linearity properties.

Typically, $DR_{LNA} \gg DR_{ADC}$ so the RX use Automatic Gain Control (AGC) and selectivity (distributed) in-between the LNA and the ADC to optimize the mapping of the wanted signal and the interference to the DR_{ADC}. For simplicity, a fixed gain setting is considered here.

A further simplified receiver model can be derived by lumping the Front End (FE), RX, and ADC into three cascaded blocks, as shown in Fig. 29.17. This model cannot replace a more rigorous analysis but will demonstrate interdependencies between the main parameters.

Focusing on the small signal co-channel noise floor, the impact of various signal and linearity impairments can be studied to arrive at a simple noise factor, or noise figure, expression.

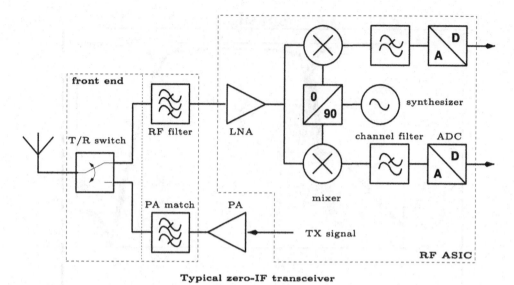

Fig. 29.16 Typical zero-IF transceiver schematic.

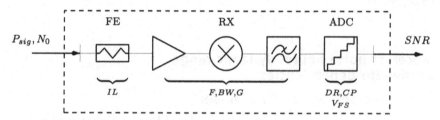

Fig. 29.17 A simplified receiver model.

29.5.2 Noise factor and noise floor

Assuming matched conditions, Friis' formula can be used to find the noise factor at the receiver input as (linear units unless noted),

$$F_{RX} = 1 + (F_{LNA} - 1) + (F_{ADC} - 1)/G$$

The RX input referred small-signal co-channel noise floor will then equal

$$N_{RX} = F_{LNA} \cdot N_0 + N_{ADC}/G$$

where $N_0 = k \cdot T \cdot BW$ and N_{ADC} are the available noise power and the ADC effective noise floor in the channel bandwidth, respectively (k and T being Boltzmann's constant and absolute temperature, respectively). The ADC noise floor is typically set by a combination of quantization, thermal and intermodulation noise, but here a flat noise floor is assumed as defined by the ADC effective number of bits.

The effective gain G from LNA input to ADC input depends on small–signal gain, AGC setting, selectivity and desensitization (saturation), but here it is assumed that the gain is set such that the antenna referred input compression point (CPi) corresponds to the ADC clipping level, that is the ADC full scale input voltage (V_{FS}).

For weak non–linearities, there is a direct mathematical relationship between CP and the third-order intercept point (IP_3) such that $IP_3 \approx CP + 10\,dB$. For higher-order non–linearities, the difference can be larger than 10 dB, but then CP is still a good estimate of the maximum signal level while intermodulation for lower signal levels may be overestimated.

29.5.3 Compression point and gain

Between the antenna and the RX there is the FE with its associated insertion loss ($IL > 1$), for example due to a T/R switch, a possible RF filter, and PCB/substrate losses. These losses have to be accounted for in the gain and noise expressions. Knowing IL, the CPi can be found that corresponds to the ADC clipping as

$$CP_i = IL \cdot N_{ADC} \cdot DR_{ADC}/G$$

The antenna referred noise factor and noise figure will then become

$$F_i = IL \cdot F_{RX} = IL \cdot F_{LNA} + CP_i/(N_0 \cdot DR_{ADC})$$

and

$$NF_i = 10 \cdot \log_{10}(F_i),$$

respectively. When comparing two designs, for example, at 2 and 30 GHz, respectively, the 30 GHz IL will be significantly higher than that of the 2 GHz. From the F_i expression it can be seen that to maintain the same noise figure (NF_i) for the two carrier frequencies, the higher FE loss at 30 GHz needs to be compensated for by improving the RX noise factor. This can be accomplished by (1) using a better LNA, (2) relaxing the input compression point, that is increasing G, or (3) increasing the DR_{ADC}. Usually a good LNA is already used at 2 GHz to achieve a low NF_i, so this option is rarely possible. Relaxing CP_i is an option but this will reduce IP3 and the linearity performance will degrade. Finally, increasing DR_{ADC} comes at a power consumption penalty ($4\times$ per extra bit). Especially wideband ADCs may have a high power consumption. That is, when BW is below some 100 MHz the $N_0 \cdot DR_{ADC}$ product (that is $BW \cdot DR_{ADC}$) is proportional to the ADC power consumption, but for higher bandwidths the ADC power consumption is proportional to $BW^2 \cdot DR_{ADC}$, thereby penalizing higher BW (see Section 29.1). Increasing DR_{ADC} is typically not an attractive option and it is inevitable that the 30 GHz receiver will have a significantly higher NF_i than that of the 2 GHz receiver.

29.5.4 Power spectral density and dynamic range

A signal consisting of many similar subcarriers will have a constant power-spectral density (PSD) over its bandwidth and the total signal power can then be found as $P = PSD \cdot BW$.

When signals of different bandwidths but similar power levels are received simultaneously, their PSDs will be inversely proportional to their BW. The antenna-referred noise floor will be proportional to BW and F_i, or $N_i = F_i \cdot k \cdot T \cdot BW$, as derived above. Since CP_i will be fixed, given by G and ADC clipping, the dynamic range, or maximum SNR, will decrease with signal bandwidth, that is $SNR_{max} \propto 1/BW$.

The above signal can be considered as additive white Gaussian noise (AWGN) with an antenna-referred mean power level (Psig) and a standard deviation (σ). Based on this assumption the peak-to-average-power ratio can be approximated as $PAPR = 20 \cdot \log_{10}(k)$, where the peak signal power is defined as $P_{sig} + k \cdot \sigma$, that is there are k standard deviations between the mean power level and the clipping level. For OFDM an unclipped PAPR of 10 dB is often assumed (that is 3σ) and this margin must be subtracted from CPi to avoid clipping of the received signal. An OFDM signal with an average power level, for example, 3σ below the clipping level will result in less than 0.2% clipping.

29.5.5 Carrier frequency and mm-wave technology aspects

Designing a receiver at, for example, 30 GHz with a 1 GHz signal bandwidth leaves much less design margin than what would be the case for a 2 GHz carrier frequency, $f_{carrier}$ with, for example, 50 MHz signal bandwidth. The IC technology speed is similar in both cases, but the design margin and performance depend on the technology being much faster than the required signal processing, which means that the 2 GHz design will have better performance.

The graph in Fig. 29.18 shows expected evolution of some transistor parameters important for mm-wave IC design, as predicted by the International Technology Roadmap for Semiconductors (ITRS). Here f_t, f_{max}, and V_{dd}/BV_{ceo} data from the ITRS 2007 targets [37] for CMOS and bipolar RF technologies are plotted vs. the calendar year when the technology is anticipated to become available. f_t is the transistor transit frequency (that is, where the RF device's extrapolated current gain is 0 dB), and f_{max} is the maximum frequency of oscillation (that is when the extrapolated power gain is 0 dB). V_{dd} is the RF/high-performance CMOS supply voltage and BV_{ceo} is the bipolar transistor's collector-emitter base open breakdown voltage limits. For example, an RF CMOS device is expected to have a maximum V_{dd} of 700 mV by 2024 (other supply voltages will be available as well, but at a lower speed).

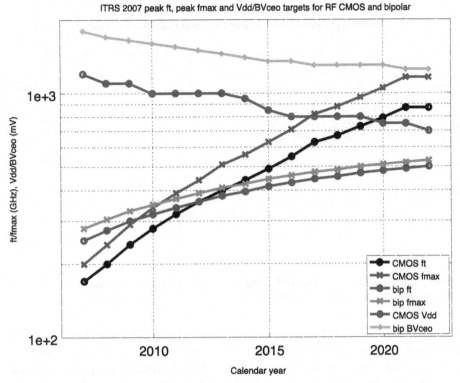

Fig. 29.18 Expected evolution over time of some transistor parameters: f_t, f_{max}, and V_{dd}/BV_{ceo} [39].

ITRS has recently been replaced by the International Roadmap for devices and Systems (IRDS). In the IRDS Outside System Connectivity 2018 update [97] expected performance data for high-speed bipolar NPN transistors is listed, as illustrated in the graph in Fig. 29.19.

The free space wavelength at 30 GHz is only 1 cm which is one tenth of what is the case for existing 3GPP bands below 6 GHz. Antenna size and path loss are related to wavelength and carrier frequency. To compensate the small physical size of a single antenna element, multiple antennas, for example array antennas, will have to be used. When beam forming is used the spacing between antenna elements will still be related to the wavelength, constraining the size of the FE and RX. Some of the implications of these frequency and size constraints are:

- The ratios $f_t/f_{carrier}$ and $f_{max}/f_{carrier}$ will be much lower at millimeter wave frequencies than for below 6 GHz applications. As the receiver gain drops with operating frequency when this ratio is less than some 10–100×, the available gain at millimeter waves will be lower and consequently the device noise factor, F_i, higher (similar to when Friis' formula was applied to a transistor's internal noise sources).

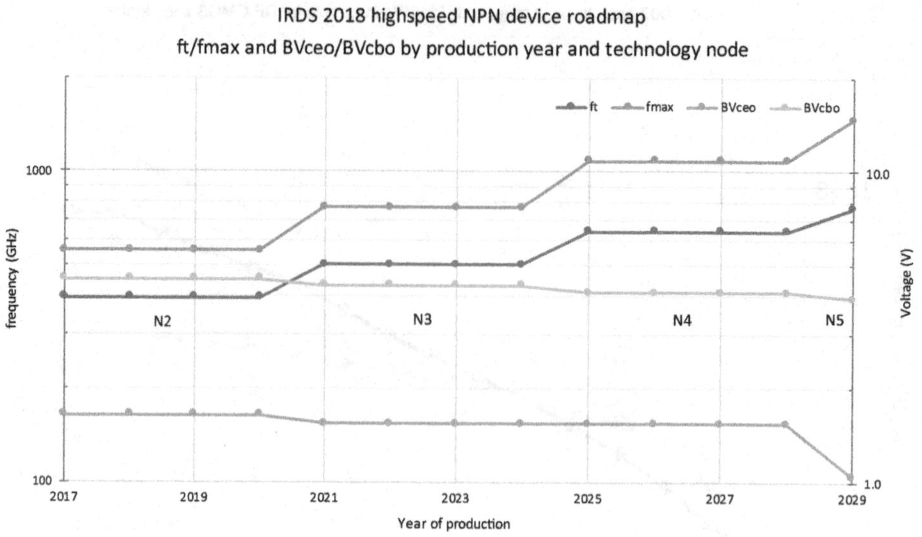

IRDS 2018 highspeed NPN device roadmap
ft/fmax and BVceo/BVcbo by production year and technology node

Fig. 29.19 Expected evolution of bipolar highspeed NPN parameters (f_t, f_{max}, BV_{ceo}, and BV_{cbo}) versus technology node (N_2–N_5) and production year [97].

- The semiconductor material's electrical breakdown voltage (E_{br}) is inversely proportional to the charge carrier saturation velocity (V_{sat}) of the device due to the Johnson limit. This can be expressed as $v_{sat} \cdot E_{br} = $ constant or $f_{max} \cdot V_{dd} = $ constant. Consequently, the supply voltage will be lower for millimeter-wave devices compared to devices in the low GHz frequency range. This will limit the CP_i and the maximum available dynamic range.
- A higher level of transceiver integration is required to save space, either as system-on-chip (SoC) or system-in-package (SiP). This will limit the number of technologies suitable for the RF transceiver and limit F_{RX}.
- RF filters will have to be placed close to the antenna elements and fit into the array antenna. Consequently, they have to be small, resulting in higher physical tolerance requirements, possibly at the cost of insertion loss and stopband attenuation. That is, IL and selectivity get worse. The filtering aspect for mm-wave frequencies is further elaborated on in Section 29.4.
- Increasing the carrier frequency from 2 GHz to 30 GHz (that is >10×) has a significant impact on the circuit design and its RF performance. For example, modern high-speed CMOS devices are velocity saturated and their maximum operating frequency is inversely proportional to the minimum channel length, or feature size. This dimension halves roughly every 4 years, as per Moore's law (stating that complexity, that is transistor density, doubles every other year). With smaller feature sizes, internal voltages must also be lowered to limit electrical fields to safe levels. Thus, designing a 30 GHz RF receiver corresponds to designing a 2 GHz receiver using about

15-year-old low-voltage technology (that is today's breakdown voltage but 15 years old f_t (see Fig. 29.18) with ITRS device targets). With such a mismatch in device performance and design margin it is not to be expected to maintain both 2 GHz performance and power consumption at 30 GHz.

The signal bandwidth at mm-wave frequencies will also be significantly higher than at 2 GHz. For an active device, or circuit, the signal swing is limited by the supply voltage at one end and by thermal noise at the other. The available thermal noise power of a device is proportional to BW/g_m, where g_m is the intrinsic device gain (transconductance). As g_m is proportional to bias current it can be seen that the dynamic range becomes the ratio

$$DR \propto V_{dd}^2 \cdot I_{bias}/BW = V_{dd} \cdot P/BW$$

or

$$P \propto BW \cdot DR/V_{dd}$$

where P is the power dissipation.

Receivers for mm-wave frequencies will have an increased power consumption due to their higher BW, aggravated by the low-voltage technology needed for speed, compared to typical 2 GHz receivers. Thus, considering the thermal challenges given the significantly reduced area/volume for mm-wave products, the complex interrelation between linearity, NF, bandwidth and dynamic range in the light of power dissipation should be considered.

29.6 Summary

This chapter gave an overview of what mm-wave technologies can offer and how to derive requirements. The need for highly integrated mm-wave systems with many transceivers and antennas will require careful and often complex consideration regarding the power efficiency and heat dissipation in small area/volume affecting the achievable performance.

Important areas presented were DAC/ADC converters, power amplifiers, and the achievable power versus efficiency as well as linearity. Receiver essential metrics are noise figure, bandwidth, dynamic range, and power dissipation and they all have complex dependencies. The mechanism for frequency generation as well as phase noise aspects were also covered. Filtering aspects for mm-wave frequencies were shown to have substantial impact in new NR bands and the achievable performance for various technologies and the feasibility of integrating such filters into NR implementations needs to be accounted for when defining RF requirements. All these aspects are accounted for throughout the process of developing the RF characteristics of NR in Frequency Range 2.

CHAPTER 30

Further 5G evolution and the first step toward 6G

The first release of 5G/NR (release 15) primarily focused on eMBB services and, to some extent, URLLC. Later releases have provided further enhancements to these areas, but also introduced new features addressing new deployment scenarios and new verticals as described in the previous chapters. With release 18 nearing its completion, 3GPP has initiated discussions on the next release, that is, release 19 (Fig. 30.1).

The content of release 19 is, at the time of this writing, not decided upon, but most likely will contain enhancements to existing features as well as new features. Studies carried out in release 18 – for example the studies on network energy efficiency and duplex enhancements – are likely to form the basis for work items in release 19. Although NR will continue to evolve for many years, release 19 will likely also be the starting point for discussions on the next generation cellular networks, commonly referred to as 6G, with release 19 looking into basic requirements. It is then expected that this will be followed by a 6G study in release 20 and the actual specification work taking place in release 21. This is needed in order to match the ITU time plan for IMT-2030, as well as to ensure commercial deployment of 6G systems in the 2030 time frame.

In the following, some potential technology components for the coming 3GPP releases are discussed. By no means are the set of components exhaustive and there is no guarantee that these technologies will be specified in 3GPP. Nevertheless, it provides some insights into the direction cellular technology may evolve in the future.

30.1 General enhancements to NR

A large part of the additions included in a certain release have, in general, been enhancements to features in the previous release and release 19 is not expected to be any different in this respect. Although some of these enhancements might be relatively small and limited in scope, one should not underestimate the benefits of gradual improvement since they do contribute to making NR an even more efficient system.

Multi-antenna transmission has been part of NR since the beginning and has been continuously evolving in every release. Enabling larger array sizes, for example, to improve coverage, is one important area to consider, in particular in FR2 and the higher part of FR1 where the physical antenna size could still remain modest despite an increase

5G/5G-Advanced
https://doi.org/10.1016/B978-0-443-13173-8.00030-X

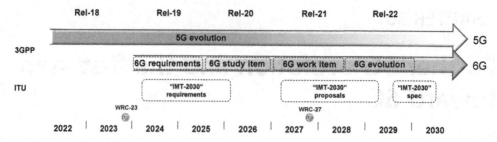

Fig. 30.1 NR evolution and the path to 6G.

in the number of antenna elements. In many cases, hybrid beamforming solutions are used as the number of radio chains and AD/DA converters otherwise becomes prohibitively large. Obtaining CSI with a large number of antenna elements is challenging and enhancements to CSI reporting might prove useful.

In FR2, a larger number of antenna elements translates into an increased number of beams. Combined with improved beam management to further speed up establishing a connection on a higher frequency band this can be useful to offload traffic from more scarce lower-frequency spectrum. To avoid the larger number of beams to lead to an increase in RRC reconfigurations, mechanisms to handle this using DCI might be useful, for example to dynamically switch QCL relations and further improve the dynamic handling of SRS. Multi-TRP enhancements, for example antenna calibration and coherent joint transmission are yet other possibilities.

Mobility is another example of a feature with enhancements in multiple releases. L1/L2-triggered mobility is introduced in release 18, although limited to mobility between cells belonging to the same gNB (or, more specifically, to the same CU). A natural area to consider in release 19 is inter-gNB L1/L2-triggered mobility.

Improved uplink performance was identified as an important part of Release 18. However, there was insufficient time to specify PRACH beam sweeping with beam indication for msg3. This can be a target for future NR evolution.

30.2 Duplex evolution

One of the much-debated topics in release 18 is duplex enhancements, in particular subband full duplex. To investigate the feasibility and potential benefits for full duplex, a study was carried out in release 18 [119], focusing on subband full duplex at the gNB and conventional time-domain separation at the device. In subband full duplex, uplink transmissions in unpaired spectrum can occur in parts of the downlink slots, see Fig. 30.2. This might be beneficial in, for example, terms of uplink latency as data can be transmitted "when needed" without having to wait for an uplink time slot. Thus, from a latency perspective, subband full duplex would come close to the performance of FDD-based deployments, despite operating in unpaired spectrum.

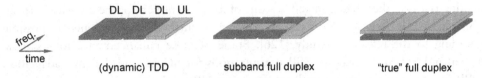

Fig. 30.2 TDD, subband full duplex, and "true" full duplex.

However, cross-link interference, especially inter-site and inter-sector interference is very challenging. Handling the interference might be possible in a low-power, non-sectorized deployment while being extremely challenging in high-power macro deployments with three-sector sites.

The release 18 study item has not come to a firm conclusion on the performance gains and the feasibility of subband full duplex. Continued work in this area is one possibility for the future. It should be noted that mechanism for handling inter-site interference problems would enable dynamic TDD to be applicable in a much wider range of scenarios than today, providing similar gains as full duplex without the need for handing the self-interference.

30.3 AI/ML

In recent years there has been significant focus on AI/ML in many areas and radio communication networks are no exception. Already today, many functions of cellular systems can be, and are, implemented using AI/ML techniques in a proprietary manner without explicit standardization support. In other words, AI/ML techniques are just yet another set of tools in the overall implementation toolbox. The set of problems benefitting from AI/ML is still under discussion. However, in general, AI/ML-based solutions are more applicable to complex problems that lack a relatively simple, but still accurate, model of the real world and for which non-linear solutions, together with training based on vast amount of data for training, are anticipated to provide major gains.

As discussed in Chapter 5, 3GPP studied the applicability of AI/ML to cellular communication in release 18 [121,125], focusing on three areas:

- CSI feedback, for example, reduced overhead and/or improved accuracy;
- Beam management, for example, beam prediction to reduce overhead and more accurate beam selection; and
- Positioning, for example, ways to improve the accuracy in different (non-line-of-sight) scenarios.

These studies are likely to form the basis for specification work in future releases.

30.4 Network energy efficiency

Network energy efficiency is a very important aspect for an operator, both from a cost perspective and from an environmental angle (reduction of CO_2 emissions). This was identified early in the 5G design and was one of the main drivers for the ultra-lean-design principle of NR (see Section 5.1.2). Nevertheless, further enhancements are possible according to the release 18 study [126]. Some of these enhancements, for example on-demand SSB transmissions, SSB optimizations for SCells, and dynamic adaptation of PRACH occasions, are possibilities for future NR evolution.

30.5 Zero-energy devices and ambient IoT

Support for massive machine-type communication is currently provided by the NB-IoT and eMTC technologies. Although device power consumption is low, these devices typically still rely on an ordinary chargeable battery as the energy source. For a truly massive number of sensors, this may not be feasible and devices which, from an end-user perspective, can operate without an ordinary battery may be required.

A "zero-energy" device is in essence a device operating without the need for explicit charging or battery replacement. Instead, the energy necessary for operation is harvested from the surroundings, for example vibrations, temperature gradients, or even the RF energy itself. Given the minuscule amounts of power possible to harvest, the amount of data possible to transmit is very small, possibly requiring rethinking many of the existing protocols and procedures. Performance-wise, the data rates and capabilities in general are below NB-IoT/eMTC, see Fig. 30.3. This is an interesting area for future 6G systems and 3GPP has already started initial discussions on such devices using the term "ambient IoT", aiming at partial support already as part of 5G-Advanced.

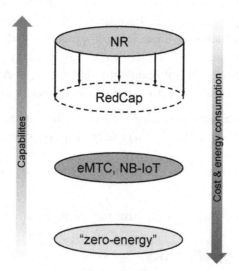

Fig. 30.3 "Zero-energy" devices/ambient IoT in relation to eMTC/NB-IoT and RedCap.

30.6 Joint communication and sensing

Positioning support has been part of cellular systems for several releases and will continue to be an important service offered by the networks. Joint communication and sensing takes this one step further. Given the large number of cellular nodes installed for communication purposes, can these nodes in addition be used for sensing purposes with no or little extra cost? This would open up a wide range of new use cases such as collecting information for digital twins, real-time updates of maps and using sensing for radio-resource management in the communication network. The sensing capability could be in the form of "radar"-like sensing building a map of the surroundings, but also sensing other interesting quantities such as precipitation, see Fig. 30.4.

Sensing is to a large extent still in the research stage with questions such as waveform design, multi-antenna techniques, coexistence between sensing and communication, and privacy aspects. It is expected to be part of future 6G systems, but initial studies on, for example, suitable channel models for sensing research, might be part of release 19.

30.7 The road to 6G

The requirements on the next-generation cellular systems, 6G, are likely to be discussed in release 19 as already mentioned. Starting 6G discussions already in release 19, given that 5G networks are still being deployed at a rapid pace, may sound early. However, targeting initial deployments of 6G around 2028 and widespread deployments in 2030, and the fact that developing 6G specifications takes time, requirement discussions need to start in release 19 as shown in Fig. 30.1.

Not only will 6G meet the expanding needs of current use cases, it also needs to address new, including not-yet-foreseen, use cases and deployment scenarios. Predicting future use cases is very difficult, if not impossible, but nevertheless is such an exercise

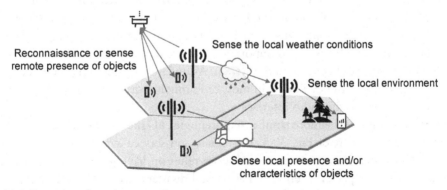

Fig. 30.4 Examples of sensing scenarios.

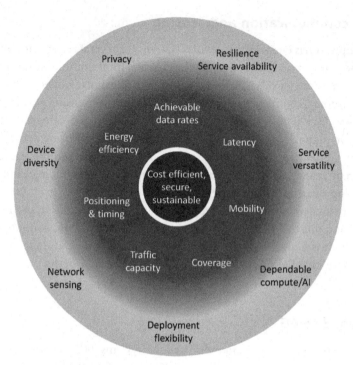

Fig. 30.5 Illustration of some requirements on 6G.

valuable and results in an expansion of the requirement space. A discussion on some possible use cases can be found in [128].

With the potential use cases in mind, the set of requirements for a future 6G system can be identified, see Fig. 30.5. Some of the requirements, for example mobility, coverage, and capacity, are recognized from earlier cellular generations although the numbers may be different. Data rates and latency are two other requirements typically found also in earlier generations. However, it should be emphasized that simply scaling the 5G requirements is not the right approach. For example, scaling up the 5G peak rate would result in data rates in the Tbit/s range which is very seldom needed and would come with a very small coverage. Instead, it is important to improve the *achievable* data rate, that is, what a typical user would expect in the system. A similar reasoning can be applied to the latency; reducing the latency jitter is in most cases much more useful than scaling down the smallest latency in the radio interface.

Apart from some of the more familiar requirements, 6G should also address new areas, shown in the outer part of Fig. 30.5. Resilience and service availability, for example, address the fact that the society more and more relies on having connectivity in place, even in case of unexpected events such as natural disasters, accidents, and human mistakes. Cellular networks may also be used for sensing, which can be seen as an expansion of the positioning services possible also in 5G. A wide range of devices, from simple zero-energy sensors to very high-end smartphone and communicating machines need to be handled, covered under the device diversity term in the figure. The possibility to quickly and in a simple way introduce new services is captured under service versatility. Dependable compute is another area, typically not discussed in previous generations. It relates to the fact that many tasks require not only communication services but also compute capabilities, and these compute resources must also be provided in a predictable manner. This requirement is also an illustration that 6G is wider in scope than 5G, which was more oriented toward communication only.

Last but not least, 6G must be able to meet these requirements in a cost efficient, sustainable, and secure manner, shown at the core of the circular figure.

Spectrum-wise, 6G is expected to address an even wider frequency range than 5G, from around 450 MHz up to the sub-THz range and including both frequency bands used for 5G as well as new frequency bands (Fig. 30.6). The sub-THz range my provide fairly large amounts of spectrum and correspondingly high data rates, but given the propagation conditions at these frequencies, the coverage will be very limited and the sub-THz frequencies will likely play a complementary role for specific scenarios rather than widespread use in general. Furthermore, the mm-wave spectrum already today provides a fair amount of spectrum suitable for denser deployments.

The other "new" spectrum range – the centimeter waves in 7–24 GHz, especially the lower range 7–15 GHz – is also an interesting frequency range for 6G, striking a good balance between the amount of spectrum potentially available the coverage. This spectrum range is currently used for other, non-cellular systems and coexistence mechanisms are therefore important.

From a coverage perspective, the lower frequency bands, below 6 GHz, will play an important role despite a very limited amount of new spectrum in that range being expected. To facilitate a smooth introduction of 6G, dynamic spectrum sharing with 5G is crucial with a corresponding impact on the choice of waveform.

Once the requirements are defined and agreed upon in release 19, the technical solutions can be discussed in release 20. As already mentioned, 6G will be able to exploit a very wide range of frequency bands. It is also foreseen that 6G will be a stand-alone technology, able to operate without relying on a 5G system being deployed it the same geographical area. Together with dynamic spectrum sharing, a stand-alone system allows the

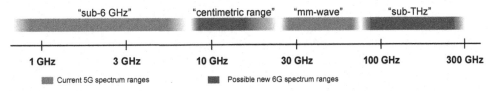

Fig. 30.6 Spectrum range for 6G.

overall 6G architecture to be simplified and avoiding architecturally complex dual connectivity solutions.

Clearly, providing a detailed list of all the 6G technology components is premature at this stage, but it is likely than many of the technology components found in 5G will be refined and reused, based on the experience of 5G deployments. Some of the technologies developed for NR in release 19, for example zero energy devices (Section 30.5) and sensing (Section 30.6), can be thought of as precursors to the full 6G solution. For a discussion of potential 6G technology components, see [128,129].

30.8 Concluding remarks

Above, some technologies which may play a role in both in the 5G evolution in release 19 has been outlined. Clearly, the NR evolution will continue for several releases after that, in parallel to the upcoming 6G standardization. As always, when trying to predict the future, there are a lot of uncertainties and new, not-yet-known requirements or technologies, which may motivate evolutions into directions not discussed above. It is clear though that NR is a very flexible platform, capable of evolving in a wide range of directions and an attractive path to future wireless communication for many years to come.

References

[1] 3GPP RP-172290, New SID Proposal: Study on Integrated Access and Backhaul for NR.
[2] 3GPP TS 37.141, E-UTRA, UTRA and GSM/EDGE; Multi-Standard Radio (MSR) Base Station (BS) Conformance Testing.
[3] 3GPP R1-163961, Final Report of 3GPP TSG RAN WG1 #84bis.
[4] 3GPP TS 38.104, NR; Base Station (BS) Radio Transmission and Reception.
[5] 3GPP TS 38.101-1, NR; User Equipment (UE) Radio Transmission and Reception. Part 1. Range 1 Standalone.
[6] 3GPP TS 38.101-2, NR; User Equipment (UE) Radio Transmission and Reception. Part 2. Range 2 Standalone.
[7] 3GPP TS 38.101-3, NR; User Equipment (UE) Radio Transmission and Reception. Part 3. Range 1and Range 2 Interworking Operation with Other Radios.
[8] 3GPP TS 38.101-4, NR; User Equipment (UE) Radio Transmission and Reception. Part 4. Performance Requirements.
[9] 3GPP RP-172021, Study on NR-Based Access to Unlicensed Spectrum.
[10] 3GPP TR 36.913, Requirements for Further Advancements for Evolved Universal Terrestrial Radio Access (E-UTRA) (LTE-Advanced) (Release 9).
[11] 3GPPTR38.803, Study on New Radio Access Technology: Radio Frequency (RF) and Coexistence Aspects (Release 14).
[12] 3GPP TS 23.402, Architecture Enhancements for Non-3GPP Accesses.
[13] 3GPP TS 23.501, System Architecture for the 5G System.
[14] 3GPP TS 36.211, Evolved Universal Terrestrial Radio Access (E-UTRA); Physical Channels and Modulation.
[15] 3GPP TS 38.331, NR; Radio Resource Control (RRC) Protocol Specification (Release 15).
[16] 3GPP TR 36.913, Requirements for Further Advancements for Evolved Universal Terrestrial Radio Access (E-UTRA) (LTE-Advanced).
[17] E. Arikan, Channel polarization: a method for constructing capacity-achieving codes for symmetric binary input memoryless channels, IEEE Trans. Inf. Theory 55 (7) (2009) 3051–3073.
[18] R.E. Best, Phases Locked Loops: Design, Simulation and Applications, sixth ed., McGraw-Hill Professional, 2007.
[19] T. Chapman, E. Larsson, P. von Wrycza, E. Dahlman, S. Parkvall, J. Sköld, HSPA Evolution: The Fundamentals for Mobile Broadband, Academic Press, 2014.
[20] D. Chase, Code combining—a maximum-likelihood decoding approach for combining and arbitrary number of noisy packets, IEEE Trans. Commun. 33 (1985) 385–393.
[21] J. Chen, Does LO noise floor limit performance in multi-gigabit mm-wave communication? IEEE Microwave Wireless Compon. Lett. 27 (8) (2017) 769–771.
[22] J.-F. Cheng, Coding performance of hybrid ARQ schemes, IEEE Trans. Commun. 54 (2006) 1017–1029.
[23] D.C. Chu, Polyphase codes with good periodic correlation properties, IEEE Trans. Inf. Theory 18 (4) (1972) 531–532.
[24] S.T. Chung, A.J. Goldsmith, Degrees of freedom in adaptive modulation: a unified view, IEEE Trans. Commun. 49 (9) (2001) 1561–1571.
[25] D. Colombi, B. Thors, C. Törnevik, Implications of EMF exposure limits on output power levels for 5G devices above 6 GHz, IEEE Antennas Wirel. Propag. Lett. 14 (2015) 1247–1249.
[26] E. Dahlman, S. Parkvall, J. Sköld, 4G LTE-Advanced Pro and the Road to 5G, Elsevier, 2016.
[27] DIGITALEUROPE, 5G Spectrum Options for Europe, (October 2017).
[28] Ericsson, Ericsson Mobility Report, November 2022.
[29] 3GPP R4-1712718, On mm-wave Filters and Requirement Impact, TSGRAN WG4 Meeting #85, Ericsson (December 2017).
[30] Federal Communications Commission, Title 47 of the Code of Federal Regulations (CFR).

[31] P. Frenger, S. Parkvall, E. Dahlman, Performance comparison of HARQ with chase combining and incremental redundancy for HSDPA, in: Proceedings of the IEEE Vehicular Technology Conference, Atlantic City, NJ, USA, October 2001, pp. 1829–1833.

[32] R.G. Gallager, Low Density Parity Check Codes, Monograph, M.I.T. Press, 1963.

[33] Global mobile Suppliers Association (GSA), The future of IMT in the 3300—4200 MHz frequency range, (June 2017).

[34] M. Hörberg, Low Phase Noise GaN HEMT Oscillator Design Based on High-Q Resonators (Ph.D. thesis), Chalmers University of Technology, 2017.

[35] IEEE, IEEE Standard for Local and metropolitan area networks Part 16: Air Interface for Broadband Wireless Access Systems Amendment 3: Advanced AirInterface, IEEE Std 802.16m-2011 (Amendment to IEEE Std 802.16-2009).

[36] IETF, Robust header compression (ROHC): framework and four profiles: RTP, UDP, ESP, and Uncompressed, RFC 3095.

[37] ITRS, Radio Frequency and Analog/Mixed-Signal Technologies for Wireless Communications, Edition International Technology Roadmap for Semiconductors (ITRS), 2007.

[38] ITU-R, Workplan, timeline, process and deliverables for the future development of IMT, ITU-R Document 5D/1297, Attachment 2.12 (July 2019).

[39] ITU-R, Framework and overall objectives of the future development of IMT-2000 and systems beyond IMT-2000, Recommendation ITU-R M.1645, June 2003.

[40] ITU-R, Unwanted emissions in the spurious domain, Recommendation ITU-R SM.329-12, September 2012.

[41] ITU-R, Future technology trends of terrestrial IMT systems, Report ITU-R M.2320, November 2014.

[42] ITU-R, Technical feasibility of IMT in bands above 6 GHz, Report ITU-R M.2376, July 2015.

[43] ITU-R, Detailed specifications of the terrestrial radio interfaces of International Mobile Telecommunications Advanced (IMT-Advanced), Recommendation ITU-R M.2012-5, February 2022.

[44] ITU-R, Frequency arrangements for implementation of the terrestrial component of International Mobile Telecommunications (IMT) in the bands identified for IMT in the Radio Regulations, Recommendation ITU-R M.1036-6, October 2019.

[45] ITU-R, IMT Vision—Framework and overall objectives of the future development of IMT for 2020 and beyond, Recommendation ITU-R M.2083, September 2015.

[46] ITU-R, Radio regulations, Edition of 2020.

[47] ITU-R, Detailed specifications of the terrestrial radio interfaces of International Mobile Telecommunications-2000 (IMT-2000), Recommendation ITU-R M.1457-15, October 2020.

[48] ITU-R, Guidelines for evaluation of radio interface technologies for IMT-2020, Report ITU-R M.2412, November 2017.

[49] ITU-R, Minimum requirements related to technical performance for IMT-2020 radio interface(s), Report ITU-R M.2410, November 2017.

[50] ITU-R, Requirements, evaluation criteria and submission templates for the development of IMT-2020, Report ITU-R M.2411, November 2017.

[51] M. Jain, et al., Practical, real-time, full-duplex wireless, in: MobiCom'11, Las Vegas, NV, USA, September, 2011.

[52] E.O. Johnson, Physical limitations on frequency and power parameters of transistors, RCA Rev. 26 (1965) 163–177.

[53] E.G. Larsson, O. Edfors, F. Tufvesson, T.L. Marzetta, Massive MIMO for next generation wireless systems, IEEE Commun. Mag. 52 (2) (2014) 186–195.

[54] D.B. Leeson, A simple model of feedback oscillator noise spectrum, Proc. IEEE 54 (2) (1966) 329–330.

[55] O. Liberg, M. Sundberg, E. Wang, J. Bergman, J. Sachs, Cellular Internet of Things: Technologies, Standards, and Performance, Academic Press, 2017.

[56] D.J.C. MacKay, R.M. Neal, Near Shannon limit performance of low density parity check codes, Electron. Lett. 33 (6) (1996) 1645–1646.

[57] 3GPP R1-040642, Comparison of PAR and Cubic Metric for Power De-rating, Motorola.

[58] B. Murmann, The race for the extra decibel: a brief review of current ADC performance trajectories, IEEE Solid-State Circuits Mag. 7 (3) (2015) 58–66.

[59] B. Murmann, ADC Performance Survey 1997-2019, Available from: http://web.stanford.edu/murmann/adcsurvey.html.

[60] M. Olsson, S. Sultana, S. Rommer, L. Frid, C. Mulligan, SAE and the Evolved Packet Core—Driving the Mobile Broadband Revolution, Academic Press, 2009.

[61] J. Padhye, V. Firoiu, D.F. Towsley, J.F. Kurose, Modelling, TCP reno performance: a simple model and its empirical validation, ACM/IEEE Trans. Netw. 8 (2) (2000) 133–145.

[62] S. Parkvall, E. Dahlman, A. Furuskär, M. Frenne, NR: the new 5G radio access technology, IEEE Commun. Stand. Mag. 1 (4) (2017) 24–30.

[63] M.B. Pursley, S.D. Sandberg, Incremental-redundancy transmission for meteor burst communications, IEEE Trans. Commun. 39 (1991) 689–702.

[64] T. Richardson, R. Urbanke, Modern Coding Theory, Cambridge University Press, 2008.

[65] C.E. Shannon, A mathematical theory of communication, Bell Syst. Tech. J. 27 (1948). 379–423, 623–656.

[66] L.A. Gerhardt, R.C. Dixon, Special issue on spread spectrum, IEEE Trans. Commun. 25 (1977) 745–869.

[67] S.B. Wicker, M. Bartz, Type-I hybrid ARQ protocols using punctured MDS codes, IEEE Trans. Commun. 42 (1994) 1431–1440.

[68] J.M. Wozencraft, M. Horstein, Digitalised communication over two-way channels, in: Fourth London Symposium on Information Theory, London, UK, September 1960.

[69] C. Mollen, E.G. Larsson, U. Gustavsson, T. Eriksson, R.W. Heath, Out-of-band radiation from large antenna arrays, IEEE Commun. Mag. 56 (4) (2018) 196–203.

[70] 3GPP TR 38.817-01, NR; General aspects for UE RF for NR (Release 15).

[71] 3GPP TR 38.817-02, NR; General aspects for BS RF for NR (Release 15).

[72] K. Zetterberg, P. Ramachandra, F. Gunnarsson, M. Amirijoo, S. Wager, T. Dudda, On heterogeneous networks mobility robustness, in: 2013 IEEE 77th Vehicular Technology Conference (VTC Spring), 2013.

[73] 3GPP RP-193229, New WID on Extending current NR operation to 71 GHz.

[74] 3GPP RP-193251, New WID on Enhancements to Integrated Access and Backhaul.

[75] 3GPP RP-193252, Work Item on NR small data transmission in INACTIVE state.

[76] 3GPP RP-133231, New WID on NR sidelink enhancements.

[77] 3GPP RP-193260, New WID on NR Dynamic spectrum sharing.

[78] 3GPP RP-193133, New WID: Further enhancements on MIMO for NR.

[79] 3GPP RP-193239, New WID: UE Power Saving Enhancements.

[80] 3GPP TR 38.811, Study on New Radio (NR) to support non-terrestrial networks (release 15).

[81] 3GPP RP-193234, Solutions for NR to support non-terrestrial networks (NTN).

[82] 3GPP RP-193248, New work Item on NR support of Multicast and Broadcast Services.

[83] 3GPP RP-133237, New SID on NR positioning Enhancements.

[84] S. Rommer, P. Hedman, M. Olsson, L. Frid, S. Sultana, C. Mulligan, 5G Core Networks: Powering Digitalization, Academic Press, 2019.

[85] 3GPP TR 38.866, Study on remote interference management for NR (Release 16).

[86] 3GPP TR 38.889, Study on NR-based access to unlicensed spectrum (Release 16).

[87] ETSI EN 301 893, 5 GHz RLAN; Harmonised Standard covering the essential requirements of article 3.2 of Directive 2014/53/EU.

[88] 3GPP TS 37.213, Physical layer procedures for shared spectrum channel access.

[89] 3GPP TR 38.855, Study on NR positioning support (Release16).

[90] 3GPP TS 38.212, NR; Multiplexing and channel coding.

[91] P. Groves, Principles of GNSS, Inertial, and Multisensor Integrated Navigation Systems, second ed., Artech House, 2013.

[92] ITU-R Working party 5D, Document IMT-2020/2-E, Revision 2, Submission, evaluation process and consensus building for IMT-2020.

[93] 3GPP TR 37.910, Study on self evaluation towards IMT-2020 submission (Release 16).

[94] 3GPP TR 38.913 V15, Study on Scenarios and Requirements for Next Generation Access Technologies; (Release 15).

[95] P. Kinget, Integrated GHz voltage controlled oscillators, in: W. Sansen, J. Huijsing, R. van de Plassche (Eds.), Analog Circuit Design, Springer, Boston, MA, 1999.

[96] H. Wang, et al., Power Amplifiers Performance Survey 2000-Present, Available from: https://gems.ece.gatech.edu/PA_survey.html.

[97] International Roadmap for Devices and Systems (IRDS), 2018 update, Outside system connectivity, IEEE 2018.

[98] 3GPP RP-140955, Revised Work Item Description: LTE Device to Device Proximity Services.

[99] 3GPP RP-193238, New SID on support of reduced-capability devices.

[100] P. Butovitsch, et al., Advanced Antenna Systems for 5G Network Deployments: Bridging the Gap Between Theory and Practice, Associated Press, 2020.

[101] 3GPP RP-162519, Revised WI proposal: LTE-based V2X Services.

[102] 3GPP RP-190984, 5G V2X with NR Sidelink.

[103] 3GPP TS 22.186, Enhancement of 3GPP support for V2X scenarios.

[104] 3GPP TS 38.133, Requirements for support of radio resource management.

[105] 3GPP TR 38.470, F1 general aspects and principles (Release 15).

[106] 3GPP TR 38.820, Study on the 7 to 24 GHz frequency range for NR (Release 16).

[107] 3GPP TR 38.840, Study on User Equipment (UE) power saving in NR.

[108] 3GPP TR 38.858, Study on Evolution of NR Duplex Operation.

[109] 3GPP RP-211574, Revised WID on support of reduced capability NR devices.

[110] Ericsson, RedCap—expanding the 5G device ecosystem for consumers and industries, https://www.ericsson.com/en/reports-and-papers/white-papers/redcap-expanding-the-5g-device-ecosystem-for-consumers-and-industries.

[111] 3GPP TR 38.875, Study on support of reduced capability NR devices.

[112] 3GPP R1-2202535, RAN1 agreements for Rel-17 NR RedCap.

[113] 3GPP RP-230796, New WID on Channel raster enhancement.

[114] 3GPP TR 38.835, Study on XR enhancements for NR.

[115] 3GPP RP-223502, New WID on XR Enhancements for NR.

[116] 3GPP RP-223545, Revised WID: NR Support for UAV (Uncrewed Aerial Vehicles).

[117] J.M.B. da Silva, G. Wikström, R.K. Mungara, C. Fischione, Full duplex and dynamic TDD: pushing the limits of spectrum reuse in multi-cell communications, IEEE Wirel. Commun. 28 (1) (2021).

[118] 3GPP R4-2218839, Additional simulation results related to SBFD adjacent channel co-existence evaluation.

[119] 3GPP RP-223041, Revised SID: Study on evolution of NR duplex operation.

[120] 3GPP RP-223540, New WID: Network energy savings for NR.

[121] 3GPP RP-221348, Revised SID: Study on Artificial Intelligence (AI)/Machine Learning (ML) for NR Air Interface.

[122] ITU-R, Future technology trends of terrestrial IMT systems towards 2030 and beyond, Report ITU-R M.2516, November 2022.

[123] ITU-R, Detailed specifications of the terrestrial radio interfaces of International Mobile Telecommunications-2020 (IMT-2020), Recommendation ITU-R M.2150-1, February 2022.

[124] ITU-R, Work plan, timeline, process and deliverables for the future development of IMT, ITU-R Document 5D/1668, Attachment 2.12 (February 2023).

[125] 3GPP TR 38.843, Study on Artificial Intelligence (AI)/Machine Learning (ML) for NR air interface.

[126] 3GPP TR 38.864, Study on network energy savings for NR.

[127] 3GPP TR 36.869, Study on low-power wake up signal and receiver for NR.

[128] 6G—Connecting a cyber-physical world, Ericsson, https://www.ericsson.com/en/reports-and-papers/white-papers/a-research-outlook-towards-6g.

[129] What to expect from 6G: Here are nine important takeaways from early global research, Ericsson, https://www.ericsson.com/en/blog/2023/2/6g-early-research-global-takeaways.

[130] 3GPP TR 38.821, Solutions for NR to support non-terrestrial networks (NTN).

[131] Ericsson, Microwave https://www.ericsson.com/en/portfolio/networks/ericsson-radio-system/ mobile-transport/microwave.

[132] 3GPP RP-222671. Revised WID: Mobile IAB (Integrated Access and Backhaul) for NR.

[133] 3GPP RP-193257, New WID on NR sidelink enhancements.

[134] 3GPP RP-210904, New WID on NR Sidelink Relay.

[135] 3GPP RP-220300, WID revision: NR sidelink evolution.

Index

Note: Page numbers followed by *f* indicate figures, *t* indicate tables, and *np* indicate footnotes.

Printed in the United States
by Baker & Taylor Publisher Services